AF614804

Methods in Molecular Biology™

Series Editor
John M. Walker
School of Life Sciences
University of Hertfordshire
Hatfield, Hertfordshire, AL10 9AB, UK

For further volumes:
http://www.springer.com/series/7651

siRNA Design

Methods and Protocols

Edited by

Debra J. Taxman

Department of Microbiology and Immunology and Lineberger Comprehensive Cancer Center, University of North Carolina, Chapel Hill, NC, USA

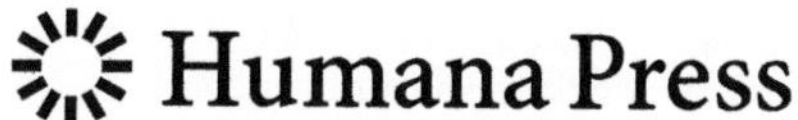

Editor
Debra J. Taxman
Department of Microbiology and Immunology
and Lineberger Comprehensive Cancer Center
University of North Carolina
Chapel Hill, NC, USA

ISSN 1064-3745 ISSN 1940-6029 (electronic)
ISBN 978-1-62703-118-9 ISBN 978-1-62703-119-6 (eBook)
DOI 10.1007/978-1-62703-119-6
Springer New York Heidelberg Dordrecht London

Library of Congress Control Number: 2012947406

Printed on acid-free paper

Humana Press is a brand of Springer
Springer is part of Springer Science+Business Media (www.springer.com)

Preface

The discovery of RNA interference (RNAi) as a methodology for gene silencing has revolutionized biological research and has provided an invaluable avenue for therapeutics. Since its unearthing in 1998, RNAi has been utilized by researchers for assessing the biological function of genes and has been the focus of numerous clinical trials. An ongoing challenge within the field of RNAi is the design of effective double-stranded RNA (dsRNA) molecules for the targeted silencing of genes. Small interfering RNA (siRNA), typically composed of two complementary 19–27 bp RNA molecules homologous to the target gene, is the most common strategy utilized for enacting RNAi. Its function is based on the sequence-specific cleavage of mRNA by the RNA-induced silencing complex (RISC), which functions endogenously to mediate gene regulation by cellular microRNA (miRNA). In addition to chemically synthesized siRNA, a number of methodologies exist for utilizing the Dicer machinery to produce siRNA within the cell from precursor RNA molecules, the most common of which is short hairpin (sh) RNA. There is debate within the field about the most advantageous methods for designing RNAi molecules for gene silencing. However, several innovative new RNAi designs have been developed to improve siRNA performance. Guidelines and protocols are provided within this volume for the selection of siRNA targeting sequences, for the strategic incorporation of chemical modifications, and for advantageous structural modifications to the classic siRNA and shRNA design. Strategies are also described for specific applications such as immunostimulatory siRNA that may provide therapeutic benefit against viral infections in mammals, the simultaneous targeting of multiple siRNAs, and siRNA-mediated crop virus resistance. The design of RNAi for gene silencing in embryonic, invertebrate, and plant systems requires a variety of unique approaches, several of which are described towards the end of this volume.

The foundation of any RNAi design protocol, regardless of the application, is the selection of the sequence to be targeted within the mRNA molecule. At this stage in understanding of RNAi methodology, targeting sequences cannot be predicted with certainty of efficacy in gene silencing (also called "knockdown" efficiency). For this reason, a validated siRNA target is often the first choice in target sequence selection. The term "validated siRNAs" can refer to those that have anywhere from 50 to 100% knockdown; however, depending on the experimental parameters, less than complete knockdown may not be sufficient for the ablation of downstream functions. Furthermore, siRNA target sequences have not been validated for all genes, and this particularly applies to nonmammalian species with partially characterized genomes. In other cases, it could be desirable to design siRNAs to selectively target one splice form of an mRNA or a homolog of a closely related gene. For any of these scenarios, a validated siRNA for the gene of interest, even if available, may not have the appropriate specificity. Moreover, most validated targets are not thoroughly tested for the so-called off-target effects that occur, in part, due to the sequence-specific activation of immune responses or degradation of nontargeted mRNAs. The ability to select an effective targeting sequence that also is specific to its target is, therefore, at the forefront of siRNA design.

To improve the success rate for selecting effective targeting sequence, algorithms have been developed based on empirical criteria, structural modeling of the RISC complex, and RNA secondary structure predictions. Chapter 1 provides an overview of the most important factors to consider when selecting a target sequence and also describes the development and utility of algorithms for target sequence selection within a historical context. Particular attention is applied to an increased understanding of the role of thermodynamics in selecting sequences that are effective in terms of the overall stability of the duplex, as well as the local internal stability, the incorporation of asymmetry that can facilitate the loading of the appropriate strand into the RISC, and the avoidance of off-target recognition of nontargeted mRNA sequences. A protocol is provided for incorporating each of these factors into the design of either siRNA or shRNA.

In Chapter 2, several analytical methodologies are considered for selecting functional siRNA target sequences using statistical modeling to predict the probability that a candidate siRNA sequence will be effective. These methodologies are evaluated by application to recently reported effective and ineffective siRNA sequences for a number of genes, and new methods for target selection are proposed. An additional newly emerging algorithm for siRNA target selection, siDirect 2.0, is described in Chapter 3. This algorithm is designed with regard to reducing off-target effects that can complicate experimental interpretation. Rationales and instructions are provided for using a Web-based server to identify candidate siRNA target sequences with reduced likelihood of producing off-target effects.

Target site accessibility is another major hurdle in siRNA site selection that is particularly important for the design of siRNA against highly structured targets, such as RNA virus genomes. In Chapter 4, a high-throughput method is described for identifying accessible target sites within highly structured RNA by measuring oligonucleotide hybridization kinetics under non-denaturing conditions.

In addition to careful target sequence selection, siRNA efficacy and specificity can be enhanced through selective chemical modification of nucleotides within the duplex. Chemical modification can reduce immunogenicity, increase the stability of the dsRNA complex, and increase the duration of knockdown. Chapter 5 reviews the recent literature describing a variety of modifications that have been used effectively in siRNA design and provides guidelines for incorporation of these modifications for specific in vitro and in vivo applications. In Chapter 6, the detailed applications and advantages are described for strategically applying substitutions using one form of helix-destabilizing non-nucleotide analog, Unlocked Nucleobase Analogs (UNA). Protocols are provided for the de novo design of optimized siRNAs containing UNA and the evaluation of these molecules for knockdown efficiency, target specificity, and low immunogenicity.

The functionality of siRNA can also be further enhanced by structural modification to the dsDNA backbone, and in Chapters 7 and 8 design protocols are provided for two such variants of the standard siRNA duplex, asymmetric siRNA (asiRNA) and fork-like siRNA (fsiRNA). These variants are designed to impart additional favorable asymmetry to the siRNA duplex to assist in RISC guide strand selection. The asiRNA design (Chapter 7) utilizes a shortened 15–16 nt sense strand and an antisense strand whose 5′ end is blunted and 3′ end has an extended overhang. Methods are provided for designing asiRNAs and for assessing their silencing activity and specificity. The fsiRNA design (Chapter 8) comprises a strategic introduction of duplex-destabilizing nucleotide substitutions at the 3′ end region of the sense strand that can enable the effective use of target sites with otherwise unfavorable sequence parameters. Nuclease sensitive sites within the fsiRNA, which are identified through rational design and experimental observation, are stabilized by introducing chemical modification.

The decision of which siRNA design strategy to choose depends on the application, and Chapter 9 provides an innovative protocol for designing dual-targeting siRNAs for applications that require the silencing of two genes simultaneously. In this design, both strands within the duplex are functional, and through a careful series of steps, each strand can be fashioned to mediate silencing activity through both siRNA-like RNA cleavage and miRNA-like translational repression mechanisms. Methods are provided for finding partially complementary dual-targeting siRNA candidates, for predicting their siRNA and miRNA activity, and for scoring and prioritizing the most promising candidates.

For the majority of RNAi experimental designs, the immunostimulatory responses that can occur as a side product of the dsRNA-mediated activation of Toll-like receptors and other RNA-sensing proteins within the cell are perceived as an undesirable off-target response. However, in specialized cases, it may be advantageous to promote immune activation during siRNA-mediated knockdown. As an example, immunostimulation may help to enhance the function of siRNAs that are antiviral or antitumorigenic. Chapter 10 reviews the current strategies for designing siRNA variants that have increased immunostimulatory potency. Methods are also provided for the assessment of immune activation by these bifunctional siRNAs.

As a cost-effective alternative to chemically synthesized siRNA, recent studies have shown that pools of siRNA molecules that target a single gene can be produced enzymatically (termed endoribonuclease-prepared siRNA, or esiRNA). The use of a pool of siRNAs has the additional advantage of diluting out the sequence-specific off-target effects of individual transfected siRNA clones. Chapter 11 provides guidelines for the in silico selection of an esiRNA target region that is most suitable for gene silencing.

As an additional alternative to chemically synthesized siRNA, a number of methodologies exist for utilizing the endogenous cellular machinery to produce siRNA within the cell from precursor molecules that are processed by the Dicer complex. Vector-expressed siRNA has particular advantage for difficult-to-transfect cells and for stable expression. Chapter 12 provides an overview of the history and application of shRNAs in mammalian systems, including a comprehensive outline of the options for shRNA effector molecule design, vector and promoter selection, and the simultaneous delivery of multiple shRNAs. In Chapter 13, methods are provided for the design and construction of multiple different forms of vector-expressed siRNA, including shRNAs and artificial miRNAs (amiRNAs) for the RNAi of single gene targets; and polycistrons, extended shRNAs (e-shRNAs), and long hairpin RNAs (lhRNAs) for the RNAi of multiple targets. Practical information is provided for the usage of viral vectors for expressing these molecules.

In Chapters 14 and 15, the design of two additional structural variants of shRNA is described. Chapter 14 describes a unique RNAi design termed bifunctional shRNA or bi-shRNA. The bi-shRNA technology harnesses both cleavage-dependent and cleavage-independent RISC loading pathways to enhance knockdown potency, thus providing increased potential for research and therapeutic applications. Methods are provided for the design and construction of bi-shRNA, the assessment of RNAi functionality, and the liposome-mediated delivery of bi-shRNA for in vivo RNAi studies. Chapter 15 describes an additional variant form of shRNA termed short shRNA, or sshRNA. Rather than being expressed from a vector, sshRNA is chemically synthesized and transfected, allowing for the incorporation of enhancing chemical modifications. Principles and procedures are provided for the design and production of right- and left-handed loop forms of sshRNA.

For mammalian oocytes and early embryos, the interferon response is nonfunctional, and therefore, long dsRNAs can be used for RNAi without concern about immunostimulatory off-target effects. Chapter 16 provides a review of the production and application of long dsRNA in mammalian cells. Currently available vectors for dsRNA expression and approaches for oocyte-specific transgenic RNAi are outlined.

In Chapter 17, the use of short and long dsRNA is described for RNAi of invertebrate model systems and human disease vectors. Similar to mammalian oocytes, the interferon response is not a critical factor in invertebrate RNAi design. Procedures are described for designing and preparing long dsRNA from genomic or cDNA sources using a bioinformatic approach that can be applied to the design of genome-wide RNAi libraries. Procedures are also described for designing short dsRNA for invertebrates and human disease vectors, which could be valuable for very short exons; and shRNA, which is useful for RNAi in invertebrate oogenesis and for long-term knockdown in cell lines. The construction and use of one such vector for shRNA in silkworm cell lines is detailed in Chapter 18.

In plants, amiRNA, has been used successfully for creating siRNA-mediated resistance to crop viruses. Chapter 19 describes the principles behind the design and production of a multi-amiRNA vector that can confer resistance to *Wheat streak mosaic virus*. The design involves replacement of five arms of the polycistronic rice miR395 with sequences targeting the viral genome and has the advantage over other historic methods of dsRNA-mediated RNAi in plants in that the target sites can be screened against matches to the plant genome to reduce off-target knockdown. Another method of choice for siRNA production in plants mimics a plant-specific class of endogenous small RNA termed trans-acting small interfering RNAs (tasiRNAs). tasiRNAs are produced from *TAS* gene-derived transcripts after they are targeted by an miRNA to promote downregulation of genes in trans. In Chapter 20, steps are described for designing and implementing the use of an innovative method for triggering tasiRNA-type transitivity to produce secondary siRNAs for target gene silencing. Instructions are provided for adapting this procedure, termed miRNA-induced gene silencing (MIGS), to the gene silencing of more than one gene.

RNAi has great potential for both research and therapeutic applications. Each application for each model system has its own set of hurdles in ensuring that the siRNA effector molecules are efficient, specific, and appropriately sustainable. However, through careful design, many of these hurdles can be overcome, ultimately improving experimental outcome and providing greater therapeutic value. The present volume provides practical information to assist in the design of siRNA and upstream siRNA-generating molecules using a variety of strategies that are relevant to a diverse array of applications in RNAi. It is my hope that the topics covered in this book will provide a broad understanding of the issues in RNAi and how they can be overcome strategically through design. This book should be of assistance to researchers, educators, clinicians, and biotech companies interested in harnessing an understanding of the power of RNAi technology.

Chapel Hill, NC, USA *Debra J. Taxman*

Contents

Preface . . . . *v*
Contributors . . . . *xi*

1 What Parameters to Consider and Which Software Tools to Use for Target Selection and Molecular Design of Small Interfering RNAs 1
Olga Matveeva

2 Methods for Selecting Effective siRNA Target Sequences Using a Variety of Statistical and Analytical Techniques 17
Shigeru Takasaki

3 Designing Functional siRNA with Reduced Off-Target Effects 57
Yuki Naito and Kumiko Ui-Tei

4 Design and Screening of siRNAs Against Highly Structured RNA Targets 69
Neda Nasheri, John Paul Pezacki, and Selena M. Sagan

5 Engineering Small Interfering RNAs by Strategic Chemical Modification 87
Jesper B. Bramsen and Jørgen Kjems

6 The Design, Selection, and Evaluation of Highly Specific and Functional siRNA Incorporating Unlocked Nucleobase Analogs 111
Narendra Vaish and Pinky Agarwal

7 The Design, Preparation, and Evaluation of Asymmetric Small Interfering RNA for Specific Gene Silencing in Mammalian Cells 135
Chanil Chang, Sun Woo Hong, Pooja Dua, Soyoun Kim, and Dong-ki Lee

8 Design of Nuclease-Resistant Fork-Like Small Interfering RNA (fsiRNA) 153
Elena L. Chernolovskaya and Marina A. Zenkova

9 Designing Dual-Targeting siRNA Duplexes Having Two Active Strands that Combine siRNA and MicroRNA-Like Targeting 169
Pål Sætrom

10 Strategies for Designing and Validating Immunostimulatory siRNAs 179
Michael P. Gantier

11 Designing Efficient and Specific Endoribonuclease-Prepared siRNAs 193
Vineeth Surendranath, Mirko Theis, Bianca H. Habermann, and Frank Buchholz

12 Short Hairpin RNA-Mediated Gene Silencing 205
Luke S. Lambeth and Craig A. Smith

13 Design of Lentivirally Expressed siRNAs 233
Ying Poi Liu and Ben Berkhout

14 Bifunctional Short Hairpin RNA (bi-shRNA): Design and Pathway to Clinical Application. 259
Donald D. Rao, Neil Senzer, Zhaohui Wang, Padmasini Kumar, Chris M. Jay, and John Nemunaitis

15 Design and Chemical Modification of Synthetic Short shRNAs as Potent RNAi Triggers 279
Anne Dallas and Brian H. Johnston

16 Production and Application of Long dsRNA in Mammalian Cells 291
Katerina Chalupnikova, Jana Nejepinska, and Petr Svoboda

17 Design of RNAi Reagents for Invertebrate Model Organisms and Human Disease Vectors 315
Thomas Horn and Michael Boutros

18 Construction of shRNA Expression Plasmids for Silkworm Cell Lines Using Single-Stranded DNA and *Bst* DNA Polymerase. 347
Hiromitsu Tanaka

19 Designing Effective amiRNA and Multimeric amiRNA Against Plant Viruses 357
Muhammad Fahim and Philip J. Larkin

20 Downregulation of Plant Genes with miRNA-Induced Gene Silencing 379
Felipe Fenselau de Felippes

Index *389*

Contributors

PINKY AGARWAL • *Evergreen Neuroscience Institute, Kirkland, WA, USA*

BEN BERKHOUT • *Laboratory of Experimental Virology, Department of Medical Microbiology, Center for Infection and Immunity Amsterdam (CINIMA), Academic Medical Center, University of Amsterdam, Amsterdam, The Netherlands*

MICHAEL BOUTROS • *Division of Signaling and Functional Genomics, German Cancer Research Center (DKFZ), Heidelberg, Germany; Department of Cell and Molecular Biology, Heidelberg University, Heidelberg, Germany*

JESPER B. BRAMSEN • *Department of Molecular Biology and Genetics, Interdisciplinary Nanoscience Center (iNANO), University of Aarhus, Aarhus, Denmark*

FRANK BUCHHOLZ • *Department of Medical Systems Biology, University Cancer Center, University Hospital and Medical Faculty Carl Gustav Carus, University of Technology Dresden, Dresden, Germany*

KATERINA CHALUPNIKOVA • *Institute of Molecular Genetics AS CR, Prague, Czech Republic*

CHANIL CHANG • *Global Research Laboratory for RNAi Medicine, Department of Chemistry, BK21 School of Chemical Materials Science, Sungkyunkwan University, Suwon, South Korea*

ELENA L. CHERNOLOVSKAYA • *Institute of Chemical Biology and Fundamental Medicine, Novosibirsk, Russia*

ANNE DALLAS • *SomaGenics, Inc., Santa Cruz, CA, USA*

POOJA DUA • *Global Research Laboratory for RNAi Medicine, Department of Chemistry, BK21 School of Chemical Materials Science, Sungkyunkwan University, Suwon, South Korea; Department of Biomedical Engineering, Dongguk University, Seoul, South Korea*

MUHAMMAD FAHIM • *Lab of Plant Developmental Molecular Genetics, School of Life Science and Biotechnology, Korea University, Seoul, South Korea*

FELIPE FENSELAU DE FELIPPES • *Department of Biology, Swiss Federal Institute of Technology (ETH), Zurich, Switzerland*

MICHAEL P. GANTIER • *Monash Institute of Medical Research, Monash University, Clayton, VIC, Australia*

BIANCA H. HABERMANN • *Max Planck Institute for Biology of Ageing, Cologne, Germany*

SUN WOO HONG • *Department of Biomedical Engineering, Dongguk University, Seoul, South Korea*

THOMAS HORN • *Department of Cell and Molecular Biology, Heidelberg University, Heidelberg, Germany; Division of Signaling and Functional Genomics, German Cancer Research Center (DKFZ), Heidelberg, Germany*

CHRIS M. JAY • *Gradalis, Inc., Dallas, TX, USA*
BRIAN H. JOHNSTON • *SomaGenics, Inc., Santa Cruz, CA, USA; Department of Pediatrics, School of Medicine, Stanford University, Stanford, CA, USA*
SOYOUN KIM • *Department of Biomedical Engineering, Dongguk University, Seoul, South Korea*
JØRGEN KJEMS • *Department of Molecular Biology and Genetics, Interdisciplinary Nanoscience Center (iNANO), University of Aarhus, Aarhus, Denmark*
PADMASINI KUMAR • *Gradalis, Inc., Dallas, TX, USA*
LUKE S. LAMBETH • *Murdoch Childrens Research Institute, Royal Childrens Hospital, Melbourne, VIC, Australia; Poultry Cooperative Research Centre, Armidale, NSW, Australia*
PHILIP J. LARKIN • *CSIRO Plant Industry, Canberra, ACT, Australia*
DONG-KI LEE • *Global Research Laboratory for RNAi Medicine, Department of Chemistry, BK21 School of Chemical Materials Science, Sungkyunkwan University, Suwon, South Korea*
YING POI LIU • *Laboratory of Experimental Virology, Department of Medical Microbiology, Center for Infection and Immunity Amsterdam (CINIMA), Academic Medical Center, University of Amsterdam, Amsterdam, The Netherlands*
OLGA MATVEEVA • *Department of Human Genetics, University of Utah, Salt Lake City, UT, USA; Novosibirsk State University, Novosibirsk, Russia*
YUKI NAITO • *Department of Biophysics and Biochemistry, Graduate School of Science, University of Tokyo, Tokyo, Japan*
NEDA NASHERI • *Department of Biochemistry, Microbiology, & Immunology, Steacie Institute for Molecular Sciences, National Research Council of Canada, and Ottawa Institute of Systems Biology, University of Ottawa, Ottawa, ON, Canada*
JANA NEJEPINSKA • *Institute of Molecular Genetics AS CR, Prague, Czech Republic*
JOHN NEMUNAITIS • *Gradalis, Inc., Dallas, TX, USA; Mary Crowley Cancer Research Centers, Dallas, TX, USA; Texas Oncology, P.A., Dallas, TX, USA; Medical City Dallas Hospital, Dallas, TX, USA*
JOHN PAUL PEZACKI • *Department of Biochemistry, Microbiology, & Immunology, Steacie Institute for Molecular Sciences, National Research Council of Canada, and Ottawa Institute of Systems Biology, University of Ottawa, Ottawa, ON, Canada*
DONALD D. RAO • *Gradalis, Inc., Dallas, TX, USA*
PÅL SÆTROM • *Departments of Computer and Information Science and Cancer Research & Molecular Medicine, Norwegian University of Science and Technology, Trondheim, Norway*
SELENA M. SAGAN • *Department of Microbiology & Immunology, School of Medicine, Stanford University, Stanford, CA, USA*
NEIL SENZER • *Gradalis, Inc., Dallas, TX, USA; Mary Crowley Cancer Research Centers, Dallas, TX, USA; Texas Oncology, P.A., Dallas, TX, USA; Medical City Dallas Hospital, Dallas, TX, USA*
CRAIG A. SMITH • *Murdoch Childrens Research Institute, Royal Childrens Hospital, Melbourne, VIC, Australia; Department of Paediatrics, University of Melbourne, Parkville, VIC, Australia; Poultry Cooperative Research Centre, Armidale, NSW, Australia*

VINEETH SURENDRANATH • *Max Planck Institute of Molecular Cell Biology and Genetics, Dresden, Germany*
PETR SVOBODA • *Institute of Molecular Genetics AS CR, Prague, Czech Republic*
SHIGERU TAKASAKI • *Toyo University, Ora-gun, Gunma, Japan*
HIROMITSU TANAKA • *Insect Mimetics Research Unit, National Institute of Agrobiological Sciences, Tsukuba, Ibaraki, Japan*
MIRKO THEIS • *Max Planck Institute of Molecular Cell Biology and Genetics, Dresden, Germany*
KUMIKO UI-TEI • *Department of Biophysics and Biochemistry, Graduate School of Science, University of Tokyo, Tokyo, Japan*
NARENDRA VAISH • *Vaish Biotech Consulting, Kirkland, WA, USA*
ZHAOHUI WANG • *Gradalis, Inc., Dallas, TX, USA*
MARINA A. ZENKOVA • *Institute of Chemical Biology and Fundamental Medicine, Novosibirsk, Russia*

Chapter 1

What Parameters to Consider and Which Software Tools to Use for Target Selection and Molecular Design of Small Interfering RNAs

Olga Matveeva

Abstract

The design of small gene silencing RNAs with a high probability of being efficient still has some elements of an art, especially when the lowest concentration of small molecules needs to be utilized. The design of highly target-specific small interfering RNAs or short hairpin RNAs is even a greater challenging task. Some logical schemes and software tools that can be used for simplifying both tasks are presented here. In addition, sequence motifs and sequence composition biases of small interfering RNAs that have to be avoided because of specificity concerns are also detailed.

Key words: siRNA efficiency, siRNA specificity, Duplex stability, Nucleotide preferences, RNA structure, Software-tools

1. Introduction

1.1. siRNA- and shRNA-Mediated Gene Silencing

Designing any molecule to mimic nature's design is not an easy task. Knowing the pathways that nature has taken and roles that the molecule plays in these pathways simplifies the mission. Small RNA interfering (RNAi) molecules have been extensively studied, yet there are substantial gaps in our knowledge of their natural function and processing. For this reason, bioinformatics of RNAi design represents a combination of solid statistical facts and educated guesses.

What is the function of small interfering RNA (siRNA) in mammals? It is well established that in plants and insects short siRNA duplexes can be generated from long double-stranded RNA by the enzyme Dicer in response to viral infections that promote viral RNA destruction (1). Whether RNAi plays a similar role as a viral infection-fighting agent in mammals is as yet unknown. siRNA

Debra J. Taxman (ed.), *siRNA Design: Methods and Protocols*, Methods in Molecular Biology, vol. 942,
DOI 10.1007/978-1-62703-119-6_1, © Springer Science+Business Media, LLC 2013

in a form that resembles a Dicer cleavage product (two paired 21 nt synthetic oligonucleotides with a 19 nt duplex region and 3′ end 2 nt overhangs) has become a popular research tool. After their introduction into a cell, siRNAs can cause specific target mRNA degradation. Thus, the biological consequences of the down-regulation of any gene can be studied.

siRNA-mediated silencing of mammalian genes uses synthetic oligonucleotides transfected into cells. An alternative approach employs the expression of short hairpin RNAs (shRNAs) in cells following delivery of expression plasmids or viral vectors (2, 3). shRNAs are artificial analogs of endogenous microRNAs (miRNAs), the vast class of small noncoding RNA molecules that regulate the stability and translation of their target mRNAs. Precursors of miRNAs (pre-miRNAs) are stable hairpins, which are encoded in plant and animal genomes. miRNAs play important regulatory roles in animals and plants. Most of these molecules appear to regulate the expression of a diverse set of genes. This regulation takes place during embryonic development, apoptosis, tissue regeneration, and so forth (for review, see ref. 4).

The approach to gene silencing based on synthesized siRNAs is fast and simple. The shRNA-based approach is more laborious and time consuming, but it is becoming increasingly popular. Compared to chemically synthesized siRNAs, the shRNA methodology offers advantages in silencing longevity and lower costs for genome-wide studies. Also, gene therapy is a particularly promising application for shRNAs. It is believed that transcription of shRNA delivers lower intracellular concentrations of siRNA-like products, compared to synthetic siRNA oligonucleotides transfected into cells. Lower intracellular concentrations achieved through the natural process of transcription for extended periods of time can yield more specific silencing effects. It is assumed that after shRNA molecules are processed in cells they enter the same enzymatic pathways as siRNAs.

One of the two strands of an siRNA duplex, the "guide strand," enters the RNA-induced silencing complex (RISC). RISC loaded with a guide strand acquires the ability to cleave RNA sequences that become base paired with the guide strand. The other duplex strand, the "passenger" strand, is the first "victim" of such newly acquired enzymatic ability. It gets cleaved and its fragments dissociate from the duplex. After passenger strand cleavage and dissociation, the guide strand becomes free from any base pairing; consequently, it becomes available for interaction with a new partner (5, 6). This new partner must feature some complementarity to the guide strand RNA sequence. Thus, an mRNA can be cleaved and destroyed if it includes a region complementary to the guide strand. In relation to the mRNA, the siRNA "guide strand" is synonymous to the "antisense strand," while the "passenger strand" is synonymous to the "sense strand."

The RISC core protein complex contains a member of the Argonaute (Ago) family of proteins (7). So far, variable number of different Ago protein-coding genes have been characterized in different organisms (8). In mammals there are eight Ago genes (9, 10). All of these eight genes have been identified in the human genome (11). Among them, only the product of Ago2 mediates RNA cleavage directed by siRNA (12).

RISC can cleave mRNA between residues base paired to nucleotides 10 and 11 of the siRNA, and the cleavage itself does not require ATP. Multiple rounds of mRNA cleavage can be guided by the same siRNA as long as it remains associated with the Ago complex. It is likely that the release of the cleaved mRNA products involves an RNA helicase, which is dependent on ATP, because several proteins associated with the RNAi pathway in *Drosophila* and other organisms contain RNA helicase/ATPase domains (13). The cleavage of the human siRNA-mediated passenger strand and of mRNA is associated solely with an Ago2-containing RISC (Ago2-RISC) (12, 14).

Because processing of artificial siRNAs and shRNAs in cells utilizes the main components of the cellular RNAi machinery, the design of new versions of these molecules should allow provision for successful interaction with RISC and mRNA targets.

1.2. History of Development of Algorithms for Predicting siRNA Silencing Efficiency

Independently selected siRNA duplexes for different mRNA target regions can have vastly different silencing efficiency. In 2003, two independent studies made very important contributions to understanding how this happens (15, 16).

The experiments performed in the first study (15) demonstrated that the rate of RISC entry might be very different for the two strands of an siRNA duplex. Both the absolute and relative 5′ terminal duplex stabilities of the two siRNA strands determine the speed of the process. In other words, RISC prefers the strand whose 5′ end more loosely pairs with its complement. Such a strand enters RISC fast with consequent efficient and fast cleavage of the target. Conversely, a strand whose 5′ end pairs tightly with its complement enters RISC slowly with an elevated chance of target cleavage failure. The statistical analysis performed in the other study (16) revealed that efficient and inefficient siRNAs differ in their 5′ antisense strand terminal duplex stability. The 5′ ends of efficient guide strands pair less stably with their complements. The findings of both studies are consistent with one another; fast RISC entry is a prerequisite for efficient siRNA guide strand cleavage. Thus, asymmetry in siRNA terminal duplex stability defines which strand enters RISC efficiently and which strand guides efficient target cleavage. The asymmetry can be evaluated by the calculation of difference in G/C or A/T nucleotide content at siRNA duplex ends; however, it is more correctly captured by the difference in 5′ terminal free energy (ddG) of the guide and passenger strand.

While asymmetry in siRNA duplex stability is an important feature defining molecular silencing efficiency, it is not the only important feature. Studies from 2004 onwards, some bioinformatic, focused on analysis of other features defining siRNA silencing efficiency as well as on ways to predict it. The logic of these studies followed a common series of steps. A database of siRNA with variable efficiency was assembled, sequence features (parameters) associated with efficient siRNAs were identified, and recommendations were formulated for how to use all of these features for experimental molecular design. The first published study resulted from analysis of more than a hundred siRNA molecules with variable efficiency (17). The list of features that were found to be associated with siRNA efficiency includes optimal G/C content, low terminal duplex stability of sense strand, lack of inverted repeats, and certain nucleotide preferences in the sense strand. A second published study that resulted from analysis of a few dozen siRNAs also suggested a list of features that included base preference of A or U at the 5′ end of the antisense strand; G or C at the 5′ end of the sense strand; at least five A/U residues in the 5′ terminal one-third of the antisense strand; and the absence of any GC stretch of more than 9 nt in length (18). The authors of both works suggested that the probability of an siRNA candidate being efficient could be evaluated according to the number of these sequence features associated with the candidate. The studies described above involved only "human intelligence" without sophisticated machine learning techniques—"artificial intelligence."

The sequence features associated with siRNA efficiency even in these early and not-so-sophisticated bioinformatics studies can be separated into at least four different categories. The first category relates to asymmetry in siRNA terminal duplex stability, the second to total siRNA duplex stability, the third to base preferences at different siRNA strand positions, and the fourth to the necessity of avoiding certain nucleotide motifs in siRNA sequences. Some of these categories are overlapping. For example, as mentioned above, asymmetry in siRNA terminal duplex stability can be roughly evaluated through certain base preferences at the 5′ and 3′ ends of siRNA strands. Total siRNA duplex stability can be roughly approximated through calculation of the total GC content of any particular duplex strand.

Computer models that can deliver siRNA design candidates with high probability of being efficient require reliable input parameters (sequence features associated with siRNA efficiency), large experimental datasets, and good data processing schemes. Almost all studies that were published in 2004 and in 2005 used comparatively small databases (no more than a few hundred data points). However, exploration of different machine learning techniques for the purpose of siRNA design begins even with small databases. The initial explorations involved Regression Trees (RT) (19, 20), Genetic Programming (GP) (21), Generalized String Kernel (GSK) combined with Support Vector Machine (SVM) (22, 23), and so forth.

A large experimental database (more than 2,000 siRNA data points) was published by Novartis in 2005 (24). This database, as well as others (25) that were compiled from experiments described in the literature, became fuel for the next round of studies that explored advanced statistical analysis and Artificial Intelligence for developing algorithms for efficient siRNA design. These studies utilized Neuronal Networking (24, 26) and supervised learning of a Radial Basis Function (RBF) network combined with Regression Trees (27), Linear Regression (28–30), SVM (31–34), Random Forest Regression (35), and RBF (36). The siRNA design algorithms utilizing large experimental databases were called second-generation algorithms.

There are a number of statistical tools which can be helpful for evaluating the performance of the algorithms listed above. When a predictive method outputs a continuous measure of siRNA efficiency, then correlation between experimental and theoretical values can be calculated. When a predictive method outputs a binary measure of efficiency, for example, siRNA is predicted to be either efficient or inefficient, then Receiver Operating Characteristic or Precision Recall Curves can be created. Both types of curves characterize an algorithm's performance from slightly different angles (37). Despite the existence of all of these statistical techniques and the expansion of approaches for siRNA design, very few studies addressed algorithm comparison issues (30, 38). A more recent work was published in 2007 (30), and it revealed that four algorithms (24, 29, 30, 39) out of 11 used for comparison fared better than the others.

Although considerable success has been reported in predicting siRNA activity, very few studies deal with the analysis of shRNA features related to their silencing efficiency (3, 25, 40). The difference between the fate of chemically synthesized siRNA and intracellular transcribed shRNA within the cell is in the transcription and enzymatic processing. shRNA has to be first transcribed and then cleaved by RNases to enter the RISC pathway, while siRNA enters this pathway without transcription or cleavage. Two popular types of shRNA molecules are currently experimentally used. The first type is miRNA-like shRNAs. Their design is based upon one particular miRNA (miR30) and they are frequently employed for loss-of-function assays (41, 42). These molecules have long (more than 21 nt) sometimes partially mismatched stems, and undergo processing by RNaseIII-like endoribonucleases (Drosha and Dicer). The second type is shRNAs with short perfectly paired stems (19–21 nt). They are more frequently used in experiments that require silencing of individual genes (2).

It is likely that the processing of miR30-based shRNAs depends on the rates of Drosha and Dicer cleavages. Both enzymes demonstrate certain nucleotide preferences for their cleavage sites. This is probably the reason why approaches that try to transfer siRNA design to miRNA-like shRNAs have a tendency to fail (40). It has been suggested, however, that the processing of shRNAs of the second type from shorter (19 nt) stems is not Dicer dependent (43). Perhaps single-strand RNases (for example, representatives of the RNase A gene superfamily) are involved in the processing of shRNAs with short stems. Thus, the ability of such shRNAs to silence genes might depend on the susceptibility of their loop sequences to RNase A cleavage. The 9 nt "UUCAAGAGA" loop was described as the optimal configuration for a potent silencing trigger (2, 44). Both types of shRNA enter the RISC after enzymatic cleavage, and at this stage the pathways of chemically synthesized and intracellularly transcribed molecules merge.

This scenario suggests that design rules for shRNA duplexes, which emerge after the enzymatic cleavage step, should be highly similar to those for siRNA duplexes. However, the precise location of enzymatic cleavage sites as well as cleavage site sequence requirements are not always well characterized. Moreover, efficient transcription considerations should also affect shRNA design rules. Thus, bioinformatics of shRNA design belongs primarily to the future. Future progress needs to be fueled by new experimental databases that permit careful analysis of transcription and cleavage preferences of all the enzymatic machinery involved in the shRNA pathway. One such database was recently published and hopefully is going to be a great asset for the design of efficient miR30-based shRNAs (40).

1.3. List of siRNA Sequence Features Associated with Specific Silencing Efficiency

The sequence features found to be associated with siRNA-specific silencing efficiency can be subdivided into at least seven categories. A list of these categories is provided below, along with the references to the studies in which the relationship between the experimental siRNA silencing efficiency and feature presence were studied. Additionally, the rationales for incorporating these features are presented in greater detail in Subheadings 1.4–1.7 below.

Category 1. Asymmetry in terminal duplex stability in siRNA (15–18, 25, 31, 34, 39, 45–47).

Category 2. Total siRNA duplex stability (a rough approximation can be made through the evaluation of GC content) (17, 21, 25, 27, 31, 34, 39, 40, 45, 47).

Category 3. Internal local siRNA duplex stability (30, 39, 40).

Category 4. Base preferences at different siRNA strand positions (17, 20, 23, 24, 31, 36, 39, 40, 45, 47–50).

Category 5. Frequency of occurrence or avoidance of certain motifs in the complete sequence and at certain positions along the siRNA (22, 29, 31, 51).

Category 6. Guide siRNA strand (31, 33, 34, 52, 53) and/or target mRNA secondary structure (33, 34, 47, 54–61).

Category 7. Negative influence of cross-hybridization potential (39, 62).

The siRNA sequence features belonging to the first two categories most strongly affect silencing efficiency. In one way or another they are included in all siRNA design algorithms.

1.4. Optimal Asymmetry in Terminal Duplex Stability

High asymmetry in terminal duplex stability is responsible for a high rate of RISC loading for the siRNA cleavage guidance strand and a low rate of RISC loading for the passenger strand. This rate difference eliminates strand competition for RISC access. Theoretically, this competition is not a problem when the RISC molecular concentration is higher than the concentration of siRNAs. However, even in such cases, RISC loading with the passenger strand is not a desirable event because RISC being loaded with the passenger strand can make an extra contribution to nonspecific cleavages. In other words, optimization of the terminal asymmetry in siRNA duplexes increases both the efficiency and specificity of silencing.

Two scenarios for interaction between the siRNA strands and RISC are possible (Fig. 1). The first scenario happens when the concentration of RISC is below that of the siRNA and the second when the concentration of RISC is equal to, or above, that of the siRNA. According to the first scenario, the competition between the strands for RISC entry happens if the 5′ ends of both strands are loosely paired and does not occur when one of the 5′ ends is tightly paired with its complement. According to the second scenario, there is no competition between the two strands regardless of how tightly their 5′ ends are paired with the complements. The absence of competition between the strands can be a consequence of RISC concentration excess or terminal asymmetry in siRNA duplexes.

If experiments are performed in RISC excess, symmetrical or asymmetrical duplexes will appear efficient (no strand competition). If experiments are done in siRNA excess only asymmetrical duplexes will appear efficient (strand competition exists but only for symmetrical duplexes). It is important to understand that sometimes it appears that asymmetry is not needed because in a RISC excess scenario symmetrical duplexes with loosely paired ends will look efficient.

Even though considerable time has passed since the 2003 discovery that asymmetry in siRNA terminal duplex stability defines which strand is loaded into RISC (15), the optimal method for calculating this asymmetry is still debatable. Mismatches in the first four nucleotides from the 5′ end of an siRNA duplex affect the rate of strand entry into RISC. This led to the conclusion that 4 nt should be considered for asymmetry calculation. However, careful statistical analysis has demonstrated that the contribution of the first nucleotide to siRNA efficiency is the strongest, while that of the fourth is the weakest (63).

Analysis of four independent experimental databases revealed that a ddG value equal to, or above, 2 kcal/mol calculated for the two terminal nucleotides of an siRNA duplex corresponds to the highest silencing activity (25). Almost all siRNA duplexes with two U/A at the 5′ end of the cleavage guidance strand and two G/C

I

RISC concentration below siRNAs
(it appears from a database analysis that asymmetry is very important)

RISC

Competition between the strands for RISC entry
Symmetrical duplexes are non efficient

guide strand
passenger strand

No competition between the strands for RISC entry
Asymmetrical duplexes are efficient

loosely paired 5' end
tightly paired 5' end

Only asymmetrical duplexes are efficient

II

RISC concentration above or equal to siRNAs
(it appears from a database analysis that asymmetry is less important)

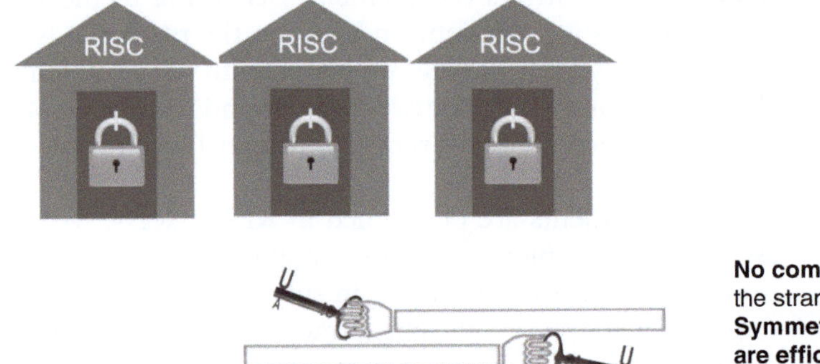

No competition between the strands for RISC entry
Symmetrical duplexes are efficient

No competition between the strands for RISC entry
Asymmetrical duplexes are efficient

Both symmetrical and asymmetrical duplexes are efficient
However, only asymmetrical duplexes are guide strand specific

Fig. 1. Two scenarios of siRNA–RISC interaction. Two scenarios for interaction between siRNA strands and RISC are possible. The first scenario (I) happens when the concentration of RISC is below that of siRNA, and the second (II) when the concentration of RISC is equal to, or above, that of siRNA. According to the first scenario, the competition between the strands for RISC entry happens if the 5′ ends of both strands are loosely paired and does not occur when one of the 5′ ends is tightly paired with its complement. According to the second scenario, there is no competition between both strands regardless of how tightly their 5′ ends are paired with the complements. The absence of competition between the strands can be a consequence of RISC concentration excess or terminal asymmetry in siRNA duplexes.

at its 3′ end belong to the category of those with optimal ddG. It was noticed that the guide strand of efficient siRNA and shRNA molecules frequently starts with U rather than with A (40). This bias is most likely explained by AGO2 binding preferences (64). Consequently, selection of siRNAs with U at the first position from the 5′ end of the cleavage guidance strand, U/A at the second position, as well as two G/C at their 3′ end should deliver the best silencing efficiency and specificity in RISC loading. That is why in Fig. 1 "U" is presented with larger symbol in comparison with "A" in the RISC entry key.

For shRNAs, transcription start nucleotide preferences have to be considered to achieve optimal duplex asymmetry. The transcription of shRNAs is usually performed from Polymerase III promoters H1 or U6 (2, 3). The H1 promoter favors adenine at the first position of a potential encoded shRNA, while a U6 promoter favors guanine.

1.5. Optimal siRNA Duplex Stability

siRNA duplex stability should be neither too high nor too low for efficient siRNA functioning and have dG values between −35 and −27 kcal/mol (25). Low duplex stability results in slow formation and short lifetime of cleavage guidance strand–target duplexes, with consequent inadequate opportunity for RNA cleavage to occur. On the other hand, the siRNA passenger strand of duplexes that are too stable may dissociate too slowly even after cleavage by RISC. The same problem can occur with target RNA cleavage products. In addition, stable siRNA duplexes are usually GC rich and GC-rich guide strands, or their target regions, are usually constituents of stable RNA secondary structures. The optimum value of duplex stability of most efficient siRNAs or shRNAs might be concentration dependent, however.

1.6. Optimal Profile of Internal Local siRNA Duplex Stability and Nucleotide Preferences

The nucleotide preferences at terminal siRNA strand positions of efficient molecules are related to terminal duplex asymmetry, so they are easily explainable. However, it is still unclear why some nucleotides are preferred at certain nonterminal positions of siRNA strands and some are avoided. Speculatively, the preference for A at the 10th position of a guidance strand has been revealed in a number of studies (17, 20, 30, 34), and may be related to RISC cleavage preference, because this cleavage occurs between residues base paired to nucleotides 10 and 11 of the siRNA. Perhaps preferential low stability (A/U enrichment) of some base pairs (6th, 7th, 12th, 13th, and 14th from the 5′ end of the guide strand) in the siRNA duplex (30, 34) is needed for easier dissociation of the cleaved RNA product and enhances RISC turnover. Some evidence was obtained that this dissociation proceeds energetically uphill because it involves an ATP-dependent RNA helicase P (10).

1.7. Specificity Problems

Early reports related to siRNA target gene silencing specificity were very optimistic; they suggested that siRNA is a highly specific tool for targeted gene knockdown (65, 66). However, later microarray studies showed that in addition to the intended target, siRNAs down regulate many unintended target transcripts (67–69). Intended and unintended transcript targets may be silenced with indistinguishable kinetics, though unintended transcripts are generally down regulated to a smaller degree (67). Off-target silencing is now widely recognized as a complication of any studies involving siRNAs (70, 71). It seems that to a large extent, target silencing effects are mediated by the participation of artificial siRNAs in natural miRNA pathways. The miRNA translational suppression pathway is directed by imperfect base pairing between the target and the miRNA guide strand (72). The specificity of this base-pairing depends on the 6–7 base "seed region" at the 5′ end of the guide strand of an miRNA. This miRNA-type seed-matched, off-target silencing by siRNAs is very common. Most of the siRNA unintended targets share sequence complementarity in their 3′ UTR regions with residues 1–8 (or contain seed motif matching nucleotides 1–6, 2–7, or 3–8) of the siRNA guide strand (67–69, 73).

The investigation of "seed region" complementarity and siRNA off-target silencing or miRNA down regulation revealed several interesting facts. It was found that the stability of a duplex between the siRNA "seed region" and an off-target region, measured as free energy dG, and the degree of gene silencing correlate with each other (74). A correlation was also discovered between the frequency of a seed hexamer being present in the 3′ UTR portion of a transcriptome and siRNA off-targeting effects. In other words, siRNAs with the most unique hexamers in their seed regions are less prone to off targeting (74). It was also found that extra complementarity within the 3′ end of a guide strand increases the miRNA targeting effect. The effect was also increased if miRNA target seed regions were located within AU-rich regions of 3′ UTRs (75). The latter observation might indicate that AU-rich mRNA regions are usually not involved in stable base-pairing within RNA secondary structure and consequently are more likely to be available for intermolecular interaction with miRNA.

siRNAs can be potent inducers of interferons (IFNs) and inflammatory cytokines both in vivo and in vitro (76–78). These findings are promoting a growing concern among researchers whether activation of an immune response can be systematically avoided during gene silencing experiments. The problem of immune-stimulatory motifs can be addressed through certain chemical modifications; however, this approach has its own limitations. For example, it cannot be used for in vivo transcribed shRNA, which along with chemically synthesized siRNA, is a popular tool for gene silencing experiments.

A basic series of steps for incorporating the design criteria presented in Subheading 1.3 and further detailed in Subheadings 1.4–1.7 is provided below. Suggested Web resources that can best help to achieve these criteria are also provided.

2. Materials

Computer with Internet access.

3. Methods

Steps for designing siRNAs according to the seven criteria categories provided in Subheading 1.3, and further discussed in Subheadings 1.4–1.7, are provided below. Step 1 can be used as a broad method for selecting siRNAs that satisfy criteria categories 1–5. The remaining steps will help to refine the selection of siRNAs with consideration of target secondary structure as well as siRNA specificity requirements to satisfy criteria categories 6–7.

1. Use one of the following Web tools for filtering out inefficient siRNAs according to the features from the categories 1–5 in Subheading 1.3: http://www5.appliedbiosystems.com/tools/siDesign/ or http://www.med.nagoya-u.ac.jp/neurogenetics/i_Score/i_score.html (see Notes 1 and 2).
2. (Optional) Use Web tools for filtering out inefficient siRNAs according to the optimal accessibility of the guide strand and target secondary structures (see Note 3): http://rna.tbi.univie.ac.at/cgi-bin/RNAxs?hakim=1, http://sfold.wadsworth.org/cgi-bin/sirna.pl, or http://rna.urmc.rochester.edu/servers/oligowalk.
3. (Optional) Use siRNA candidates that are in overlapping pools in outputs from steps 1 and 2.
4. Filter out siRNA candidates that are vulnerable to nonspecific silencing efficiency due to the presence of immune-stimulating motifs if you are using any software other than http://www5.appliedbiosystems.com/tools/siDesign/. Three sequence motifs, "UGUGU," "GUCCUUCAA," and "AUCGAU(N)nGGGG," should be included in an immune-motif avoidance list. In addition, U-rich sequences or sequences with biased nucleotide content, such as (G + U) >> (C + A) or with A + U or G + U rich motifs, should also be avoided (see Note 4).
5. Filter out siRNA candidates with the motif "UGGC," which is associated with reduced cell viability (see Note 5).
6. Filter out siRNA candidates with motifs of low sequence complexity such as "GGGG," "UUUU," "CCCC," and "AAAA" (see Note 6).
7. Use Washington University BLAST (WU-BLAST) to filter out the least specific siRNAs: http://informatics-eskitis.griffith.edu.au/SpecificityServer/ (see Notes 7–9).
8. Use for your experiments siRNA candidates with the top scores from step 1 that pass all the filters listed above.

4. Notes

1. A large number of Web services for siRNA design are currently available. However, very few of them use the most relevant updated information, so I have recommended two Web sites that are most committed to predicting efficient siRNAs:

   http://www5.appliedbiosystems.com/tools/siDesign/ (34). This Web service filters out siRNA candidates with miRNA "seed" matches, and toxic and immunostimulatory motifs.

   http://www.med.nagoya-u.ac.jp/neurogenetics/i_Score/i_score.html (27) This Web site allows the calculation of siRNA efficiency scores based upon nine independent siRNA design tools. In addition it provides thermodynamic information related to the dG value of an siRNA duplex, dG of guide strand secondary structure, and dinucleotide dG values at the 5′ and 3′ ends. However, the Web service does not filter out siRNAs with continuous runs of identical nucleotides or other non-desirable siRNA sequence motifs. It does not consider the BLASTN score.
2. Two independent studies suggested that the ability of the siRNA guide strand to cross hybridize with multiple off-target regions (39, 62) diminishes this strand's ability to silence its own target. Recently, a downloadable software was described that can perform siRNA design steps that can substitute for steps 1, 2, and 7 (79).
3. There are reported correlations between the openness of the secondary structures of the siRNA guide strand and target mRNA. I have recommended three Web tools that offer design of siRNAs based upon consideration of these sequence features. It has to be noted that for http://sfold.wadsworth.org/cgi-bin/sirna.pl the restriction in mRNA target length is 250 nt. Two other Web sites can deal with longer mRNAs. However, all the listed Web services do not filter out siRNAs with continuous runs of identical nucleotides or some other non-desirable siRNA sequence motifs.
4. Varieties of different specialized cellular receptors have evolved for recognition of pathogen-associated molecular patterns (PAMPs) that trigger immune responses when their ligands (viral or bacterial components) become present (80, 81). Some of these receptors are restricted to immune cells only, while others are present in all cells. Toll-like receptors (TLRs) are type I transmembrane proteins involved in the functioning of innate immunity by recognizing pathogen-specific molecules including DNA and RNA. The human TLR7 and TLR8 immune response appears to be stimulated strongest with U-rich sequences (82–84). AU-rich motifs mostly were

shown to activate TLR8 and GU-rich motifs both TLR7 and TLR8 receptors (85). A- and U-rich sequences stimulate TLR8, which triggers the production of both IFNα and TNFα from monocytes (84, 85), while G- and U-rich sequences trigger TLR7, thereby causing IFNα production from pDCs (86). Certain siRNA sequences are particularly immuno-stimulatory, such as "GUCCUUCAA" (77) and "UGUGU" (78, 87). Experimental evidence demonstrates that the latter motif triggers induction of IFN type 1 and nonspecific gene down regulation through TLR7. However, some siRNA sequences, independent of their GU or AU enrichment, are potent inducers of IFN-alpha production. One study showed that the presence of the CpG motif "AUCGAU" in RNA oligonucleotides together with a poly-G tail stimulates monocytes to produce large amounts of IL-12 (88).

5. A strong correlation has been found between the presence of motif "UGGC" in an siRNA cleavage guidance strand and reduced cell viability (89).

6. Clear answers are elusive about why certain motifs are preferred or avoided in efficient siRNA molecules. These parameters are poorly studied. However, it is widely believed that it is desirable to avoid targeting mRNA regions with low sequence complexity, particularly those represented by continuous runs of four identical nucleotides. It is assumed that the "UUUU" motif should be avoided during design of shRNAs because it represents an RNA III polymerase termination site.

7. NCBI BLASTN allows the fast discrimination of siRNA design candidates according to their homology with potential off-target hits. It can be also used for simple homology searches. For BLAST searching, it is recommended to discard any candidate with a score above 30 (34). However, BLASTN default parameters have been demonstrated to be inappropriate for siRNA design and other parameters have been suggested (90).

8. In general, siRNA off-target silencing is a function of many characteristics of duplexes between the guide strand and multiple regions in a transcriptome that are partially complementary to this strand (off-target hits) (91). So far, there is no ideal software for careful screening and characterization of these imperfect mismatched duplexes.

9. It is preferable to design siRNAs without "seed regions" present in known miRNAs. miRNA-like off targeting that involves mRNA translation inhibition can lead to gene-nonspecific down regulation (92). The list of these regions can be extracted from the sequences of mature miRNAs. A database of these sequences is available from http://www.mirbase.org/ftp.shtml. However, the siRNA design tool available from http://www5.appliedbiosystems.com/tools/siDesign/ performs this function automatically.

Acknowledgement

The author is very grateful to John Atkins for careful reading of this manuscript and very constructive comments. The work was supported in part by Fred Hutchinson Cancer Center grant DK056465 to animal core facility and by Russian Ministry of Science and Education grant 11.G34.31.0034 to Novosibirsk State University.

References

1. Van Rij RP, Andino R (2006) The silent treatment: RNAi as a defense against virus infection in mammals. Trends Biotechnol 24:186–193
2. Brummelkamp TR, Bernards R, Agami R (2002) A system for stable expression of short interfering RNAs in mammalian cells. Science 296:550–553
3. Paddison PJ et al (2002) Short hairpin RNAs (shRNAs) induce sequence-specific silencing in mammalian cells. Genes Dev 16:948–958
4. Bartel DP (2009) MicroRNAs: target recognition and regulatory functions. Cell 136:215–233
5. Matranga C et al (2005) Passenger-strand cleavage facilitates assembly of siRNA into Ago2-containing RNAi enzyme complexes. Cell 123:607–620
6. Rand TA et al (2005) Argonaute2 cleaves the anti-guide strand of siRNA during RISC activation. Cell 123:621–629
7. Hutvagner G, Simard MJ (2008) Argonaute proteins: key players in RNA silencing. Nat Rev Mol Cell Biol 9:22–32
8. Höck J, Meister G (2008) The Argonaute protein family. Genome Biol 9:210–210
9. Meister G, Tuschl T (2004) Mechanisms of gene silencing by double-stranded RNA. Nature 431:343–349
10. Carmell MA et al (2002) The Argonaute family: tentacles that reach into RNAi, developmental control, stem cell maintenance, and tumorigenesis. Genes Dev 16:2733–2742
11. Sasaki T et al (2003) Identification of eight members of the Argonaute family in the human genome small star, filled. Genomics 82:323–330
12. Meister G et al (2004) Human Argonaute2 mediates RNA cleavage targeted by miRNAs and siRNAs. Mol Cell 15:185–197
13. Witold F (2005) RNAi: the nuts and bolts of the RISC machine. Cell 122:17–20
14. Liu J, Carmell MA et al (2004) Argonaute2 is the catalytic engine of mammalian RNAi. Science 305:1437–1441
15. Schwarz DS et al (2003) Asymmetry in the assembly of the RNAi enzyme complex. Cell 115:199–208
16. Khvorova A et al (2003) Functional siRNAs and miRNAs exhibit strand bias. Cell 115:209–216
17. Reynolds A et al (2004) Rational siRNA design for RNA interference. Nat Biotechnol 22: 326–330
18. Ui-Tei K et al (2004) Guidelines for the selection of highly effective siRNA sequences for mammalian and chick RNA interference. Nucleic Acids Res 32:936–948
19. Chalk AM et al (2004) Improved and automated prediction of effective siRNA. Biochem Biophys Res Commun 319:264–274
20. Jagla B et al (2005) Sequence characteristics of functional siRNAs. RNA 11:864–872
21. Saetrom P (2004) Predicting the efficacy of short oligonucleotides in antisense and RNAi experiments with boosted genetic programming. Bioinformatics 20:3055–3063
22. Teramoto R et al (2005) Prediction of siRNA functionality using generalized string kernel and support vector machine. FEBS Lett 579: 2878–2882
23. Jia P et al (2006) Demonstration of two novel methods for predicting functional siRNA efficiency. BMC Bioinform 7:271
24. Huesken D et al (2005) Design of a genome-wide siRNA library using an artificial neural network. Nat Biotechnol 23:995–1001
25. Matveeva OV et al (2010) Optimization of duplex stability and terminal asymmetry for shRNA design. PLoS One 5:e10180
26. Ge G et al (2005) Prediction of siRNA knockdown efficiency using artificial neural network models. Biochem Biophys Res Commun 336: 723–728
27. Takasaki S, Kawamura Y, Konagaya A (2006) Selecting effective siRNA sequences by using radial basis function network and decision tree learning. BMC Bioinform 7(Suppl 5):S22

28. Ichihara M et al (2007) Thermodynamic instability of siRNA duplex is a prerequisite for dependable prediction of siRNA activities. Nucleic Acids Res 35:e123
29. Vert J-P et al (2006) An accurate and interpretable model for siRNA efficacy prediction. BMC Bioinform 7:520
30. Matveeva O et al (2007) Comparison of approaches for rational siRNA design leading to a new efficient and transparent method. Nucleic Acids Res 35:e63
31. Ladunga I (2007) More complete gene silencing by fewer siRNAs: transparent optimized design and biophysical signature. Nucleic Acids Res 35:433–440
32. Peek AS, Behlke MA (2007) Design of active small interfering RNAs. Curr Opin Mol Ther 9:110–118
33. Lu ZJ, Mathews DH (2008) Efficient siRNA selection using hybridization thermodynamics. Nucleic Acids Res 36:640–647
34. Wang X et al (2009) Selection of hyperfunctional siRNAs with improved potency and specificity. Nucleic Acids Res 37:e152
35. Jiang P et al (2007) RFRCDB-siRNA: improved design of siRNAs by random forest regression model coupled with database searching. Comput Methods Programs Biomed 87:230–238
36. Takasaki S (2009) Methods for selecting effective siRNA sequences by using statistical and clustering techniques. Methods Mol Biol 487:1–39
37. Davis J, Goadrich M (2006) The relationship between Precision-Recall and ROC curves. ACM Press, New York, NY, pp 233–240
38. Saetrom P, Snøve O Jr (2004) A comparison of siRNA efficacy predictors. Biochem Biophys Res Commun 321:247–253
39. Shabalina SA, Spiridonov AN, Ogurtsov AY (2006) Computational models with thermodynamic and composition features improve siRNA design. BMC Bioinform 7:65
40. Fellmann C et al (2011) Functional identification of optimized RNAi triggers using a massively parallel sensor assay. Mol Cell 41:733–746
41. Silva JM et al (2008) Profiling essential genes in human mammary cells by multiplex RNAi screening. Science 319:617–620
42. Schlabach M et al (2008) Cancer proliferation gene discovery through functional genomics. Science 319:620–624
43. Siolas D et al (2005) Synthetic shRNAs as potent RNAi triggers. Nat Biotechnol 23:227–231
44. Boudreau RL, Monteys AM, Davidson BL (2008) Minimizing variables among hairpin-based RNAi vectors reveals the potency of shRNAs. RNA 14:1834–1844
45. Amarzguioui M, Prydz H (2004) An algorithm for selection of functional siRNA sequences. Biochem Biophys Res Commun 316: 1050–1058
46. Walton SP, Wu M, Gredell JA, Chan C (2010) Designing highly active siRNAs for therapeutic applications. FEBS J 277:4806–4813
47. Gong W et al (2006) Integrated siRNA design based on surveying of features associated with high RNAi effectiveness. BMC Bioinform 7:516
48. Takasaki S, Kotani S, Konagaya A (2004) An effective method for selecting siRNA target sequences in mammalian cells. Cell Cycle 3: 790–795
49. Holen T (2006) Efficient prediction of siRNAs with siRNA rules 1.0: an open-source JAVA approach to siRNA algorithms. RNA 12: 1620–1625
50. Takasaki S (2009) Selecting effective siRNA target sequences by using Bayes' theorem. Comput Biol Chem 33:368–372
51. Peek AS (2007) Improving model predictions for RNA interference activities that use support vector machine regression by combining and filtering features. BMC Bioinform 8:182
52. Patzel V et al (2005) Design of siRNAs producing unstructured guide-RNAs results in improved RNA interference efficiency. Nat Biotechnol 23:1440–1444
53. Köberle C, Kaufmann SHE, Patzel V (2006) Selecting effective siRNAs based on guide RNA structure. Nat Protoc 1:1832–1839
54. Bohula EA et al (2003) The efficacy of small interfering RNAs targeted to the type 1 insulin-like growth factor receptor (IGF1R) is influenced by secondary structure in the IGF1R transcript. J Biol Chem 278:15991–15997
55. Yoshinari K, Miyagishi M, Taira K (2004) Effects on RNAi of the tight structure, sequence and position of the targeted region. Nucleic Acids Res 32:691–699
56. Heale BS et al (2005) siRNA target site secondary structure predictions using local stable substructures. Nucleic Acids Res 33:e30
57. Brown KM, Chu C-Y, Rana TM (2005) Target accessibility dictates the potency of human RISC. Nat Struct Mol Biol 12:469–470
58. Overhoff M et al (2005) Local RNA target structure influences siRNA efficacy: a systematic global analysis. J Mol Biol 348:871–881
59. Schubert S et al (2005) Local RNA target structure influences siRNA efficacy: systematic analysis of intentionally designed binding regions. J Mol Biol 348:883–893
60. Shao Y et al (2007) Effect of target secondary structure on RNAi efficiency. RNA 13: 1631–1640

61. Tafer H et al (2008) The impact of target site accessibility on the design of effective siRNAs. Nat Biotechnol 26:578–583
62. Alsheddi T et al (2008) siRNAs with high specificity to the target: a systematic design by CRM algorithm. Mol Biol (Mosk) 42:163–171
63. Tilesi F et al (2009) Design and validation of siRNAs and shRNAs. Curr Opin Mol Ther 11: 156–164
64. Frank F, Sonenberg N, Nagar B (2010) Structural basis for 5(prime)-nucleotide base-specific recognition of guide RNA by human AGO2. Nature 465:818–822
65. Elbashir SM et al (2001) Duplexes of 21-nucleotide RNAs mediate RNA interference in cultured mammalian cells. Nature 411:494–498
66. Amarzguioui M et al (2003) Tolerance for mutations and chemical modifications in a siRNA. Nucleic Acids Res 31:589–595
67. Jackson AL et al (2003) Expression profiling reveals off-target gene regulation by RNAi. Nat Biotechnol 21:635–637
68. Jackson AL et al (2006) Widespread siRNA "off-target" transcript silencing mediated by seed region sequence complementarity. RNA 12:1179–1187
69. Birmingham A et al (2006) 3′ UTR seed matches, but not overall identity, are associated with RNAi off-targets. Nat Meth 3: 199–204
70. Echeverri CJ et al (2006) Minimizing the risk of reporting false positives in large-scale RNAi screens. Nat Meth 3:777–779
71. Vankoningsloo S et al (2008) Gene expression silencing with "specific" small interfering RNA goes beyond specificity—a study of key parameters to take into account in the onset of small interfering RNA off-target effects. FEBS J 275: 2738–2753
72. Doench JG, Sharp PA (2004) Specificity of microRNA target selection in translational repression. Genes Dev 18:504–511
73. Lin X et al (2005) siRNA-mediated off-target gene silencing triggered by a 7 nt complementation. Nucleic Acids Res 33:4527–4535
74. Anderson EM et al (2008) Experimental validation of the importance of seed complement frequency to siRNA specificity. RNA 14: 853–861
75. Grimson A et al (2007) MicroRNA targeting specificity in mammals: determinants beyond seed pairing. Mol Cell 27:91–105
76. Sioud M (2005) Induction of inflammatory cytokines and interferon responses by double-stranded and single-stranded siRNAs is sequence-dependent and requires endosomal localization. J Mol Biol 348:1079–1090
77. Hornung V et al (2005) Sequence-specific potent induction of IFN-alpha by short interfering RNA in plasmacytoid dendritic cells through TLR7. Nat Med 11:263–270
78. Judge AD et al (2005) Sequence-dependent stimulation of the mammalian innate immune response by synthetic siRNA. Nat Biotechnol 23:457–462
79. Mysara M et al (2011) MysiRNA-designer: a workflow for efficient siRNA design. PLoS One 6:e25642
80. Armant MA, Fenton MJ (2002) Toll-like receptors: a family of pattern-recognition receptors in mammals. Genome Biol 3:reviews3011.1–reviews3011.6
81. Akira S, Hemmi H (2003) Recognition of pathogen-associated molecular patterns by TLR family. Immunol Lett 85:85–95
82. Heil F et al (2004) Species-specific recognition of single-stranded RNA via toll-like receptor 7 and 8. Science 303:1526–1529
83. Sioud M (2006) Single-stranded small interfering RNA are more immunostimulatory than their double-stranded counterparts: a central role for 2′-hydroxyl uridines in immune responses. Eur J Immunol 36:1222–1230
84. Goodchild A et al (2009) Sequence determinants of innate immune activation by short interfering RNAs. BMC Immunol 10:40
85. Forsbach A et al (2008) Identification of RNA sequence motifs stimulating sequence-specific TLR8-dependent immune responses. J Immunol 180:3729–3738
86. Gantier MP et al (2008) TLR7 is involved in sequence-specific sensing of single-stranded RNAs in human macrophages. J Immunol 180: 2117–2124
87. Stewart CR et al (2011) Immunostimulatory motifs enhance antiviral siRNAs targeting highly pathogenic avian influenza H5N1. PLoS One 6:e21552
88. Sugiyama T et al (2005) CpG RNA: identification of novel single-stranded RNA that stimulates human CD14+CD11c+ monocytes. J Immunol 174:2273–2279
89. Fedorov Y et al (2006) Off-target effects by siRNA can induce toxic phenotype. RNA 12: 1188–1196
90. Birmingham A et al (2007) A protocol for designing siRNAs with high functionality and specificity. Nat Protoc 2:2068–2078
91. Dahlgren C et al (2008) Analysis of siRNA specificity on targets with double-nucleotide mismatches. Nucleic Acids Res 36:e53
92. Alemán LM, Doench J, Sharp PA (2007) Comparison of siRNA-induced off-target RNA and protein effects. RNA 13:385–395

Chapter 2

Methods for Selecting Effective siRNA Target Sequences Using a Variety of Statistical and Analytical Techniques

Shigeru Takasaki

Abstract

Short interfering RNA (siRNA) has been widely used for studying gene function in mammalian cells but varies markedly in its gene silencing efficacy. Although many design rules/guidelines for effective siRNAs based on various criteria have been reported recently, there are only a few consistencies among them. This makes it difficult to select effective siRNA sequences in mammalian genes. This chapter first reviews the recently reported siRNA design guidelines and then proposes new methods for selecting effective siRNA sequences from many possible candidates by using decision tree learning, Bayes' theorem, and average silencing probability on the basis of a large number of known effective siRNAs. These methods differ from the previous score-based siRNA design techniques and can predict the probability that a candidate siRNA sequence will be effective. Evaluation of these methods by applying them to recently reported effective and ineffective siRNA sequences for a number of genes indicates that they would be useful for many other genes. They should, therefore, be of general utility for selecting effective siRNA sequences for mammalian genes. The chapter also describes another method using a hidden Markov model to select the optimal functional siRNAs and discusses the frequencies of combinations of two successive nucleotides as an important characteristic of effective siRNA sequences.

Key words: siRNA, siRNA design, RNA interference, Gene silencing, Estimation of gene silencing, Decision tree learning, Bayes' theorem, Average gene silencing, Hidden Markov model

1. Introduction

RNA interference (RNAi) silences gene expression by introducing double-stranded RNA homologous to the target mRNA. It has been widely used for studying gene functions, but many practical obstacles need to be overcome before it becomes an established tool for use in mammalian systems (1–6). One of the important problems is designing effective short interfering RNA

Debra J. Taxman (ed.), *siRNA Design: Methods and Protocols*, Methods in Molecular Biology, vol. 942,
DOI 10.1007/978-1-62703-119-6_2,

(siRNA) sequences for target genes. The effectiveness of the siRNA responsible for RNA interference varies widely depending on the target sequence positions (sites) selected from the target gene (7, 8). We therefore need useful criteria for gene silencing efficacy when we design siRNA sequences (9, 10).

Schwarz et al. and Khvorova et al. showed that the 5′ end of the antisense strand might be incorporated into the RNA-induced silencing complex. Strand incorporation may depend on weaker base-pairing, and an A–T terminus may thus lead to more strand incorporation than a G–C terminus (11, 12). Other factors reported to be related to gene silencing efficacy are GC content, point-specific nucleotides, specific motif sequences, and secondary structures of mRNA. Several siRNA design rules/guidelines using efficacy-related factors have been reported (13–17).

Although the effectiveness of siRNA sequences seems to be determined largely by their nucleotide sequences, there are few consistencies among the reported rules (18–23). This implies that they might result in the generation of many candidate target sequences, making it difficult to select the effective ones. In addition, the previously reported rules cannot estimate the probability that a candidate siRNA will actually silence the target gene. What are therefore needed are not only methods for selecting high-potential siRNA candidates but also methods for estimating the probability that the selected candidates will indeed silence their target genes. Furthermore, there is in RNAi a risk of off-target regulation: a possibility that the siRNA will silence other genes whose sequences are similar to those of the target gene. When we use gene silencing for studying gene functions, we have to first somehow select high-potential siRNA candidate sequences and then eliminate possible off-target ones (24).

This chapter first reviews the recently reported siRNA design guidelines and clarifies their problems. It then describes prediction methods for selecting effective siRNA target sequence from many possible candidate sequences by using decision tree learning, Bayes' theorem, and average silencing probability of a large number of siRNA sequences known to be effective (25–32). They are quite different from the previous score-based siRNA design techniques and can predict the probability that a candidate siRNA sequence will be effective. The results obtained when applying these statistical methods to recently reported effective and ineffective siRNA sequences for various genes showed that they are accurate and thus imply that they would be useful for selecting siRNA sequences silencing many other genes. This chapter also describes another method using a hidden Markov model (HMM) to select the optimal functional siRNAs (25) and discusses the frequencies of combinations of two successive nucleotides as an important characteristic of effective siRNA sequences.

2. siRNA Sequence Selection Problems

To use RNAi as a biological tool for mammalian cell experiments, we first need to identify target sequences causing gene degradation. They have so far been identified by using a trial-and-error method (3, 8), but siRNAs extracted from different regions of the same gene have varied remarkably in their effectiveness. The difficulty of using the trial-and-error method to select target sequences causing gene silencing increases when the coding regions are long, as they are in mammalian cells. This is because the number of candidates increases with the length of the coding region.

2.1. The Reported Guidelines for Designing siRNA Sequences

The earliest guidelines for siRNA sequence design were proposed by Elbashir et al. (4, 8, 33). They suggested that the target mRNA is silenced effectively by siRNA duplexes 21 nucleotides long: 19-nt base-paired sequences with 2-nt overhangs at the 3′ ends. Many siRNA design guidelines/rules have been reported since then, and this chapter considers the following five (herein designated guidelines G1–G5).

Reynolds et al. (18) analyzed 180 siRNAs systematically, targeting every other position of two 197-base regions of firefly luciferase and human cyclophilin B mRNA (90 siRNAs per gene), and reported eight criteria for improving siRNA selection.

Guideline G1

1. G/C content 30–52%.
2. at least three As or Ts at positions 15–19.
3. absence of internal repeats.
4. an A at position 19.
5. an A at position 3.
6. a T at position 10.
7. a base other than G or C at position 19.
8. a base other than G at position 13.

Ui-Tei et al. (19) examined 72 siRNAs targeting six genes and reported four rules for effective siRNA designs.

Guideline G2

1. an A or T at position 19.
2. a G or C at position 1.
3. at least five T or A residues from positions 13 to 19.
4. no GC stretch more than 9 nt long.

Amarzguioui and Prydz (20) analyzed 46 siRNAs targeting four genes and reported six rules for effective siRNA designs.

Table 1
Effective and ineffective nucleotides specified in the individual guidelines

	Position	1	3	6	10	11	13	16	19
G1	Preferred		A		T		A/C/T		A/T
G2	Preferred	G/C							A/T
	Unpreferred	A/T							G/C
G3	Preferred	G/C		A			T	C	A/T
	Unpreferred	T			T				G
G4	Preferred	G/C			A/T				A/T
G5	Preferred					C/G	A	G	T
	Unpreferred			C		A/T			G

Position: Nucleotide position from 1 to 19 (5′ to 3′, cDNA form)
Preferred: Effective (positive), unpreferred: ineffective (negative)

Guideline G3

1. a G or C at position 1.
2. an A at position 6.
3. a base other than T at position 10.
4. a T at position 13.
5. a C at position 16.
6. an A or T at position 19.

Jagla et al. (22) tested 601 siRNAs targeting one exogenous and three endogenous genes and reported four rules.

Guideline G4

1. an A or T at position 19.
2. an A or T at position 10.
3. a G or C at position 1.
4. more than three A/Ts between positions 13 and 19.

Hsieh et al. (21) examined 138 siRNAs targeting 22 genes and reported five position-specific characteristics.

Guideline G5

1. a T at position 19.
2. a C or G at position 11.
3. a G at position 16.
4. an A at position 13.
5. a base other than C at position 6.

These guidelines are summarized in Table 1.

Other methods for scoring, screening, and designing functional siRNAs have also been reported recently. Chalk et al. (13) reported the following seven rules ("Stockholm rules") based on thermodynamic properties: (1) total hairpin energy <1, (2) antisense 5′ end binding energy <9, (3) sense 5′ end binding energy in range 5–9 exclusive, (4) GC between 36% and 53%, (5) middle (7–12) binding energy <13, (6) energy difference <0, and (7) energy difference between −1 and 0. The score of an siRNA candidate is incremented by one point for each rule fulfilled and is thus between 0 and 7.

Huesken et al. (23) reported a method for screening functional siRNAs by using an artificial neural network. This network was first trained by 2,182 randomly selected siRNAs targeted to 34 genes and was used in the design of a genome-wide siRNA collection with two potent siRNAs per gene.

Teramoto et al. (14) and Ladunga (34) have reported functional siRNA selection methods using support vector machines (SVMs). Teramoto et al. used a generalized string kernel (GSK) combined with an SVM. siRNA sequences were represented as vectors in a multidimensional feature space according to the number of subsequences in each siRNA and were classified as effective or ineffective (14). Ladunga used an SVM with polynomial kernels and constrained optimization models from 572 sequence, thermodynamic, accessibility, and self-hairpin features over 2,200 published siRNAs (23, 34). As the key to SVM success is to collect many useful features of effective siRNA sequences, the usefulness of methods using SVMs may depend on the selected siRNAs.

Holen recently reported siRNA rules based on apparent overrepresentation or underrepresentation of certain nucleotides in certain positions of the Novartis data set (35). The criteria for an siRNA candidate depend on the positive and negative scores computed for each position by using a scoring table generated by the percentage overrepresentation or underrepresentation of individual nucleotides for each position in the large Novartis data set (23). Although the method was evaluated by using other reported siRNA sets, which of the candidate siRNAs actually silence genes is not clear. In addition, the original scores in the scoring table are based on the percentage overrepresentation or underrepresentation of certain nucleotides in certain positions and thus may vary drastically depending on what sets of siRNAs are used. This makes it difficult to evaluate the scores computed for siRNA candidates.

Although secondary structures of siRNA sequences are also thought to be important in predicting siRNA efficacy, there are conflicting results concerning the effects of secondary structures on siRNA functionality. Some studies have suggested that the secondary structure of the siRNA plays a role in determining the efficacy of gene silencing (36–38), but others did not find any correlation between the functionality of the siRNA and the secondary structures of the target mRNA (7, 18, 20). This issue therefore requires further study (39, 40).

Table 2
Features of individual siRNA design rules/algorithms

siRNA design rules	Citation	No. of genes	No. of siRNAs	Description	Technique
Reynolds et al.	(18)	2	197	Sequence features	
Ui-Tei et al.	(19)	6	72	Sequence features	
Amarzguioui et al.	(20)	4	46	Sequence features	
Hsieh et al.	(21)	22	138	Sequence features	
Hesken et al.	(23)	34	2,128	Sequence motifs	Neural network
Jagla et al.	(22)	4	601	Sequence features	Decision tree
Holen	(34)	34	400	Sequence features	Percentage
Saetrom	(35)	40	581	Sequence motifs	Genetic programming
Teramoto et al.	(14)	2	94	Sequence motifs	Support vector machine
Ladunga	(41)	34	2,252	Position features	Support vector machine
Chalk et al.	(13)	92	398	Binding energy	Regression tree
Takasaki et al.	(29–31)	490	833	Sequence features	Statistics, SOM

The features of various siRNA design rules are summarized in Table 2. In addition, other design rules using a combination of the above techniques have also been used to obtain efficacious siRNAs based on public/open siRNA data sets (42–54).

2.2. Problems with the Previous Guidelines

Among the problems with the reported guidelines is the problem of inconsistencies with regard to the nucleotide frequencies of each position. Although some guidelines have the same preferred and unpreferred nucleotides at positions 1 and 19, there are few consistencies at other positions (Table 1). These results indicate that though some rules from the guidelines are suitable for identifying effective sequences for some genes, they might be unsuitable for others. Because the previous guidelines are based on the analyses of specific genes, it could be inferred that they are not always effective for many other genes. Therefore, if these guidelines were used to select siRNA target sequence candidates for other mammalian genes, many ineffective sequences might be selected as candidates. This is due to the prevalence of long coding regions in mammalian genes. This poses an additional problem because experimentally evaluating whether the selected sequences provide effective gene degradation is a costly and time-consuming task. Still another problem is that the previously reported methods cannot estimate the probability that a candidate siRNA will actually silence the target gene. Even if a high-scoring siRNA were obtained using the

Table 3
Relations between attributes (positions and nucleotides of 19 nt sequences) and training instances (no. of effective and ineffective siRNAs)

Position	1				2				3				⋮	19			
Nucleotide	A	G	C	T	A	G	C	T	A	G	C	T	⋮	A	G	C	T
No. of nt (effect)	100	464	173	96	250	225	202	156	265	191	173	204	⋮	259	173	194	207
No. of nt (ineffect)	264	157	182	244	209	194	214	230	179	217	251	200	⋮	71	270	361	145

No. of nt (effect): No. of nucleotides in effective siRNAs
No. of nt (ineffect): No. of nucleotides in ineffective siRNAs

reported methods, it would be difficult to estimate the probability that it would actually accomplish the expected gene degradation. To overcome the problems of the previous guidelines, Takasaki et al. recently reported new scoring methods using the statistical and clustering techniques listed in Table 2 (28–32).

3. Methods for siRNA Sequence Selections

To design effective siRNA sequences for target genes, decision tree learning, Bayes's theorem, and average silencing probability of a large number of effective siRNA sequences were used in the proposed methods (26, 27). These methods mainly consist of two phases, one is to learn the relations between individual siRNA sequences and their gene silencing efficacies by using known data, and the other is to predict gene silencing probabilities for new candidate siRNA sequences by using the learned relations. The learning phase is carried out by training and validation phases by supplying many known effective and ineffective siRNA sequences. The prediction phase is where predictions of gene silencing probabilities are actually computed for new candidate siRNA sequences.

3.1. Prediction by the Decision Tree Learning Method

3.1.1. Model for Decision Tree Learning

Individual positions and nucleotides of 19 sequences listed in Table 3 were used as attributes. 833 effective and 847 ineffective siRNA sequences were used as training instances.

To carry out the supervised learning for effective siRNA classifications by using decision tree learning, the training instances are partitioned into two sets, one for the growth of the decision tree (training data) and other for the decision tree pruning (testing data). The processes of the classifications were carried out in two phases: the growth and pruning of the decision tree.

3.1.2. The Growth of the Decision Tree

The algorithm, in outline, is as follows:

1. If all the instances belong to a single class, there is nothing to do (except create a leaf node labeled with the name of that class).
2. Otherwise, for each attribute that has not already been used, calculate the information gain that would be obtained by using that attribute on the particular set of instances classified to this branch node. The information gain can be computed in the following way (55):

$$I(p,u) = -\frac{p}{p+u}\log_2\left(\frac{p}{p+u}\right) - \frac{u}{p+u}\log_2\left(\frac{u}{p+u}\right) \quad (1)$$

where p is the total number of nucleotides for this attribute in effective (preferred) siRNA sequences and u is the total number of nucleotides in ineffective (unpreferred) siRNA sequences.

The entropy $H(L)$ associated with the attribute L is

$$H(L) = \sum_{i=1}^{v} \frac{p_i + u_i}{p+u} I(p_i, u_i) \quad (2)$$

where v is a kind of nucleotide, i.e., $i = 1 = \mathrm{A}$, $2 = \mathrm{G}$, $3 = \mathrm{C}$, and $4 = \mathrm{T}$, p_i and u_i are, respectively, a number of the corresponding nucleotides in effective and ineffective siRNA sequences, and L is the attribute (position) listed in Table 3. The information gain is therefore obtained as follows:

$$gain(L) = I(p,u) - H(L) \quad (3)$$

3. Use the attribute (position) with the greatest information gain as a branch node and for each nucleotide of L, create a new descendant of the node.
4. If the information gain becomes less than the specified criterion, stop the growth of the decision tree and create leaf nodes; otherwise continue to build the tree.

3.1.3. Decision Tree Pruning

Working backwards from the bottom of the tree, the subtree starting at each nonterminal node is examined. If the error (misclassification) rate on the testing data improves by pruning it, the subtree is removed. The process continues until no improvement can be made by pruning a subtree.

The predictions of gene silencing for new candidate siRNA sequences are carried out by the decision tree method described above. This article used the recently reported effective and ineffective siRNA sequences as the evaluation data for predictions (see Subheading 3.5).

3.2. Prediction Analysis of siRNA Target Sequences Based on Bayes' Theorem

Bayes' theorem tells us the probability that a hypothesis is true given the prior probability that we would have assigned to the hypothesis, and both conditional probabilities A and B. In this case, the hypothesis is that an siRNA candidate will effectively silence mammalian genes, the prior probability is the best guess, i.e., the empirical information, the conditional probability A is that an siRNA candidate will effectively silence genes if the individual nucleotides at each of the positions of the candidate belonged to the set of nucleotide frequently found in effective siRNAs, and the conditional probability B is that the siRNA candidate will effectively silence genes even if the corresponding nucleotides belonged to the set of nucleotides frequently found in ineffective siRNAs. The sets of nucleotide frequencies for effective and ineffective siRNAs can be estimated from a large number of known effective and ineffective siRNA sequences in the literature.

3.2.1. Prediction of Gene Degradation Ratio for a Given siRNA Candidate

Given a candidate siRNA sequence **X** (**X** = X_1, X_2, …$X_{19,}$ where X_i is a nucleotide) for a specified target gene, Bayes' theorem (25) could be used to predict the following gene silencing probability $P(eff|X)$ for that sequence:

$$P(\text{eff}|X) = \frac{P^{\text{eff}}\,P(X\mid\text{eff})}{P(X)} \tag{4}$$

where P^{eff} is the prior probability of more than 80% gene reduction by siRNA sequences. This probability is obtained, as empirical knowledge, from many siRNA experiments and is known to be approximately 0.1–0.2 in mammalian genes (9, 10, 18). This prior probability represents the best guess that we can make about an siRNA candidate sequence before we have seen information about the sequence itself. $P(X|\text{eff})$ is the probability that **X** would cause effective gene silencing if X_1, X_2, …X_{19} belonged to the set of nucleotides frequently found in effective siRNA sequences, for example, from 833 effective siRNAs in the literature (see Subheading 3.5.1). $P(X|\text{eff})$ is therefore computed by the product of individual frequency ratios of the positional nucleotides in the following way:

$$P(X|\text{eff}) = \prod_{i=1}^{19} q_{x_i^n}^{\text{eff}} \tag{5}$$

where i is the nucleotide position and n is the kind of nucleotide (A, C, G, or T). That is, $x_i^A = A$, $x_i^C = C$, $x_i^G = G$, and $x_i^T = T$ and $q_{x_i^n}^{\text{eff}}$ is the frequency ratio of ith position nucleotide n of the candidate **X**. Individual $q_{x_i^n}^{\text{eff}}$ can be derived from the corresponding nucleotide frequency ratios in the set of effective siRNA sequences described before. As $P(X)$ is the probability that X will result in effective gene silencing, it can be computed by summation of both probabilities derived from frequency ratio sets of known effective and not-effective siRNAs in the following way:

$$P(X) = P^{\text{eff}}\,P(X|\text{eff}) + P^{\text{inf}}\,P(X|\text{inf}) \tag{6}$$

where P^{inf} is also the prior probability and is equal to $1 - P^{\text{eff}}$.

$P(X|\text{inf})$ is the probability that X will result in effective gene silencing if X_1, X_2, …X_{19} belongs to the set of nucleotides frequently found in not-effective sequences. $P(X|\text{inf})$ is therefore computed by the product of individual frequency ratios of the positional nucleotides in the following way:

$$P(X|\text{inf}) = \prod_{i=1}^{19} q_{x_i^n}^{\text{inf}} \tag{7}$$

where $q_{x_i^n}^{\text{inf}}$ is the frequency ratio of *i*th position nucleotide *n* (A, C, G, or T) of the candidate X. Individual $q_{x_i^n}^{\text{inf}}$ can be obtained from the corresponding nucleotide frequency ratios in the set of ineffective siRNA sequences (see Subheading 3.5.2).

Using Eqs. 5–7, we can express Eq. 4 as follows:

$$P(\text{eff}|X) = \frac{P^{\text{eff}} P(X|\text{eff})}{P^{\text{eff}} P(X|\text{eff}) + P^{\text{inf}} P(X|\text{inf})} = \frac{P^{\text{eff}} \prod_{i=1}^{19} q_{x_i^n}^{\text{eff}}}{P^{\text{eff}} \prod_{i=1}^{19} q_{x_i^n}^{\text{eff}} + P^{\text{inf}} \prod_{i=1}^{19} q_{x_i^n}^{\text{inf}}} \tag{8}$$

This $P(\text{eff}|X)$ is the gene silencing probability we want. It is called the posterior probability that **X** will result in gene silencing because it is our best guess after we have seen the siRNA candidate sequence.

Suppose, for example, that we have a candidate siRNA sequence CCATCAACACCGAGTTCAA for some target gene. What is the probability of gene silencing by this sequence? If we assume that $p^{\text{eff}} = 0.1$ and $p^{\text{inf}} = 0.9$ and the observed frequencies of effective ($q_{x_i^n}^{\text{eff}}$) and ineffective ($q_{x_i^n}^{\text{inf}}$) nucleotides at each of the positions are given as $q_{x_i^n}^{\text{eff}} = (0.208, 0.242, 0.318, 0.197, 0.248, 0.294, 0.247, 0.229, 0.271, 0.275, 0.256, 0.247, 0.259, 0.276, 0.286, 0.23, 0.235, 0.313, 0.312)$ and $q_{x_i^n}^{\text{inf}} = (0.215, 0.253, 0.211, 0.207, 0.266, 0.231, 0.273, 0.253, 0.235, 0.242, 0.269, 0.257, 0.226, 0.266, 0.215, 0.21, 0.262, 0.182, 0.084)$ from the sets of nucleotide frequencies in the effective and ineffective siRNA sequences, Eq. 8 predicts that the gene silencing probability of this candidate is 0.63 (63%). This indicates that although we have only information about 10% gene silencing by this siRNA candidate as the prior probability (the best guess) before we have seen the candidate, we can predict a gene silencing probability six times higher after we have seen it. This is useful information for selecting effective siRNA candidates. In the case of $p^{\text{eff}} = 0.2$ and $p^{\text{inf}} = 0.8$, the gene silencing probability of the candidate is predicted to be 79.3%.

3.2.2. Evaluation Criteria Used in the Proposed Method

To make the estimated results using Bayes' theorem easily understood, the ratio of the result estimated for a new siRNA candidate to the prior probability (best guess: empirical information) is first considered. This ratio *PR* is therefore defined as follows:

$$PR = \frac{ER}{PP} \tag{9}$$

where *PP* is the prior probability and *ER* is the result estimated by Bayes' theorem.

If $PR > 1$, it indicates that the level of gene silencing will be higher than the usual level of silencing. Conversely, a $PR < 1$ indicates that the level of silencing will be lower than the usual level.

Now let us consider the normalized ratio (NR) obtained when we divide the result estimated for a new siRNA candidate by the average of the results estimated for a large number of known effective siRNAs. We do this because the results estimated for the known effective siRNAs could be considered a standard criterion for new siRNA candidates. The ratio *NR* is therefore defined as follows:

$$NR = \frac{ER}{AP} \tag{10}$$

where AP is the average of the probabilities predicted for the known effective siRNAs.

This *NR*, therefore, indicates the gene silencing potential of the siRNA candidates relative to that of the known effective siRNAs. If $NR \geq 1$, the level of gene silencing expected to be obtained with the siRNA candidate is the same as or higher than the level of silencing obtained with the known effective siRNAs. That is, *NR* indicates that the candidate sequence has a high potential for gene silencing. If, on the other hand, $NR < 1$, it indicates that the gene silencing expected to be obtained with the candidate sequence is lower than the level of silencing obtained with the known effective siRNAs.

3.2.3. Verification Models of the Proposed Method

We can see from Eq. 8 that the accuracy of the gene silencing probability predicted by the proposed method is greatly dependent on individual $q_{x_i^n}^{\mathrm{eff}}$ and $q_{x_i^n}^{\mathrm{inf}}$ values. Generally, the larger the sets of effective and ineffective siRNAs, the higher the prediction accuracy, and a set of $q_{x_i^n}^{\mathrm{inf}}$ would be a complementary set of $q_{x_i^n}^{\mathrm{eff}}$. It is difficult, however, to obtain complete sets of effective and ineffective siRNAs because of the difficulty of the siRNA verification for all genes. Individual $q_{x_i^n}^{\mathrm{eff}}$ and $q_{x_i^n}^{\mathrm{inf}}$ values were therefore generated on the basis of a large number of effective and ineffective siRNAs in the literature. We make two assumptions about the occurrence of nucleotides in the sets of effective siRNA sequences from which $q_{x_i^n}^{\mathrm{eff}}$ is determined. One is that the nucleotides occur independently at individual positions (see Table 4a), and the other is that the occurrence of individual nucleotides at individual positions depends on the nucleotides at other positions. A typical occurrence dependency is, for example, the Markov chain dependency (the simple (first) Markov model). That is, the nucleotide occurrences at the present position are dependent on the nucleotides at the previous position. The probability of the simple Markov model $M(q_{x_i^n}^{\mathrm{eff}})$ for the nucleotide x^n at the present position i is therefore expressed as follows:

$$M(q_{x_i^n}^{\mathrm{eff}}) = P(x_i^n \,|\, q_{x_{i-1}^n}^{\mathrm{eff}}) \tag{11}$$

Table 4
Probabilities of individual nucleotide occurrences at each position

(a) Probabilities of independent nucleotide occurrences in 833 effective siRNAs

	1	2	3	4	5	6	7	8	9	10	11	12	13	14	15	16	17	18	19
A	0.12	0.3	0.318	0.218	0.271	0.294	0.247	0.298	0.271	0.229	0.25	0.283	0.259	0.282	0.27	0.232	0.288	0.313	0.312
G	0.557	0.27	0.229	0.291	0.262	0.239	0.304	0.253	0.224	0.257	0.279	0.247	0.255	0.276	0.242	0.287	0.24	0.239	0.208
C	0.208	0.242	0.208	0.294	0.248	0.208	0.263	0.229	0.244	0.275	0.256	0.234	0.247	0.217	0.202	0.251	0.235	0.178	0.233
T	0.115	0.187	0.245	0.197	0.218	0.259	0.186	0.22	0.261	0.239	0.216	0.235	0.239	0.224	0.286	0.23	0.236	0.27	0.248

(b) Probabilities of dependent nucleotide occurrences—the simple Markov model

		1′–2		2′–3		3′–4		4′–5		5′–6		6′–7		7′–8		8′–9		9′–10	
A	A	0.12	0.16	0.3	0.328	0.318	0.2	0.218	0.247	0.271	0.292	0.294	0.229	0.247	0.282	0.298	0.278	0.271	0.204
	G		0.35		0.272		0.336		0.341		0.265		0.363		0.311		0.286		0.323
	C		0.28		0.192		0.264		0.22		0.212		0.257		0.214		0.254		0.257
	T		0.21		0.208		0.2		0.192		0.23		0.151		0.194		0.181		0.217
G	A	0.557	0.33	0.27	0.364	0.229	0.241	0.291	0.314	0.262	0.394	0.239	0.342	0.304	0.316	0.253	0.313	0.224	0.337
	G		0.284		0.169		0.293		0.194		0.206		0.241		0.241		0.218		0.193
	C		0.213		0.227		0.251		0.26		0.22		0.246		0.213		0.223		0.203
	T		0.172		0.24		0.215		0.231		0.179		0.171		0.229		0.246		0.267
C	A	0.208	0.358	0.242	0.371	0.208	0.277	0.294	0.331	0.248	0.329	0.208	0.295	0.263	0.37	0.2229	0.33	0.244	0.261
	G		0.116		0.149		0.127		0.171		0.13		0.185		0.137		0.131		0.192
	C		0.295		0.198		0.318		0.245		0.203		0.243		0.224		0.199		0.271
	T		0.231		0.282		0.277		0.253		0.338		0.277		0.269		0.34		0.276
T	A	0.115	0.198	0.187	0.167	0.245	0.172	0.197	0.146	0.218	0.137	0.259	0.144	0.186	0.187	0.22	0.153	0.261	0.134
	G		0.396		0.353		0.368		0.409		0.368		0.389		0.361		0.246		0.304
	C		0.25		0.218		0.353		0.268		0.192		0.301		0.284		0.301		0.359
	T		0.156		0.263		0.108		0.177		0.302		0.167		0.168		0.301		0.203

10′–11		11′–12		12′–13		13′–14		14′–15		15′–16		16′–17		17′–18		18′–19	
0.229	0.267	0.25	0.274	0.283	0.195	0.259	0.245	0.282	0.264	0.27	0.218	0.232	0.249	0.288	0.3	0.313	0.3
	0.356		0.284		0.314		0.329		0.306		0.329		0.275		0.288		0.235
	0.257		0.25		0.263		0.236		0.179		0.258		0.218		0.154		0.231
	0.12		0.192		0.229		0.19		0.251		0.196		0.259		0.254		0.235
0.257	0.271	0.279	0.358	0.247	0.364	0.255	0.373	0.276	0.33	0.242	0.267	0.287	0.285	0.24	0.37	0.239	0.387
	0.271		0.263		0.204		0.274		0.226		0.231		0.238		0.2		0.166
	0.285		0.19		0.204		0.193		0.165		0.249		0.247		0.185		0.196
	0.173		0.19		0.228		0.16		0.278		0.151		0.23		0.245		0.236
0.275	0.306	0.256	0.305	0.234	0.282	0.247	0.325	0.217	0.331	0.202	0.222	0.251	0.421	0.235	0.352	0.178	0.412
	0.131		0.155		0.169		0.146		0.16		0.107		0.148		0.133		0.108
	0.262		0.239		0.241		0.214		0.204		0.187		0.167		0.204		0.23
	0.301		0.3		0.308		0.316		0.304		0.231		0.263		0.306		0.23
0.239	0.146	0.216	0.172	0.235	0.204	0.239	0.181	0.224	0.144	0.286	0.151	0.23	0.188	0.236	0.228	0.27	0.183
	0.382		0.294		0.321		0.357		0.262		0.396		0.307		0.325		0.272
	0.216		0.267		0.281		0.226		0.273		0.236		0.313		0.173		0.263
	0.256		0.267		0.194		0.236		0.321		0.276		0.193		0.274		0.277

(c) Probabilities of independent nucleotide occurrences in 847 ineffective siRNAs

	1	2	3	4	5	6	7	8	9	10	11	12	13	14	15	16	17	18	19
A	0.312	0.247	0.211	0.247	0.254	0.231	0.273	0.26	0.235	0.295	0.251	0.243	0.226	0.235	0.203	0.261	0.262	0.182	0.084
G	0.185	0.229	0.256	0.262	0.237	0.259	0.236	0.254	0.286	0.279	0.231	0.257	0.323	0.266	0.293	0.26	0.255	0.301	0.319
C	0.215	0.253	0.296	0.285	0.266	0.321	0.283	0.253	0.298	0.242	0.269	0.256	0.247	0.244	0.289	0.269	0.262	0.319	0.426
T	0.288	0.272	0.236	0.207	0.243	0.189	0.208	0.234	0.182	0.184	0.248	0.43	0.204	0.255	0.215	0.21	0.221	0.198	0.171

(continued)

Table 4
(continued)

(d) Probabilities of deductive nucleotide occurrences from 833 effective siRNAs

	1	2	3	4	5	6	7	8	9	10	11	12	13	14	15	16	17	18	19
A	0.293	0.233	0.227	0.261	0.243	0.235	0.251	0.234	0.243	0.257	0.25	0.239	0.247	0.239	0.243	0.256	0.237	0.229	0.229
G	0.148	0.243	0.257	0.236	0.246	0.254	0.232	0.249	0.259	0.248	0.24	0.251	0.248	0.241	0.253	0.238	0.253	0.254	0.264
C	0.264	0.253	0.264	0.235	0.251	0.264	0.246	0.257	0.252	0.242	0.248	0.255	0.251	0.261	0.266	0.25	0.255	0.274	0.256
T	0.295	0.271	0.252	0.268	0.261	0.247	0.271	0.26	0.246	0.254	0.261	0.255	0.254	0.259	0.238	0.257	0.255	0.243	0.251

Equation 2.11 tells us that the probability of the nucleotide occurrence x_i^n at the present position i is determined under the condition of the effective nucleotide $q_{x_{i-1}^n}^{\text{eff}}$ at the previous position $i-1$. Suppose, for example, that we have a sequence CCATCAACACCGAGTTCAA as a candidate siRNA sequence for some target gene. In this case, the second nucleotide "C" may occur depending on the first nucleotide "C." From Table 4b (the simple Markov model table) one sees that the probability that C occurs at position 2 under the condition that there is a 0.208 probability that C occurs at position 1 is 0.295. One similarly sees that the probability that the third nucleotide is A is 0.371 under the condition that there is a 0.242 probability that the second nucleotide is C. Two types of evaluations were therefore considered for the proposed prediction method: (a) one for independent nucleotide occurrences at individual positions and (b) one for nucleotide occurrences dependent on the previous nucleotide occurrences at individual positions.

The set from which $q_{x_i^n}^{\text{inf}}$ is determined could also be assumed to have two types of nucleotide occurrences: independent nucleotide occurrences at individual positions (see Table 4c) and the deductive nucleotide occurrences induced from the effective siRNAs. That is, individual nucleotides could occur as deductive complement ratios of the effective siRNAs. The probabilities of deductive nucleotide occurrences $D(q_{x_i^n}^{\text{inf}})$ at the position i are computed as follows:

$$D(q_{x_i^n}^{\text{inf}}) = \frac{(1 - q_{x_i^n}^{\text{eff}})}{3} \tag{12}$$

Suppose, for example, the deductive probability of the nucleotide "A" occurrence at the position 3. As $q_{x_3^A}^{\text{eff}}$ is 0.318 from Table 4a, $D(q_{x_3^A}^{\text{inf}})$ can be computed as 0.227 by using Eq. 12. The deductive probabilities of individual nucleotides at positions 1–19 are shown in Table 4d. Comparing Table 4a, d, it is clear that the probabilities of individual nucleotide occurrences shown in Table 4d are deductive complement ratios of the effective nucleotides shown in Table 4a. Consequently there could be two cases: (c) one with independent nucleotide occurrences and (d) one with the deductive nucleotide occurrences. The evaluation of the proposed method was therefore carried out in four groups: the combinations of the sets of (a) and (c), (a) and (d), (b) and (c), and (b) and (d). They are, respectively, listed in Table 5 as Cases 1, 2, 3, and 4. Individual nucleotide frequencies at each of the positions of (a), (b), (c), and (d) were generated on the basis of 833 effective siRNA sequences and 847 ineffective siRNA sequences. They are listed in Table 4a–d.

Table 5
The combinations of the proposed method evaluations

	(a) Independent	(b) Dependent
(c) Independent	Case 1	Case 3
(d) Deductive	Case 2	Case 4

Independent: Independent nucleotide occurrences at individual positions
Dependent: Nucleotide occurrences dependent on the simple Markov model
Deductive: Deductive nucleotide occurrences from effective siRNAs

3.2.4. A Procedure for Selecting Useful siRNA Sequences Using Bayes' Theorem

Useful siRNA sequences could be selected in the following way.

Estimate the gene silencing potential of candidate siRNA sequences.

- Carry out gene silencing predictions using Bayes' theorem based on the four groups listed in Table 5 (or carry out only the Case 3 because it yielded the most accurate predictions in the evaluations described later).
- Select the siRNA candidates whose gene silencing probability is predicted to be more than 40% when the prior probability is 10%.
- Compare candidate sequences with the sequences of known effective siRNAs.
- Carry out the normalized analyses of the siRNA candidates by using a large number of known siRNAs.
- Select the candidates having a normalized ratio >1.

(Users can easily utilize the proposed method by using Eq. 8 and Table 4.)

3.3. Procedure for Selecting Effective siRNAs Based on the Average Silencing Probability

Many effective gene silencing siRNA sequences have been reported recently and can be used to predict how new siRNA candidates will function. If the probability of individual nucleotide occurrences at positions from 1 to 19 in the effective siRNA population is obtained, it can be used to calculate the probability that candidate siRNAs will be effective. In addition, if the average probability of a large number of effective siRNA sequences is computed, it could be considered a measure of the potential effectiveness of siRNAs and used to evaluate whether or not a candidate siRNA is likely to silence its target gene. If the probability of the candidate siRNA were greater than the average probability of a large number of effective siRNAs, it would indicate a high likelihood of gene silencing. To calculate this measure, 833 effective siRNA sequences reported in the literature (PubMed) were collected and nucleotide occurrences at positions from 5′ to 3′ in the cDNA were summarized.

The probability of individual nucleotide occurrence frequencies f_p^N at individual positions can be computed as follows:

$$f_p^N = \frac{\sum_{i=1}^{I}\{A,G,C,T\}}{I} \qquad (13)$$

where N is the kind of nucleotide (A, G, C, or T), p is the position in the cDNA (1, 2, ...,19 from 5′ to 3′), and I is the number of the effective siRNA sequences (e.g., 833).

Then the probability OF_i of each effective siRNA sequence is calculated in the following way:

$$OF_i = \prod_{p=1}^{19} f_{ip}^N \qquad (14)$$

where i is the sequence identification number of the effective siRNAs (i.e., $i = 1, 2, \ldots, I$).

The average sequence probability A_E for the effective siRNAs is therefore computed as follows:

$$A_E = \frac{\sum_{i=1}^{I} OF_i}{I} \qquad (15)$$

A_E could be considered a criterion for candidate siRNA. That is, if the probability of the candidate sequence were greater than A_E, it would indicate a high likelihood of gene silencing. On the other hand, if the probability of the candidate sequence were remarkably lower than A_E, it would indicate a low likelihood of effectiveness.

3.4. siRNA Sequence Selection Based on a Hidden Markov Model

As an siRNA sequence X basically consists of 19 nucleotides, it can be described as $\mathbf{X} = X_1, X_2, \ldots X_{19}$, where X_i indicates the nucleotide A, C, G, or T at position i. Furthermore, this sequence can be expressed as state diagrams of nucleotides A, C, G, and T from the positions 1 to 19 shown in Fig. 1. As shown in Fig. 1, if the state at position 1 is, for example, the nucleotide C, it can be transmitted to all the states A, C, G, or T at position 2. Likewise, these nucleotide state transitions proceed from the positions 1 to 19. In relations between the state diagrams (top) and the frequency ratios (bottom) as shown in Fig. 1, although what states are allocated to the individual positions of effective siRNA sequences are unknown in the intermediate processes, the ratios of the individual nucleotide occurrences are obtained as shown in the bottom of Fig. 1. Therefore, the transmission of the individual nucleotides A, C, G, and T from the positions 1 to 19 can be considered a hidden Markov process. If the state diagrams of effective siRNAs were expressed as an HMM, the optimal states (nucleotides) for maximizing the state transition probability could be solved as a decoding problem by using the Viterbi algorithm (25).

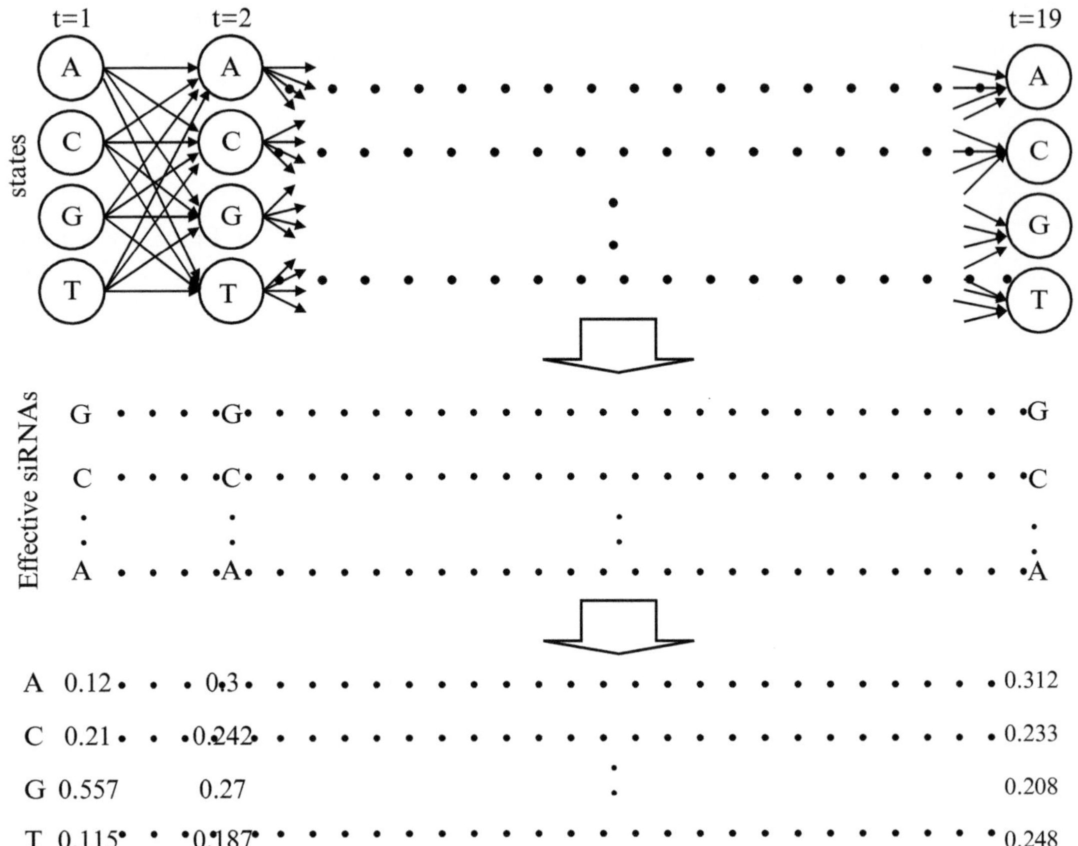

Fig. 1. State diagram of the hidden Markov model. A set of 833 effective siRNA sequences from the literature is shown in the *middle part*, and the frequency ratios of individual nucleotides at each position in those sequences are shown in the *bottom*. The ratios of the nucleotides A, C, G, and T at position 2, for example, are, respectively, 250/833 (=0.3), 202/833 (=0.242), 225/833 (=0.27), and 156/833 (=0.187).

3.4.1. The Viterbi Algorithm for Selection of the Optimal siRNA Nucleotide

The Viterbi algorithm for selecting the optimal siRNA sequence is expressed as follows:

1. Initialization for individual states i=A, C, G, T.

$$\delta_1 = \prod_i b_i(O_1) \tag{16}$$

$$\varphi_1(i) = 0,$$

where $\prod_i$ is the initial state probability distribution for the state i and $bi(o_1)$ is the output of the state i at the sequence position 1.

2. Recursive computations for the sequence positions t=1, 2, …, 18 and the individual states j=A, C, G, T.

$$\delta_{t+1}(j) = \max_i(\delta_t(i)a_{ij})b_j(O_{t+1}) \tag{17}$$

$$\varphi_{t+1}(j) = \arg\max_i(\delta_t(i)a_{ij}) \tag{18}$$

where a_{ij} is the state transition probability from the state i to j.

3. Termination of the recursive computations.

$$\hat{P} = \max_i \delta_{19}(i) \quad (19)$$

$$\hat{q}_{19} = \arg\max_i \delta_{19}(i) \quad (20)$$

4. Optimal state generation for sequence positions $t=18, 17, \ldots, 1$.

$$\hat{q}_t = \varphi_{t+1}(\hat{q}_{t+1}) \quad (21)$$

3.4.2. Nucleotide Occurrence Models

Two types of nucleotide occurrences from positions 1 to 19 were assumed. One is that the nucleotides occur independently at individual positions as listed in Table 4a, and the other is that the occurrence of individual nucleotides at individual positions depends on the nucleotides at other positions. A typical occurrence dependency is, for example, the Markov chain dependency (the simple (first) Markov model). That is, the nucleotide occurrences at the present position depend on the nucleotides at the previous position. The probability of the simple Markov model for the nucleotide at the present position i ($i=2, 3, \ldots, 19$) is determined under the condition of the effective nucleotide at the previous position $i-1$ as listed in Table 4b. Suppose, for example, that we have the sequence GACTCAACACCGAGTTCAA as a candidate siRNA sequence for some target gene. In this case, the probability of the second nucleotide being A may depend on the first nucleotide being G. From Table 4b (the simple Markov model table) one sees that the probability that A occurs at position 2 under the condition that there is a 0.557 probability that G occurs at position 1 is 0.33. One similarly sees that the probability that the third nucleotide is C is 0.192 under the condition that there is a 0.3 probability that the second nucleotide is A.

3.4.3. Evaluation Criteria Used in the Proposed Method

To make the results estimated using the average silencing probability easily understood, the ratio of the result estimated for a new siRNA candidate to the average sequence probability A_E is considered. This is because the results estimated for the known effective siRNAs could be considered a standard criterion for candidate new siRNAs. This normalized ratio NR is therefore defined as follows:

$$NR = \frac{ER}{A_E}, \quad (22)$$

where ER is the result estimated by the average silencing probability method and A_E is the average of the probabilities predicted for the known effective siRNAs.

This NR therefore indicates the gene silencing potential of the siRNA candidates relative to that of the known effective siRNAs. If $NR \geq 1$, the level of gene silencing expected to be obtained with the siRNA candidate is the same as or higher than the level of silencing obtained with the known effective siRNAs. That is, NR

indicates that the candidate sequence is likely to silence its target gene. If, on the other hand, $NR < 1$, the gene silencing expected to be obtained with the candidate sequence is lower than the level of silencing obtained with the known effective siRNAs.

3.5. Training and Testing Data

3.5.1 Effective siRNA Sequences

Two kinds of known effective siRNA sequences were used for the training data. One kind was 833 effective siRNA sequences (more than 80% effective at gene silencing at the protein level) from 490 different cDNAs in the published references of the PubMed database (5, 6, 29). The other was the 636 top-ranked effective siRNA sequences (more than 0.832 (normalized inhibitory activity)) from 34 genes (23).

3.5.2. Ineffective siRNA Sequences

Because ineffective siRNAs are rarely published, we used the 847 worst-ranked ineffective siRNAs (<0.612 (normalized inhibitory activity)) from Huesken et al. (23) for the training data of the proposed methods.

3.5.3. Testing Data

The recently reported effective and ineffective siRNAs were used as the testing data. These testing data were not included in the training data.

Reynolds et al. analyzed 90 siRNAs systematically, targeting every other position of 197-base regions of human *cyclophilin B* mRNA (GeneBank accession no. M60875) (18). For simplicity, human *cyclophilin B* is symbolized throughout the present article as MG1. From the 90 analyzed siRNA sequences we selected as effective ones the 25 top-ranked sequences for which the MG1 target gene silencing was >80% and selected as ineffective ones the 25 worst-ranked sequences for which MG1 target gene silencing was <50% with the standard deviation 10%.

Ui-Tei et al. reported 38 effective and 24 ineffective sequences for six genes: *firefly luciferase* (*PRL-TK*), *vimentin*, *Oct 4*, *EGFP*, *ECFP*, and *DsRed* (19). For simplicity, in the rest of this article all six of these genes are symbolized as MG2.

Amarzguioui and Prydz reported 21 effective and 25 ineffective siRNA sequences for four genes: *hTF* (accession no. M16553), *mTF* (accession no. M26071), *PSK* (accession no. J272212), and *CSK* (accession no. NM_004383) (20). For simplicity, in the rest of this article these four genes are symbolized as MG3.

Takasaki et al. reported seven effective and seven ineffective siRNA sequences for the *cyclin B1* (accession no. NM_031966) (28). For simplicity, in the rest of this article this gene is symbolized as MG4.

Huesken et al. reported 37 siRNAs for *TC10* (accession no. BD135193), *UBE2I* (accession no. NM_003345), and *CDC34* (accession no. NM_004359) (23). The 12 top-ranked effective and 12 worst-ranked ineffective siRNA sequences were selected for these genes. For simplicity, they are symbolized as MG5 in the rest of this article.

4. Evaluations of the Proposed Methods

4.1. Evaluation of the Decision Tree Learning Method

The decision tree diagram shown in Fig. 2 was obtained by the learning of the decision tree using 833 effective and 847 ineffective sequences. Then the prediction probabilities of gene silencing were computed for MG1 to MG5 by the learned decision tree diagram. The distributions of the predicted probabilities for MG1 to MG5 are shown in Fig. 3a–e. We also calculated the average for them.

The distributions of the predicted probabilities by decision tree learning, as a whole, indicated that there are differences between the effective and ineffective siRNA sequences as shown in Fig. 3a–e. There were clear distinctions between the distributions of effective and ineffective siRNAs for MG1 to MG5 except MG3. The entire average predicted probability of 103 effective siRNA sequences for these genes was 68.9%, whereas that of 93 ineffective siRNA sequences was 34.7%.

4.2. Evaluation of the Bayes' Theorem Method

The average probability predicted for these 833 siRNAs can be considered as a standard criterion for the effectiveness of siRNAs targeting other genes. The four cases listed in Table 5 were evaluated while computing the predicted gene silencing probabilities of

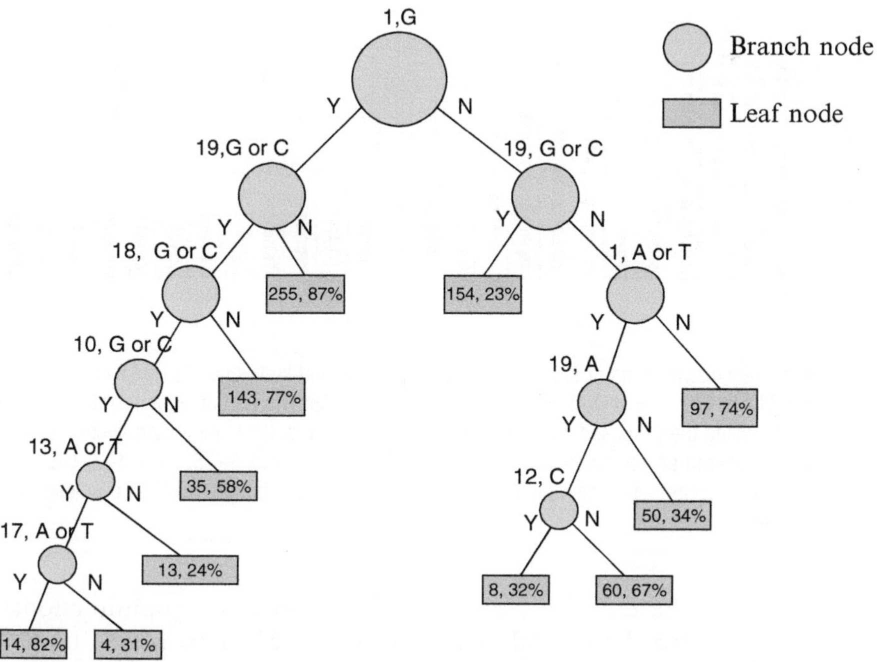

Fig. 2. Decision tree diagram for known 833 effective and 847 ineffective siRNA sequences. The *top* of the branch node indicates the position and nucleotide attribute, e.g., "19, G or C" means that the position of cDNA is 19 and the nucleotide at the position shows G or C. The *bottom* of the branch node shows yes (Y) and no (N). The leaf node indicates the number of effective siRNA sequences and its percentage, e.g., "255, 87%" means that the number of effective siRNA sequences is 255 and its percentage is 87% (=255/292).

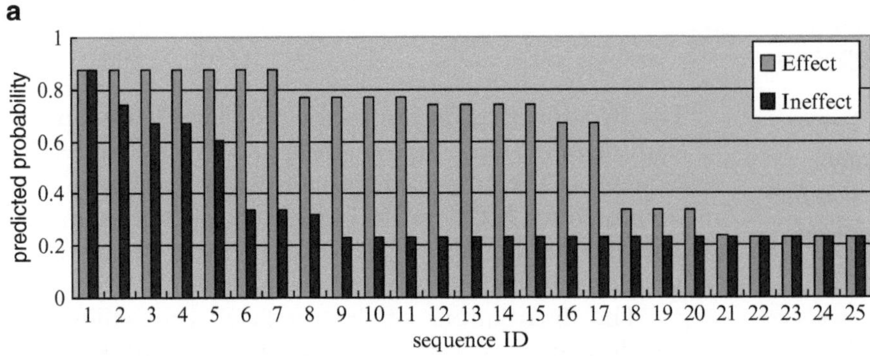

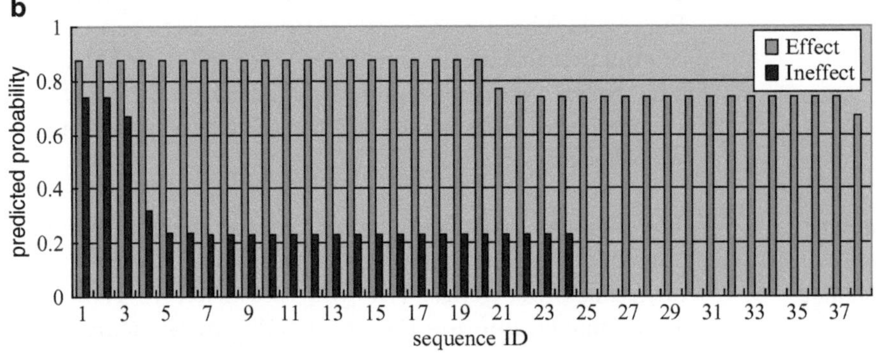

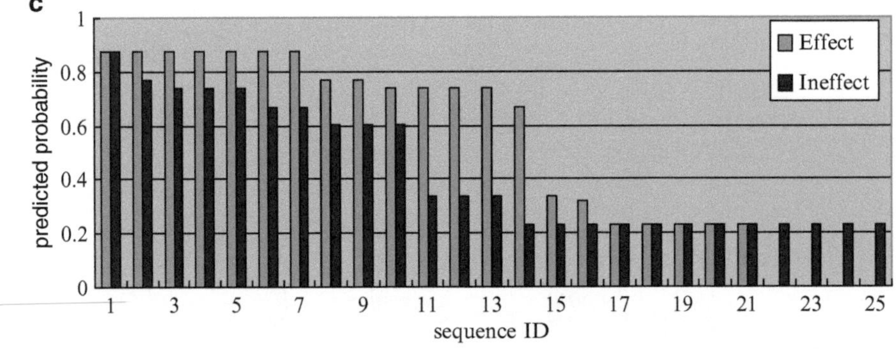

Fig. 3. Prediction probability distributions of siRNA sequences effective and ineffective for MG1 to MG5 by the proposed decision tree method. Effect: Effective siRNAs. Ineffect: Ineffective siRNAs. The probabilities for effective and ineffective siRNAs are computed by using the proposed decision tree method, and the siRNAs are sorted according to the predicted probabilities. They are numbered as sequence IDs (MG1: 1–25, MG2: 1–38, MG3: 1–25, MG4: 1–7, MG5: 1–12). (**a**) MG1 prediction distribution. (**b**) MG2 prediction distribution. (**c**) MG3 prediction distribution. (**d**) MG4 prediction distribution. (**e**) MG5 prediction distribution.

the effective (functional) and ineffective (nonfunctional) siRNAs for the recently reported genes (MG1 to MG5) under the prior probabilities of $P^{eff} = 0.1$ and 0.2.

Because the probabilities predicted for individual siRNA sequences for the reported genes varied, the average predicted probability was used as an evaluation measure in the following verification. The reasonability of this is discussed later.

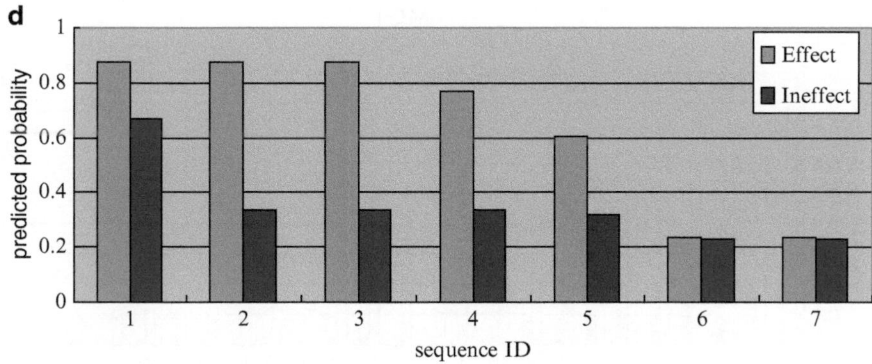

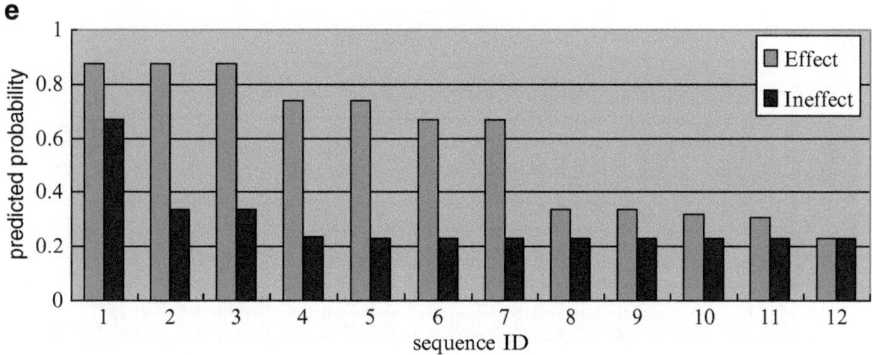

Fig. 3. (continued)

4.2.1. Case 1: Combination of Both Independent Nucleotide Occurrences

In Case 1, the average probability predicted that 833 effective siRNAs would be effective in silencing their target genes (Ae-833) was computed to be 29.9% by using Eq. 8 under the prior probability $P^{\mathrm{eff}} = 0.1$ (see Subheading 3.2). Although this probability might seem low for effective siRNAs, Bayes' theorem estimated it to be 2.99 times the prior probability. Therefore, this could be considered a standard criterion for the effectiveness of siRNAs. The distributions of the probabilities predicted for the effective and ineffective siRNA sequences for MG1 to MG5 are shown in Fig. 4. In the case of MG1, as shown in Fig. 4a, it is clear that there are big differences between the distributions for effective and ineffective siRNA sequences. Most of the effective siRNAs (88%), for example, have predicted probabilities of gene silencing that are >10%, whereas siRNAs more than 70% ineffective have predicted probabilities of effective gene silencing that are <10%. The average probability of effective silencing predicted for the effective siRNAs for MG1 (Ae-MG1) was 32.2%, whereas that predicted for the ineffective ones (Ai-MG1) was 6.7%. This means, according to Eq. 9, that the *PR* for sequences effective for MG1 is 3.22. That is, Bayes' theorem estimated a probability more than three times as high as the prior probability. The *PR* for Ai-MG1, in contrast, is 0.67, which indicates that the silencing obtained with sequences

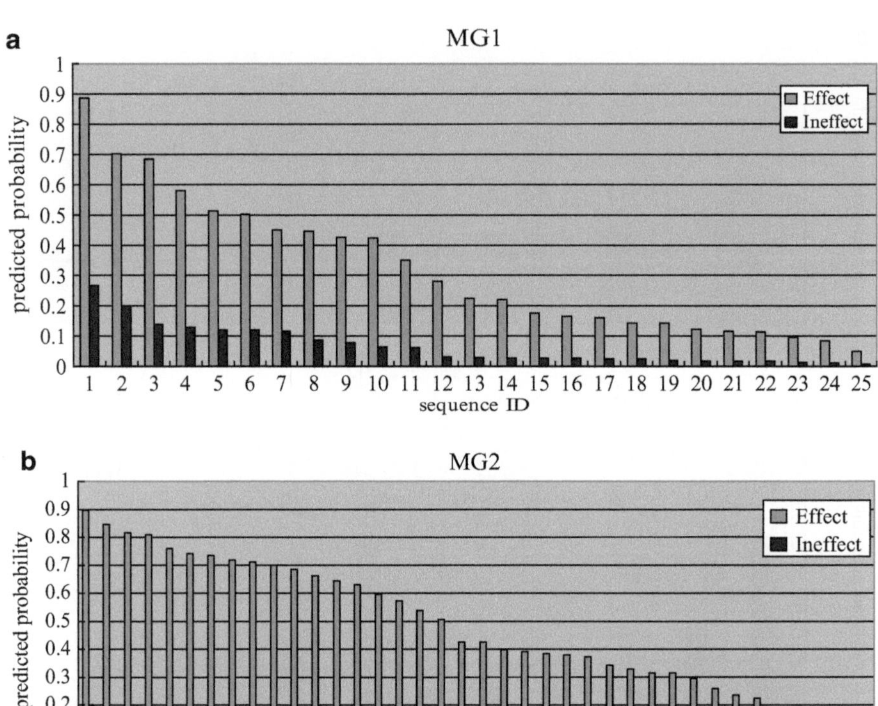

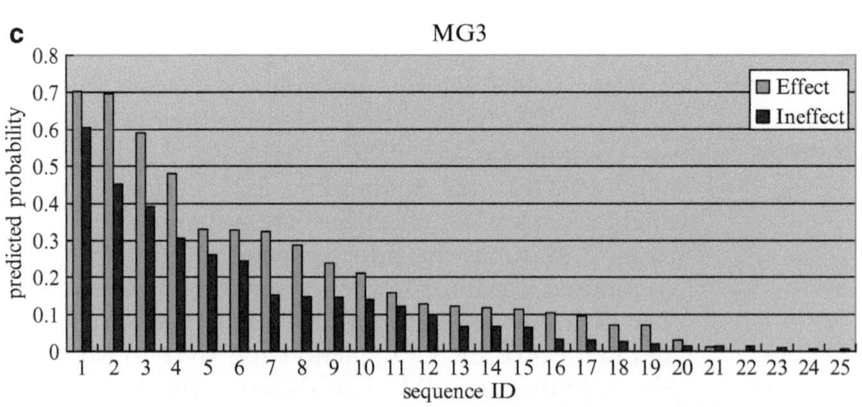

Fig. 4. Distributions of probabilities predicted for siRNA sequences that are effective and ineffective for MG1 to MG5 (results of Case 1 evaluation). Effect: Effective siRNAs. Ineffect: Ineffective siRNAs. The probabilities for effective and ineffective siRNAs were calculated by using Eq. 8, and the siRNAs are sorted according to the predicted probabilities. They are numbered as sequence IDs (MG1: 1–25, MG2: 1–38, MG3: 1–25, MG4: 1–7, MG5: 1–12). (**a**) Distribution predicted for MG1. (**b**) Distribution predicted for MG2. (**c**) Distribution predicted for MG3. (**d**) Distribution predicted for MG4. (**e**) Distribution predicted for MG5.

ineffective for MG1 sequences would be about two-third the usual level of silencing.

MG2 has remarkably distinct distributions. As shown in Fig. 4b, the gene silencing probabilities predicted for most siRNAs effectively silencing MG2 are high, whereas those predicted for

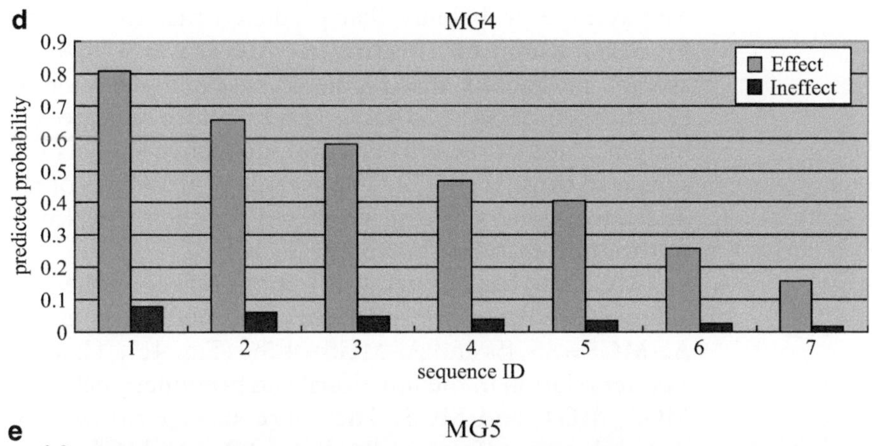

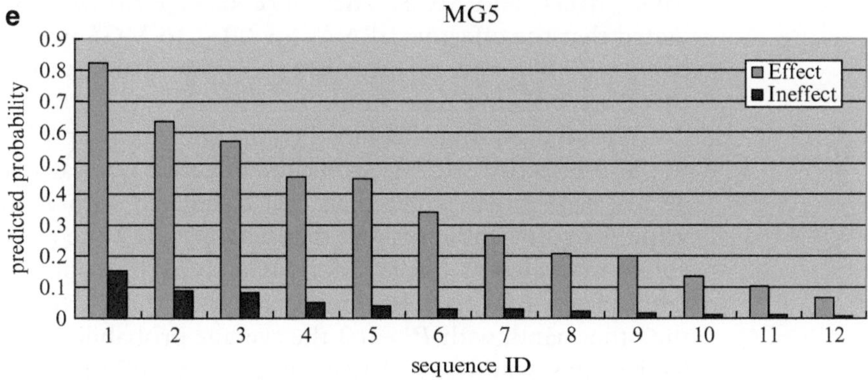

Fig. 4. (continued)

most of the ineffective ones are low. The average probability that predicted that the effective siRNAs for MG2 (Ae-MG2) would be effective was 47.9%, whereas the average probability that predicted that the ineffective ones would be effective (Ai-MG2) was 4.2%. These effective and ineffective estimation ratios are therefore better than those of MG1.

On the other hand, as shown in Fig. 4c, although the gene silencing probabilities predicted for most of the siRNAs effectively silencing MG3 are larger than those predicted for the ineffective ones, the differences are not so big as the differences between the probabilities predicted for the siRNAs effective and ineffective for MG1 and MG2. The average probability that predicted that the effective siRNAs for MG3 would be effective (Ae-MG3) was 24.8%, whereas the average probability that predicted that the ineffective ones would be effective (Ai-MG3) was 13.8%. This indicates that the *PR*s for siRNAs effective and ineffective for silencing MG3 (Ae-MG3 and Ai-MG3) are 2.48 and 1.38, so there might not be much difference between them. The average probabilities Ae-MG3 and Ai-MG3 reflect distribution features shown in Fig. 4c.

As shown in Fig. 4d, in contrast, the siRNAs effective and ineffective in silencing MG4 have remarkably distinct distributions.

The average probability that predicted that the effective siRNAs for MG4 would be effective (Ae-MG4) was 47.6%, whereas the average probability that predicted that ineffective ones would be effective (Ai-MG4) was 4.4%. This therefore indicates that the estimation ratios of the effective and ineffective siRNAs for MG4 are similar to those for MG2. The average probabilities of effective and ineffective gene silencing predicted for these siRNAs also reflect distribution features shown in Fig. 4d.

MG5 also has distinct distributions reflecting the average probabilities predicted for effective and ineffective gene silencing: Ae-MG5 = 35.4% and Ai-MG5 = 4.5% (Fig. 4e). There is therefore a better relation in the functional and nonfunctional ratios of MG1, MG2, MG4, and MG5. The entire average probability that predicted that the effective siRNAs for MG1 to MG5 would be effective was 37.9%, whereas the entire average probability that predicted that the ineffective ones would be effective was 7.7%.

With the prior probability $P^{eff} = 0.2$ the average probability that the 833 effective siRNAs would be effective was calculated to be 43.5%, 1.45 times larger than that calculated with $P^{eff} = 0.1$. The values listed in Table 6 show that with $P^{eff} = 0.2$ the average probability that siRNAs effective for MG1 to MG5 would be effective ranges from 1.32 to 1.46 times larger than that with $P^{eff} = 0.1$. On the other hand, with $P^{eff} = 0.2$ the average probability that the ineffective ones would be effective ranges from 1.67 to 2.1 times larger than that with $P^{eff} = 0.1$.

4.2.2. Case 2: Combination of Independent and Deductive Nucleotide Occurrences

In Case 2, Ae-833 was computed to be 25.1% under $P^{eff} = 0.1$. Ae-MG1 was 27.9%, whereas Ai-MG1 was 5.8%. These predictions mean that *PR*s of Ae-MG1 and Ai-MG1 are, respectively, 2.79 and 0.58. Ae-MG2 was 26.7%, whereas Ai-MG2 was 5.7%. As the *PR*s of Ae-MG2 and Ai-MG2 are, respectively, 2.67 and 0.57, they indicate levels similar to those of MG1. Because Ae-MG3 and Ai-MG3 were, respectively, 20.2% and 11.6%, their *PR*s are, respectively, 2.02 and 1.16. As Ae-MG4 and Ai-MG4, in contrast, were, respectively, 51.1% and 2.3%, their *PR*s are, respectively, 5.11 and 0.23. These predicted probabilities therefore indicate remarkably clearer distinctions than the prior probability. Because Ae-MG5 and Ai-MG5 were, respectively, 17.9% and 5.4%, their *PR*s are, respectively, 1.79 and 0.54. The entire average probability that predicted that the effective siRNAs for MG1 to MG5 would be effective was 26.3%, whereas the entire average probability that predicted that the ineffective ones would be effective was 7.1%. As a whole, the prediction accuracy of Case 2 is lower than that of Case 1. This indicates that the deductive nucleotide occurrences would not contribute to the prediction accuracy.

In the case of $P^{eff} = 0.2$, the average probability that 833 effective siRNAs would be effective was computed to be 38.5%, 1.54 times larger than that computed with $P^{eff} = 0.1$. The values listed in

Table 6
Average probabilities of effective silencing predicted for siRNAs effective and ineffective for various genes

			Case 1		Case 2		Case 3		Case 4	
Genes	Eff/Ineff	No. of seqs	$P^{eff}=0.1$	$P^{eff}=0.2$	$P^{eff}=0.1$	$P^{eff}=0.2$	$P^{eff}=0.1$	$P^{eff}=0.2$	$P^{eff}=0.1$	$P^{eff}=0.2$
MG 1	Effect	25	32.2	47	27.9	41.9	36	49.7	32.7	46.3
	Ineffect	25	6.7	13.1	5.8	11.6	9.5	15.5	7.7	13.7
MG 2	Effect	38	47.9	63.3	26.7	41.6	41	53.9	25	36.4
	Ineffect	24	4.2	8.2	5.7	10.4	1.4	2.9	2	4.2
MG 3	Effect	21	24.8	38	20.2	32.2	32.7	45.5	27.9	40.1
	Ineffect	25	13.8	22.9	11.7	20.8	15	23.1	13.9	22
MG 4	Effect	7	47.6	64	51.1	68.9	65.1	77.7	70.2	82.7
	Ineffect	7	4.4	9.2	2.3	5	5.9	12	3.1	6.7
MG 5	Effect	12	35.4	50.7	17.9	30	36.1	49.8	19.8	31.4
	Ineffect	12	4.5	9.3	5.5	11	5.1	9.1	5.9	10.4
siRNAs	Effect	833	29.9	43.5	25.1	38.5	38.4	51.6	33.7	46.9

Effect: Effective siRNAs, ineffect: ineffective siRNAs, P^{eff}: prior probability
No. of seqs: Number of siRNA sequences

Table 6 show that the average probability that effective siRNAs for MG1 to MG5 would be effective ranges from 1.35 to 1.68 times larger than that in case of $P^{eff}=0.1$. On the other hand, the average probability that the ineffective ones would be effective ranges from 1.8 to 2.2 times larger than that in case of $P^{eff}=0.1$.

4.2.3. Case 3: Combination of Markov Model and Independent Nucleotide Occurrences

In Case 3, Ae-833 was computed to be 38.4% under the assumption that $P^{eff}=0.1$. This is a higher probability than the corresponding Ae-833 computed in Case 1. Ae-MG1 was 36%, whereas Ai-MG1 was 9.5%. As the *PR*s of Ae-MG1 and Ai-MG1 are, respectively, 3.6 and 0.95, Ae-MG1 indicates better accuracy than that in Case 1 and Ai-MG1 shows worse accuracy than that in Case 1. Ae-MG2 was 40.95%, whereas Ai-MG2 was 1.37%. So the MG2 of Case 3 indicates that Ae-MG2 has a worse accuracy than that in Case 1, whereas Ai-MG2 has an accuracy three times better than that in Case 1. Because Ae-MG3 and Ai-MG3 were, respectively, 32.7% and 15%, their *PR*s are, respectively, 3.27 and 1.5. So the MG3 of Case 3 indicates that Ae-MG3 is better than that in Case 1, whereas Ai-MG3 is little bit worse than that in Case 1. As Ae-MG4 and Ai-MG4 in Case 3 were, respectively, 65.1% and 5.86%, Ae-MG4 is better than that in Case 1 and Ai-MG4 is a little bit worse than that in Case 1. Because Ae-MG5 and Ai-MG5 are, respectively, 36.1% and 5.1%, their *PR*s are similar to those in Case 1. As a whole, the *PR*s in Case 3 indicate better accuracy than do those in Case 1.

With $P^{eff}=0.2$ the average probability that 833 effective siRNAs would be effective was computed to be 51.6%, 1.34 times larger than that computed with $P^{eff}=0.1$. The values listed in Table 6 show that with $P^{eff}=0.2$ the average probability that siRNAs effective for MG1 to MG5 would be effective ranges from 1.19 to 1.39 times larger than that with $P^{eff}=0.1$. On the other hand, with $P^{eff}=0.2$ the average probability that the ineffective ones would be effective ranges from 1.5 to 2.1 times larger than that with $P^{eff}=0.1$.

4.2.4. Case 4: Combination of Markov Model and Deductive Nucleotide Occurrences

In Case 4, Ae-833 was computed to be 33.7% under $P^{eff}=0.1$. This is 4.7% lower than the corresponding Ae-833 computed in Case 3. Ae-MG1 was 32.7%, whereas Ai-MG1 was 7.7%. The *PR*s of Ae-MG1 and Ai-MG1 are therefore, respectively, 3.27 and 0.77. Ae-MG2 was 25%, whereas Ai-MG2 was 2%. Because Ae-MG3 and Ai-MG3 were, respectively, 27.9% and 13.9%, their *PR*s are 2.79 and 1.39. As Ae-MG4 and Ai-MG4 in Case 4 were, respectively, 70.2% and 3.1%, they are distinguished more clearly in this case than in any other. Because Ae-MG5 and Ai-MG5 were, respectively, 19.8% and 5.94%, their *PR*s are 1.98 and 0.594. As a whole, although the evaluation results of MG1 to MG5 in Case 4 showed better accuracy than those in Case 2, they showed worse accuracy than those in Cases 1 and 3.

In the case of $P^{eff}=0.2$, the average probability that 833 effective siRNAs would be effective was computed to be 46.9%, 1.39 times larger than that computed with $P^{eff}=0.1$. The values listed in

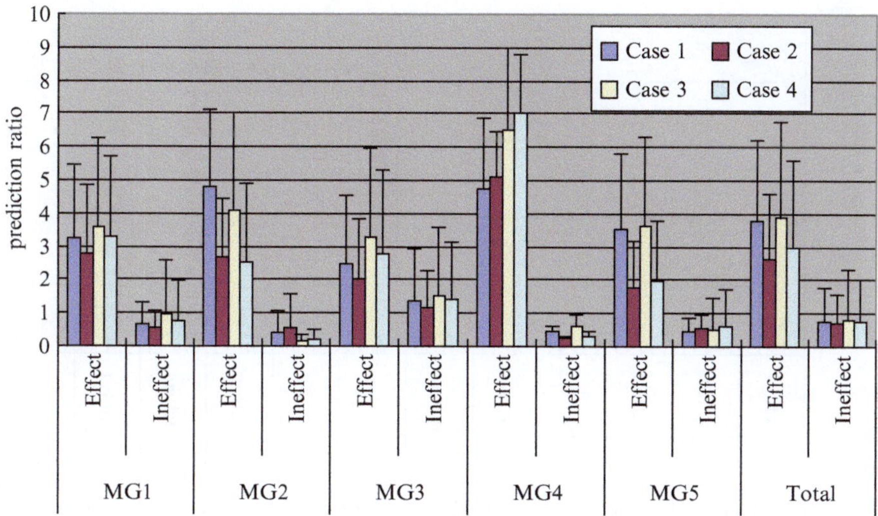

Fig. 5. Relations between the predicted probability and the prior probability. Effect: Effective siRNAs. Ineffect: Ineffective siRNAs. The prediction ratio graphed here is the ratio of the average of the probabilities estimated for the siRNAs to the prior probability $P^{\text{eff}} = 0.1$. Case 1 is that in which both the effective and ineffective siRNAs have independent nucleotide occurrences at the individual positions. Case 2 is that in which the effective siRNAs have independent and ineffective siRNAs have deductive nucleotide occurrences at the individual positions. Case 3 is that in which the effective siRNAs have dependent nucleotide occurrences at the individual positions (the simple Markov model) and the ineffective siRNAs have independent nucleotide occurrences at the individual positions. Case 4 is that in which the effective siRNAs have dependent nucleotide occurrences and the ineffective siRNAs have deductive nucleotide occurrences at the individual positions.

Table 6 show that the average probability that siRNAs effective for MG1 to MG5 would be effective ranges from 1.18 to 1.59 times larger than that in case of $P^{\text{eff}} = 0.1$. On the other hand, the average probability that the ineffective ones would be effective ranges from 1.6 to 2.1 times larger than that in case of $P^{\text{eff}} = 0.1$.

The ratios of the average estimated probabilities to the prior probabilities of the reported genes are shown for Cases 1–4 in Fig. 5, where it is clear that the ratios computed in Case 3 show the most distinct differences between the siRNAs effective and ineffective for silencing MG1 to MG5 except MG2. This means that the combination of Markov model and independent nucleotide occurrences might yield the most accurate predictions when Bayes' theorem is used to select siRNA sequences effective for gene silencing. With regard to gene classes, MG1 and MG5 show distinctions between the effective and ineffective siRNAs more clearly than MG3 does, and MG2 and MG4 show distinctions remarkably clearly. These results therefore imply that there are some differences in the individual nucleotide frequencies at each position of the siRNAs effective for these gene classes. Although MG3 shows differences between the effective and ineffective siRNAs, the ratios of the predicted probabilities for effective ones to ineffective ones are <2.2. This implies that there is no big difference between the individual nucleotide frequencies of the siRNAs effective and ineffective for silencing this class of genes.

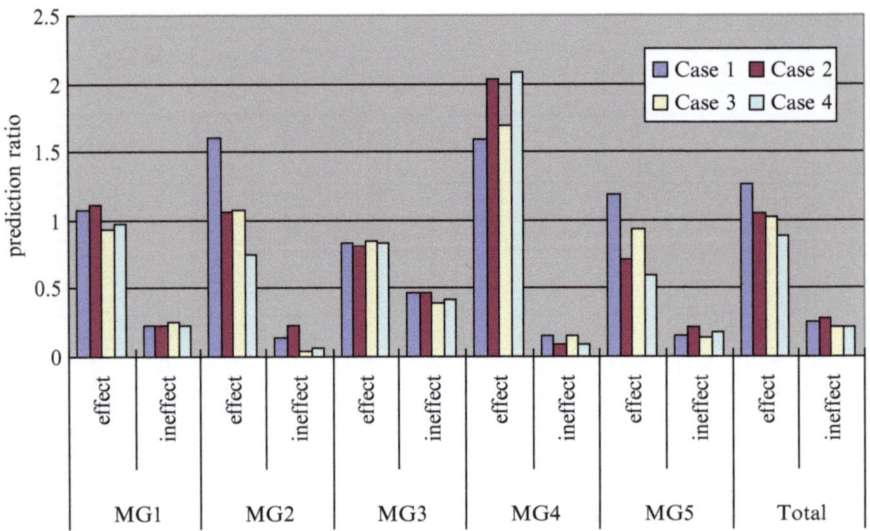

Fig. 6. Normalized relations among the siRNAs for MG1 to MG5. Effect: Effective siRNAs. Ineffect: Ineffective siRNAs. The prediction ratio graphed here is the ratio of the average estimated for the designated siRNAs to the average for the 833 siRNAs known to be effective. Case 1 is one in which both the effective and ineffective siRNAs have independent nucleotide occurrences at the individual positions under the assumption that $P^{\mathrm{eff}} = 0.1$ when the 833 siRNAs known to be effective are used as normalization data. Case 2 is also one in which P^{eff} is assumed to be 0.1 when the 833 siRNAs known to be effective are used as normalization data, but in this case the effective siRNAs have independent nucleotide occurrences and ineffective siRNAs have deductive nucleotide occurrences. Case 3 is one in which the effective siRNAs have dependent (simple Markov model) nucleotide occurrences and the ineffective siRNAs have independent nucleotide occurrences at the individual positions. Case 4 is one in which the effective siRNAs have dependent nucleotide occurrences and the ineffective siRNAs have deductive nucleotide occurrences at the individual positions.

4.2.5. Comparative Analysis by the Normalization

Because the average predicted probability that 833 effective siRNAs would be effective (Ae-833) could be considered a standard criterion for the other gene functionality as described earlier, we calculated *NR*s for MG1 to MG5 by using Eq. 10 (see Subheading 3.2). As a whole, individual *NR*s result in clearer distinctions between the effective and ineffective siRNAs. As shown in Fig. 6, in all Cases 1–4 the average *NR*s of siRNAs effective for MG1 are about 1. These results therefore indicate that all 833 siRNAs effective for MG1 are about equally effective. Because the average *NR*s of the 833 siRNAs effective for MG2 that are calculated in three Cases 1–3 are more than 1.1, they indicate that these siRNAs are more likely to be effective than the 833 siRNAs known to be effective. In contrast, the average *NR*s of siRNAs effective for MG3 are about 0.83 in all cases, whereas those of the siRNAs ineffective for MG3 are in the ranges of 0.39–0.47. Therefore, although there are differences between the average *NR*s for siRNAs effective and ineffective for MG3, the ratios of between them are 1.73–2.18 and do not indicate differences as clearly as do the average *NR*s for siRNAs effective and ineffective for MG1 and MG2. Meanwhile, in all cases the average *NR*s of siRNAs effective for MG4 are about 1.6 times higher than the Ae-833 standard criterion. These results imply that

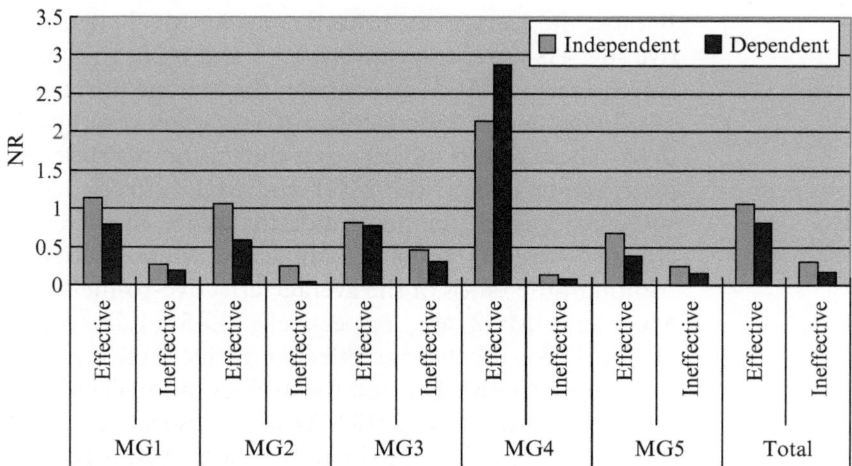

Fig. 7. Normalized ratios based on 833 effective siRNAs. Effective: Effective siRNAs. Ineffective: Ineffective siRNAs. NR: Normalized ratio calculated by Eq. 22. Independent: Independent occurrences at individual positions. Dependent: The simple Markov model.

the siRNAs effective for MG4 are potentially more effective than the 833 siRNAs known to be effective. The average *NR*s of siRNAs effective for MG5 range from 0.59 to 1.18, whereas those of ineffective siRNAs for MG5 are 0.13–0.22. The average *NR*s of all the siRNAs effective for MG1 to MG5 are more than 1 in three Cases 1, 2, and 3, indicating that these siRNAs are potentially more effective than the 833 siRNAs known to be effective. On the other hand, the average *NR*s of the siRNAs ineffective for MG1 to MG5 are in the ranges of 0.21–0.28. They therefore contribute to the clearer differences between the average *NR*s of the siRNAs effective and ineffective for MG1, MG2, MG4, and MG5 (Fig. 6). This makes the selection of candidate effective siRNA sequences easier.

4.3. Evaluation of the Average Silencing Probability Method

The proposed average silencing probability method was evaluated by first computing A_E for 833 effective siRNA sequences and then using Eq. 14 to compute the individual probabilities of the effective and ineffective siRNAs. Because there were ups and downs in the individual ratios of the effective and ineffective siRNAs, the average of them were calculated. The relations between the normalized average ratios of the effective and ineffective siRNAs for the recently reported genes are shown in Fig. 7.

4.3.1. Evaluation Using Nucleotide Frequencies Based on 833 Effective siRNAs

Case 1: Independent Nucleotide Occurrences at Individual Positions

The average normalized ratio *NR* for the MG1 effective siRNAs was 1.14, whereas that for the ineffective ones was 0.26. This indicates that as the *NR* for the sequences of MG1 effective siRNAs are 1.14 times higher than that for the 833 effective siRNAs, it shows the higher level potential in gene silencing. On the other hand, as the *NR* for the sequences of MG1 ineffective ones shows 0.256 times, i.e., one-fourth, compared to the *NR* for the 833 effective siRNAs, it implies one-fourth (low) level potential of gene silencing.

Because the average normalized ratios for MG2 effective and ineffective siRNAs were, respectively, 1.06 and 0.25, they indicate a similar tendency of MG1. In contrast, the average normalized ratios for MG3 effective and ineffective siRNAs were, respectively, 0.82 and 0.46. These results indicate that there is no big difference between them (compared to the MGl and MG2 effective and ineffective siRNAs). That is, the nucleotide frequency characteristics of MG3 effective siRNAs resemble those of MG3 ineffective siRNAs. Although the ratios of the average effective-to-ineffective ratios for MGl and MG2 are, respectively, 4.45 (1.14/0.26) and 4.08 (1.06/0.25), the average effective-to-ineffective ratio for MG3 is 1.78 (0.82/0.46). Because the average normalized ratios of MG4 effective and ineffective siRNAs were, respectively, 2.14 and 0.13, the ratio of the effective to ineffective siRNAs was 16.5 (2.14/0.13). The *NR* of the effective siRNAs for MG4 therefore implies a high likelihood (2.2 times) of gene silencing compared to that of the 833 effective siRNAs, whereas the *NR* of the ineffective ones shows quite low likelihood (0.13 times). On the other hand, because the normalized ratios for MG5 were, respectively, 0.68 and 0.24, the ratio of the effective to ineffective siRNAs was 2.83. The entire normalized ratio that effective siRNAs for MGl to MG5 would be effective was 1.06, whereas the entire normalized ratio that the ineffective ones would be effective was 0.297. These evaluation results for the independent nucleotide occurrences indicate that the proposed prediction method based on the effective siRNA sequences is useful for selecting candidate siRNAs for target genes.

Case 2: Dependent Nucleotide Occurrences Based on the Simple Markov Model

As shown in Fig. 7, in Case 2 as a whole the average normalized ratios of the effective and ineffective siRNAs for MG1, MG2, MR3, and MG5 were lower than those in Case 1. In contrast, the normalized ratio of MG4 effective siRNAs was higher than that in Case 1 and the normalized ratio of MG4 ineffective ones was lower than that in Case 1. There is, however, a similar tendency in the ratios of the effective-to-ineffective average ratios for MGl, MG2, MG3, and MG5. The average normalized ratio of the MGl effective siRNAs was 0.79, whereas that of the ineffective ones was 0.19. The average ratio of the effective siRNAs is thus about four times larger than that of the ineffective ones. As the average normalized ratios of the MG2 effective and ineffective siRNAs were, respectively, 0.59 and 0.04, the average ratio of the effective siRNAs was about 14 times larger than that of the ineffective ones. On the other hand, the average normalized ratios of the MG3 effective and ineffective siRNAs were, respectively, 0.77 and 0.31. Although the average ratio of the effective siRNAs was only about 2.5 times larger than that of the ineffective ones and this ratio was lower than the corresponding ratios for the MGl and MG2 siRNAs, there was still a clear difference between the average normalized ratios of the MG3 effective and ineffective siRNAs. Similarly, the normalized ratios of the MG5 effective and ineffective siRNAs

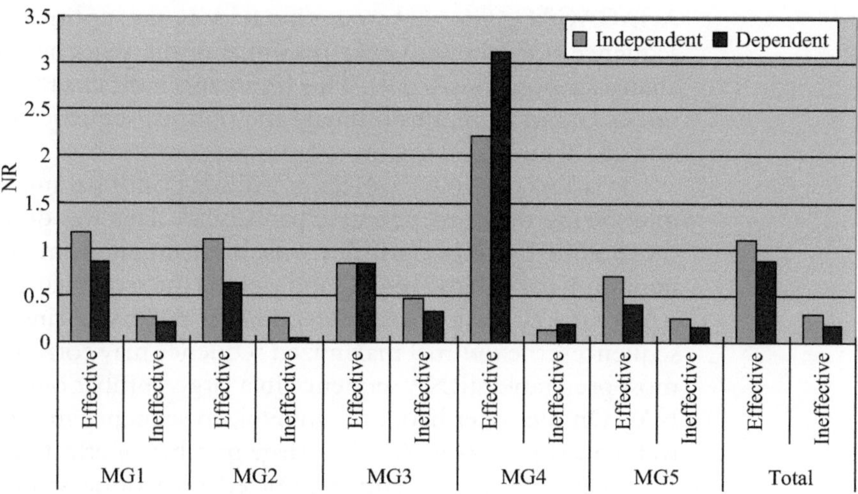

Fig. 8. Normalized ratios based on 636 effective siRNAs.

were, respectively, 0.37 and 0.15. Therefore, the average ratio of effective siRNAs was approximately 2.5 times larger than that of the ineffective ones. On the other hand, because the average normalized ratios of MG4 (cyclin B1) effective and ineffective siRNAs were, respectively, 2.88 and 0.08, the difference between them was remarkably large (36-fold). The *NR* of the effective siRNAs for MG4 therefore indicated the higher likelihood of gene silencing compared to that of the 833 siRNAs, whereas the *NR* for the ineffective ones showed the quite low likelihood. These evaluation results for the dependent nucleotide occurrences based on the simple Markov model indicate that the proposed prediction method is useful for selecting candidate siRNAs for target genes.

4.3.2. Evaluation Using Another Large Number of Known siRNAs

Gene silencing probabilities were also evaluated using the nucleotide frequencies at individual positions in 636 other effective siRNAs. The independent (Case 1) and dependent (Case 2) nucleotide frequencies at individual positions for these siRNAs and the probabilities that the effective and ineffective siRNAs would be effective for the reported genes are shown in Fig. 8. Although there were ups and downs in *NR*s predicted for MG1 to MG5 using either the 833 or 636 effective siRNAs, the total *NR*s predicted are similar for both cases. That is, the *NR*s based on the 833 effective siRNAs are, respectively, 1.06 and 0.81 for the independent and dependent cases, and those based on the 636 effective siRNAs are, respectively, 1.1 and 0.87 for the independent and dependent cases. This implies that the proposed method using the average silencing probabilities could be useful for many other genes.

4.4. Evaluation of the HMM Method

The Viterbi algorithm was carried out for the state diagram of the HMM shown in Fig. 1. As a result, the siRNA sequence GAAGA

AGAGAGAGAGCAGA was obtained as the optimal nucleotide sequence (i.e., the sequence maximizing the sequence state probability for positions 1–19). This result also indicates that the nucleotides G and A might dominate the optimal sequence in reported sets of effective siRNAs.

It is also possible to select individual positional nucleotides for minimizing the sequence state probability. This was done by using the modified Viterbi algorithm, i.e., by changing from maximum to minimum in the Eqs. 16–21, and yielded the sequence TTTTTATT AATCGCGTTCG. From the point of gene silencing by siRNA sequences, the optimal maximized sequence may correspond to the most preferable siRNA sequence in a large number of effective siRNAs. On the other hand, the minimized sequence may correspond to the least preferable one in a large number of effective siRNAs.

These maximized and minimized nucleotide sequences were then compared with the upper and lower level significant nucleotides obtained using the previously proposed statistical significance testing for 833 effective siRNA sequences (30). One sees in Table 7 that the maximized nucleotide obtained using the Viterbi algorithm corresponds to the upper level nucleotides obtained using the significance testing, and the minimized nucleotide sequence corresponds to the lower level one obtained using the significance testing. Interestingly, there are many coincidences between the maximized and minimized nucleotides and the upper and lower level significant nucleotides. Between the maximized nucleotides and the upper level ones there are 13 coincidences (at positions 1, 2, 3, 4, 6, 7, 8, 9, 12, 14, 16, 17, and 19), and between the minimized nucleotides and the lower level ones there are 11 coincidences (at positions 1, 2, 4, 5, 7, 10, 11, 14, 15, 18, and 19). There are six coincidence positions in both relations: at positions 1, 2, 4, 7, 14, and 19. The positions 1, 2, and 19 correspond to around the 5′ and 3′ terminal points. This implies that these positions play important roles in gene silencing.

4.4.1. Evaluation for MG1 to MG5 Based on a Large Number of Ineffective siRNAs

It is also possible to clarify the probability of how siRNA candidates are effective on the basis of a large number of ineffective siRNAs. 847 known siRNAs were selected as ineffective ones (see Subheading 3.5). The probabilities of individual nucleotide occurrence frequencies at individual positions are listed in Table 4c. The relations among *NR*s of effective and ineffective siRNAs for MG1 to MG5 computed by using the Eqs. 14, 15, and 22 are shown in Fig. 9. In the case of using the 847 known ineffective siRNAs, *NR*s of effective siRNAs for MG1 to MG5 are <1, whereas those of ineffective ones are more than 1 as shown in Fig. 9. The *NR* of the total effective siRNAs is 0.67, whereas that of the ineffective ones is 2.37.

Comparing Fig. 9 with Fig. 7, it is clear that the corresponding *NR*s of effective and ineffective siRNAs for MG1 to MG5 are, respectively, reverse relations. This depends on what set of siRNAs,

Table 7
Relations between the maximized and minimized nucleotide sequences and the upper and lower level significant nucleotides

	1	2	3	4	5	6	7	8	9	10	11	12	13	14	15	16	17	18	19
Upper level	G	A	A/T	C/G	C	A/T	G/C	A	C/G	C	C	A/C		A	T/A	T/C	A	T/A	A
Maximized	G	A	A	G	A	A	G	A	G	A	G	A	G	A	G	C	A	G	A
Coincidence	=	=	=	=		=	=	=	=			=		=		=	=		=
Lower level	A/T	T	G	T/A	T	G/C	T/A	C	G	A/G	T			C	G/C	A	G	C/G	G
Minimized	T	T	T	T	T	A	T	T	A	A	T	C	G	C	G	T	T	C	G
Coincidence	=	=		=	=		=			=	=			=	=			=	=

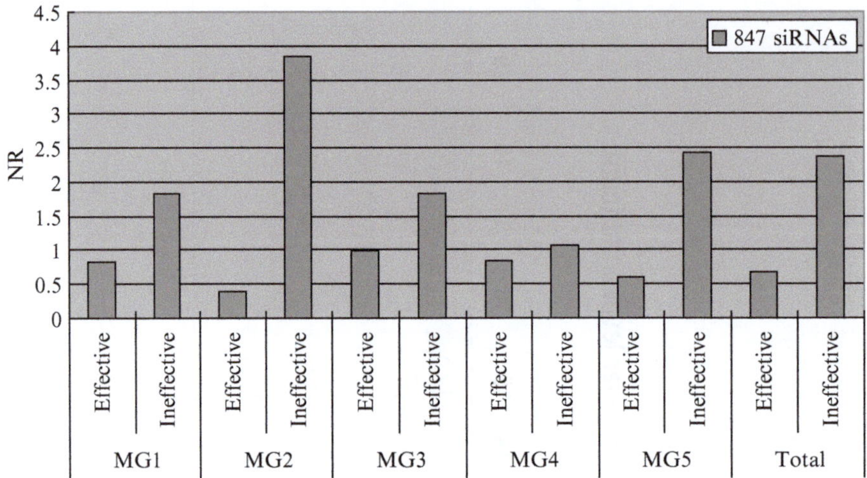

Fig. 9. Normalized ratios based on 847 ineffective siRNAs.

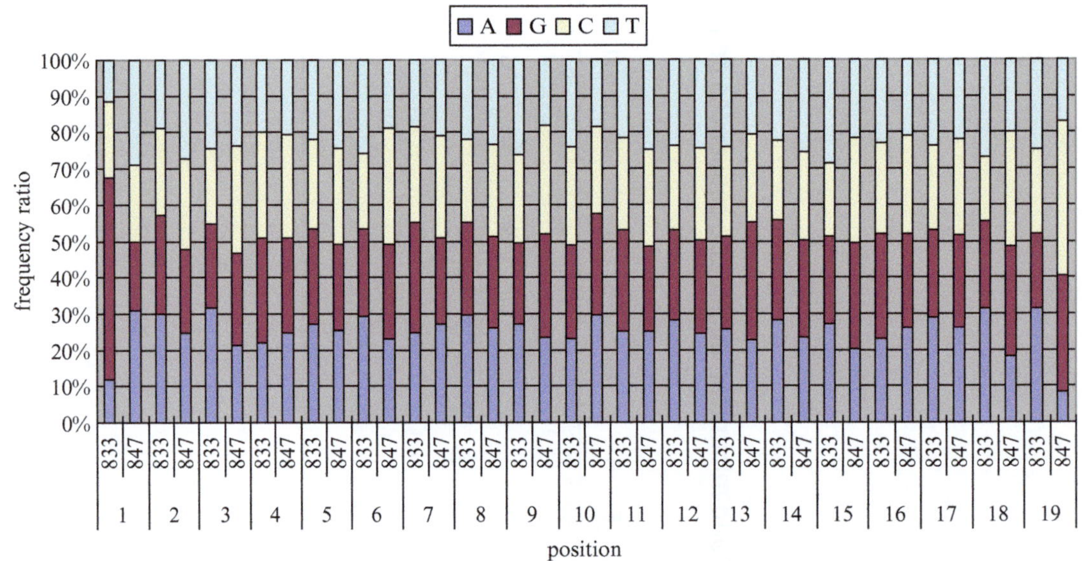

Fig. 10. Relations of nucleotide occurrence frequencies between 833 effective and 847 ineffective siRNAs.

i.e., 833 or 847 siRNAs, is used. There are differences in the nucleotide occurrence frequencies between both sets of siRNAs as shown in Fig. 10. Especially, there are big differences at positions 1 and 19. These results are also useful for designing effective siRNA sequences.

4.4.2. Characteristics for the Combinations of Two Successive Nucleotides

From the relations between two successive nucleotides determined by using the first Markov model for 833 known effective siRNAs it is possible to analyze the frequencies of combinations of two successive nucleotides in the sense strand. The relations among the

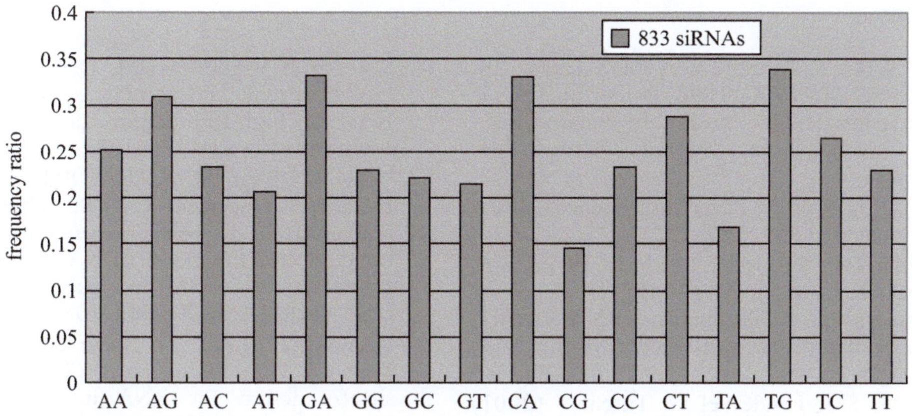

Fig. 11. Frequency ratios of combinations of two successive nucleotides in 833 effective siRNAs.

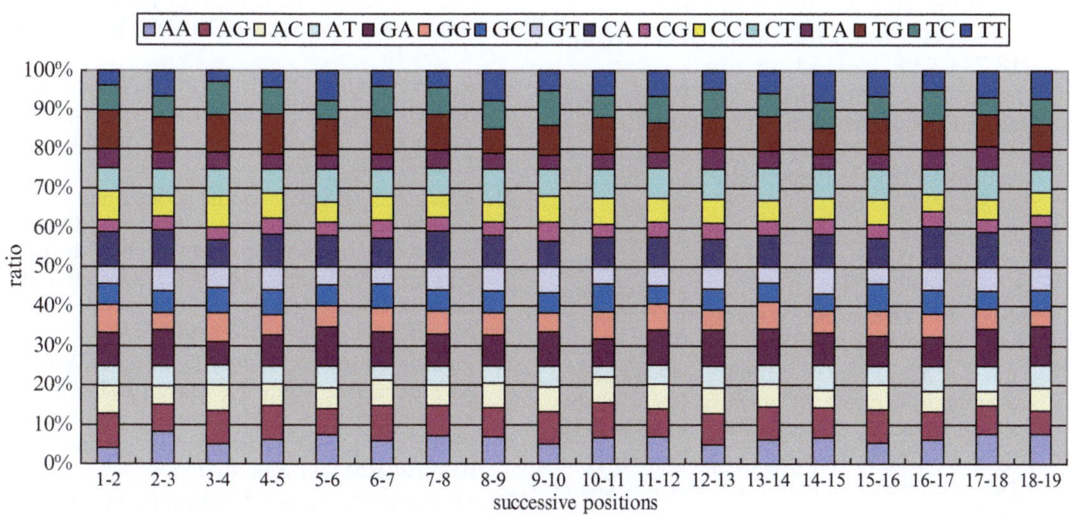

Fig. 12. Frequency ratios of two-nucleotide combinations in two successive positions from 5′ to 3′ for 833 effective siRNAs.

frequency ratios of two successive nucleotides for 833 known effective siRNAs are shown in Fig. 11, where it is clear that there are ups and downs in the frequency ratios of combinations of two nucleotides. Two-nucleotide combinations with high frequency ratios are TG (34%), GA (33%), CA (33%), and AG (31%), whereas combinations with low ones are CG (15%) and TA (17%).

It is also possible to calculate the frequency ratios of two-nucleotide combinations between two successive positions from 5′ to 3′ of the sense strand. They are shown in Fig. 12. When designing effective siRNAs for the target genes, it is also necessary to consider these characteristics of the frequencies of two-nucleotide combinations.

References

1. Fire A, Xu S, Montgomery MK, Kostas SA, Driver SE, Mello CC (1998) Potent and specific genetic interference by double-stranded RNA in *Caenorhabditis elegans*. Nature 391: 806–811
2. Sharp PA (2001) RNA interference—2001. Genes Dev 15:485–490
3. Elbashir SM, Harborth J, Lendeckel W, Yalcin A, Weber K, Tuschl T (2001) Duplexes of 21-nucleotide RNAs mediate RNA interference in mammalian cell culture. Nature 411: 494–498
4. Elbashir SM, Lendeckel W, Tuschl T (2001) RNA interference is mediated by 21- and 22-nucleotide RNAs. Genes Dev 15:188–200
5. Dykxhoorn DM, Navia CD, Sharp PA (2003) Killing the messenger: short RNAs that silence gene expression. Nat Rev 4:457–467
6. Hannon GJ (2002) RNA interference. Nature 418:244–251
7. Holen T, Amarzguioui M, Wiiger MT, Babaie E, Prydz H (2002) Positional effects of short interfering RNAs targeting the human coagulation trigger tissue factor. Nucleic Acids Res 30:1757–1766
8. Elbashir SM, Martinez J, Patkaniowska A, Lendeckel W, Tuschl T (2001) Functional anatomy of siRNAs for mediating efficient RNAi in *Drosophila melanogaster* embryo lysate. EMBO J 20:6877–6888
9. Kumar R, Conklin DS, Mittal V (2003) High-throughput selection of effective RNAi probes for gene silencing. Genome Res 13:2333–2340
10. Mittal V (2004) Improving the efficiency of RNA interference in mammals. Nat Rev Genet 5:355–365
11. Schwarz DS, Hutvagner G, Du T, Xu Z, Aronin N, Zamore PD (2003) Asymmetry in the assembly of the RNAi enzyme complex. Cell 115:199–208
12. Khvorova A, Reynolds A, Jayasena SD (2003) Functional siRNAs and miRNAs exhibit strand bias. Cell 115:209–216
13. Chalk AM, Wahlestedt C, Sonnhammer ELL (2004) Improved and automated prediction of effective siRNA. Biochem Biophys Res Commun 319:264–274
14. Teramoto R, Aoki M, Kimura T, Kanaoka M (2005) Prediction of siRNA functionality using generalized string kernel and support vector machine. FEBS Lett 579:2878–2882
15. Naito Y, Yamada T, Ui-Tei K, Morishita S, Saigo K (2004) siDirect: highly effective, target-specific siRNA design software for mammalian RNA interference. Nucleic Acids Res 32:W124–W129
16. Santoyo J, Vaguerizas JM, Dapozo J (2004) Highly specific and accurate selection of siRNAs for high-throughput functional assays. Bioinformatics 21:1376–1382
17. Truss M, Swat M, Kielbasa SM, Schafer R, Herzed H, Hagemeier C (2005) HuSiDa—the human siRNA database: an open-access database for published functional siRNA sequences and technical details of efficient transfer into recipient cells. Nucleic Acids Res 33:D108–D111
18. Reynolds A, Leake D, Boese Q, Scaringe S, Marshall WS, Khvorova A (2004) Rational siRNA design for RNA interference. Nat Biotechnol 22:326–330
19. Ui-Tei K, Naito Y, Takahashi F, Haraguchi T, Ohki-Hamazaki H, Juni A, Ueda R, Saigou K (2004) Guidelines for the selection of highly effective siRNA sequences for mammalian and chick RNA interference. Nucleic Acids Res 32: 936–948
20. Amarzguioui M, Prydz H (2004) An algorithm for selection of functional siRNA sequences. Biochem Biophys Res Commun 316: 1050–1058
21. Hsieh AC, Bo R, Monola J, Vazquez F, Bare O, Khvorova A, Scaringe S, Sellers WR (2004) A library of siRNA duplexes targeting the phosphoinositide 3-kinase pathway: determinants of gene silencing for use in cell-based screens. Nucleic Acids Res 32:893–901
22. Jagla B, Aulner N, Kelly PD, Song D, Volchuk A, Zatorski A, Shum D, Mayer T, De Angelis DA, Ouerfelli O, Rutishauser U, Rothman JE (2005) Sequence characteristics of functional siRNAs. RNA 11:864–872
23. Huesken D, Lange J, Mikanin C, Weiler J, Asselbergs F, Warner J, Meloon B, Engel S, Rosenberg A, Cohen D, Labow M, Reinhardt M, Natt F, Hall J (2005) Design of a genome-wide siRNA library using an artificial neural network. Nat Biotechnol 23:995–1001
24. Snove O Jr, Nedland M, Fjeldstad SH, Humberset H, Birkeland OR, Grunfeld T, Saetrom PO (2004) Designing effective siRNAs with off-target control. Biochem Biophys Res Commun 325:769–773
25. Durbin R, Eddy SR, Krogh A, Mitchison G (1998) Biological sequence analysis—probabilistic models of proteins and nucleic acids. Cambridge University Press, Cambridge
26. Takasaki S (2009) Selecting effective siRNA target sequences by using Bayes' theorem. Comput Biol Chem 33:368–372
27. Takasaki S, Kawamura Y, Konagaya A (2006) Selecting effective siRNA sequences by using radial basis function network and decision tree learning. BMC Bioinform 7(Suppl 5):S22

28. Takasaki S, Kotani S, Konagaya A (2004) An effective method for selecting siRNA target sequences in mammalian cells. Cell Cycle 3: 790–795
29. Takasaki S, Kotani S, Konagaya A (2005) Selecting effective siRNA target sequences for mammalian genes. RNA Biol 2:21–27
30. Takasaki S, Kawamura Y, Konagaya A (2006) Selecting effective siRNA sequences based on the self-organizing map and statistical techniques. Comput Biol Chem 30:169–178
31. Takasaki S, Konagaya A (2006) Comparative analyses for selecting effective siRNA sequences. Chem-Bioinform J 6:69–84
32. Takasaki S, Kawamura Y (2007) Using radial basis function networks and significance testing to select effective siRNA sequences, *Comput.* Stat Data Anal 51:6476–6487
33. Elbashir SM, Harborth J, Weber K, Tuschl T (2002) Analysis of gene function in somatic mammalian cells using small interfering RNAs. Methods 26:199–213
34. Ladunga I (2007) More complete gene silencing by fewer siRNAs: transparent optimized design and biophysical signature. Nucleic Acids Res 35:433–440
35. Holen T (2006) Efficient prediction of siRNAs with siRNA rules 1.0: an open-source JAVA approach to siRNA algorithms. RNA 12: 1620–1625
36. Heale BSE, Sifer HS, Bowers C, Rossi JJ (2005) siRNA target site secondary structure predictions using local stable substructures. Nucleic Acids Res 33:e-30
37. Luo KQ, Chang DC (2004) The gene silencing efficacy of siRNA is strongly dependent on the local structure of mRNA at the target region. Biochem Biophys Res Commun 318:303–310
38. Bohula EA, Salisbury AJ, Sohail M, Playford MP, Riedemann J, Southern EM, Macaulay VM (2003) The efficacy of small interfering RNAs targeted to the type I insulin-like growth factor receptor (IGFIR) is influenced by secondary structure in the IGFIR transcript. J Biol Chem 278:15991–15997
39. Chan CY, Carmack CS, Long DD, Maliyekkel A, Shao Y, Roninson IB, Ding Y (2009) A structural interpretation of the effect of GC-content on efficiency of RNA interference. BMC Bioinform 10(Suppl 1):S33
40. Vig K, Lewis N, Moore EG, Pillai S, Dennis VA, Singh SR (2009) Secondary RNA structure and its role in RNA interference to silence the respiratory syncytial virus fusion protein gene. Mol Biotechnol 43:200–211
41. Saetrom P, Snove O Jr (2004) A comparison of siRNA efficacy predictors. Biochem Biophys Res Commun 321:247–253
42. Shabalina SA, Spiridonov AN, Ogurtsov AY (2006) Computational models with thermodynamic and composition features improve siRNA design. BMC Bioinform 7:65
43. Vert J, Foveau N, Lajaunie C, Vandenbrouck Y (2006) An accurate and interpretable model for siRNA efficacy prediction. BMC Bioinform 7:520
44. Matveeva O, Nechipurenko Y, Rossi L, Moore B, Sactrom P, Ogurtsov AY, Atkins JF, Shabalina SA (2007) Comparison of approaches for rational siRNA design leading to a new efficient and transparent method. Nucleic Acids Res 35:e63
45. Lu ZJ, Mathews DH (2008) Efficient siRNA selection using hybridization thermodynamics. Nucleic Acids Res 36:640–647
46. Wang X, Wang X, Varma RK, Beauchamp L, Magdaleno S, Sendera TJ (2009) Selection of hyperfunctional siRNAs with improved potency and specificity. Nucleic Acids Res 37:e152
47. Klingelhoefer JW, Moutsianas L, Holmes C (2009) Approximate Bayesian feature selection on a large meta-dataset offers novel insights on factors that effect siRNA potency. Bioinformatics 25:1594–1601
48. Gong W, Ren Y, Zhou H, Wang Y, Kang S, Li T (2008) siDRM: an effective and generally applicable online siRNA design tool. Bioinformatics 24:2405–2406
49. Patzel V (2007) In silico selection of active siRNA. Drug Discov Today 12:139–148
50. Tafer H, Ameres SL, Obemosterer G, Gebeshuber CA, Schroeder R (2008) The impact of target site accessibility on the design of effective siRNAs. Nat Biotechnol 26: 578–583
51. Walton SP, Wu M, Gredell JA, Chan C (2010) Designing highly active siRNAs for therapeutic applications. FEBS J 277:4806–4813
52. Ahmed F, Raghava GP (2011) Designing of highly effective complementary and mismatch siRNAs for silencing a gene. PLoS One 6: e23443
53. Chaudhary A, Srivastava S, Garg S (2011) Development of a software tool and criteria evaluation for efficient design of small interfering RNA. Biochem Biophys Res Commun 404: 313–320
54. Katoh T, Suzuki T (2007) Specific residues at every third position of siRNA shape its efficient RNAi activity. Nucleic Acids Res 35:e27
55. Quinlan JR (1986) Induction of decision trees. Mach Learn 1:81–106
56. Saetrom P (2004) Predicting the efficacy of short oligonucleotides in antisense and RNAi experiments with boosted genetic programming. Bioinformatics 20:3055–3063

Chapter 3

Designing Functional siRNA with Reduced Off-Target Effects

Yuki Naito and Kumiko Ui-Tei

Abstract

RNA interference (RNAi) mediated by small interfering RNA (siRNA) is now widely used to knock down gene expression in a sequence-specific manner, making it a powerful tool not only for studying gene functions but also for therapeutic applications. siRNA decreases the expression level of the intended target gene with complete complementarity by cleaving its mRNA. However, the efficacy of each siRNA widely varies depending on its sequence in mammalian cells; only a limited fraction of randomly designed siRNAs is functional. Moreover, off-target silencing effects arise when the siRNA has partial complementarity in the seed region with unintended genes. Here, we describe the rational designing of functional, off-target effect-reduced siRNAs using siDirect 2.0 Web server (http://siDirect2.RNAi.jp/). By using the default parameters, siDirect 2.0 can design at least one qualified siRNA for >94% of human mRNA sequences in the RefSeq database.

Key words: RNAi, siRNA, shRNA, siDirect

1. Introduction

RNA interference (RNAi) mediated by double-stranded RNA is a powerful tool not only for studying gene functions but also for therapeutic purposes, including antiviral treatments (1, 2). RNAi is induced by small interfering RNA (siRNA), a duplex of 21-nucleotide (nt) RNAs with 2-nt 3′ overhangs. The siRNA incorporated into cells is loaded onto an RNAi effector complex called the RNA-induced silencing complex (RISC) (3, 4). One of the two strands of an siRNA duplex binds to the target mRNA in a sequence-specific manner, leading to the silencing of gene expression by the cleavage of the target mRNA. Thus, RNAi is induced by siRNA with complete complementarity with target mRNA. However, RNAi efficacies are widely varied depending on siRNA sequences, and a limited fraction of siRNAs is functional in mammalian cells. Previously, we reported that functional siRNA showed asymmetrical sequences, as described below (in Subheading 2,

Debra J. Taxman (ed.), *siRNA Design: Methods and Protocols*, Methods in Molecular Biology, vol. 942,
DOI 10.1007/978-1-62703-119-6_3, © Springer Science+Business Media, LLC 2013

Fig. 1a), suggesting that siRNA performances depend on the direction from which a terminal siRNA could unwind to a single strand.

Meanwhile, siRNAs can give rise to the down-regulation of other genes with partial complementarity with the siRNA guide strand. This phenomenon is referred to as an "off-target" effect, which represents siRNA silencing effects on non-targeted genes and is induced through mRNA degradation (5–9). Our and others' microarray-based off-target effect profilings using cells transfected with siRNA showed that the transcripts with 3′ UTRs containing 7-nt-long sequences complementary to the seed region at positions 2–8 of the siRNA guide strand were prominently reduced (Fig. 1b, (5–7)). Furthermore, we revealed that the capability of siRNAs to induce seed-dependent off-target effect is highly correlated with the thermodynamic stability of the duplex between the seed region of the siRNA guide strand and its target mRNA (10). An siRNA with lower seed–target duplex stability should minimize seed-dependent off-target silencing (Fig. 1c, d).

Previously, we released a highly effective, target-specific siRNA design software, siDirect, in which siRNA sequences were selected using our guidelines established by extensive experiments to clarify the relationship between siRNA sequences and RNAi activities (Fig. 1a) (11). By using siDirect or other software developed by several groups (12–18), RNAi became a versatile experimental tool used widely and routinely. However, a fundamental concern to avoid off-target effects remained unresolved. To overcome this problem, we developed an improved version of siDirect, siDirect 2.0, which reduces off-target silencing by selecting siRNAs with low thermodynamic stabilities in the seed region followed by the elimination of unrelated mRNAs with nearly perfect match (19). By using the default parameters, at least one functional, off-target-reduced siRNA could be designed for >94% of the human mRNA sequences in RefSeq (Table 1; see Note 1).

2. Algorithms

An overall flow of designing functional siRNA with reduced off-target effects using siDirect 2.0 for human RNAi is shown in Fig. 2. All possible 23-mer subsequences, corresponding to the complementary sequence of the 21-nt guide strand and the 2-nt 3′ overhang of the passenger strand within the human mRNA sequences, are generated and filtered in the three selection steps described below.

2.1. Selecting Functional siRNAs

Among all human 23-mer subsequences, highly functional siRNA sequences are selected using our algorithm (20) (Fig. 2a, see

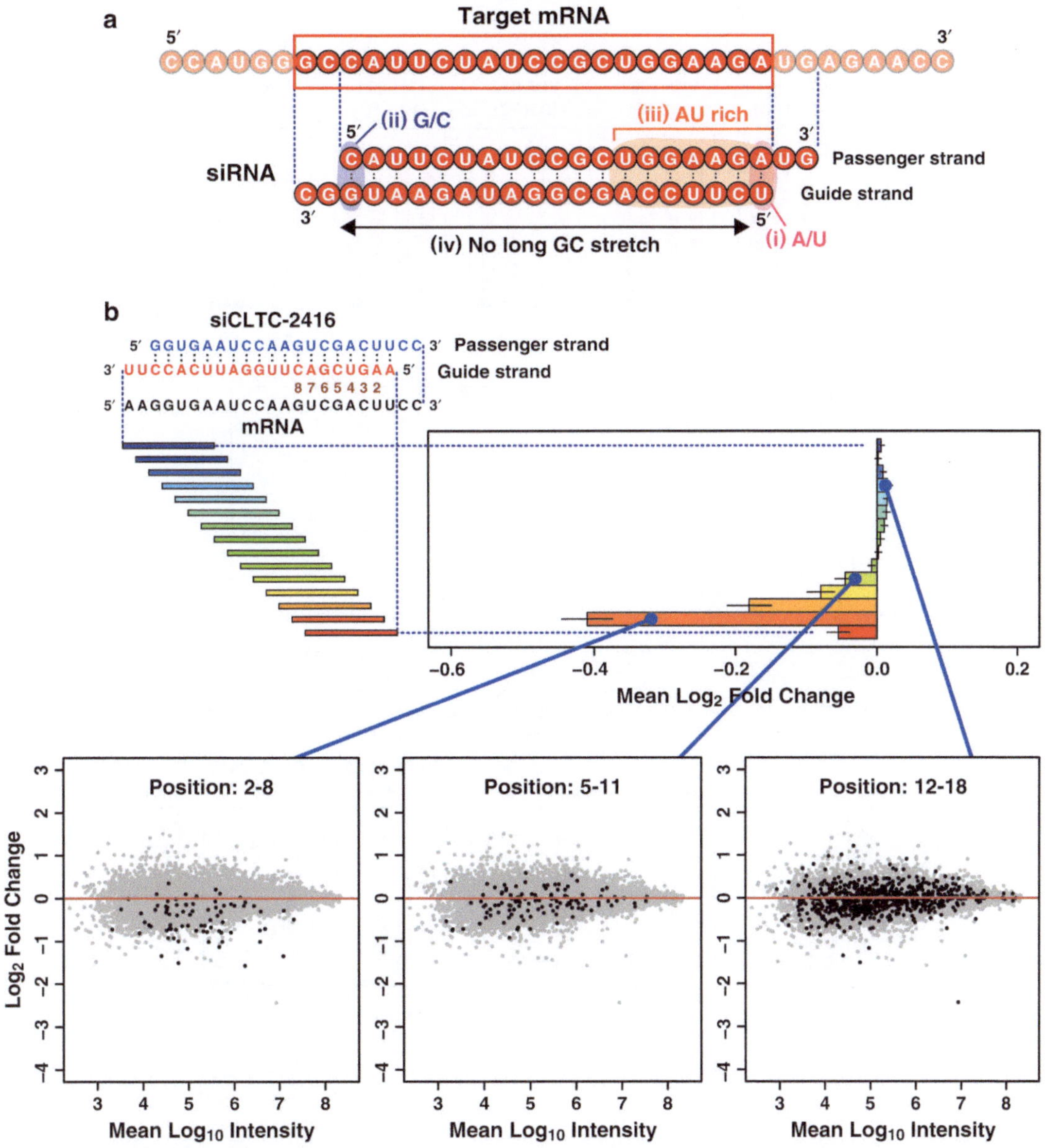

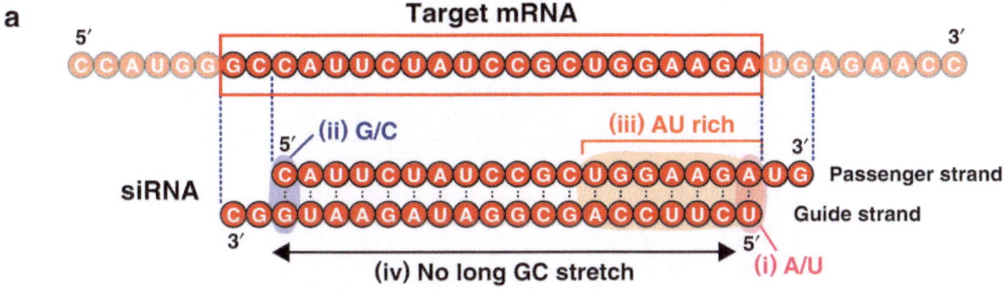

Fig. 1. Functional siRNA sequence and seed-dependent off-target effects depending on the thermodynamic stability in the seed–target duplex. (**a**) Schematic representation of a functional siRNA duplex. The target mRNA and the functional siRNA design algorithm presented by Ui-Tei et al. (20) is shown (see Text). (**b**) Microarray analysis of cells transfected with siRNA. siRNA (siCLTC-2416) and target mRNA sequences used in this experiment (*upper left*). Transcripts possessing the complements to a 7-nt region of the siCLTC-2416 guide strand were divided into 15 groups based on the positions of the complementary regions. The silencing effects of each group were shown as log2 fold change to mock transfection. The MA plots show the expression profile of three groups (positions 2–8, 5–11, and 12–18, respectively). The vertical axis indicates the log2 fold change, and the horizontal axis indicates the mean signal intensity of the probes for each gene before and after siRNA transfection. Black dots indicate transcripts with 7-nt complementarity. *Gray dots*, unrelated transcripts. Complementarity between the seed region (nucleotides 2–8) of an siRNA and the target mRNA is the key determinant in off-target silencing. (**c**) Off-target silencing activities of 12 siRNAs were examined in HeLa cells using the psiCHECK seed-match reporter assay (10). The line plot illustrates the distribution of Tm values of all possible 7-nt seed sequences ($4^7 = 16{,}384$). Twelve siRNAs are mapped on the line plot by the calculated Tm of the seed–target duplex. The silencing activity is highly correlated with the calculated Tm of the duplex between the seed region of the siRNA guide strand and its target mRNA. (**d**) Expression profiling of transcripts downregulated by the transfection of siVIM-270 with high seed Tm (28.5°C) and siVIM-812 with low seed Tm (8.7°C). Transcripts with one or more sequence complementary to the siRNA seed in the 3′ UTR are plotted in *black*. Transcripts with no seed complementarity are plotted in *gray*. The inset represents the mean log2 fold change of the transcripts with one or more seed complementarity (*black*) and with no seed complementarity (*gray*).

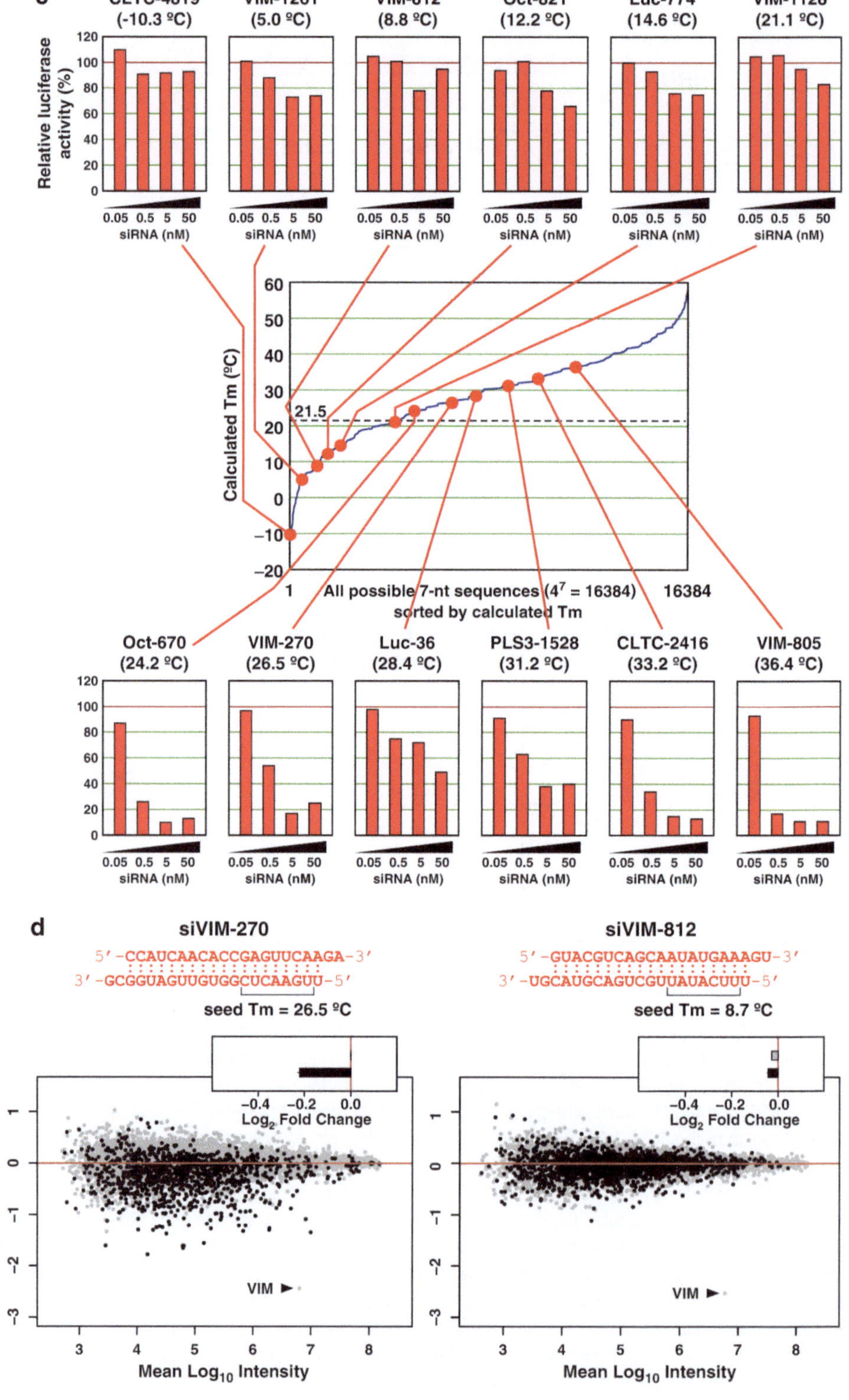

Fig. 1. (continued)

Table 1
The proportion of selectable siRNAs according to the Tm values in the seed–target duplexes. The default parameters: Tm in the seed–target duplex = 21.5°C, minimum mismatches against off-targets = 2

Tm in the seed–target duplex (°C)	Number of minimum mismatches against off-targets	Selectable siRNAs among all 23-mers in RefSeq (%)[a]	Human mRNAs for which at least one siRNA is selectable[b]
≤21.5	0	2.99	25,426 (99.1%)
≤21.5	1	2.74	24,765 (96.6%)
≤21.5	2	2.08	24,283 (94.7%)
≤21.5	3	0.20	19,807 (77.2%)
≤21.5	4	0.00	38 (0.15%)
≤15.0	0	1.13	24,447 (95.3%)
≤15.0	1	1.02	22,898 (89.3%)
≤15.0	2	0.70	21,821 (85.1%)
≤15.0	3	0.04	12,049 (47.0%)
≤15.0	4	0.00	2 (0.01%)
≤10.0	0	0.52	23,161 (90.3%)
≤10.0	1	0.46	20,616 (80.4%)
≤10.0	2	0.27	18,656 (72.7%)
≤10.0	3	0.01	4,739 (18.5%)
≤10.0	4	0.00	0 (0.00%)

[a]The total number of siRNA (56,375,087) is set to 100%
[b]100% indicates 25,651 mRNAs in RefSeq release 30

Note 2). In brief, siRNA sequences that satisfy the following conditions simultaneously are selected: (1) A/U at the 5′ terminus of the guide strand; (2) G/C at the 5′ terminus of the passenger strand; (3) at least 4 A/U residues in the 5′ terminal 7 bp of the guide strand; and (4) G/C stretches longer than 9 bp should be avoided (Fig. 1a). Our experimental validation using luciferase reporter assays showed that 98% of the siRNAs satisfying the above conditions reduce the expression of luciferase reporter to below 33% (21).

2.2. Reducing Seed-Dependent Off-Target Effects

To avoid off-target silencing, siRNAs with low thermodynamic stability in the seed region are selected. Because thermodynamic stability shows a strong relation to melting temperature (Tm) (10), the Tm of the seed–target duplex is calculated using the nearest

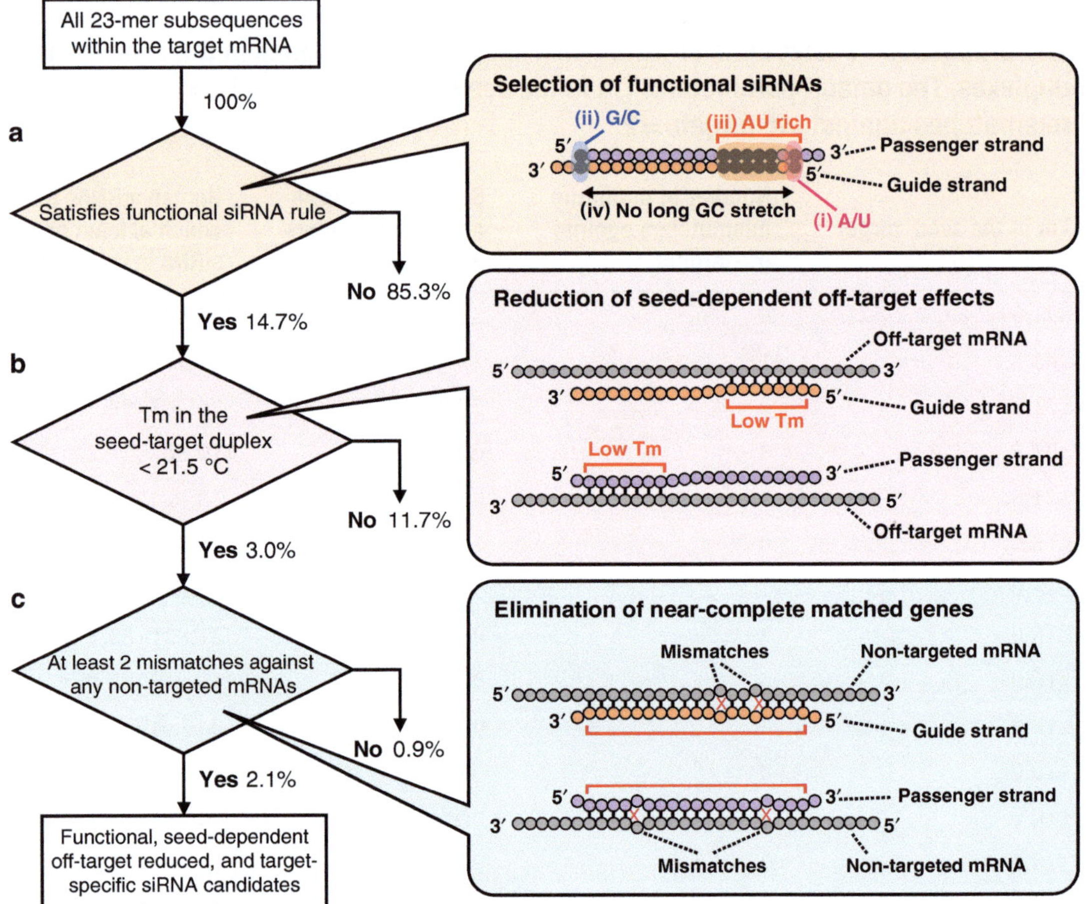

Fig. 2. Overall flow of steps for selecting functional siRNAs with reduced off-target effects. (**a**) Functional siRNA sequences are selected by using the algorithm described in ref. (20). (**b**) siRNAs with seed–target Tm below 21.5°C for both strands are selected. (**c**) Nucleotides positioned in the 2–20 nt region of both strands of the siRNAs are examined to find near-perfect matches with unrelated genes. siRNAs that have at least two mismatches to any other non-targeted transcripts are selected. The percentages in this figure denote the proportions of selected ("Yes") or unselected ("No") siRNA candidates calculated using all 23-mer subsequences (56,375,087; 100%) generated from human mRNAs in RefSeq release 30.

neighbor method and the thermodynamic parameters for the formation of RNA duplex (22) as follows (Fig. 2b):

$$\mathrm{Tm} = \{(1{,}000 \times \Delta H) / (A + \Delta S + R \ln(C_{\mathrm{T}} / 4))\} \\ -273.15 + 16.6 \log[\mathrm{Na}^{+}],$$

where ΔH (kcal/mol) is the sum of the nearest neighbor enthalpy change, A (cal/ml/K) is the helix initiation constant (−10.8), ΔS (kcal/mol/K) is the sum of the nearest neighbor entropy change (22), R (cal/deg/mol) is the gas constant (1.987), and C_{T} (mol/L) is the total molecular concentration of the strand (100 μM). $[\mathrm{Na}^{+}]$ is fixed at 100 mM.

As shown in our previous report, a calculated Tm of 21.5°C may be a benchmark to discriminate almost all off-target-free seed sequences from the off-target-positive ones (10), and thus is used as the default parameter in siDirect 2.0. Furthermore, the passenger strands of functional siRNAs also take part in the seed-dependent off-target gene silencing. Thus, siRNAs whose seed–target Tms are below 21.5°C for both strands are selected (see Note 3).

2.3. Eliminating Near-Perfect Matched Genes

Even when the Tm value of the seed–target duplex is sufficiently low, target gene silencing can still take place if the non-seed region is completely complementary (e.g., VIM is potently downregulated by siVIM-812 with a low seed Tm in Fig. 1d). Therefore, in the third step, siRNAs that have near-perfect matches to any other non-targeted transcripts are eliminated. In siDirect 2.0, off-target searches are performed for 19-mer sequences at positions 2–20 of both strands of the siRNA duplex (Fig. 2c) because these 19 nucleotides are responsible for target mRNA recognition. Because the widely used BLAST tends to overlook near-perfect match candidates frequently, we used a fast and sensitive algorithm (23). By default, siRNA sequences that have at least two mismatches to any other non-targeted transcripts are selected (see Note 4).

3. Methods

3.1. Designing siRNAs for Human/Mouse/Rat Genes

Highly functional, off-target-reduced siRNAs for most human, mouse, and rat genes can be designed using the Web-based online software, siDirect 2.0 (http://siDirect2.RNAi.jp/). In this section, the procedure for selecting siRNA for human claudin 17 (CLDN17) is presented as an example.

1. Open "http://siDirect2.RNAi.jp/" with a Web browser (Fig. 3a).
2. Enter the accession number for human claudin 17 (NM_012131) into the upper blank box (Fig. 3a, 1).
3. Click "retrieve sequence" to get the nucleotide sequence from GenBank (Fig. 3a, 2).
4. Or you can directly paste the cDNA sequence of claudin 17 in a nucleotide sequence box (Fig. 3a, 3). Accepted input types are FASTA or plain nucleotide sequence up to 10 kb.
5. Click the "Options: [click here]" button to set optional parameters if necessary (Fig. 3b). Functional siRNA selection algorithms (Fig. 3b, 5), maximum Tm of the seed–target duplex (Fig. 3b, 6), and nonredundant database for the specificity search (Fig. 3b, 7) can be selected. "Other options" can specify the target range, contiguous bases to avoid, the GC content, and custom sequence patterns to include/exclude (Fig. 3b, 8).

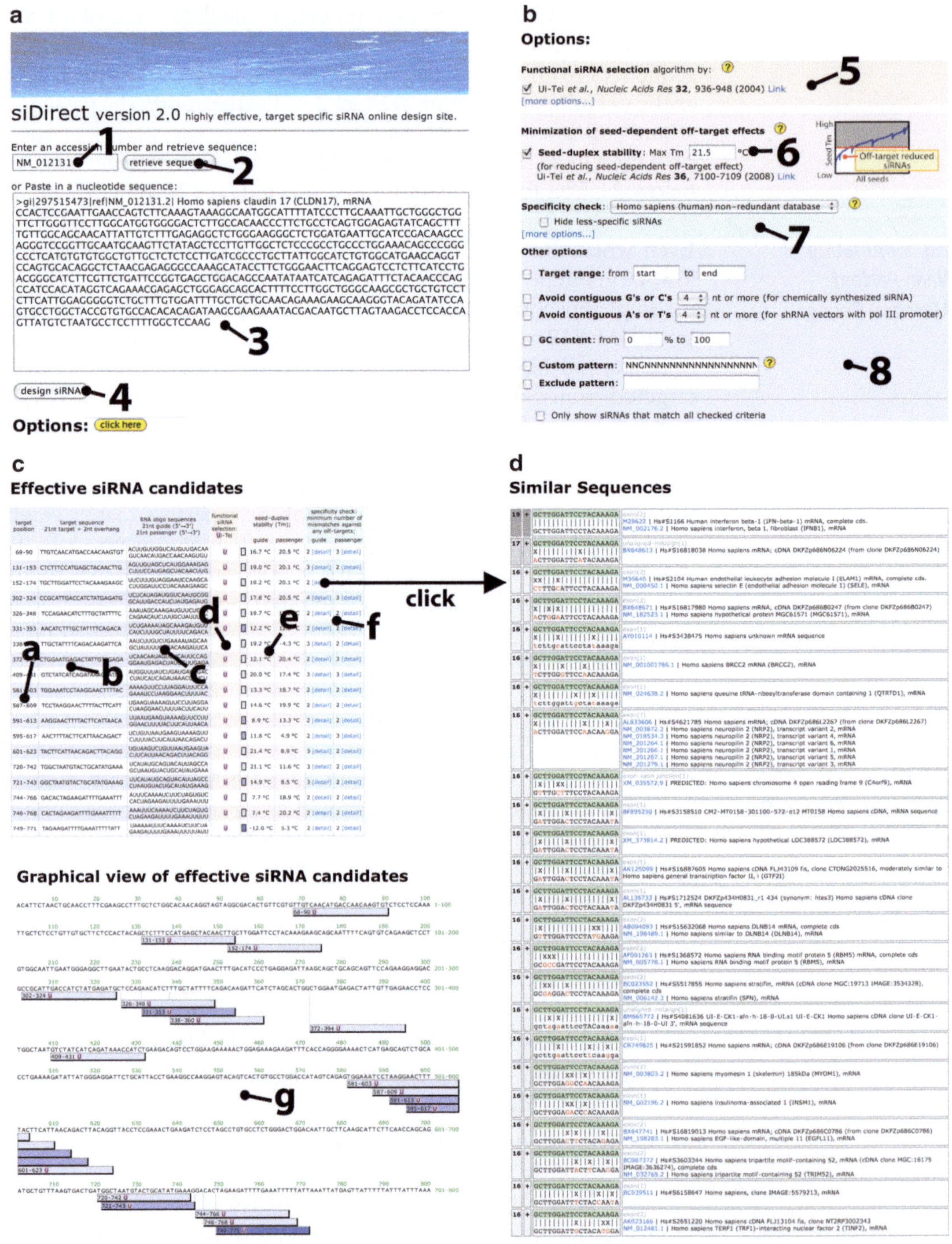

Fig. 3. Screenshots of siDirect 2.0 (http://siDirect2.RNAi.jp/). (**a**) Top page. (**b**) Optional parameters. (**c**) Result page. (**d**) Detailed list of off-target gene candidates with near-complete matches. The alignment between each off-target sequence and the siRNA sequence allows visualization of the positions of mismatches. See Subheading 3.1 for a detailed description of the features 1–8 and a–g.

6. Click "design siRNA" (Fig. 3a, 4).
7. Results (Fig. 3c): (a) siRNA target positions; (b) siRNA target sequences (21 nt+2 nt overhang); (c) siRNA guide strand sequences (21nt) and passenger strand sequences (21nt); (d) siRNA efficacy predictions. siRNAs with "U", "R", and "A" satisfy the functional siRNA design algorithms of Ui-Tei et al. (20), Reynolds et al. (24), and Amarzguioui et al. (25), respectively. "R" and "A" are displayed if specified in the option (Fig. 3b, 5). (e) Calculated Tms of the seed regions of siRNA guide and passenger strands, respectively. Selecting an siRNA with lower seed Tm reduces off-target effects; (f) Homology search results against human mRNAs. The minimum number of mismatches against off-target transcripts is shown. (g) Graphical view of the designed siRNAs. Click on each siRNA to show the off-target list.
8. A tab-delimited siRNA list is provided at the bottom of the results page. Users can directly copy–paste a result into a text-editing program such as Excel.
9. List of off-target candidates for individual siRNAs (Fig. 3d). The alignment between each off-target candidate and the siRNA sequence clarifies the locations of mismatches. Hits with a complete match (i.e., 19/19 matches), one mismatch (18/19 matches), two mismatches (17/19 matches), or three mismatches (16/19 matches) are shown.
10. Select the appropriate siRNAs from the siDirect 2.0 results.

3.2. Designing shRNA Expression Constructs with pol III Promoters

Functional short hairpin RNAs (shRNAs) can be also designed using siDirect 2.0 in the same manner. However, additional sequence conditions are required for designing shRNAs transcribed by a pol III promoter. The procedure for designing shRNA for human vimentin (VIM) is presented as an example.

1. Open "http://siDirect2.RNAi.jp/" with a Web browser.
2. Enter the accession number for human vimentin (NM_003380).
3. Click "retrieve sequence" to get the nucleotide sequence from GenBank.
4. Or you can directly paste in a nucleotide sequence. Accepted input types are FASTA or plain nucleotide sequence up to 10 kb.
5. Click "Options: [click here]" (Fig. 4a).
6. Select "Avoid contiguous A's or T's 4 nt or more" (Fig. 4a, 1). These sequences should be avoided because shRNA templates containing TTTT may terminate the transcription by RNA polymerase III.
7. Select "Custom pattern" and input "NNGNNNNNNNNN NNNNNNNNNNN" in the field (Fig. 4a, 2). This fixes the

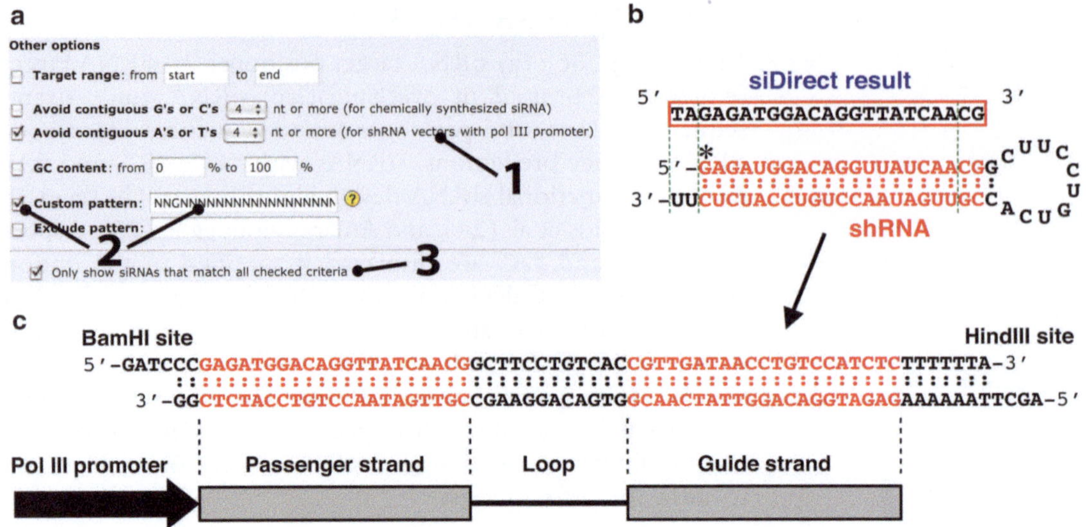

Fig. 4. Designing siRNA for shRNA expression constructs with pol III promoters. (**a**) Optional parameters for pol III-driven shRNAs. Avoiding AAAA/TTTTT prevents transcription termination by pol III. The third letter (i.e., shRNA transcription start site) should be G. (**b**) siRNA designed using siDirect 2.0 is predicted to be cleaved out by cellular Dicer from the illustrated hairpin-structured RNA. The double-stranded siRNA sequence is shown in *red characters*, and the loop and termination regions are shown in *black characters*. An *asterisk* indicates the transcription start site. (**c**) The 67-nt single-stranded complementary DNA oligonucleotides are annealed and the resultant double-stranded DNA is inserted into BamHI and HindIII sites located downstream of RNA pol III promoter in the vector, such as pSUPER.retro.puro (OligoEngine) or pENTR.H1 (Invitrogen).

third letter of the target sequence (i.e., the shRNA transcription start) to "G". A purine is required for the RNA polymerase III transcription start site (an asterisk in Fig. 4b).

8. Select "Only show siRNAs that match all checked criteria" (Fig. 4a, 3).
9. Click "design siRNA". See step 7 of Subheading 3.1 for a detailed description of the result window.
10. Design the oligonucleotide sequence from the siDirect result (Fig. 4b, c).

4. Notes

1. To demonstrate the effects of varying selection stringency, we performed a genome-wide design of siRNAs for human mRNAs in RefSeq release 30 using various parameters (Table 1). As the stringency of selection is increased, the number of siRNA candidates is decreased. If no siRNA candidates are found for a given set of criteria, try relaxing the parameters, including the specificity check and seed-duplex stability. If the

parameters need to be relaxed further, it is recommended that the detailed list of off-target gene candidates is manually investigated (Fig. 3d) and that siRNAs are selected that do not have off-target hits to unrelated transcripts.

2. The proportion of functional siRNA sequences selected by this algorithm is 14.7% of all human 23-mer sequences generated from RefSeq 30 (Fig. 2a).
3. After this step, 3.0% of all human 23-mer sequences remained available (Fig. 2b).
4. Finally, 2.1% of all human 23-mer sequences remained available (Fig. 2c). These can be used as functional, off-target-reduced siRNAs.

Acknowledgement

This work was supported by grants from the Ministry of Education, Culture, Sports, Science and Technology (MEXT) of Japan to YN and KU-T.

References

1. Boutros M, Ahringer J (2008) The art and design of genetic screens: RNA interference. Nat Rev Genet 9:554–566
2. Castanotto D, Rossi JJ (2009) The promises and pitfalls of RNA-interference-based therapeutics. Nature 457:426–433
3. Hutvagner G, Simard MJ (2008) Argonaute proteins: key players in RNA silencing. Nat Rev Mol Cell Biol 9:22–32
4. Jinek M, Doudna JA (2009) A three-dimensional view of the molecular machinery of RNA interference. Nature 457:405–412
5. Lin X, Ruan X, Anderson MG, McDowell JA, Kroeger PE, Fesik SW, Shen Y (2005) siRNA-mediated off-target gene silencing triggered by a 7 nt complementation. Nucleic Acids Res 33:4527–4535
6. Birmingham A, Anderson EM, Reynolds A, Ilsley-Tyree D, Leake D, Fedorov Y, Baskerville S, Maksimova E, Robinson K, Karpilow J, Marshall WS, Khvorova A (2006) 3′ UTR seed matches, but not overall identity, are associated with RNAi off-targets. Nat Methods 3:199–204
7. Jackson AL, Burchard J, Schelter J, Chau BN, Cleary M, Lim L, Linsley PS (2006) Widespread siRNA "off-target" transcript silencing mediated by seed region sequence complementarity. RNA 12:1179–1187
8. Jackson AL, Bartz SR, Schelter J, Kobayashi SV, Burchard J, Mao M, Li B, Cavet G, Linsley PS (2003) Expression profiling reveals off-target gene regulation by RNAi. Nat Biotechnol 21:635–637
9. Scacheri PC, Rozenblatt-Rosen O, Caplen NJ, Wolfsberg TG, Umayam L, Lee JC, Hughes CM, Shanmugam KS, Bhattacharjee A, Meyerson M, Collins FS (2004) Short interfering RNAs can induce unexpected and divergent changes in the levels of untargeted proteins in mammalian cells. Proc Natl Acad Sci USA 101:1892–1897
10. Ui-Tei K, Naito Y, Nishi K, Juni A, Saigo K (2008) Thermodynamic stability and Watson-Crick base pairing in the seed duplex are major determinants of the efficiency of the siRNA-based off-target effect. Nucleic Acids Res 36:7100–7109
11. Naito Y, Yamada T, Ui-Tei K, Morishita S, Saigo K (2004) siDirect: highly effective, target-specific siRNA design software for mammalian RNA interference. Nucleic Acids Res 32:W124–W129

12. Yuan B, Latek R, Hossbach M, Tuschl T, Lewitter F (2004) siRNA Selection Server: an automated siRNA oligonucleotide prediction server. Nucleic Acids Res 32:W130–W134
13. Ding Y, Chan CY, Lawrence CE (2004) Sfold web server for statistical folding and rational design of nucleic acids. Nucleic Acids Res 32:W135–W141
14. Yiu SM, Wong PW, Lam TW, Mui YC, Kung HF, Lin M, Cheung YT (2005) Filtering of ineffective siRNAs and improved siRNA design tool. Bioinformatics 21:144–151
15. Gong W, Ren Y, Xu Q, Wang Y, Lin D, Zhou H, Li T (2006) Integrated siRNA design based on surveying of features associated with high RNAi effectiveness. BMC Bioinformatics 7:516
16. Vert JP, Foveau N, Lajaunie C, Vandenbrouck Y (2006) An accurate and interpretable model for siRNA efficacy prediction. BMC Bioinformatics 7:520
17. Ladunga I (2007) More complete gene silencing by fewer siRNAs: transparent optimized design and biophysical signature. Nucleic Acids Res 35:433–440
18. Park YK, Park SM, Choi YC, Lee D, Won M, Kim YJ (2008) AsiDesigner: exon-based siRNA design server considering alternative splicing. Nucleic Acids Res 36:W97–W103
19. Naito Y, Yoshimura J, Morishita S, Ui-Tei K (2009) siDirect 2.0: updated software for designing functional siRNA with reduced seed-dependent off-target effect. BMC Bioinformatics 10:392
20. Ui-Tei K, Naito Y, Takahashi F, Haraguchi T, Ohki-Hamazaki H, Juni A, Ueda R, Saigo K (2004) Guidelines for the selection of highly effective siRNA sequences for mammalian and chick RNA interference. Nucleic Acids Res 32:936–948
21. Naito Y, Saigo K, Ui-Tei K (2008) Evaluation of published rational siRNA design algorithms using firefly *luciferase* gene as a reporter. In: Lyland RT, Browning IB (eds) RNA interference research progress. Nova, New York, pp 3–11
22. Freier SM, Kierzek R, Jaeger JA, Sugimoto N, Caruthers MH, Neilson T, Turner DH (1986) Improved free-energy parameters for predictions of RNA duplex stability. Proc Natl Acad Sci USA 83:9373–9377
23. Yamada T, Morishita S (2005) Accelerated off-target search algorithm for siRNA. Bioinformatics 21:1316–1324
24. Reynolds A, Leake D, Boese Q, Scaringe S, Marshall WS, Khvorova A (2004) Rational siRNA design for RNA interference. Nat Biotechnol 22:326–330
25. Amarzguioui M, Prydz H (2004) An algorithm for selection of functional siRNA sequences. Biochem Biophys Res Commun 316:1050–1058

Chapter 4

Design and Screening of siRNAs Against Highly Structured RNA Targets

Neda Nasheri, John Paul Pezacki, and Selena M. Sagan

Abstract

RNA silencing is an invaluable tool to interrogate gene function. The cytoplasmic delivery of small interfering RNAs (siRNAs) complementary to a gene of interest results in cleavage and degradation of the target mRNA. Given the potential to target virtually any RNA, siRNA-based therapeutics may revolutionize the treatment of disease. Target site accessibility is a significant barrier to the design and efficacy of siRNAs, particularly against highly structured targets such as the genomes of positive-sense RNA viruses. Here, we describe a bead-based approach to screen for target site accessibility of siRNAs designed against highly structured target RNAs and demonstrate that this approach can be used to assess target site accessibility in vitro and predict potent target sites for siRNAs in cell culture against a highly structured RNA target.

Key words: Short-interfering RNA, Target site accessibility, Highly structured target RNA, Hepatitis C virus, Huh-7

1. Introduction

RNA silencing is an evolutionarily conserved pathway whereby small double-stranded RNA duplexes sequence-specifically suppress the expression of a target gene(s) (1). The introduction of small interfering RNA (siRNA) duplexes into the cytoplasm of cells results in their incorporation into the RNA-induced silencing complex (RISC), which uses one strand of the siRNA (the guide strand) to target complementary RNAs for endonucleolytic cleavage (2, 3). The cleaved target RNA is then further degraded by cellular 5′ and 3′ exonucleases (4).

Many factors contribute to the overall efficiency of RNA silencing. Numerous studies have led to the establishment of parameters that, when incorporated into a rational siRNA design

Debra J. Taxman (ed.), *siRNA Design: Methods and Protocols*, Methods in Molecular Biology, vol. 942,
DOI 10.1007/978-1-62703-119-6_4, © Springer Science+Business Media, LLC 2013

algorithm, increase the probability of selecting an effective siRNA (i.e., one capable of silencing gene expression by >50%) (5, 6). Firstly, siRNAs should have a G+C content between 30 and 52% (5–7). High G+C content can inhibit duplex unwinding, whereas low G+C content is associated with decreased functionality, likely due to a lower target affinity and specificity (6, 8). The G+C content also typically correlates with accessibility (5, 6); however, relying on G+C content alone may result in a large number of predicted false negatives (9). Designed siRNAs should have a low internal stability at the 3′ terminus of the sense (passenger) strand since this will determine which strand of the duplex will be incorporated into RISC (6, 10, 11). In addition, siRNAs should lack inverted repeat sequences because palindromes can foster internal secondary structures that may result in poor RISC loading (6). Lastly, several sense (passenger) strand base preferences have been described which correlate with increased siRNA efficacy including an A base at position 3, a U base at position 10, the absence of a G base at position 13, and the presence of an A/U base at position 19 (6). The preference for a U at position 10 likely reflects the RISC endonuclease's preference to cleave the 3′ end of U>A>G>>>C (5, 6) and the A/U at position 19 is likely a reflection of strand selection (6, 10, 11).

Despite a rational approach to siRNA design, only a fraction of siRNAs are effective at reducing the expression of their RNA targets and the efficiency of different siRNAs directed against the same target often varies significantly (7, 12, 13). Additional parameters must therefore affect the efficiency of siRNAs, such as target site accessibility or the presence of RNA-binding proteins. These parameters are not taken into account by many currently available siRNA design algorithms, despite the fact that target site accessibility is known to hamper efficient knockdown (7, 14–18). A few algorithms now take into account local secondary structure prediction in siRNA design and can improve the probability of designing effective siRNAs (9, 19–21). However, the prediction accuracy of currently available computational algorithms is still very low, generating numerous alternative structures (22, 23). In addition, long-range RNA–RNA interactions, common to challenging RNA targets such as RNA viruses, are not yet amenable to accurate modeling, and hence reliable results of analysis on target RNA secondary structure prediction methods cannot be guaranteed for such targets (23). Also, for long RNA targets, the demands on the computer are high, and in lieu of the long running time needed to predict the secondary structure of large RNAs, many siRNA design algorithms still rely solely on the siRNA sequence characteristics described above. This may not be an important parameter for the design of siRNAs targeting the ORFs of relatively unstructured cellular mRNAs, but for more challenging targets, such as the complex and highly structured genomes of positive-sense RNA viruses, target site accessibility may be an important consideration.

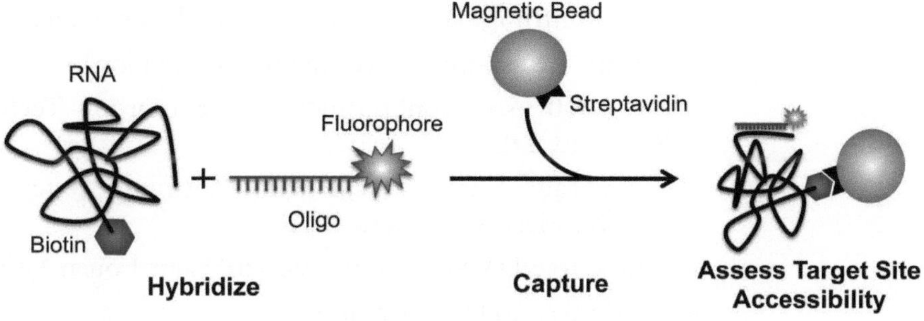

Fig. 1. Schematic diagram of bead-based oligo screening assay. Biotin-conjugated RNAs of interest are hybridized to fluorescently labeled oligos under native or denaturing conditions and captured on magnetic streptavidin-conjugated microbeads. Target site accessibility is then assessed by fluorescence microscopy.

Herein, we describe a high-throughput, bead-based approach to screen for effective siRNAs against challenging (large, highly structured) target RNAs, such as the hepatitis C virus (HCV) genome (Fig. 1). In vitro hybridization reactions are carried out under native conditions meant to preserve the native structure of the target RNA and are hence a measure of hybridization speed (kinetics) rather than duplex stability (thermodynamics). Assessment of hybridization of oligos to a target RNA under native conditions thus reflects the accessibility of the target RNA. We demonstrate that this approach can be used to assess target site accessibility and is able to predict potent siRNA target sites in cell culture against the highly structured HCV RNA genome. This approach also has the potential to be applied more broadly to the identification of novel protein–RNA and RNA–RNA interactions, and has recently been applied to the identification of novel microRNA–target RNA interactions (24).

2. Materials

2.1. Selecting siRNA Target Sites and Control Oligonucleotides

1. To select siRNA target sites as well as control oligonucleotides, Internet access is required.

2.2. Oligonucleotide preparation

1. RNA and/or DNA oligonucleotides can be synthesized commercially (see Note 1). For oligo screening, 5′ end fluorophore-labeled single-stranded RNA or DNA oligos can be used; however, duplexed siRNAs must be used for siRNA knockdown (see Note 2).

2.3. In Vitro Transcription and End-Labeling of Target RNA

1. Linearized plasmid DNA (see Note 3), or PCR product containing RNA of interest with a T7 RNA polymerase promoter.
2. MEGAscript™ T7 kit (Ambion, Austin, TX).
3. MEGAclear™ kit (Ambion).

4. Microcentrifuge 5415 D (Eppendorf, Hauppauge, NY).
5. 1.5 mL RNase-free microfuge tubes (Ambion).
6. ND-1000 spectrophotometer (NanoDrop Technologies, Rockland, DE).
7. 5′-EndTag Nucleic Acid End Labeling system (Vector Laboratories, Burlingame, CA).
8. RNase-free DMSO (Sigma-Aldrich, Saint Louis, MO).
9. RNase-free EDTA (Ambion).
10. EZ-Link Maleimide-PEG11-Biotin (Thermo Fisher Scientific, Rockford, IL), prepare a 200 mM stock in RNAse-free dry DMSO and store at −20°C in a desiccator.
11. Pierce® RNA 3′ End Biotinylation Kit (Thermo Fisher Scientific).
12. Owl EasyCast™ B1 (Thermo Fisher Scientific).
13. RNase free 10× Tris/Borate/EDTA (TBE) buffer (Ambion).
14. RNase free Agarose (Ambion).
15. SYBR® Safe DNA gel stain (Invitrogen, Burlington, ON).
16. Gel Loading Buffer II (Ambion).
17. UV transilluminator (UVP, Upland, CA).

2.4. Magnetic Microbead Conjugation and Oligo Screening

1. Labquake® Tube Shaker/Rotators (Cole-Parmer, Vernon Hills, IL).
2. Streptavidin-coated Dynabeads® M-280 (Invitrogen).
3. 2× Bind and wash buffer (B&W): 10 mM Tris–HCl, pH 7.5 + 2 M NaCl.
4. Solution A: 0.1 M NaOH + 0.05 M NaCl.
5. Hybridization buffer (20 mM HEPES, pH 7.8, 50 mM KCl, 10 mM $MgCl_2$, 1 mM DTT, 0.625 mg/mL salmon sperm DNA, 50 μg/mL yeast tRNA).
6. SUPER-In RNase Inhibitor (Ambion).
7. MagneSphere® Technology Magnetic Separation Stand (Promega, Madison, WI).

2.5. Imaging and Data Analyses

1. 1.0 mm Micro slides (VWR, West Chester, PA).
2. 18 mm Circle coverslips (VWR).
3. Olympus 1 × 81 inverted confocal microscope (Olympus Optical Co LTD, Japan).
4. Image-Pro® Plus 4.5 software (Media Cybernetics, Inc., Silver Spring, MD).

2.6. Tissue Culture and Transfection of siRNA Duplexes

1. Dulbecco's modified Eagle's medium (DMEM) (Gibco/BRL, Bethesda, MD) supplemented with 100 nM MEM non-essential amino acids (Gibco), 50 U/mL penicillin, 50 μg/mL

streptomycin (Pen–Strep, Gibco), and 10% fetal bovine serum (FBS, CANSERA, Rexdale, ON).

2. 10× Trypsin (Sigma Aldrich, St. Louis, MO) diluted to 1× with PBS.
3. Lipofectamine™ RNAiMAX (Invitrogen).
4. Opti-MEM® reduced serum medium (Invitrogen).

2.7. Northern Blot and Quantitative PCR Analyses

1. RNeasy extraction kit (Qiagen, Mississauga, ON).
2. Cell scraper (BD Biosciences, Franklin Lakes, NJ).
3. 20-Gauge needle (0.9 mm diameter) (BD Biosciences).
4. BrightStar® Biotinylated RNA Millenium™ Markers (Ambion).
5. NorthernMAX® kit (Ambion).
6. Hybond XL nylon membrane (GE Healthcare, Piscataway, NJ).
7. Chemiluminescent Nucleic Acid Detection Module (Pierce, Rockford, IL).
8. VWR 2700 Mini hybridization incubator (VWR).
9. Hand-held UV lamp (Spectroline, Westbury, NY).
10. Superscript II kit (Invitrogen).
11. Microseal 96-well PCR plates (Bio-Rad, Hercules, CA).
12. iQ SYBR Green Supermix (Bio-Rad).
13. iQ5 iCycler (Bio-Rad).
14. Biotin-11-UTP and biotin-11-CTP (Perkin Elmer, Boston, MA).

2.8. Western Blot Analyses

1. RIPA buffer (50 mM Tris–HCl (pH 6.8), 2% SDS, 10% glycerol, 100 mM DTT, 0.1% bromophenol blue).
2. Bio-Rad DC protein assay (Bio-Rad).
3. Spectramax® (Molecular Devices).
4. Softmax® pro plus (Molecular Devices).
5. Mini Trans-Blot® electrophoretic transfer cell (Bio-Rad).
6. Whatman filter paper (GE Healthcare).
7. Autoradiography film (Mandel Scientific, Guelph, ON).
8. Film processor (Konica, SRX-101a).

3. Methods

3.1. Designing siRNAs and Control Oligonucleotides

1. Establish whether the gene of interest has one or more splice variants. Decide whether you want to target a specific splice variant or all potential forms of the gene. Select regions of the mRNA for targeting accordingly (see Note 4).

2. Select several potential target sites within your gene depending on which splice forms of the gene are to be targeted (see Note 5). If an siRNA against the gene of interest has been functionally validated commercially or in the literature, it may be useful to test the same target site in your system. Often siRNAs that have shown efficacy in one cell system have similar efficacy in other cell systems.
3. There are several siRNA design Web sites available (both commercial and noncommercial, see Note 6). Updates to design algorithms are ongoing based on new findings in the field, and hence it is important to use the most current design algorithm available.
4. Eliminate potential off-target effects of your siRNAs by performing a BLAST search and eliminating sequences that have a perfect match ≥16 nt to other genes in the genome of the same species (http://www.ncbi.nlm.nih.gov/BLAST/) (see Note 7).
5. Avoid target sites that have known single-nucleotide polymorphisms (SNPs) (http://www.ncbi.nlm.nih.gov/projects/SNP/).
6. Select siRNAs that target different regions of your gene of interest, have the fewest amount of BLAST matches, and that do not overlap with SNPs (see Notes 8 and 9).
7. Also design control siRNAs or oligonucleotides, including a non-targeting siRNA (see Note 10). If possible, design a control oligonucleotide against a region of known secondary structure within your gene of interest (see Note 11).

3.2. In Vitro Transcription and End-Labeling of Target RNA

1. Assemble the in vitro transcription reaction at room temperature as follows (see Notes 12 and 13):
   (a) Dilute 2 μg of template DNA in 8 μL of RNase-free water (if the template DNA is too dilute, precipitate the DNA and resuspend in a smaller volume).
   (b) Prepare an NTP mastermix by combining equal volumes of the ribonucleotides (ATP, CTP, GTP, and UTP) and add 8 μL of the mix to the template solution (avoid multiple freeze–thaw of NTPs).
   (c) Vortex the 10× reaction buffer and add 2 μL to the DNA–ribonucleotide mix (see Note 14).
   (d) Add 2 μL of the T7 enzyme mix and microfuge briefly (see Note 14).
2. Incubate the reaction at 37°C for 4 h.
3. To digest the template DNA, add 1 μL TURBO DNase, and mix well (the reaction may be viscous) and incubate at 37°C for a further 15 min.

4. To clean up the in vitro transcription reaction, use the procedure outlined in the MEGAclear™ kit (see Note 15).
5. Determine the RNA concentration by measurement of the absorbance at 260 nm with an ND-1000 spectrophotometer.
6. Examine the RNA integrity by agarose gel electrophoresis (see Note 16).
7. Precipitate 20 μg of RNA according to the following protocol (see Note 17):
   (a) Add 1 volume of 5 M LiCl to the RNA solution and mix well.
   (b) Incubate at –20°C for ≥30 min.
   (c) Pellet the RNA by centrifugation at 16,000 × *g* for 30 min at 4°C.
   (d) Carefully remove and discard the supernatant (contains free nucleotides, enzyme, and small RNAs (<200 nt)). Do not disturb the RNA pellet (contains the large RNA, >200 nt).
   (e) Wash the RNA pellet with 70% ethanol to remove residual salt.
   (f) Air-dry the RNA pellet for 5–10 min.
   (g) Resuspend the RNA in 8 μL RNase-free water.
8. Denature the RNA by incubation at 80°C for 5 min.
9. Immediately chill on ice for 2 min.
10. For 5′ end-labeling reactions, assemble the reaction using the 5′-EndTag Nucleic Acid End Labeling system as follows:
    (a) 8 μL target RNA (from above).
    (b) 1 μL universal reaction buffer.
    (c) 1 μL alkaline phosphatase.
    (d) 1 μL SUPER-In RNase Inhibitor.
11. Mix and incubate for 30 min at 37°C.
12. Microfuge briefly and add:
    (a) 2 μL universal reaction buffer.
    (b) 1 μL ATPγS.
    (c) 2 μL T4 polynucleotide kinase.
    (d) 5 μL RNase-free water.
13. Mix and incubate for a further 30 min at 37°C.
14. Add 1 μL of EZ-Link Maleimide-PEG11-Biotin (200 mM stock).
15. Add RNase-free EDTA to a final concentration of 0.5 mM.
16. Mix and incubate for 30 min at 65°C.

17. To clean up the end-labeling reaction, use the procedure outlined in the MEGAclear™ kit.
18. Alternatively, 3′-end label the RNA using the Pierce® RNA 3′ End Biotinylation Kit. Assemble the RNA ligation reaction as follows (see Note 18):
    (a) 8 μL target RNA (from above).
    (b) 3 μL 10× RNA Ligase Reaction Buffer.
    (c) 1 μL RNase inhibitor.
    (d) 1 μL Biotinylated Cytidine (Bis)phosphate.
    (e) 2 μL T4 RNA Ligase.
    (f) 15 μL 30% PEG.
19. Incubate the reactions at 16°C overnight (16 h).
20. To clean up the end-labeling reaction, use the procedure outlined in the MEGAclear™ kit.

3.3. Magnetic Microbead Preparation, Conjugation of RNA, and Oligo Screening

1. For oligo screening under native conditions, dilute 1 μg of biotinylated RNA in 20 μL hybridization buffer. Add fluorophore-conjugated single-stranded oligos to final concentrations ranging from 0.01 to 1 μM.
2. For oligo screening under denaturing conditions, prepare the reactions as outlined above for native conditions, but heat-denature the hybridization reactions at 80°C for 15 min, followed by immediately placing the reactions on ice for 2 min.
3. Incubate reactions at 37°C for 30 min.
4. For each sample set, include the following control reactions: Beads only (no biotinylated RNA, no labeled oligo), oligo only (no biotinylated RNA), and biotinylated RNA only (no labeled oligo).
5. To prepare the magnetic microbeads, resuspend the streptavidin-coated Dynabeads® M-280 by gently swirling the vial. Use approximately 5×10^6 (10 μL) beads per sample and transfer them to 1.5 mL microfuge tubes. Place the tubes on a magnet for 2 min and remove the supernatant with a pipette while the tubes are on the magnet. Remove the tubes from the magnet and resuspend the beads in 100 μL of 1× B&W wash buffer. To ensure that the beads are RNase free, wash twice with 100 μL of solution A. Finally, equilibrate the beads by washing them with 100 μL of 1× hybridization buffer.
6. Remove the supernatant from the last wash and add the hybridization reactions to the beads.
7. Place the tubes on a rotator and incubate at 37°C for 30 min.
8. Wash the beads three times with 100 μL of hybridization buffer.

9. Resuspend the beads in 10 μL of PBS and pipette them onto a 1.0 mm micro slide.
10. Gently apply a coverslip on the beads, trying not to introduce any air bubbles.
11. Proceed with imaging immediately (see Note 19).

3.4. Image Acquisition and Data Analysis

1. Images are taken using a confocal or a conventional fluorescent microscope with 40× magnification objective lens. Optimal exposure time and emission filter vary depending on the microscope and the fluorophore used for the assay. However, it is important to remain consistent with the optimum condition across all samples (see Note 20).
2. The quantification of signal intensities from the microbeads is performed using Image-Pro® Plus 4.5 software by calculating the integrated optical density (IOD) from individual beads. Exclude out-of-focus bead clusters.
3. Net signal intensities are obtained by local-ring background subtraction (net = raw − background). The net signal intensity for each bead is normalized to the area of the bead (net IOD/area) and averaged for approximately 100 beads per image. For normalization purposes, the net IODs from the control hybridization reactions carried out in the absence of biotinylated RNA are subtracted from the net IODs.
4. For each oligo, standard curves for native and denatured target RNA are generated from the linear least squares fit of the net IODs vs. oligo concentration. Target site accessibility is assessed by calculating the ratio of the slope of the oligo hybridization to native target RNA over the slope of the oligo hybridization to denatured target RNA (Fig. 2). Since hybridizations carried out under denaturing conditions indicate that the RNA is in an open conformation, ratios close to 1 between the native and denatured hybridization conditions indicate highly accessible sites within the native target RNA. Those oligos with ratios close to 1 therefore represent promising candidates for effective siRNA knockdown in cell culture (see Note 21).

3.5. Transfection of Adherent Cells with siRNA Duplexes

1. Plate cells into 60-mm dishes and grow until approaching 70–80% confluence. For assessment of HCV knockdown, we used the human hepatoma-derived Huh-7 cells that stably express an HCV replicon (pFK-I389neo/luc/NS3-3′/5.1, genotype 1b, Con1 isolate) (13, 25).
2. For each dish of cells, dilute 2–6 μL of Lipofectamine™ 2000 into 500 μL OptiMEM® reduced serum medium and incubate for 5 min at room temperature.
3. Dilute the siRNA duplexes (10–100 nM) into 500 μL OptiMEM® reduced serum medium.

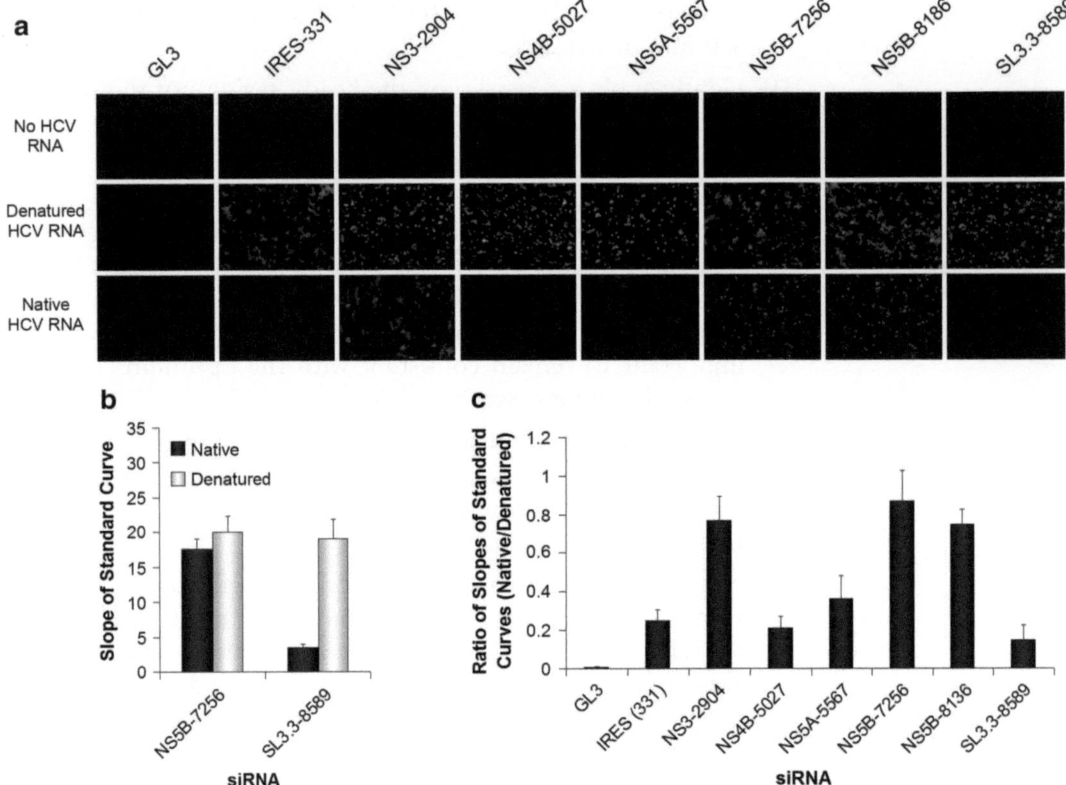

Fig. 2. Bead imaging and assessment of target site accessibility. Biotin-conjugated HCV replicon RNA (pFK-I389neo/NS3-3′/5.1, genotype 1b, Con1 isolate, Genbank accession # AJ242654 (29)) was hybridized to single-stranded fluorophore-conjugated RNA oligos directed against the HCV replicon RNA under native and denaturing conditions and captured on streptavidin-coated magnetic beads. (**a**) Representative fluorescent images of the beads after hybridization with 100 nM RNA oligos under native and denaturing conditions (oligos are denoted by the region of the genome they are directed against as well as the nucleotide # based on Genbank accession # AJ242654) (13). "No HCV RNA" represents hybridization reactions carried out in the absence of HCV replicon RNA to assess nonspecific binding of the RNA oligos to the beads. (**b**) The slopes of standard curves generated by quantifying net IODs from individual beads ($n = 100$) from images of increasing RNA oligo concentrations were used to generate a bar graph of the positive-control (NS5B-7256) and negative-control (SL3.3-8589) oligo hybridizations. Error bars represent SE. (**c**) Ratio of slopes from native and denaturing hybridization of the HCV-specific RNA oligos. Error bars represent SD of at least three independent replicates. Reprinted from ref. 13 with permission from Elsevier © 2010.

4. Combine the diluted Lipofectamine™ 2000 (step 2) with the diluted siRNA (step 3). Mix and incubate at room temperature for 20 min to allow transfection complexes to form.
5. Aspirate media from cells and add 1 mL transfection complexes to each dish.
6. Incubate at 37°C/5% CO_2.
7. At 4 h post transfection, add 1 mL media supplemented with 20% FBS.
8. Incubate at 37°C/5% CO_2 for 24–72 h (see Note 22).

3.6. Confirmation of Knockdown of HCV RNA by Quantitative RT-PCR and Northern Blot Analyses

1. Isolate total RNA from knockdown cells at 24–72 h post transfection (see Note 22). Also isolate RNA from controls, including mock-transfected and cells transfected with control siRNAs targeting an irrelevant gene (see Note 10). Use an RNeasy purification kit to purify total RNA from a confluent 60-mm or 10-cm dish. Determine the RNA concentration by measurement of the absorbance at 260 nm with an ND-1000 spectrophotometer and examine RNA integrity by agarose gel electrophoresis.
2. For quantitative reverse-transcription PCR (qRT-PCR) analyses, design primers that target the same splice forms as the siRNAs and should span an intron/exon junction if possible (see Note 23). Primers should lie 100–200 nt apart and should have a T_m of approximately 57°C. Primer design programs such as Beacon Designer™ or Primer3 (Whitehead Institute, MIT) can assist in designing appropriate qRT-PCR primers.
3. Prepare cDNA as follows:
   (a) Combine 1 μL random primer mix (0.5 μg/μL), 500 μg total RNA, 1 μL dNTPs (10 mM), and RNase-free water up to 12 μL.
   (b) Heat mixture to 65°C for 5 min and quick chill on ice for at least 1 min.
   (c) Add 4 μL 5× first-strand buffer, 2 μL 0.1 M DTT, and 1 μL SUPER-In RNase Inhibitor.
   (d) Incubate at 42°C for 2 min.
   (e) Add 1 μL Superscript II.
4. Incubate at 42°C for 1 h.
5. Inactivate by heating at 70°C for 15 min.
6. For quantitative PCR, prepare a mastermix of SYBR green and primers: 10 μL 2× SYBR Green PCR Master Mix, 1 μL forward primer (10 μM), 1 μL reverse primer (10 μM), and 7 μL water. Mix well. Include reactions against the target gene as well as a reference gene (e.g., 18S rRNA) for each sample (see Note 24).
7. Pipet 19 μL of the mix to the corresponding wells of a 96-well PCR plate. Pipette 1 μL of cDNA to the corresponding wells. Each PCR reaction should be carried out in triplicate. Also include a no-template control (mastermix only).
8. Seal the plate with microseal adhesive sealer and centrifuge the plate for 5 min at 1,500 × *g*.
9. Run on an iCycler instrument (Bio-Rad) or a similar thermocycler with the following program:
   (a) 95°C × 10 min.
   (b) 95°C × 30 s.

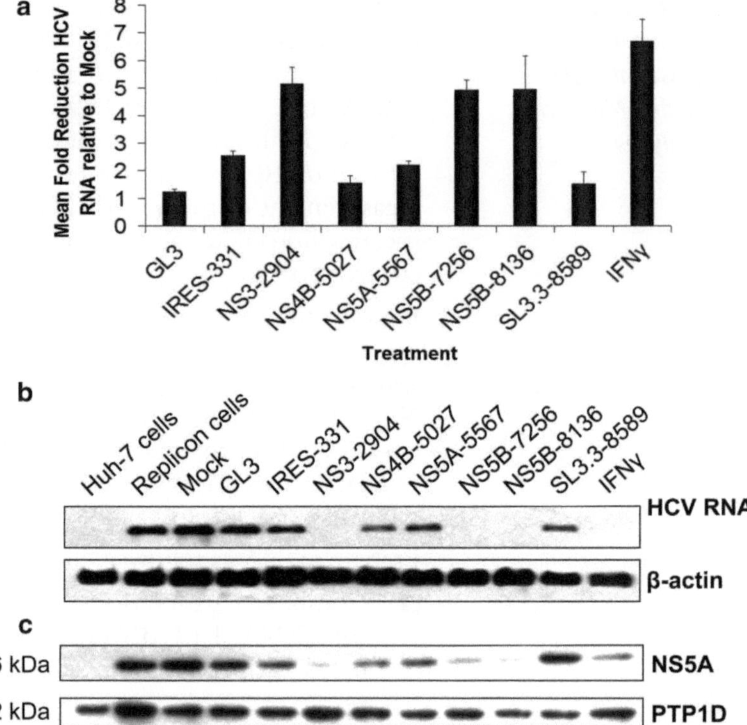

Fig. 3. Confirmation of knockdown of HCV in Huh-7 cells stably replicating HCV replicon RNA (pFK-I389neo/luc/NS3-3′/5.1, genotype 1b, Con1 isolate) (25) transfected with siRNA duplexes for 48 h. (**a**) Quantitative PCR analysis of siRNA knockdown in cell culture. Data are represented as mean fold reduction in HCV replicon RNA relative to mock-treated samples. Results represent mean fold reduction in HCV replicon RNA relative to mock-treated samples performed in triplicate and error bars represent SD. All samples were standardized to 18S rRNA expression. (**b**) Northern blot analysis of HCV replicon RNA in HCV replicon-harboring cells transfected with HCV-specific siRNAs. Samples from parental Huh-7 cells (*lane 1*) and replicon-harboring cells (*lane 2*) are shown. Mock-transfected samples (*lane 3*, transfection reagent only) and samples transfected with the negative control (GL3, *lane 4*) and HCV-specific siRNAs (*lanes 5–11*) are shown. IFNγ was used as a positive control for knockdown of HCV RNA and proteins. (**c**) Western blot analysis of HCV NS5A protein levels (and PTP1D as a control) in HCV replicon-harboring cells transfected with HCV-specific siRNAs (13). The samples analyzed by western blot are identical to those described for (**b**). Reproduced from ref. 13 with permission from Elsevier © 2010.

60°C × 30 s.

72°C × 1 min.

(c) Repeat step 9b for 40 cycles.

10. Determine the relative amounts of mRNA for your target gene (Fig. 3a) using the $2^{-\Delta\Delta Ct}$ method (26). Standardize values to a reference gene (see Note 24).

11. For northern blot analyses, design a 300–500 nt negative-sense probe complementary to the target RNA (HCV RNA, nts

6,648–7,770, Genbank accession #AJ242654) and a housekeeping gene as a control (such as β-actin, Genbank accession #X00351). Prepare the DNA for transcription using a primer containing a T7 promoter and synthesize probes using the MEGAscript™ T7 kit. In vitro transcriptions were performed as described above (Subheading 3.2) with the inclusion of biotin-11-UTP and biotin-11-CTP at a molar ratio of 1:3 with unlabeled UTP and unlabeled CTP, respectively.

12. Dilute 1–10 μg of total RNA in formamide loading dye, heat to 65°C for 15 min, and load onto a 0.5–2% agarose-MOPS gel. Include an RNA ladder (such as biotinylated RNA millennium marker). Run the gel at 100 V in 1× MOPS running buffer for 1.5–2 h depending on the size of the RNA of interest.

13. Transfer the gel to a nylon membrane (Hybond XL, Amersham). Prepare the transfer apparatus as follows: 3 cm thick dry paper towel stack, three dry Whatmann papers, two wet Whatmann papers, one wet nylon membrane (make sure that there are no bubbles), align the notches, make sure that the gels are in upright position (roll out any bubbles, by a disposable pipette), three wet Whatmann papers, three wet filter paper bridges onto the stack, and reaching into the reservoir containing 100 mL transfer buffer (avoid bubbles between the layers by rolling out with a plastic pipette). Insert parafilm around the stack to prevent short circuit of the buffer. Position the gel casting tray on top of the stack and add a 200 g weight. Transfer for 2–4 h at RT.

14. Remove the membranes from the stack, rinse briefly (10 s) in 1× gel running buffer, and blot excess liquid (do not dry membrane). Using a handheld UV light source, cross-link the membrane with long-range UV light (385 nm) for 15 min.

15. Prehybridize membrane with 10 mL preheated ULTRAhyb for 30 min at 68°C in hybridization buffer. Mix 22 ng of RNA probe in 1 mL of ULTRAhyb (for a final concentration of 2 ng/mL in a total of 11 mL ULTRAhyb solution) and boil at 95°C for 5 min to denature. Add the denatured probe (1 mL) to the hybridization tube and incubate overnight (16 h) at 68°C.

16. Wash the membrane with 15 mL low-stringency wash solution #1 twice for 5 min at RT and with 15 mL high-stringency wash solution #2 twice for 15 min at 68°C.

17. Warm blocking buffer and 4× Wash Buffer between 37 and 50°C, and block membranes with 20 mL blocking buffer for 15 min at RT. Prepare conjugate/blocking buffer solution: 66.7 μL stabilized streptavidin–HRP conjugate and 20 mL blocking buffer. Decant blocking buffer and add the 20 mL conjugate/blocking buffer and incubate for 15 min at RT. Decant the conjugate/blocking buffer and wash with 20 mL 1× wash solution in water four times for 5 min at RT.

18. Decant washes and incubate membranes with 30 mL substrate equilibration buffer for 5 min at RT. Mix equal volumes of luminol/enhancer solution with stable peroxide solution and place on membrane. Incubate for 5 min at RT. Blot excess liquid. Wrap membrane in plastic wrap and expose to film (Fig. 3b).

3.7. Confirmation of Knockdown by Western Blot Analyses

1. Prepare whole cell lysates from knockdown cells using RIPA buffer. Also prepare lysates from controls, including mock-transfected and cells transfected with control siRNAs targeting an irrelevant gene. Measure protein concentration using the Bio-Rad DC protein assay and a spectrophotometer.
2. Dilute 10–60 μg protein in SDS-gel loading dye, and heat for 5 min at 95°C prior to loading onto a 8–12% SDS-PAGE gel. After electrophoresis, transfer proteins to a Hybond-P PVDF membrane, and probe membrane with antibodies specific for the gene of interest (Fig. 3c).

4. Notes

1. RNA oligonucleotides can be synthesized from many commercial sources including Ambion (http://www.ambion.com), Dharmacon (http://www.dharmacon.com), and Invitrogen (http://www.invitrogen.com).
2. 5′ End-labeled oligonucleotides for imaging experiments can contain a range of fluorophores; in our experience Cy3 (or Dy547) as well as Cy5 dyes have performed well. Duplexing of siRNAs for cell culture experiments can be performed as indicated by the manufacturer.
3. Plasmid DNA for in vitro transcription should be linearized using a restriction enzyme that will cleave to generate the appropriate 3′ end of the gene of interest by run-off transcription. It is important to confirm complete digestion by analyzing the linearized DNA by agarose gel electrophoresis.
4. Consideration of the splice variants of the gene of interest is crucial to the selection of appropriate target sites for siRNA. If a specific isoform is to be targeted, often one is restricted to a smaller region of the gene that will be unique to that splice variant. When general gene knockdown is desired, one must also be careful to ensure that the targeted region is present in all splice variants of the gene. This also applies to pathogens, such as viruses, where one may want to target a specific viral mRNA or RNA gene segment.
5. The position of the target sites (5′ UTR, coding region, or 3′ UTR) within the appropriate splice variant does not appear to

have a general effect on the efficacy of the siRNA. However, local secondary structure and RNA-binding proteins have been known to decrease the efficacy of siRNAs. Hence, regions of known secondary structure or those occupied by RNA-binding proteins should be avoided. In addition, it is generally assumed that UTR-binding proteins may interact with the region around the start codon, and hence target sites are typically chosen at least 50–100 nt downstream of the start codon.

6. Many Web sites offer algorithms for siRNA target site selection including http://www.ambion.com, http://www.dharmacon.com, and http://rnaidesigner.invitrogen.com. It is often helpful to select target sites that are predicted by multiple algorithms, as this may increase the likelihood of predicting efficacious siRNAs.
7. When targeting pathogens, such as viruses, it is important to perform a BLAST search against both the viral genome as well as the host to eliminate off-target effects.
8. Using multiple siRNAs favors knockdown efficiency while reducing the off-target effects as each can be used at a lower concentration. Additionally, demonstrating functional effects with two siRNAs independently reinforces the outcome due to the statistical unlikelihood of the knockdown being due to shared off-target effects.
9. For in vitro screening of siRNA target sites using the native bead-based screening assay, it may be more cost effective to use DNA oligonucleotides. In addition, siRNAs can be designed with thymidine overhangs (2 nt, 3′ overhangs) as this can reduce the cost of RNA synthesis and may allow for greater nuclease resistance in culture medium as well as within transfected cells (27). Furthermore, the sequence of the 2 nt, 3′ overhang does not appear to significantly contribute to target recognition (28).
10. Experimental controls include mock-transfection, transfection of a scrambled siRNA, or of an siRNA targeting another gene entirely. It is favorable to use an siRNA targeting another gene entirely as one can then be sure that the control siRNA is able to enter a functional RISC. For these purposes, it is often preferable to use an established siRNA known to function in knockdown of the targeted gene, such as GL2 (5′-CGU ACG CGG AAU ACU UCG A-3′) or GL3 luciferase (5′-CUU ACG CUG AGU ACU UCG A-3′) (27).
11. For in vitro screening of siRNA target sites using the native bead-based screening assay, it is helpful to have a control oligonucleotide that is against a region of established secondary structure within the gene of interest. This will act as a negative control as it will have decreased target site accessibility in the

native state. This is particularly important for challenging targets such as the highly structured genomes of positive-sense RNA viruses, but may be less important for ORFs of relatively unstructured cellular mRNAs.

12. All solutions, buffers, and reactions should be RNase free and prepared with RNase-free water.
13. The MEGAscript Kit is designed to function best with templates that code for RNA transcripts of approximately 500 nt and longer. All kit reagents should be microfuged briefly before opening to prevent loss and/or contamination of material that may be present around the rim of the tube.
14. To prevent the spermidine in the 10× reaction buffer from precipitating the template DNA, assemble the reaction at RT. The T7 enzyme mix as well as the NTP stock solutions should be kept on ice.
15. The yield is often higher if the water for elution is heated at 80°C before adding to the MEGAclear™ columns and incubated for 1 min prior to centrifuging.
16. The concentration of agarose can vary from 0.6 to 2% according to the size of the RNA of interest.
17. Lithium chloride selectively precipitates large RNA (>200 nt) such as mRNA, while small RNAs (<200 nt), free nucleotides, and salts will remain in solution.
18. Thaw all Pierce® RNA 3′ End Biotinylation kit components on ice except the DMSO and 30% PEG. Thaw DMSO at RT and warm the 30% PEG to 37°C for 5 min or until the solution becomes fluid.
19. In order to prevent the beads from drying out, place them on slides just prior to imaging.
20. The optimum exposure time represents the shortest exposure period in which a strong signal is captured in the positive control while no signal can be detected from the negative control.
21. Assessment of target site accessibility does not guarantee that an RNA oligo will be an effective siRNA in cell culture due to the presence of RNA-binding proteins among other parameters; however, the screening approach can increase the likelihood that the siRNA will be effective against the gene of interest.
22. Expand the cells as needed. For siRNA knockdown of most targets it is necessary to wait 48–72 h to give the cells a chance to actively reduce gene expression.
23. Designing primers that span an intron/exon junction reduces the background signal from contaminating cellular DNA in

the sample. Alternatively, to reduce contaminating cellular DNA, an on-the-column DNase digestion can be performed during RNeasy kit purification.

24. All samples must be normalized to an endogenous control to account for variability in the initial concentration and quality of the total RNA as well as the conversion efficiency of the reverse transcription reaction. We routinely use 18S rRNA (5′-GCG ATG CGG CGG CGT TAT TC-3′ and 5′-CAA TCT GTC AAT CCT GTC CGT GTC C-3′) and GAPDH (5′-AGG CTG TGG GCA AGG TCA TCC-3′ and 5′-AGT GGG TGT CGC TGT TGA AGT CA-3′) as endogenous controls (13). Other housekeeping genes can also be used.

References

1. Fire A, Xu S, Montgomery MK, Kostas SA, Driver SE, Mello CC (1998) Potent and specific genetic interference by double-stranded RNA in *Caenorhabditis elegans*. Nature 391:806–811
2. Liu J, Carmell MA, Rivas FV, Marsden CG, Thomson JM, Song JJ, Hammond SM, Joshua-Tor L, Hannon GJ (2004) Argonaute2 is the catalytic engine of mammalian RNAi. Science 305:1437–1441
3. Meister G, Landthaler M, Patkaniowska A, Dorsett Y, Teng G, Tuschl T (2004) Human Argonaute2 mediates RNA cleavage targeted by miRNAs and siRNAs. Mol Cell 15:185–197
4. Orban TI, Izaurralde E (2005) Decay of mRNAs targeted by RISC requires XRN1, the Ski complex, and the exosome. RNA 11: 459–469
5. Elbashir SM, Harborth J, Weber K, Tuschl T (2002) Analysis of gene function in somatic mammalian cells using small interfering RNAs. Methods 26:199–213
6. Reynolds A, Leake D, Boese Q, Scaringe S, Marshall WS, Khvorova A (2004) Rational siRNA design for RNA interference. Nat Biotechnol 22:326–330
7. Holen T, Amarzguioui M, Wiiger MT, Babaie E, Prydz H (2002) Positional effects of short interfering RNAs targeting the human coagulation trigger tissue factor. Nucleic Acids Res 30:1757–1766
8. Birmingham A, Anderson E, Sullivan K, Reynolds A, Boese Q, Leake D, Karpilow J, Khvorova A (2007) A protocol for designing siRNAs with high functionality and specificity. Nat Protoc 2:2068–2078
9. Heale BS, Soifer HS, Bowers C, Rossi JJ (2005) siRNA target site secondary structure predictions using local stable substructures. Nucleic Acids Res 33:e30
10. Schwarz DS, Hutvagner G, Du T, Xu Z, Aronin N, Zamore PD (2003) Asymmetry in the assembly of the RNAi enzyme complex. Cell 115:199–208
11. Khvorova A, Reynolds A, Jayasena SD (2003) Functional siRNAs and miRNAs exhibit strand bias. Cell 115:209–216
12. Novina CD, Murray MF, Dykxhoorn DM, Beresford PJ, Riess J, Lee SK, Collman RG, Lieberman J, Shankar P, Sharp PA (2002) siRNA-directed inhibition of HIV-1 infection. Nat Med 8:681–686
13. Sagan SM, Nasheri N, Luebbert C, Pezacki JP (2010) The efficacy of siRNAs against hepatitis C virus is strongly influenced by structure and target site accessibility. Chem Biol 17:515–527
14. Vickers TA, Koo S, Bennett CF, Crooke ST, Dean NM, Baker BF (2003) Efficient reduction of target RNAs by small interfering RNA and RNase H-dependent antisense agents. A comparative analysis. J Biol Chem 278: 7108–7118
15. Bohula EA, Salisbury AJ, Sohail M, Playford MP, Riedemann J, Southern EM, Macaulay VM (2003) The efficacy of small interfering RNAs targeted to the type 1 insulin-like growth factor receptor (IGF1R) is influenced by secondary structure in the IGF1R transcript. J Biol Chem 278:15991–15997
16. Xu Y, Zhang HY, Thormeyer D, Larsson O, Du Q, Elmen J, Wahlestedt C, Liang Z (2003) Effective small interfering RNAs and phosphorothioate antisense DNAs have different preferences for target sites in the luciferase mRNAs. Biochem Biophys Res Commun 306: 712–717

17. Kretschmer-Kazemi Far R, Sczakiel G (2003) The activity of siRNA in mammalian cells is related to structural target accessibility: a comparison with antisense oligonucleotides. Nucleic Acids Res 31:4417–4424
18. Brown KM, Chu CY, Rana TM (2005) Target accessibility dictates the potency of human RISC. Nat Struct Mol Biol 12:469–470
19. Overhoff M, Alken M, Far RK, Lemaitre M, Lebleu B, Sczakiel G, Robbins I (2005) Local RNA target structure influences siRNA efficacy: a systematic global analysis. J Mol Biol 348:871–881
20. Tafer H, Ameres SL, Obernosterer G, Gebeshuber CA, Schroeder R, Martinez J, Hofacker IL (2008) The impact of target site accessibility on the design of effective siRNAs. Nat Biotechnol 26:578–583
21. Varekova RS, Bradac I, Plchut M, Skrdla M, Wacenovsky M, Mahr H, Mayer G, Tanner H, Brugger H, Withalm J et al (2008) A new program for analyzing RNA interference. Comput Methods Programs Biomed 90:89–94, http://www.rnaworkbench.com
22. Schubert S, Grunweller A, Erdmann VA, Kurreck J (2005) Local RNA target structure influences siRNA efficacy: systematic analysis of intentionally designed binding regions. J Mol Biol 348:883–893
23. Luo KQ, Chang DC (2004) The gene-silencing efficiency of siRNA is strongly dependent on the local structure of mRNA at the targeted region. Biochem Biophys Res Commun 318:303–310
24. Nasheri N, Singaravelu R, Goodmurphy M, Lyn RK, Pezacki JP (2011) Competing roles of microRNA-122 recognition elements in hepatitis C virus RNA. Virology 410: 336–344
25. Sagan SM, Rouleau Y, Leggiadro C, Supekova L, Schultz PG, Su AI, Pezacki JP (2006) The influence of cholesterol and lipid metabolism on host cell structure and hepatitis C virus replication. Biochem Cell Biol 84:67–79
26. Livak KJ, Schmittgen TD (2001) Analysis of relative gene expression data using real-time quantitative PCR and the 2(−Delta Delta C(T)) Method. Methods 25:402–408
27. Elbashir SM, Harborth J, Lendeckel W, Yalcin A, Weber K, Tuschl T (2001) Duplexes of 21-nucleotide RNAs mediate RNA interference in cultured mammalian cells. Nature 411:494–498
28. Elbashir SM, Lendeckel W, Tuschl T (2001) RNA interference is mediated by 21- and 22-nucleotide RNAs. Genes Dev 15: 188–200
29. Lohmann V, Korner F, Koch J, Herian U, Theilmann L, Bartenschlager R (1999) Replication of subgenomic hepatitis C virus RNAs in a hepatoma cell line. Science 285: 110–113

Chapter 5

Engineering Small Interfering RNAs by Strategic Chemical Modification

Jesper B. Bramsen and Jørgen Kjems

Abstract

Synthetic small interfering RNAs (siRNAs) have revolutionized functional genomics in mammalian cell cultures due to their reliability, efficiency, and ease of use. This success, however, has not fully translated into siRNA applications in vivo and in siRNA therapeutics where initial optimism has been dampened by a lack of efficient delivery strategies and reports of siRNA off-target effects and immunogenicity. Encouragingly, most aspects of siRNA behavior can be addressed by careful engineering of siRNAs incorporating beneficial chemical modifications into discrete nucleotide positions during siRNA synthesis. Here, we review the literature (Subheadings 1–3) and provide a quick guide (Subheading 4) to how the performance of siRNA can be improved by chemical modification to suit specific applications in vitro and in vivo.

Key words: RNA interference, RNAi, Small interfering RNA, siRNA, siRNA design, Off-target effect, Gene silencing, Knockdown, Chemical modification

1. Introduction

1.1. Nucleic Acids in Gene Silencing Technology

Conceptionally, the high predictability and specificity of nucleic acid base pairing provides an ideal framework for universal and specific gene silencing technologies (GSTs). The first indication that nucleic acids can bedrock GSTs came from pioneering work in the 1970–1980s where synthetic antisense oligonucleotides (ASOs) were engineered to inhibit specific protein production upon base-pairing to target RNAs (1). Since then, pivotal efforts have been invested to chemically engineer ASO designs, which initially suffered from unpredictable activity, specificity, and low biostability (2), into a modern clinical trial-grade GST encompassing phosphorodiamidate morpholino oligomers (PMOs) (3), "Gapmer" oligonucleotides (4), LNA ASOs (5, 6), and LNA antimiRs (7) (also see http://clinicaltrials.gov). The discovery in the late 1990s that endogenous double-stranded RNAs (dsRNAs) are natural

Debra J. Taxman (ed.), *siRNA Design: Methods and Protocols*, Methods in Molecular Biology, vol. 942,
DOI 10.1007/978-1-62703-119-6_5,

triggers of gene silencing through RNA interference (RNAi) provided researches with an even more powerful and physiologic nucleic acid-based GST (8, 9); the observation that synthetic small 21mer dsRNA, coined small interfering RNA (siRNAs), triggered specific gene silencing in mammalian cell cultures with no apparent side-effects pointed to obvious therapeutic potential (10).

1.2. siRNAs Need Chemical Engineering

Indeed, synthetically prepared siRNAs are now the preferred gene silencing tool in vitro, a success that emanates from their consistency, high efficiency, and ease of use; most experimenters will be able to successfully apply an unmodified siRNA to obtain the desired gene KD in short term cell culture experiments. Such benefits originate from harnessing endogenous RNAi pathways to effectuate gene silencing; upon introduction of synthetic 21mer siRNAs into the cell cytoplasm they are protected by incorporation into an RNA-induced silencing complex (RISC) (11) by a RISC loading complex (RLC) (12) containing the RNase III enzyme Dicer (13). By sensing the thermodynamic asymmetry of siRNA duplex ends (14, 15), RLC loads the siRNA guiding antisense strand into a cleavage-competent RISC containing Argonaute 2 (Ago2) (16), whereas the passenger sense strand is cleaved and released (17, 18). Subsequently, Ago2-RISC will efficiently guide and effectuate multiple rounds of target RNA cleavage to ensure potent gene knockdown (KD) (19).

Although siRNA technology has become increasing popular in vivo (20–23), serious concerns regarding siRNA delivery and safety have surfaced, and major biotech companies such as Roche and Pfizer have recently downsized their RNAi research (24, 25). In essence, siRNAs are very unlike familiar drugs by being relatively big, labile, and highly negatively charged, which severely complicates efficient intracellular delivery in vivo (26), and unmodified siRNAs will inherently trigger off-target effects (27) and can be immunogenic in immunocompetent cells (28, 29). However, RNAi is still a young therapeutic platform that needs maturing to succeed, just as for ASO technology, which currently is showing important progress in clinical trials (for more details http://clinicaltrials.gov). As described throughout Subheadings 1–3 of this chapter, researchers have for a decade successfully improved most aspects of siRNA performance by introducing chemically modified nucleotides into discrete positions of the siRNA during its chemical synthesis. These modifications have indeed proven capable of changing siRNA pharmacokinetic properties, enhancing siRNA activity, and reducing off-target effects and immunogenicity (for a short siRNA design guide see Subheading 4). Hence, siRNAs may very well still blossom into a most powerful GST for in vivo studies, if appropriately engineered.

2. Tools and Tolerances in siRNA Design

2.1. Choosing an siRNA Template

By synthesis, siRNAs can be designed to structurally mimic several of the RNA intermediates in the cellular RNAi pathway that process both endogenous microRNAs (miRNAs) and siRNAs from longer double-stranded RNA (dsRNA) substrates (30). The canonical synthetic siRNA design mimics natural Dicer cleavage products and comprises two 21 nucleotide (nt) RNA strands annealed to form a 19-bp dsRNA duplex stem and 2 nt 3′-overhangs at both ends (here referred to as 21mer siRNA) (31, 32). One strand, denoted the antisense strand (AS) or guide strand, is designed antisense to the RNA target and design principles outlined below ensure its preferential loading into RISC. Likewise, the complementary strand, denoted the sense strand (SS) or passenger strand, is designed to ensure preferential AS RISC uptake upon which the SS is cleaved and degraded. The similar Dicer-independent 23mer siRNA design has been widely used with equal or slightly higher efficiency (33, 34) yet may be slightly more immunogenic in certain settings (35) (see Subheading 3.4). Longer siRNA designs aim to improve siRNA activity by enhancing the RLC uptake of exogenous so-called Dicer-substrate siRNAs (D-siRNAs), which are subsequently processed into 21mer siRNAs by Dicer. Both 27mer siRNAs (34) and short hairpin RNAs (shRNAs) (36) exhibit high silencing activity, yet an enhanced immunogenicity of these longer dsRNA species seem to limit their usage, at least in vivo if unmodified (33). In this regard, the shorter 19mer siRNA seems non-immunogenic in vivo in mice (37) and generally exhibits only marginally reduced silencing activity. A number of siRNA designs have challenged the canonical, symmetrical dsRNA structure of siRNAs; Asymmetric siRNA (asiRNA) uses a shortened 15 nt passenger strand to abrogate passenger strand function while preserving, or even enhancing, siRNA activity (38). Similarly, the small internally segmented interfering RNAs (sisiRNAs) utilize two short 9–11 nt locked nucleic acid (LNA)-modified sense strands to abrogate sense strand function and enhance the tolerance for additional chemical modification (39). Fork-siRNAs (also called fork-like siRNAs or fsiRNAs) containing mismatches into the 3′-part of the sense strand have been shown to exhibit enhanced silencing activity (40, 41), but chemical modification of single-stranded regions in asymmetric siRNA design appears to be needed to ensure sufficient nuclease resistance, at least in vivo (42). Other functional, but less used siRNA designs include Dicer-independent short shRNAs (sshRNAs: typically RNA stems ≤19 bp) (43–45), blunt 19mer siRNAs (46, 47), single-stranded siRNAs (ss-siRNAs) (48–50), dumbbell-shaped nanocircular RNAs (51, 52), and self-dimerizing single-oligo RISC substrates (53).

2.2. Tolerances for siRNA Modification

Chemically modified nucleotide analogues have long been utilized in ASO technologies to increase ASO stability, specificity and efficiency (54) and subsequently also in siRNA with promise of similar improvements. However, as modified siRNA must be compatible with endogenous RNAi pathways, the tolerance for chemical modification is quite naturally dependent on the position and nature of the particular modification type (size, charge, etc.). In general, the entire SS as well as the 3′-proximal part and 3′ overhang of the AS seem to be most tolerant to chemical modification, whereas the 5′-phosphate, the 5′-proximal part and central positions of the AS are more sensitive, especially to multiple or bulky modifications (55–60). This intolerance, identified empirically, is fully in line with solved structures of RNAi proteins. The guide strand 5′-phosphate group is bound by the Ago2 MID domain (61) and modifications at this location often compromise binding by Ago2 (62, 63). Also, the initial interactions between the guide strand and target RNA are mediated only by the 5′-proximal seed region exposed by the RISC (61), and Ago2-mediated target RNA cleavage requires forming of an alpha-helical duplex structure between the AS and the target spanning both the seed region (positions 2–8) and around the cleavage site (opposite of guide strand position 10/11) (64).

2.3. Tools for Chemical Engineering

Three chemical modification types are most popular in siRNA design: (1) modification of the phosphodiester backbone, (2) modification of the ribose 2′-OH group, and (3) modification of the ribose ring and nucleoside bases.

2.3.1. Backbone Modification

Backbone modifications are typically used to enhance oligonucleotide/siRNA nuclease resistance; the most widely used phosphate backbone alteration is the phosphorothioate (PS) modification where one of the non-bridging phosphate oxygen atoms is replaced with a sulfur atom. Introducing PS modifications will additively enhance oligonucleotide nuclease resistance, however, will also increase their toxicity and reduce silencing activity (56–59, 65–67). Therefore, only partial PS-modification is widely used in ASO (68) and siRNA design today, typically in combination with other modifications types, to help enhance siRNA nuclease resistance (7, 69). An alternative is the boranophosphate linkage, which is more nuclease-resistant and less toxic compared to its PS counterparts (67), and functional boranophosphate-modified single-stranded siRNAs have been reported (50). Finally, the phosphonoacetate linkage is potentially interesting for siRNA design as it is completely resistant to nuclease degradation and electrochemically neutral (if esterified) (70), which allows modified oligonucleotides to be taken up by cells in the absence of delivery reagents (71).

2.3.2. 2′-OH Modification

Modification of the ribose 2′-OH group is the most diverse and popular class of modification in siRNA design and is generally used to enhance siRNA nuclease resistance and hybridization properties. Two types of 2′-modifications are utilized: (1) substitution of the ribose 2′-OH group with for example 2′-O-methyl (2′-OMe), 2′-fluoro (2′-F), 2′-methoxyethyl (2′-O-MOE), etc. (2) locking of the 2′oxygen via intramolecular linkages to create so-called bridged nucleic acids (BNAs).

The small electronegative 2′-substituted 2′-OMe and 2′-F are among the most extensively tested; They both enhance siRNA nuclease resistance and duplex thermostability (T_m increase of ~0.5–0.7°C and ~0.7–1°C per modification, respectively) and are generally well tolerated at most duplex positions (46, 47, 56, 58, 65, 66, 72). The impact of more extensive or full 2′-OMe or 2′-F modification is sequence dependent; in many cases siRNA potency is reduced (31, 47, 56, 57), and in other cases not (65, 73), and alternating 2′-OMe and 2′-F substitutions have produced fully substituted, nuclease-resistant, and extremely potent siRNAs (74). The 2′-OMe modification is particularly useful in siRNAs; the first successful KD of an endogenous gene in vivo using a systemic delivery strategy suitable for therapeutics was performed using a 2′-OMe/PS-modified siRNA (69), and very notably, careful 2′-OMe modification can reduce siRNA immunogenicity and off-target effects (see Subheadings 3.4 and 3.5).

More bulky 2′-modifications such as 2′-O-MOE, 2′-O-allyl, and others may enhance siRNA nuclease resistance but are tolerated at fewer positions within the siRNA duplex, typically in 3′-overhangs and strand 3′-ends likely as they destabilize the helical dsRNA structure essential to RNAi (46, 55, 58, 75). Such destabilization can, however, be intentionally employed to alter siRNA thermodynamic asymmetry and consequently siRNA activity; for example, 2′-aminoethyl modifications have been reported to enhance siRNA function when inserted into the passenger strand 3′-end (55).

Bridged nucleic acids (BNAs) are a class of conformationally locked nucleotide analogues with interesting properties in siRNA design. Here, the 2′-oxygen can be connected to the 4′-carbon via a methylene bridge as in LNA (76) and carbocyclic-LNA (55, 77) or via an ethylene bridge as in ENA (60) and carbocyclic-ENA (55, 77) or to the 1′-carbon as in oxetane (OXE) (55, 78). LNA has been extensively tested in siRNAs (39, 55, 57, 66, 79) as it enhances nuclease resistance in vitro (55, 57) and in vivo (80, 81). LNA dramatically increase the thermostability (2–10°C per LNA monomer) upon incorporation into RNA duplexes (82) and, in effect, only relatively few (4–6) simultaneous LNA modifications are tolerated in siRNA design (57, 66, 79). Typically, LNA can be used to enhance siRNA potency and specificity (39, 55, 79) by modulating the thermodynamic profile of the siRNA duplex and subsequent RISC strand selection or to build novel siRNA designs that rely on only short stretches of base pairing such as the sisiRNAs (39).

2.3.3. Ribose Ring Modification

Substituting the ribose 5-carbon sugar (pentose) with a 6-carbon sugar (hexose) has formed the basis for successful modifications in siRNA design such as in ANA, HNA, 2′-F-ANA, and CeNA nucleotides, which are based on anitrol, hexitol, arabinose, and cyclohexenyl, respectively (55, 83–87). ANA, HNA, and CeNA modifications at strand 3′-ends can modestly enhance siRNA serum stability, activity, and silencing duration (84, 86, 87). Similarly, an siRNA consisting of a fully 2′-F-ANA-modified SS and 3′-end modified AS displayed higher potencies and serum stability than unmodified siRNAs (83). 4′-Thio-modified nucleotides (4′-S RNA) contain a sulfur atom instead of the 4′-carbon of the ribose ring. Modification of the guide strand with 4′-S nucleotides is only tolerated at certain positions but enhances nuclease resistance and may enhance siRNA potency (88, 89), albeit in a sequence-dependent fashion (90).

A more radical modification type is unlocked nucleic acids (UNAs), which are acyclic derivatives of RNA lacking the C2′–C3′-bond of the RNA ribose ring. Incorporation of UNA residues induces additive instability by 5–8°C per UNA monomer, and as a consequence, extensive UNA modification will not support standard siRNA strand annealing (91, 92). In effect, internal UNA modification of the duplex stem can lower siRNA serum stability, whereas 3′-overhang modifications can significantly enhance stability; siRNAs modified with UNA residues only in their 3′-overhangs exhibit prolonged biostability upon intravenous injection in mice (even compared to extensively LNA-modified siRNA) and elicited efficient gene KD in contrast to unmodified siRNAs (92). Minimal UNA modification (1–2 residues per duplex) can be strategically used to introduce local destabilization into the siRNA duplex to improve the potency of extensively (e.g., LNA-) modified siRNAs that are otherwise too stable or rigid to support RNAi (92) or to modulate strand selection by the RISC; UNA-modification of the passenger strand 5′-terminus (an extra UNA as pos. "–1") and of the 3′-end overhangs favors guide strand selection during RISC loading to enhance activity and specificity (55, 92, 93). Owing to its strong destabilizing properties, UNA is not well tolerated in the guide strand seed region (94), yet very notably, incorporation of a single UNA modification at position 7 of the guide strand can efficiently reduce siRNA off-targeting while preserving on-target activity for multiple sequences tested (95).

2.3.4. Base Modifications

A number of modified nucleotide bases, such as 5-bromo-, 5-iodo-, 2-thio-, 4-thio, dihydro, and pseudo-uracil, have been tested in siRNA design, but are not yet widely used. These may modulate base pairing potential to enhance siRNA potency and specificity (96); 5-bromo- and 5-iodo-uracil slightly reduced siRNA potency (56), whereas 2-thio- and pseudo-uracil was reported to enhance siRNA potency (97) and to reduce cellular immune responses (98).

3. Enhancing siRNA Performance by Chemical Engineering

3.1 Maximizing siRNA Activity

Maximizing siRNA potency (reducing half maximal inhibitory concentration, IC_{50}) will minimize the siRNA dose required for efficient gene silencing to reduce both cost and potential adverse side effects such as immunological responses, flooding of endogenous RNAi pathways, or bioaccumulation of synthetic siRNA breakdown products. siRNA activity is highly sequence dependent (99–105) as the chosen sequence dictates both the target site accessibility within the mRNA and the thermodynamic properties of the siRNA duplex that are major determinants for their recognition by RNAi proteins (such as efficiency of uptake, strand selection and silencing kinetics). Large siRNA chemical modification screens show that careful chemical engineering of siRNAs can enhance their activity beyond unmodified siRNAs (55). Typically, conservative modification levels, especially of siRNA strand 3′-ends, can lead to modest improvements in siRNA potency, often less than twofold (55, 83–87, 106). In some cases dramatic improvements in siRNA potency is reported even upon extensive modification. Allerson et al. reported a 500-fold increase in siRNA potency using fully substituted siRNAs with alternating 2′-OMe/2′-F-modifications (74) which may be attributed to enhanced, but sequence specific, RISC uptake (107). Although interesting, such examples of dramatic improvements by extensive modification have proven too sequence specific to be applicable as general siRNA design rules. Instead, chemical modifications that enhance RISC uptake and favor the preferential loading of the siRNA guide strand into Ago2-RISC seem to be a reliable, sequence-independent strategy to enhance siRNA potency as described below.

Target site accessibility: Efficient siRNAs can be predicted in silico to exclude target sequences located in stable secondary structures (108–112) or in regions that are occupied by RNA-binding proteins. In effect, efficient target sites are often found in AU-rich regions (113), and therefore the corresponding siRNAs often have low GC-contents of 30–50% (101, 114), especially in the guide strand "seed" region (101, 104). Improvement of siRNA activity on poorly accessible targets by chemical engineering has not been reported, likely as the hybridization energy between the target and the siRNA/miRNA guide strand is reported as a poor predictor of silencing activity (115). Also, most seed modifications tend to reduce guide strand activity (55, 95), and studies preferentially identify potent unmodified siRNAs before continuing with chemical modification.

siRNA duplex-end modification: Chemically engineering 5′-ends and 3′-overhangs of the two siRNA strands can enhance siRNA kinetics and favor guide strand loading into the RISC to improve silencing activity and specificity. The essential binding of the siRNA guide strand 5′-end by the Ago2 MID domain requires a free 5′-mono-phosphate group (49, 62). Therefore, modification of the 5′-phosphate by for example a single 5′-O-methyl or an additional single UNA (position −1) in the passenger sense strand will disfavor its selection by the RISC and favor guide strand selection even of thermodynamically unfavorable siRNAs (93, 116). The siRNA 3′-overhangs are bound within the PAZ domain of Dicer/Ago2 with only very limited sequence preference (117), and even blunt-ended siRNAs are only slightly less potent (47, 118). Although 3′-overhang modification is generally well tolerated, it does affect PAZ binding affinity slightly and can therefore be utilized to ensure preferential loading of the guiding strand into the RISC; both chemically modified overhangs that are favored and those that are disfavored during strand selection by the RISC have been identified, and these overhangs can be incorporated into the guide and passenger strands of the siRNA, respectively (55, 119). Notably, the industry-standard and popular two nucleotide DNA-overhang dTdT, which only marginally reduces siRNA activity, seems to significantly reduce silencing longevity (120).

siRNA duplex stem modifications: Because the siRNA strand having the least thermostable 5′-end is preferentially utilized as the guiding strand in the RISC (14, 15), several siRNA modification strategies and designs (such as fsiRNAs; see above) aim to engineer optimal thermodynamic asymmetry within the siRNA duplex stem. Many modification types can be employed to modulate siRNA thermodynamic properties as they either enhance (e.g., such as LNA (79), 2-thiouracil (97), 2′-F) or decrease duplex stability (e.g., oxetane, ethylamino, UNA, dihydrouracil, or PS (55, 79, 121)). Typically, the 5′-end and 3′-end of the passenger strand is stabilized or destabilized by modification, respectively, in order not to introduce harmful modifications into the siRNA guiding seed region (whereas stabilization of the guide strand 3′-end is better tolerated (55)). Modest thermostabilization of the entire siRNA duplex stem may favor guide strand loading into cleavage-competent Ago2-RISC to enhance silencing activity and specificity as the non-catalytic Ago proteins Ago1, -3, and –4 form less functional RISCs with thermostable siRNAs (122).

3.2. Enhancing siRNA Stability

The high susceptibility of RNA to nuclease degradation suggests that siRNAs may need chemical modification to withstand the RNase-A-like activities dominant in most extracellular fluids. Indeed, unmodified and unshielded siRNAs are degraded within minutes in vivo (123) and in popular mimics of “extracellular environments” such as blood serum, whole blood, purified RNases, sweat, etc. (56, 57, 92). Furthermore, chemical stabilization was found early to be critical for successful silencing by naked siRNA upon low pressure intravenous injection in mice, a strategy relevant to siRNA therapeutics (124).

The benefit of nuclease-resistant siRNAs seems to originate from effects prior to siRNA internalization and RISC incorporation (125). Once inside the cytoplasm, siRNAs are more stable, some likely protected by RISC incorporation (126), and silencing can persist for 30–90 days in slowly or nondividing cells (125, 127). In agreement, high pressure hydrodynamic intravenous injection (HDI) of siRNA in mice, which ensures rapid cytoplasmic delivery to hepatocytes, can produce silencing, regardless of siRNA stabilization (128, 129). Concurrently, in culture experiments, where siRNAs are typically delivered intracellularly by shielding transfection reagents, siRNA nuclease resistance poses little concern, and efficient silencing is typically seen for 2–7 days depending on the cell type, rate of cell-division (125) and siRNA sequence (120). For in vivo applications of naked siRNAs (or when using only partly shielding delivery vehicles) initial efforts to stabilize siRNAs aimed to maximize modification levels, preferably generating fully modified siRNAs, by substituting internucleotide phosphate linkages (typically by PS) or ribose 2′-OH groups (typically by 2′-F, 2′-OMe, LNA). A great number of ribose 2′-modifications have been used to increase nuclease resistance by either full, partial or 3′-overhang modification of the siRNA duplex; In most cases full modification will dramatically reduce siRNA function, yet some fully modified siRNAs, especially those using DNA, 2′-OMe and 2′-F substitutions, are reported to be both highly stable and potent (74, 124, 130). Typically a mix of modifications are used to create modified functional siRNAs; a fully modified siRNA (having a passenger strand with 2′-F substitutions for all pyrimidine positions, DNA in all purine positions, 5′ and 3′-inverted abasic end caps and a guide strand with 2′-F substitutions in all pyrimidine positions, 2′-OMe substitutions of all purines and a single PS-modification at the 3′-terminal linkage) produced dose-dependent gene silencing upon intravenous injection in mice (124). Also, a PS/2′-OMe-stabilized, cholesterol-conjugated siRNA was reported to efficiently silence target gene expression in mouse livers (69).

As extensive chemical modification of siRNAs generally reduces their activity (55), more recent strategies modify or shield only RNAse hyper-sensitive nucleotide positions to greatly improve siRNA stability while preserving their potency (131, 132). Modifications of siRNA 3′-overhangs are often well tolerated, can rather safely be introduced to resist 3′-exonuclease attack and will typically modestly improve serum stability (55–59, 65, 66). Further stabilization requires modification of the siRNA duplex stem. Most dsRNA-specific endo-ribonucleases are pyrimidine-specific and preferentially cleave single-stranded UpA, UpG, and CpA dinucleotide motifs transiently exposed by spontaneous thermal fluctuations (131–134). Therefore, siRNA nuclease resistance can be greatly improved by selected or full modification of siRNA pyrimidines (e.g., by 2′-OMe) (131, 132) or by shielding these hypersensitive sites by thermodynamic stabilization at selected positions, e.g., by the introduction of LNA (55, 57, 66, 79, 81) or 4′-thioribose (88).

3.3. Enhancing siRNA Silencing Duration

Enhancing siRNA silencing duration may prove essential for therapeutic applications. Quite simply, siRNA silencing duration seem to depend on siRNA potency (120) and modifications that enhance potency have been reported to enhance silencing duration, such as ANA modification (86) or using the potent 27mer siRNA design (135). Yet, even equally potent siRNAs (evaluated at 48 h) differ significantly in silencing duration in a sequence-specific manner (120). Several studies report that chemical stabilization of siRNAs can lead to enhanced silencing persistence in vitro; FANA, HNA, CeNA, 2′-OMe, or 2′-F modification have been shown capable of slightly enhancing silencing duration (56, 83, 84, 86, 87, 131). However, it remains uncertain if siRNA stabilization in general will prolong siRNA survival in the cytoplasm; siRNA stability does not immediately influence silencing duration after cationic lipid transfection in cell culture or in mice after HDI of naked siRNAs (125, 128), and double-stranded siRNA species were found to be stable inside cells, even without RISC incorporation (136). Quite notably, the industry-standard DNA overhangs, typically dTdT, significantly reduce silencing duration irrespective of the siRNA sequence (120).

3.4. Abrogating siRNA Immunogenicity

Exogenous siRNAs were initially believed to be non-immunogenic due to their structural mimicry of endogenous siRNA/miRNA species (10). However, it is now clear that siRNAs can initiate innate immune responses upon binding to classes of so-called pattern recognition receptors (PRRs) leading to an interferon response and shutdown of protein synthesis (137). Generally, siRNA sequence and duplex length are the main determinants of the immunogenic potential of particular siRNAs; however, its elicitation is highly dependent on cell type, siRNA entry route, and concentration (33).

Sequence-specific immune responses towards siRNAs are mediated by the Toll-like receptors 7 and 8 (TLR7/8), which are transmembrane receptors found in the endosomes of immune cell populations (138). Hence, TL7/8 activation poses concerns in vivo and in cell cultures of immune cells (such as peripheral blood monocytic cell (PBMC) preparations) upon endosomal delivery but only rarely in typical cell cultures of nonimmune origin. TLR7/8 recognizes single-stranded RNA, as exposed from the siRNA duplex by random thermal fluctuations, in a sequence-specific manner, and particularly immune-stimulatory sequence motifs such as GUCCUUCAA (29), UGUGU (139) UGGC (140) and GU (141) have been identified in siRNA. Recent studies suggest that simply the presence of uridines correlates with TLR7/8 activation (142), thus complicating siRNA design as it is practically impossible to exclude uridines from the primary siRNA sequence. Encouragingly, TLR7/8 activation may be largely avoided by the use of delivery agents that exclude siRNA endosomal delivery (such as electroporation (143)) or by chemically modifying or shielding immune-stimulatory sequences/nucleotides to make them unrecognizable to TLR7/8. Notably, the incorporation of several chemical modification types, especially 2′-modified nucleotides (DNA, 2′-OMe, 2′-F, LNA) have been shown to abrogate siRNA immune activation, especially at high levels of modification (124, 144–146). Yet, only limited modification may be needed; modification of only uridines with either 2'-F or 2'-OMe (147) or deoxynucleotide residues (148) abrogated siRNA immunogenicity while generally preserving siRNA potency. Interestingly, 2′-OMe-modified RNA may serve as a potent antagonist of TLR7 activation and can even reduce the immune response towards unmodified siRNA when co-transfected (149). In accordance, alternating 2′-OMe modification of the SS has been proposed as a universal approach to avoid TLR7 activation by siRNA without reducing guide strand potency (150). Alternatively, as the strength of the hybridization between the siRNA strands correlates negatively with immunostimulatory activity (151) enhancing siRNA thermostability, e.g., by moderate LNA-modification may reduced siRNA immunogenicity by making single-stranded immunogenic sequence-motifs inaccessible to TLR7/8 recognition. In this regard, it should be noted that unpurified siRNA generally contain contaminants of single-stranded RNA from the annealing reaction, and it is therefore highly recommended to purify the duplex to avoid TLR7/8 activation.

Longer siRNA duplexes are detected primarily by transmembrane TLR3 and the cytoplasmic protein kinase R (PKR) in a sequence nonspecific manner (152). TLR3 is expressed on the surface and in endosomes primarily in the dendritic subpopulation of leukocytes but notably also in many primary cell types and popular cell lines (33, 35). Studies suggest that 21mer siRNAs may be immunologically safe as tested in several cell lines at high concentrations whereas 25mer (or longer) siRNAs triggered a concentration-dependent, presumably TLR3-mediated interferon-response in HeLa S3, DU 145, and MCF7 cells, but not in HeLa and HEK293 cells (33). Similarly, Kariko et al. found sequence-independent, low level TLR3 activation by 23mer siRNAs in kidney 293 (HEK293) or keratinocyte (HaCaT) cell lines and higher activation in human primary dendritic cells or macrophages (35). Very notably, TLR3-activation by the canonical 21mer siRNA has been reported upon intraocular injection in mice regardless of its sequence and 2′-OMe-modification (37). In contrast, shorter 19mer siRNAs were safe (37), suggesting that shorter siRNAs or chemical modifications are needed to fully avoid TLR3 responses in vivo.

Whereas TLR3 is restricted to membranes, sequence-independent sensing of long RNA structures occurs in the cytoplasm via PKR. This kinase was originally described to respond only to dsRNA longer than 30 bp (153), and therefore siRNAs were considered immunologically safe in typical cell cultures (10). Long 25–30mer siRNAs did not trigger PKR in HEK293 cells (34), yet canonical 21mer siRNAs have been shown to bind or trigger modest PKR activation in murine microglial N9 cells (154), human T98G cells (155–157), and HeLa cells (156). The impact of modest PKR activation needs to be further established, but precautions typically are not taken in cell culture experiments. Yet, the site-specific introduction of purine N2-benzyl modifications in the siRNA passenger strand has been shown to reduce PKR activation in HeLa cells (156). Also, using a 47-bp dsRNA, it was shown that PKR activation may be reduced by slightly disrupting the dsRNA structure by introduction of GU wobble base pairs or by introduction of 2′-deoxyuridine, 4-thiouridine, and 2-thiouridine (s2U) modifications, whereas other modifications, e.g., 2′-F and PS did not (158).

RIG-I is another major sensor of cytoplasmic dsRNA that recognizes poly-uridine rich dsRNA and is particularly sensitive to the nature of dsRNA ends; the standard 21mer siRNA design having two 2 nt 3′-overhangs is tolerated, whereas blunt 21–27mer siRNA and 5′-end triphosphates trigger RIG-I activation (155). For in vivo applications, however, it seems that the potent responses towards 21mer siRNAs by TLR3 and TLR7/8 poses more concerns and must be addressed by chemical modification. Induction of nonspecific responses by longer siRNAs is most likely difficult to avoid (33). In fact, Reynold et al. found induction of an IFN-like gene expression profile by siRNAs in cell lines that otherwise exhibit little or no overt signs of cell stress (e.g., cell death) (33).

3.5. Reducing siRNA Off-Targets Effects

The great success of siRNAs relies on their potency and reliability conferred by harnessing endogenous RNAi pathways to achieve highly sequence-specific cleavage of intended mRNA targets (159, 160). In turn, however, the shared handling of exogenous siRNAs and endogenous microRNAs (miRNAs) by RNAi proteins inherently forces siRNAs to behave as miRNAs and trigger unintended silencing of hundreds of endogenous genes sharing only limited sequence complementarity (27, 161, 162). These so-called off-target effects can result in toxic phenotypes (140) and compromise the interpretation, outcome and safety of the particular siRNA application. During a large-scale siRNA library screen, Lin et al. even found that the most efficient siRNAs function through off-targeting rather than target cleavage (163). miRNA-target regulation, and therefore siRNA off-targeting, is identified as mRNA destabilization and/or translational inhibition upon base pairing between the guide strand seed region (nucleotide position 2–7/8 counting from its 5′-end) and the complementary sites in the target 3′-UTRs (161, 163). All investigated siRNAs trigger off-target effects, and although siRNA sequences with low seed target frequency can be predicted in silico, off-targeting cannot be fully avoided. Some reduction of siRNA off-targeting can be achieved by utilizing siRNA pools to minimize the contribution of the individual siRNAs to off-targeting while preserving on-target activity. More promising, a number of studies have successfully reduced siRNA off-targeting by chemically modifying the seed region of the guide strand; 2′-OMe modification of position 2 of the guide strand was initially reported to reduce off-targeting and only minimally reduce silencing activity (72). More recent studies have aimed to destabilize seed-target interaction by substituting position 1–8 with DNA (164) or incorporate the strongly destabilizing UNA-modification at position 7 (95). Particularly, UNA modification at position 7 seems to dramatically reduce off-targeting with minimal loss of on-target activity for the siRNA sequences tested so far.

Proper siRNA design aims to strongly favor selection of the intended guide strand during RISC loading (see above), thereby minimizing the contribution of the passenger strand to off-targeting (165). However, chemical modifications can fully abrogate the contribution of the passenger strand to off-targeting by abrogating either its function or its incorporation into RISC; Whereas the sisiRNA design utilizes two shorter sense strands incapable of RNAi function (39), LNA modification of the passenger strand 5′ duplex end disfavors passenger strand incorporation into the RISC (79). Alternatively, passenger strand incorporation can be abrogated by chemically blocking the passenger 5′-phosphate, e.g., by 5′-OMe modification (116) or addition of a UNA residue (93).

4. Recap Guide: Building Better siRNAs

When engineering siRNAs, researchers have to consider the target species/cells, delivery vehicle, and entry route and incorporate modifications accordingly. For typical siRNA applications in cell culture using commercial transfection reagents, potent unmodified siRNAs can relatively easily be designed and chemical modification is only needed to reduce miRNA-like off-targeting, or to avoid RIG-I/PKR/TLR3-mediated siRNA immunogenicity in certain cell types. For in vivo applications, researchers have to consider additional factors; unshielded siRNA should be chemically stabilized, potency should be maximized in order to minimize the required siRNA dose and prevalent TLR-3-7/8-mediated immunogenicity should be abrogated. Below we provide a quick guide

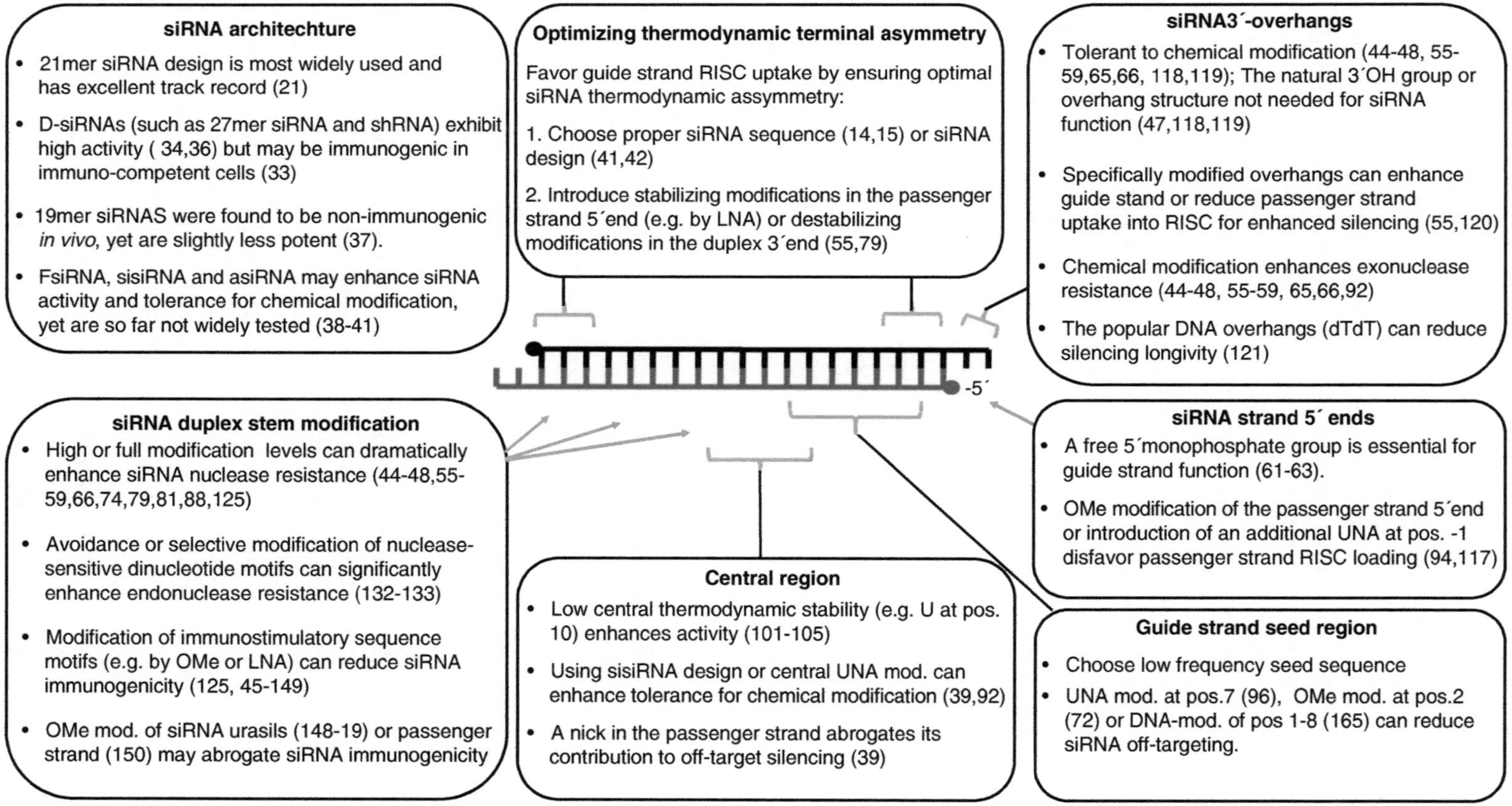

siRNA architechture

- 21mer siRNA design is most widely used and has excellent track record (21)
- D-siRNAs (such as 27mer siRNA and shRNA) exhibit high activity (34,36) but may be immunogenic in immuno-competent cells (33)
- 19mer siRNAS were found to be non-immunogenic *in vivo*, yet are slightly less potent (37).
- FsiRNA, sisiRNA and asiRNA may enhance siRNA activity and tolerance for chemical modification, yet are so far not widely tested (38-41)

Optimizing thermodynamic terminal asymmetry

Favor guide strand RISC uptake by ensuring optimal siRNA thermodynamic assymmetry:

1. Choose proper siRNA sequence (14,15) or siRNA design (41,42)

2. Introduce stabilizing modifications in the passenger strand 5´end (e.g. by LNA) or destabilizing modifications in the duplex 3´end (55,79)

siRNA3´-overhangs

- Tolerant to chemical modification (44-48, 55-59,65,66, 118,119); The natural 3´OH group or overhang structure not needed for siRNA function (47,118,119)
- Specifically modified overhangs can enhance guide stand or reduce passenger strand uptake into RISC for enhanced silencing (55,120)
- Chemical modification enhances exonuclease resistance (44-48, 55-59, 65,66,92)
- The popular DNA overhangs (dTdT) can reduce silencing longivity (121)

-5´

siRNA duplex stem modification

- High or full modification levels can dramatically enhance siRNA nuclease resistance (44-48,55-59,66,74,79,81,88,125)
- Avoidance or selective modification of nuclease-sensitive dinucleotide motifs can significantly enhance endonuclease resistance (132-133)
- Modification of immunostimulatory sequence motifs (e.g. by OMe or LNA) can reduce siRNA immunogenicity (125, 45-149)
- OMe mod. of siRNA urasils (148-19) or passenger strand (150) may abrogate siRNA immunogenicity

Central region

- Low central thermodynamic stability (e.g. U at pos. 10) enhances activity (101-105)
- Using sisiRNA design or central UNA mod. can enhance tolerance for chemical modification (39,92)
- A nick in the passenger strand abrogates its contribution to off-target silencing (39)

siRNA strand 5´ ends

- A free 5´monophosphate group is essential for guide strand function (61-63).
- OMe modification of the passenger strand 5´end or introduction of an additional UNA at pos. -1 disfavor passenger strand RISC loading (94,117)

Guide strand seed region

- Choose low frequency seed sequence
- UNA mod. at pos.7 (96), OMe mod. at pos.2 (72) or DNA-mod. of pos 1-8 (165) can reduce siRNA off-targeting.

Fig. 1. Overview of key chemical modification types that enhance siRNA performance. A canonical 21mer siRNA duplex is shown. Upper strand (*black*) represents the passenger, sense strand, whereas the lower strand represents the guide, antisense strand (see Subheading 4 for details).

on how to fulfill this goal by siRNA engineering (key beneficial modification types are illustrated in Fig. 1).

4.1. Choosing the siRNA Architecture

The canonical 21mer siRNA design is by far the most popular and has an excellent track record for most experimental conditions, yet has been shown to induce TLR3-mediated immune responses upon intraocular injection in mice, where only 19mer siRNAs were both potent and immunologically safe (37). Longer Dicer substrates siRNAs such as blunt 27mer siRNAs and shRNAs may produce more potent gene silencing (34, 36), yet can induce interferon responses via PKR, TLR3, or RIG-I in certain cell types in cell culture and in vivo (33). asiRNA, fsiRNA, and sisiRNA are reported to enhance siRNA activity and specificity but require chemical modification for structural integrity, especially in vivo.

4.2. Choosing siRNA Sequences

A number of design rules for unmodified siRNA have so far been deduced from the experimental testing of large siRNA sets (99–105); siRNA sequences should be chosen to create thermodynamically asymmetric siRNA where the 5′-end of the guide strand is least stable (14, 15). The middle of the siRNA duplex should be thermodynamically loose and the tenth base of the guide strand should be an A or U (101, 103, 104). The siRNAs should have relatively low GC-content (30–50%) and have accessible target sites not trapped in secondary structures (101, 108–111, 114) or occupied by proteins (166). siRNA should also be designed to avoid known immunostimulatory sequence motifs and the exclusion of UpA, UpG, and CpA dinucleotides may enhance nuclease resistance. Other nucleotide preferences at specific positions within the siRNA duplex have been reported (99–104), but seem not to be widely used in siRNA design.

4.3. Modifying the siRNA Duplex Ends

siRNA 3′-overhangs are tolerant to modification and are ideal for chemical engineering to enhance serum stability, activity, delivery or to allow siRNA tracking. Most 3′-overhang modification types will slightly enhance nuclease resistance (such as DNA, LNA, UNA, OMe, fluor, etc.) and some overhangs will even enhance the siRNA potency of the introduced guide strand; a 3-nt guide strand 3′-overhang composed of 5-LNA-LNA-RNA-3′ simultaneously enhances siRNA activity and serum stability (55). Conversely, the popular DNA overhang, typically dTdT, will reduce silencing duration and should be avoided (120). Passenger strand 3′-ends overhangs should similarly be modified for stability, and disfavored chemical modifications (such as UNA, etc.) can be incorporated to enhance guide strand activity (55), whereas the 5′-end of the guide strand is typically left unmodified as it has essential interactions with the RISC. Conversely, the 5′-end of the passenger strand can be modified by for example UNA or OMe to prevent its contribution to silencing and enhance guide strand activity (93, 116).

4.4. Modifying the siRNA Duplex Stem

Modification of the stem can severely compromise siRNA activity and although some fully modified siRNAs are highly active, the potential sequence dependency of such design needs to be fully established. Instead it may be desirable to keep modification levels at a minimum, especially for the guide strand, and only introduce key modification types: Modification of nuclease sensitive sites, either dinucleotide motifs or simply all pyrimidines, by for example 2′-OMe or 2′-F will greatly enhance siRNA nuclease resistance (131). Such 2′-OMe modification may simultaneously abrogate TLR7/8 mediated immunogenicity (147). Alternatively, siRNA serum stability can be significantly enhanced by chemical stabilization using for example LNA, which can also reduce siRNA immunogenicity (145). siRNA activity is enhanced by optimizing thermodynamic asymmetry; introduction of stabilizing modifications (e.g., LNA) in the passenger strand 5′-end and destabilizing modifications in its 3′-end such as UNA or oxetane can enhance siRNA activity (55, 79, 92). Finally, the introduction of a single UNA-modification at the guide strand position 7 is a potent strategy to reduce siRNA off-target activity (95).

References

1. Zamecnik PC, Stephenson ML (1978) Inhibition of Rous sarcoma virus replication and cell transformation by a specific oligodeoxynucleotide. Proc Natl Acad Sci U S A 75:280–284
2. Stein CA, Krieg AM (1994) Problems in interpretation of data derived from in vitro and in vivo use of antisense oligodeoxynucleotides. Antisense Res Dev 4:67–69
3. Cirak S, Arechavala-Gomeza V, Guglieri M, Feng L, Torelli S, Anthony K, Abbs S, Garralda ME, Bourke J, Wells DJ et al (2011) Exon skipping and dystrophin restoration in patients with Duchenne muscular dystrophy after systemic phosphorodiamidate morpholino oligomer treatment: an open-label, phase 2, dose-escalation study. Lancet 378:595–605
4. Monia BP, Lesnik EA, Gonzalez C, Lima WF, McGee D, Guinosso CJ, Kawasaki AM, Cook PD, Freier SM (1993) Evaluation of 2′-modified oligonucleotides containing 2′-deoxy gaps as antisense inhibitors of gene expression. J Biol Chem 268:14514–14522
5. Gupta N, Fisker N, Asselin MC, Lindholm M, Rosenbohm C, Orum H, Elmen J, Seidah NG, Straarup EM (2010) A locked nucleic acid antisense oligonucleotide (LNA) silences PCSK9 and enhances LDLR expression in vitro and in vivo. PLoS One 5:e10682
6. Wahlestedt C, Salmi P, Good L, Kela J, Johnsson T, Hokfelt T, Broberger C, Porreca F, Lai J, Ren K et al (2000) Potent and nontoxic antisense oligonucleotides containing locked nucleic acids. Proc Natl Acad Sci U S A 97: 5633–5638
7. Elmen J, Lindow M, Schutz S, Lawrence M, Petri A, Obad S, Lindholm M, Hedtjarn M, Hansen HF, Berger U et al (2008) LNA-mediated microRNA silencing in non-human primates. Nature 452:896–899
8. Fire A, Xu S, Montgomery MK, Kostas SA, Driver SE, Mello CC (1998) Potent and specific genetic interference by double-stranded RNA in Caenorhabditis elegans. Nature 391:806–811
9. Lee RC, Feinbaum RL, Ambros V (1993) The C. elegans heterochronic gene lin-4 encodes small RNAs with antisense complementarity to lin-14. Cell 75:843–854
10. Elbashir SM, Harborth J, Lendeckel W, Yalcin A, Weber K, Tuschl T (2001) Duplexes of 21-nucleotide RNAs mediate RNA interference in cultured mammalian cells. Nature 411:494–498
11. Hammond SM, Boettcher S, Caudy AA, Kobayashi R, Hannon GJ (2001) Argonaute2, a link between genetic and biochemical analyses of RNAi. Science 293:1146–1150
12. Maniataki E, Mourelatos Z (2005) A human, ATP-independent, RISC assembly machine fueled by pre-miRNA. Genes Dev 19:2979–2990
13. Bernstein E, Caudy AA, Hammond SM, Hannon GJ (2001) Role for a bidentate ribo-

nuclease in the initiation step of RNA interference. Nature 409:363–366

14. Khvorova A, Reynolds A, Jayasena SD (2003) Functional siRNAs and miRNAs exhibit strand bias. Cell 115:209–216
15. Schwarz DS, Hutvagner G, Du T, Xu Z, Aronin N, Zamore PD (2003) Asymmetry in the assembly of the RNAi enzyme complex. Cell 115:199–208
16. Martinez J, Tuschl T (2004) RISC is a 5′ phosphomonoester-producing RNA endonuclease. Genes Dev 18:975–980
17. Leuschner PJ, Ameres SL, Kueng S, Martinez J (2006) Cleavage of the siRNA passenger strand during RISC assembly in human cells. EMBO Rep 7:314–320
18. Matranga C, Tomari Y, Shin C, Bartel DP, Zamore PD (2005) Passenger-strand cleavage facilitates assembly of siRNA into Ago2-containing RNAi enzyme complexes. Cell 123:607–620
19. Hutvagner G, Zamore PD (2002) A microRNA in a multiple-turnover RNAi enzyme complex. Science 297:2056–2060
20. Behlke MA (2008) Chemical modification of siRNAs for in vivo use. Oligonucleotides 18:305–319
21. Behlke MA (2006) Progress towards in vivo use of siRNAs. Mol Ther 13:644–670
22. Lares MR, Rossi JJ, Ouellet DL (2010) RNAi and small interfering RNAs in human disease therapeutic applications. Trends Biotechnol 28:570–579
23. Higuchi Y, Kawakami S, Hashida M (2010) Strategies for in vivo delivery of siRNAs: recent progress. BioDrugs 24:195–205
24. http://www.fiercebiotech.com/story/roche-details-rd-cuts-new-buyout-plans-global-restructuring/2010-11-17
25. http://biopharmconsortium.com/blog/2011/02/15/pfizer-makes-massive-rd-cuts-and-exits-rnai-and-regenerative-medicine-therapeutics/
26. Meade BR, Dowdy SF (2009) The road to therapeutic RNA interference (RNAi): tackling the 800 pound siRNA delivery gorilla. Discov Med 8:253–256
27. Jackson AL, Bartz SR, Schelter J, Kobayashi SV, Burchard J, Mao M, Li B, Cavet G, Linsley PS (2003) Expression profiling reveals off-target gene regulation by RNAi. Nat Biotechnol 21:635–637
28. Kariko K, Bhuyan P, Capodici J, Weissman D (2004) Small interfering RNAs mediate sequence-independent gene suppression and induce immune activation by signaling through toll-like receptor 3. J Immunol 172:6545–6549
29. Judge AD, Sood V, Shaw JR, Fang D, McClintock K, MacLachlan I (2005) Sequence-dependent stimulation of the mammalian innate immune response by synthetic siRNA. Nat Biotechnol 23:457–462
30. Carthew RW, Sontheimer EJ (2009) Origins and mechanisms of miRNAs and siRNAs. Cell 136:642–655
31. Elbashir SM, Martinez J, Patkaniowska A, Lendeckel W, Tuschl T (2001) Functional anatomy of siRNAs for mediating efficient RNAi in Drosophila melanogaster embryo lysate. EMBO J 20:6877–6888
32. Elbashir SM, Lendeckel W, Tuschl T (2001) RNA interference is mediated by 21- and 22-nucleotide RNAs. Genes Dev 15:188–200
33. Reynolds A, Anderson EM, Vermeulen A, Fedorov Y, Robinson K, Leake D, Karpilow J, Marshall WS, Khvorova A (2006) Induction of the interferon response by siRNA is cell type- and duplex length-dependent. RNA 12:988–993
34. Kim DH, Behlke MA, Rose SD, Chang MS, Choi S, Rossi JJ (2005) Synthetic dsRNA Dicer substrates enhance RNAi potency and efficacy. Nat Biotechnol 23:222–226
35. Kariko K, Bhuyan P, Capodici J, Ni H, Lubinski J, Friedman H, Weissman D (2004) Exogenous siRNA mediates sequence-independent gene suppression by signaling through toll-like receptor 3. Cells Tissues Organs 177:132–138
36. Siolas D, Lerner C, Burchard J, Ge W, Linsley PS, Paddison PJ, Hannon GJ, Cleary MA (2005) Synthetic shRNAs as potent RNAi triggers. Nat Biotechnol 23:227–231
37. Kleinman ME, Yamada K, Takeda A, Chandrasekaran V, Nozaki M, Baffi JZ, Albuquerque RJC, Yamasaki S, Itaya M, Pan YZ et al (2008) Sequence- and target-independent angiogenesis suppression by siRNA via TLR3. Nature 452:591–597
38. Sun X, Rogoff HA, Li CJ (2008) Asymmetric RNA duplexes mediate RNA interference in mammalian cells. Nat Biotechnol 26:1379–1382
39. Bramsen JB, Laursen MB, Damgaard CK, Lena SW, Babu BR, Wengel J, Kjems J (2007) Improved silencing properties using small internally segmented interfering RNAs. Nucleic Acids Res 35:5886–5897
40. Hohjoh H (2004) Enhancement of RNAi activity by improved siRNA duplexes. FEBS Lett 557:193–198

41. Ohnishi Y, Tokunaga K, Hohjoh H (2005) Influence of assembly of siRNA elements into RNA-induced silencing complex by fork-siRNA duplex carrying nucleotide mismatches at the 3′- or 5′-end of the sense-stranded siRNA element. Biochem Biophys Res Commun 329:516–521
42. Petrova Kruglova NS, Meschaninova MI, Venyaminova AG, Zenkova MA, Vlassov VV, Chernolovskaya EL (2010) 2′-O-methyl-modified anti-MDR1 fork-siRNA duplexes exhibiting high nuclease resistance and prolonged silencing activity. Oligonucleotides 20:297–308
43. Ge Q, Ilves H, Dallas A, Kumar P, Shorenstein J, Kazakov SA, Johnston BH (2009) Minimal-length short hairpin RNAs: the relationship of structure and RNAi activity. RNA 16(1): 106–17
44. Ge Q, Dallas A, Ilves H, Shorenstein J, Behlke MA, Johnston BH (2009) Effects of chemical modification on the potency, serum stability, and immunostimulatory properties of short shRNAs. RNA 16:118–130
45. Chu CY, Rana TM (2008) Potent RNAi by short RNA triggers. RNA 14:1714–1719
46. Prakash TP, Allerson CR, Dande P, Vickers TA, Sioufi N, Jarres R, Baker BF, Swayze EE, Griffey RH, Bhat B (2005) Positional effect of chemical modifications on short interference RNA activity in mammalian cells. J Med Chem 48:4247–4253
47. Czauderna F, Fechtner M, Dames S, Aygun H, Klippel A, Pronk GJ, Giese K, Kaufmann J (2003) Structural variations and stabilising modifications of synthetic siRNAs in mammalian cells. Nucleic Acids Res 31:2705–2716
48. Holen T, Amarzguioui M, Babaie E, Prydz H (2003) Similar behaviour of single-strand and double-strand siRNAs suggests they act through a common RNAi pathway. Nucleic Acids Res 31:2401–2407
49. Martinez J, Patkaniowska A, Urlaub H, Luhrmann R, Tuschl T (2002) Single-stranded antisense siRNAs guide target RNA cleavage in RNAi. Cell 110:563–574
50. Hall AH, Wan J, Spesock A, Sergueeva Z, Shaw BR, Alexander KA (2006) High potency silencing by single-stranded boranophosphate siRNA. Nucleic Acids Res 34: 2773–2781
51. Abe N, Abe H, Nagai C, Harada M, Hatakeyama H, Harashima H, Ohshiro T, Nishihara M, Furukawa K, Maeda M et al (2011) Synthesis, structure, and biological activity of dumbbell-shaped nanocircular rnas for rna interference. Bioconjug Chem 22(10):2082–2092
52. Abe N, Abe H, Ito Y (2007) Dumbbell-shaped nanocircular RNAs for RNA interference. J Am Chem Soc 129:15108–15109
53. Lapierre J, Salomon W, Cardia J, Bulock K, Lam JT, Stanney WJ, Ford G, Smith-Anzures B, Woolf T, Kamens J et al (2011) Potent and systematic RNAi mediated silencing with single oligonucleotide compounds. RNA 17:1032–1037
54. Prakash TP (2011) An overview of sugar-modified oligonucleotides for antisense therapeutics. Chem Biodivers 8:1616–1641
55. Bramsen JB, Laursen MB, Nielsen AF, Hansen TB, Bus C, Langkjær N, Babu BR, Højland T, Abramov M, Van Aerschot A et al (2009) A large-scale chemical modification screen identifies design rules to generate siRNAs with high activity, high stability and low toxicity. Nucleic Acids Res 37:2867–2881
56. Chiu YL, Rana TM (2003) siRNA function in RNAi: a chemical modification analysis. RNA 9:1034–1048
57. Braasch DA, Jensen S, Liu Y, Kaur K, Arar K, White MA, Corey DR (2003) RNA interference in mammalian cells by chemically-modified RNA. Biochemistry 42:7967–7975
58. Amarzguioui M, Holen T, Babaie E, Prydz H (2003) Tolerance for mutations and chemical modifications in a siRNA. Nucleic Acids Res 31:589–595
59. Harborth J, Elbashir SM, Vandenburgh K, Manninga H, Scaringe SA, Weber K, Tuschl T (2003) Sequence, chemical, and structural variation of small interfering RNAs and short hairpin RNAs and the effect on mammalian gene silencing. Antisense Nucleic Acid Drug Dev 13:83–105
60. Hamada M, Ohtsuka T, Kawaida R, Koizumi M, Morita K, Furukawa H, Imanishi T, Miyagishi M, Taira K (2002) Effects on RNA interference in gene expression (RNAi) in cultured mammalian cells of mismatches and the introduction of chemical modifications at the 3′-ends of siRNAs. Antisense Nucleic Acid Drug Dev 12:301–309
61. Wang Y, Sheng G, Juranek S, Tuschl T, Patel DJ (2008) Structure of the guide-strand-containing argonaute silencing complex. Nature 456:209–213
62. Nykanen A, Haley B, Zamore PD (2001) ATP requirements and small interfering RNA structure in the RNA interference pathway. Cell 107:309–321
63. Lima WF, Wu H, Nichols JG, Sun H, Murray HM, Crooke ST (2009) Binding and cleavage specificities of human Argonaute2. J Biol Chem 284:26017–26028

64. Meister G, Landthaler M, Patkaniowska A, Dorsett Y, Teng G, Tuschl T (2004) Human Argonaute2 mediates RNA cleavage targeted by miRNAs and siRNAs. Mol Cell 15: 185–197
65. Choung S, Kim YJ, Kim S, Park HO, Choi YC (2006) Chemical modification of siRNAs to improve serum stability without loss of efficacy. Biochem Biophys Res Commun 342:919–927
66. Grunweller A, Wyszko E, Bieber B, Jahnel R, Erdmann VA, Kurreck J (2003) Comparison of different antisense strategies in mammalian cells using locked nucleic acids, 2′-O-methyl RNA, phosphorothioates and small interfering RNA. Nucleic Acids Res 31: 3185–3193
67. Hall AH, Wan J, Shaughnessy EE, Ramsay Shaw B, Alexander KA (2004) RNA interference using boranophosphate siRNAs: structure–activity relationships. Nucleic Acids Res 32:5991–6000
68. Aboul-Fadl T (2005) Antisense oligonucleotides: the state of the art. Curr Med Chem 12:2193–2214
69. Soutschek J, Akinc A, Bramlage B, Charisse K, Constien R, Donoghue M, Elbashir S, Geick A, Hadwiger P, Harborth J et al (2004) Therapeutic silencing of an endogenous gene by systemic administration of modified siRNAs. Nature 432:173–178
70. Sheehan D, Lunstad B, Yamada CM, Stell BG, Caruthers MH, Dellinger DJ (2003) Biochemical properties of phosphonoacetate and thiophosphonoacetate oligodeoxyribonucleotides. Nucleic Acids Res 31: 4109–4118
71. Yamada CM, Dellinger DJ, Caruthers MH (2007) Synthesis and biological activity of phosphonocarboxylate DNA. Nucleosides Nucleotides Nucleic Acids 26:539–546
72. Jackson AL, Burchard J, Leake D, Reynolds A, Schelter J, Guo J, Johnson JM, Lim L, Karpilow J, Nichols K et al (2006) Position-specific chemical modification of siRNAs reduces "off-target" transcript silencing. RNA 12:1197–1205
73. Kraynack BA, Baker BF (2006) Small interfering RNAs containing full 2′-O-methylribonucleotide-modified sense strands display Argonaute2/eIF2C2-dependent activity. RNA 12:163–176
74. Allerson CR, Sioufi N, Jarres R, Prakash TP, Naik N, Berdeja A, Wanders L, Griffey RH, Swayze EE, Bhat B (2005) Fully 2′-modified oligonucleotide duplexes with improved in vitro potency and stability compared to unmodified small interfering RNA. J Med Chem 48:901–904
75. Odadzic D, Bramsen JB, Smicius R, Bus C, Kjems J, Engels JW (2008) Synthesis of 2′-O-modified adenosine building blocks and application for RNA interference. Bioorg Med Chem 16:518–529
76. Wengel J, Petersen M, Nielsen KE, Jensen GA, Hakansson AE, Kumar R, Sorensen MD, Rajwanshi VK, Bryld T, Jacobsen JP (2001) LNA (locked nucleic acid) and the diastereoisomeric alpha-L-LNA: conformational tuning and high-affinity recognition of DNA/RNA targets. Nucleosides Nucleotides Nucleic Acids 20:389–396
77. Srivastava P, Barman J, Pathmasiri W, Plashkevych O, Wenska M, Chattopadhyaya J (2007) Five- and six-membered conformationally locked 2′,4′-carbocyclic ribo-thymidines: synthesis, structure, and biochemical studies. J Am Chem Soc 129:8362–8379
78. Pradeepkumar PI, Amirkhanov NV, Chattopadhyaya J (2003) Antisense oligonuclotides with oxetane-constrained cytidine enhance heteroduplex stability, and elicit satisfactory RNase H response as well as showing improved resistance to both exo and endonucleases. Org Biomol Chem 1:81–92
79. Elmén J, Thonberg H, Ljungberg K, Frieden M, Westergaard M, Xu Y, Wahren B, Liang Z, Ørum H, Koch T et al (2005) Locked nucleic acid (LNA) mediated improvements in siRNA stability and functionality. Nucleic Acids Res 33:439–447
80. Glud SZ, Bramsen JB, Dagnaes-Hansen F, Wengel J, Howard KA, Nyengaard JR, Kjems J (2009) Naked siLNA-mediated gene silencing of lung bronchoepithelium EGFP expression after intravenous administration. Oligonucleotides 19:163–168
81. Mook OR, Baas F, de Wissel MB, Fluiter K (2007) Evaluation of locked nucleic acid-modified small interfering RNA in vitro and in vivo. Mol Cancer Ther 6:833–843
82. Petersen M, Wengel J (2003) LNA: a versatile tool for therapeutics and genomics. Trends Biotechnol 21:74–81
83. Dowler T, Bergeron D, Tedeschi AL, Paquet L, Ferrari N, Damha MJ (2006) Improvements in siRNA properties mediated by 2′-deoxy-2′-fluoro-beta-D-arabinonucleic acid (FANA). Nucleic Acids Res 34:1669–1675
84. Fisher M, Abramov M, Van Aerschot A, Rozenski J, Dixit V, Juliano RL, Herdewijn P (2009) Biological effects of hexitol and altritol-modified siRNAs targeting B-Raf. Eur J Pharmacol 606:38–44
85. Watts JK, Choubdar N, Sadalapure K, Robert F, Wahba AS, Pelletier J, Pinto BM, Damha MJ (2007) 2′-Fluoro-4′-thioarabino-modified oligonucleotides: conformational switches

linked to siRNA activity. Nucleic Acids Res 35:1441–1451

86. Fisher M, Abramov M, Van Aerschot A, Xu D, Juliano RL, Herdewijn P (2007) Inhibition of MDRl expression with altritol-modified siRNAs. Nucleic Acids Res 35: 1064–1074

87. Nauwelaerts K, Fisher M, Froeyen M, Lescrinier E, Aerschot AV, Xu D, DeLong R, Kang H, Juliano RL, Herdewijn P (2007) Structural characterization and biological evaluation of small interfering RNAs containing cyclohexenyl nucleosides. J Am Chem Soc 129:9340–9348

88. Dande P, Prakash TP, Sioufi N, Gaus H, Jarres R, Berdeja A, Swayze EE, Griffey RH, Bhat B (2006) Improving RNA interference in mammalian cells by 4′-thio-modified small interfering RNA (siRNA): effect on siRNA activity and nuclease stability when used in combination with 2′-O-alkyl modifications. J Med Chem 49:1624–1634

89. Hoshika S, Minakawa N, Kamiya H, Harashima H, Matsuda A (2005) RNA interference induced by siRNAs modified with 4′-thioribonucleosides in cultured mammalian cells. FEBS Lett 579:3115–3118

90. Hoshika S, Minakawa N, Shionoya A, Imada K, Ogawa N, Matsuda A (2007) Study of modification pattern-RNAi activity relationships by using siRNAs modified with 4′-thioribonucleosides. Chembiochem 8: 2133–2138

91. Langkjær N, Pasternak A, Wengel J (2009) UNA (unlocked nucleic acid): a flexible RNA mimic that allows engineering of nucleic acid duplex stability. Bioorg Med Chem 17:5420–5425

92. Laursen MB, Pakula MM, Gao S, Fluiter K, Mook OR, Baas F, Langklaer N, Wengel SL, Wengel J, Kjems J et al (2010) Utilization of unlocked nucleic acid (UNA) to enhance siRNA performance in vitro and in vivo. Mol Biosyst 6:862–870

93. Vaish N, Chen F, Seth S, Fosnaugh K, Liu Y, Adami R, Brown T, Chen Y, Harvie P, Johns R et al (2011) Improved specificity of gene silencing by siRNAs containing unlocked nucleobase analogs. Nucleic Acids Res 39:1823–1832

94. Kenski DM, Cooper AJ, Li JJ, Willingham AT, Haringsma HJ, Young TA, Kuklin NA, Jones JJ, Cancilla MT, McMasters DR et al (2009) Analysis of acyclic nucleoside modifications in siRNAs finds sensitivity at position 1 that is restored by 5′-terminal phosphorylation both in vitro and in vivo. Nucleic Acids Res 38:660–671

95. Bramsen JB, Pakula MM, Hansen TB, Bus C, Langkjaer N, Odadzic D, Smicius R, Wengel SL, Chattopadhyaya J, Engels JW et al (2010) A screen of chemical modifications identifies position-specific modification by UNA to most potently reduce siRNA off-target effects. Nucleic Acids Res 38:5761–5773

96. Peacock H, Kannan A, Beal PA, Burrows CJ (2011) Chemical modification of siRNA bases to probe and enhance RNA interference. J Org Chem 76:7295–7300

97. Sipa K, Sochacka E, Kazmierczak-Baranska J, Maszewska M, Janicka M, Nowak G, Nawrot B (2007) Effect of base modifications on structure, thermodynamic stability, and gene silencing activity of short interfering RNA. RNA 13:1301–1316

98. Hornung V, Ellegast J, Kim S, Brzozka K, Jung A, Kato H, Poeck H, Akira S, Conzelmann KK, Schlee M et al (2006) 5′-Triphosphate RNA is the ligand for RIG-I. Science 314:994–997

99. Ladunga I (2007) More complete gene silencing by fewer siRNAs: transparent optimized design and biophysical signature. Nucleic Acids Res 35:433–440

100. Huesken D, Lange J, Mickanin C, Weiler J, Asselbergs F, Warner J, Meloon B, Engel S, Rosenberg A, Cohen D et al (2005) Design of a genome-wide siRNA library using an artificial neural network. Nat Biotechnol 23:995–1001

101. Reynolds A, Leake D, Boese Q, Scaringe S, Marshall WS, Khvorova A (2004) Rational siRNA design for RNA interference. Nat Biotechnol 22:326–330

102. Shabalina SA, Spiridonov AN, Ogurtsov AY (2006) Computational models with thermodynamic and composition features improve siRNA design. BMC Bioinformatics 7:65

103. Jagla B, Aulner N, Kelly PD, Song D, Volchuk A, Zatorski A, Shum D, Mayer T, De Angelis DA, Ouerfelli O et al (2005) Sequence characteristics of functional siRNAs. RNA 11:864–872

104. Ui-Tei K, Naito Y, Takahashi F, Haraguchi T, Ohki-Hamazaki H, Juni A, Ueda R, Saigo K (2004) Guidelines for the selection of highly effective siRNA sequences for mammalian and chick RNA interference. Nucleic Acids Res 32:936–948

105. Li W, Cha L (2007) Predicting siRNA efficiency. Cell Mol Life Sci 64:1785–1792

106. Deleavey GF, Watts JK, Alain T, Robert F, Kalota A, Aishwarya V, Pelletier J, Gewirtz AM, Sonenberg N, Damha MJ (2010) Synergistic effects between analogs of DNA and RNA

improve the potency of siRNA-mediated gene silencing. Nucleic Acids Res 38(13): 4547–4557

107. Koller E, Propp S, Murray H, Lima W, Bhat B, Prakash TP, Allerson CR, Swayze EE, Marcusson EG, Dean NM (2006) Competition for RISC binding predicts in vitro potency of siRNA. Nucleic Acids Res 34:4467–4476

108. Brown KM, Chu CY, Rana TM (2005) Target accessibility dictates the potency of human RISC. Nat Struct Mol Biol 12:469–470

109. Schubert S, Grunweller A, Erdmann VA, Kurreck J (2005) Local RNA target structure influences siRNA efficacy: systematic analysis of intentionally designed binding regions. J Mol Biol 348:883–893

110. Overhoff M, Alken M, Far RK, Lemaitre M, Lebleu B, Sczakiel G, Robbins I (2005) Local RNA target structure influences siRNA efficacy: a systematic global analysis. J Mol Biol 348:871–881

111. Tafer H, Ameres SL, Obernosterer G, Gebeshuber CA, Schroeder R, Martinez J, Hofacker IL (2008) The impact of target site accessibility on the design of effective siRNAs. Nat Biotechnol 26:578–583

112. Shao Y, Chan CY, Maliyekkel A, Lawrence CE, Roninson IB, Ding Y (2007) Effect of target secondary structure on RNAi efficiency. RNA 13:1631–1640

113. Yuan B, Latek R, Hossbach M, Tuschl T, Lewitter F (2004) siRNA selection server: an automated siRNA oligonucleotide prediction server. Nucleic Acids Res 32:W130–W134

114. Holen T (2005) Mechanisms of RNAi: mRNA cleavage fragments may indicate stalled RISC. J RNAi Gene Silencing 1:21–25

115. Marin RM, Vanicek J (2011) Efficient use of accessibility in microRNA target prediction. Nucleic Acids Res 39:19–29

116. Chen PY, Weinmann L, Gaidatzis D, Pei Y, Zavolan M, Tuschl T, Meister G (2008) Strand-specific 5′-O-methylation of siRNA duplexes controls guide strand selection and targeting specificity. RNA 14:263–274

117. Ma JB, Ye K, Patel DJ (2004) Structural basis for overhang-specific small interfering RNA recognition by the PAZ domain. Nature 429:318–322

118. Ghosh P, Dullea R, Fischer JE, Turi TG, Sarver RW, Zhang C, Basu K, Das SK, Poland BW (2009) Comparing 2-nt 3′ overhangs against blunt-ended siRNAs: a systems biology based study. BMC Genomics 10 (Suppl 1):S17

119. Sano M, Sierant M, Miyagishi M, Nakanishi M, Takagi Y, Sutou S (2008) Effect of asymmetric terminal structures of short RNA duplexes on the RNA interference activity and strand selection. Nucleic Acids Res 36:5812–5821

120. Strapps WR, Pickering V, Muiru GT, Rice J, Orsborn S, Polisky BA, Sachs A, Bartz SR (2010) The siRNA sequence and guide strand overhangs are determinants of in vivo duration of silencing. Nucleic Acids Res 38:4788–4797

121. Li ZY, Mao H, Kallick DA, Gorenstein DG (2005) The effects of thiophosphate substitutions on native siRNA gene silencing. Biochem Biophys Res Commun 329:1026–1030

122. Petri S, Dueck A, Lehmann G, Putz N, Rudel S, Kremmer E, Meister G (2011) Increased siRNA duplex stability correlates with reduced off-target and elevated on-target effects. RNA 17:737–749

123. Gao S, Dagnaes-Hansen F, Nielsen EJ, Wengel J, Besenbacher F, Howard KA, Kjems J (2009) The effect of chemical modification and nanoparticle formulation on stability and biodistribution of siRNA in mice. Mol Ther 17:1225–1233

124. Morrissey DV, Lockridge JA, Shaw L, Blanchard K, Jensen K, Breen W, Hartsough K, Machemer L, Radka S, Jadhav V et al (2005) Potent and persistent in vivo anti-HBV activity of chemically modified siRNAs. Nat Biotechnol 23:1002–1007

125. Bartlett DW, Davis ME (2007) Effect of siRNA nuclease stability on the in vitro and in vivo kinetics of siRNA-mediated gene silencing. Biotechnol Bioeng 97:909–921

126. Hoerter JA, Krishnan V, Lionberger TA, Walter NG (2011) siRNA-like double-stranded RNAs are specifically protected against degradation in human cell extract. PLoS One 6:e20359

127. Song E, Lee SK, Dykxhoorn DM, Novina C, Zhang D, Crawford K, Cerny J, Sharp PA, Lieberman J, Manjunath N et al (2003) Sustained small interfering RNA-mediated human immunodeficiency virus type 1 inhibition in primary macrophages. J Virol 77:7174–7181

128. Layzer JM, McCaffrey AP, Tanner AK, Huang Z, Kay MA, Sullenger BA (2004) In vivo activity of nuclease-resistant siRNAs. RNA 10:766–771

129. Heidel JD, Hu S, Liu XF, Triche TJ, Davis ME (2004) Lack of interferon response in animals to naked siRNAs. Nat Biotechnol 22:1579–1582

130. Blidner RA, Hammer RP, Lopez MJ, Robinson SO, Monroe WT (2007) Fully 2′-deoxy-2′-fluoro substituted nucleic acids induce RNA interference in mammalian cell culture. Chem Biol Drug Des 70:113–122

131. Volkov AA, Kruglova NS, Meschaninova MI, Venyaminova AG, Zenkova MA, Vlassov VV, Chernolovskaya EL (2009) Selective protection of nuclease-sensitive sites in siRNA prolongs silencing effect. Oligonucleotides 19:191–202

132. Turner JJ, Jones SW, Moschos SA, Lindsay MA, Gait MJ (2007) MALDI-TOF mass spectral analysis of siRNA degradation in serum confirms an RNAse A-like activity. Mol Biosyst 3:43–50

133. Sorrentino S (1998) Human extracellular ribonucleases: multiplicity, molecular diversity and catalytic properties of the major RNase types. Cell Mol Life Sci 54:785–794

134. Qiu L, Moreira A, Kaplan G, Levitz R, Wang JY, Xu C, Drlica K (1998) Degradation of hammerhead ribozymes by human ribonucleases. Mol Gen Genet 258:352–362

135. Hefner E, Clark K, Whitman C, Behlke MA, Rose SD, Peek AS, Rubio T (2008) Increased potency and longevity of gene silencing using validated Dicer substrates. J Biomol Tech 19:231–237

136. Raemdonck K, Remaut K, Lucas B, Sanders NN, Demeester J, De Smedt SC (2006) In situ analysis of single-stranded and duplex siRNA integrity in living cells. Biochemistry 45:10614–10623

137. Sioud M, Sorensen DR (2003) Cationic liposome-mediated delivery of siRNAs in adult mice. Biochem Biophys Res Commun 312:1220–1225

138. Heil F, Hemmi H, Hochrein H, Ampenberger F, Kirschning C, Akira S, Lipford G, Wagner H, Bauer S (2004) Species-specific recognition of single-stranded RNA via toll-like receptor 7 and 8. Science 303:1526–1529

139. Sioud M (2006) Single-stranded small interfering RNA are more immunostimulatory than their double-stranded counterparts: a central role for 2′-hydroxyl uridines in immune responses. Eur J Immunol 36:1222–1230

140. Fedorov Y, Anderson EM, Birmingham A, Reynolds A, Karpilow J, Robinson K, Leake D, Marshall WS, Khvorova A (2006) Off-target effects by siRNA can induce toxic phenotype. RNA 12(7):1188–1196

141. Diebold SS, Kaisho T, Hemmi H, Akira S, Reis e Sousa C (2004) Innate antiviral responses by means of TLR7-mediated recognition of single-stranded RNA. Science 303:1529–1531

142. Diebold SS, Massacrier C, Akira S, Paturel C, Morel Y, Reis e Sousa C (2006) Nucleic acid agonists for Toll-like receptor 7 are defined by the presence of uridine ribonucleotides. Eur J Immunol 36:3256–3267

143. Gantier MP, Tong S, Behlke MA, Xu D, Phipps S, Foster PS, Williams BR (2008) TLR7 is involved in sequence-specific sensing of single-stranded RNAs in human macrophages. J Immunol 180:2117–2124

144. Judge AD, Bola G, Lee AC, MacLachlan I (2006) Design of noninflammatory synthetic siRNA mediating potent gene silencing in vivo. Mol Ther 13:494–505

145. Hornung V, Guenthner-Biller M, Bourquin C, Ablasser A, Schlee M, Uematsu S, Noronha A, Manoharan M, Akira S, de Fougerolles A et al (2005) Sequence-specific potent induction of IFN-alpha by short interfering RNA in plasmacytoid dendritic cells through TLR7. Nat Med 11:263–270

146. Kariko K, Buckstein M, Ni H, Weissman D (2005) Suppression of RNA recognition by Toll-like receptors: the impact of nucleoside modification and the evolutionary origin of RNA. Immunity 23:165–175

147. Cekaite L, Furset G, Hovig E, Sioud M (2007) Gene expression analysis in blood cells in response to unmodified and 2′-modified siRNAs reveals TLR-dependent and independent effects. J Mol Biol 365:90–108

148. Flatekval GF, Sioud M (2009) Modulation of dendritic cell maturation and function with mono- and bifunctional small interfering RNAs targeting indoleamine 2,3-dioxygenase. Immunology 128:e837–e848

149. Robbins M, Judge A, Liang L, McClintock K, Yaworski E, MacLachlan I (2007) 2′-O-Methyl-modified RNAs act as TLR7 antagonists. Mol Ther 15:1663–1669

150. Hamm S, Latz E, Hangel D, Muller T, Yu P, Golenbock D, Sparwasser T, Wagner H, Bauer S (2009) Alternating 2′-O-ribose methylation is a universal approach for generating non-stimulatory siRNA by acting as TLR7 antagonist. Immunobiology 215: 559–569

151. Goodchild A, Nopper N, King A, Doan T, Tanudji M, Arndt GM, Poidinger M, Rivory LP, Passioura T (2009) Sequence determinants of innate immune activation by short interfering RNAs. BMC Immunol 10:40

152. Alexopoulou L, Holt AC, Medzhitov R, Flavell RA (2001) Recognition of double-

stranded RNA and activation of NF-kappaB by Toll-like receptor 3. Nature 413:732–738

153. Manche L, Green SR, Schmedt C, Mathews MB (1992) Interactions between double-stranded RNA regulators and the protein kinase DAI. Mol Cell Biol 12:5238–5248

154. Zhang Z, Weinschenk T, Guo K, Schluesener HJ (2006) siRNA binding proteins of microglial cells: PKR is an unanticipated ligand. J Cell Biochem 97:1217–1229

155. Marques JT, Devosse T, Wang D, Zamanian-Daryoush M, Serbinowski P, Hartmann R, Fujita T, Behlke MA, Williams BR (2006) A structural basis for discriminating between self and nonself double-stranded RNAs in mammalian cells. Nat Biotechnol 24:559–565

156. Puthenveetil S, Whitby L, Ren J, Kelnar K, Krebs JF, Beal PA (2006) Controlling activation of the RNA-dependent protein kinase by siRNAs using site-specific chemical modification. Nucleic Acids Res 34:4900–4911

157. Sledz CA, Holko M, de Veer MJ, Silverman RH, Williams BR (2003) Activation of the interferon system by short-interfering RNAs. Nat Cell Biol 5:834–839

158. Nallagatla SR, Bevilacqua PC (2008) Nucleoside modifications modulate activation of the protein kinase PKR in an RNA structure-specific manner. RNA 14:1201–1213

159. Du Q, Thonberg H, Wang J, Wahlestedt C, Liang Z (2005) A systematic analysis of the silencing effects of an active siRNA at all single-nucleotide mismatched target sites. Nucleic Acids Res 33:1671–1677

160. Dahlgren C, Zhang HY, Du Q, Grahn M, Norstedt G, Wahlestedt C, Liang Z (2008) Analysis of siRNA specificity on targets with double-nucleotide mismatches. Nucleic Acids Res 36:e53

161. Birmingham A, Anderson EM, Reynolds A, Ilsley-Tyree D, Leake D, Fedorov Y, Baskerville S, Maksimova E, Robinson K, Karpilow J et al (2006) 3′ UTR seed matches, but not overall identity, are associated with RNAi off-targets. Nat Methods 3:199–204

162. Lim LP, Lau NC, Garrett-Engele P, Grimson A, Schelter JM, Castle J, Bartel DP, Linsley PS, Johnson JM (2005) Microarray analysis shows that some microRNAs downregulate large numbers of target mRNAs. Nature 433:769–773

163. Lin X, Ruan X, Anderson MG, McDowell JA, Kroeger PE, Fesik SW, Shen Y (2005) siRNA-mediated off-target gene silencing triggered by a 7 nt complementation. Nucleic Acids Res 33:4527–4535

164. Ui-Tei K, Naito Y, Zenno S, Nishi K, Yamato K, Takahashi F, Juni A, Saigo K (2008) Functional dissection of siRNA sequence by systematic DNA substitution: modified siRNA with a DNA seed arm is a powerful tool for mammalian gene silencing with significantly reduced off-target effect. Nucleic Acids Res 36:2136–2151

165. Clark PR, Pober JS, Kluger MS (2008) Knockdown of TNFR1 by the sense strand of an ICAM-1 siRNA: dissection of an off-target effect. Nucleic Acids Res 36:1081–1097

166. Kretschmer-Kazemi Far R, Sczakiel G (2003) The activity of siRNA in mammalian cells is related to structural target accessibility: a comparison with antisense oligonucleotides. Nucleic Acids Res 31:4417–4424

Chapter 6

The Design, Selection, and Evaluation of Highly Specific and Functional siRNA Incorporating Unlocked Nucleobase Analogs

Narendra Vaish and Pinky Agarwal

Abstract

The efficient and specific silencing of genes via RNA interference (RNAi) for functional genomics and therapeutics depends on careful consideration of the factors that affect the functionality of small interfering RNA (siRNA). These factors include (1) the length of sequence available for siRNA targeting of an mRNA, (2) the structural and thermodynamic properties of target and siRNA sequences, (3) the mechanisms of siRNA off-target effects, and (4) the susceptibility of siRNA degradation when exposed to nucleases in serum and inside cells. Incorporation of Unlocked Nucleobase analogs (UNAs) in the siRNA design offers an attractive approach to design highly efficacious siRNAs with dramatically reduced off-target activity. Here, we describe methods and principles pertaining to the design, selection and screening of optimal siRNAs containing UNA.

Key words: RNA interference, Gene silencing, Small interfering RNA, siRNA design, Unlocked nucleobase analog, Modified nucleotide

1. Introduction

RNA interference (RNAi) is a biological response to double-stranded RNA and involves sequence-specific downregulation of target genes. The mammalian RNAi pathway also employs non-coding microRNA (miRNA) to regulate protein synthesis. Synthetic siRNA technology is a powerful gene intervention tool for functional genomics and therapeutics. Since the discovery that RNAi could be used to silence mammalian genes (1, 2), tens of thousands of studies involving siRNA have been published and several siRNA drug candidates have entered clinical trials (3–5). Despite becoming a prominent technology for functional genomics in a short span of time the infidelity of siRNA in silencing a target gene infers the limitations of gene silencing by this method.

Debra J. Taxman (ed.), *siRNA Design: Methods and Protocols*, Methods in Molecular Biology, vol. 942,
DOI 10.1007/978-1-62703-119-6_6,

In general, siRNAs are composed of two complementary RNA strands of 19 nucleotides (nt) with the option of a 2 nt 3′-overhang. Conceptually, the design of siRNAs appears straightforward. However, given the limitation of the siRNA target to the gene transcript (a ~200–10,000 nt region) and the stringent requirement for favorable sequence and thermodynamic properties of siRNA duplexes, designing highly functional siRNA can be cumbersome, especially for small transcripts. In addition, considering that there are 42,115 protein coding genes, 42,208 RefSeqs, 122,727 UniGenes (a set of transcript sequences that appear to arise from the same transcription locus), and 48.9 million SNPs in the NCBI human database (http://www.ncbi.nlm.nih.gov), designing a specific siRNA for a given gene is quite challenging. Requirements for multispecies targeting siRNA for cross species studies and drug development imposes an additional layer of complexity.

Because siRNAs can also act as microRNAs (miRNAs) for the translational repression and eventual degradation of mRNA via processing bodies (P-bodies), a single siRNA has potential to silence hundreds of nontarget genes (also called off-target silencing). miRNAs are endogenous noncoding RNAs of 19–21 base pairs (bp) that mammalian cells employ to regulate the translation of the transcriptome. The antisense strand, also called the guide strand, of an miRNA duplex regulates the translation of multiple genes, primarily in specific biological pathways. While the exact mechanism of target recognition by miRNAs is still elusive, the miRNA seed region (positions 2–8 from the 5′-end of the guide strand) plays a central role in recognizing complementary sequences in the 3′-UTR (3′-untranslated region) of mRNAs. Bioinformatic analysis suggests that a random 7 nt sequence occurs, on average, once in every 16,384 bp, and even the most infrequent 7-nt seed sequence has sequence complementarity within the 3′-UTR of at least 17 known mRNAs (6). Thus, it is apparent that it is nearly impossible to design specific siRNAs targeting a single gene. It has been shown biochemically that off-target silencing is a fundamental property of siRNAs that cannot be always predicted and easily eliminated (7, 8). Unlocked nucleobase analogs (UNAs) are helix destabilizing non-nucleotide analogs for which the C2′–C3′ bond of the ribose ring is absent (Fig. 1a). UNA incorporation in the seed region of the guide strand is demonstrated to confer the elimination of seed region mediated off-targeting (9, 10). Combined placement of UNA at the 3′-ends of both strands, the 5′-end of the passenger strand and within the seed region demonstrated more than tenfold reduction in the global off-target signature of an siRNA in a microarray study (10).

In addition to the sequence specific off-targeting, siRNAs can also induce proinflammatory cytokines upon recognition by endosomal or cytosolic receptors such as TLRs (11–14) and RIG-I (15–18), thus complicating the interpretation of biochemical experiments

Fig. 1. (**a**) Structure of unlocked nucleobase analogs. (**b**) Structure of an siRNA containing strategically placed UNA residues. *Circles* represent UNA residues.

(19) and hampering drug development (14, 20). For this reason, cytokine induction by double-stranded siRNAs should be carefully evaluated. Specific chemical modification can be employed to reduce the cytokine stimulatory potential of siRNAs (21). Chemical modifications of ribose sugars utilizing 2'-O-methyl (2′-OMe) and 2'-flouro (2′-F) nucleoside analogs can also be highly valuable in improving the stability of the siRNA duplex when exposed to nucleases in serum and cytosol (22, 23). Chemical modification of siRNA can improve the in vivo efficacy of siRNA and long term duration of silencing compared to the unmodified control siRNA (22, 24). Because UNA does not contain the C2′–C3′ bond of the ribose sugar, the incorporation of UNA at the 3′-ends of the siRNA might be useful in resisting siRNA degradation by 3′-exonucleases (25).

Herein, we describe methods for designing highly effective siRNAs with incorporated UNAs. Protocols are provided for selecting an optimal targeting sequence for the siRNA and for screening to ensure knockdown efficacy. Strategies are described for the addition of UNA to the 3′-overhang to increase the siRNA specificity, to the 5′-end to increase the strand selection bias during assembly of the RNA-induced silencing complex (RISC), and within the seed region to decrease miRNA-like off-target activity.

2. Materials

2.1. General

1. A personal computer with Microsoft Excel.
2. Access to the World Wide Web and the National Center for Biotechnology Information (NCBI).
3. Vector NTI or Web based access to ClustalW (http://www.genome.jp/tools/clustalw/).
4. Blastn Web access (http://blast.ncbi.nlm.nih.gov/Blast.cgi) or Smith–Waterman Web access (http://mobyle.pasteur.fr/).
5. Access to a custom script such as Perl script for the parsing of target messenger RNAs into 19 nt sequences; or MID function in Excel can also be utilized to parse the sequences.

6. An algorithm for the prediction of relevant oligo parameters such as melting temperature, thermodynamic properties, secondary structure, etc. Some examples are DINAMelt (http://mfold.rna.albany.edu/?q=DINAMelt), Vector NTI, Primer 3.0 (http://frodo.wi.mit.edu/primer3/), HPCDispatcher (http://www1.infosci.coh.org/hpcdispatcher/siRNA.aspx), and Oligo7 (http://www.oligo.net).
7. RNA synthesis capability and access to UNA phosphoramidites or a commercial source (such as IDT, Ambion/ABI, Sigma-Proligo, Thermo Fisher-Dharmacon, etc.). Note that UNA modified oligos can be purchased from Ribotask (http://www.ribotask.dk) and TriLink BioTechnologies (http://www.trilink-biotech.com) or can be accessed through Marina Biotech.
8. Tissue culture related equipment and instruments: Shakers, laminar flow hood, incubator with CO_2 supply, (VWR, Westchester, PA).
9. Sterile single and multichannel pipettes and aerosol resistant pipette tips (Rainin, Oakland, CA); aerosol resistant pipette tips can also be purchased from Molecular BioProducts (San Diego, CA).
10. Tissue culture flasks with filter lid, T75 and T150 (BD Falcon, Franklin Lakes, NJ).
11. 6-Well and 96-well tissue culture plates (BD Falcon, Franklin Lakes, NJ).
12. 96-Well sterile block, 0.5 mL (VWR, Westchester, PA).
13. Reagent Reservoir: 25–100 mL (VWR, Westchester, PA).
14. Cell culture medium, buffers and supplements: Dulbecco's modified Eagle media with high glucose (DMEM) and phosphate buffered saline (PBS), OptiMEM, HEPES buffer (100×), L-Glutamine (Gibco-Invitrogen, Carlsbad, CA), 100× Nonessential amino acids (Media Tech, Manassas, VA).
15. Heat inactivated fetal bovine serum (Gibco-Invitrogen, Carlsbad, CA).
16. 0.25% Trypsin–EDTA (Invitrogen, Carlsbad, CA).
17. Penicillin–streptomycin (100 U/mL) (Invitrogen, Carlsbad, CA).
18. Lipofectamine-RNAiMAX and Lipofectamine 2000 (Invitrogen, Carlsbad, CA).
19. Light and inverted phase contrast microscope (Olympus, Center Valley, PA).

2.2. Synthesis of Reporter Plasmids Containing siRNA and miRNA Targets

1. psiCHECK II vector: GenBank accession number AY535007 (Promega, Madison, WI).
2. DNA oligo synthesis capability or access to appropriate vendor.

3. *Xho I* and *Not I* restriction enzymes and buffer (New England Biolabs, Ipswich, MA).
4. DNA cloning capability.
5. One Shot® PIR1 Chemically Competent *Escherichia coli* (Invitrogen, Carlsbad, CA).
6. Maxiprep kits (Qiagen, Valencia, CA) to isolate the plasmid DNA.

2.3. Dual Luciferase Assay

1. HeLa cells (ATCC, Manassas, VA).
2. 12-Channel wand with 20-gauge needles, 13 mm long (V&P Scientific Inc., San Diego, CA).
3. Sterile eppendorf tubes and 15 and 50 mL conical tubes (VWR, Westchester, PA).
4. Costar Black clear bottom 96-well plates (Fisher Scientific, PA).
5. siRNA buffer (Dharmacon-Thermo Fisher, Lafayette, CO).
6. siRNA.
7. Target gene(s) cloned in a psiCheck-II plasmid construct.
8. Dual-luciferase reporter assay system (Promega, Madison, WI).
9. Victor luminometer (Perkin Elmer, MA).

2.4. Quantitative Real-Time PCR Assay

1. RNase-free water (Invitrogen, Carlsbad, CA).
2. DNase I solution (Invitrogen, Carlsbad, CA).
3. 95–100% Ethanol.
4. RNase AWAY® Reagent (Invitrogen, Carlsbad, CA).
5. 1 M Tris–HCl, pH 8.4.
6. $MgCl_2$.
7. KCl.
8. Invitrogen PURELink 96 RNA Isolation Kit (Invitrogen, Carlsbad, CA).
9. β-Mercaptoethanol (β-ME) (Invitrogen, Carlsbad, CA).
10. SuperScript III First-Strand Synthesis System (Invitrogen, Carlsbad, CA).
11. TaqMan Universal PCR Master Mix without AmpErase UNG (Applied Biosystem, Foster City, CA) or SYBR Green FastMix, ROX (Quanta Biosciences Inc., Gaithersburg, MD).
12. Vacuum manifold and vacuum supply.
13. FastPrep-24 (Sample preparation system; MP Biomedicals, Solon, OH).
14. 7900 Real-Time PCR System (Applied Biosystems, Foster City, CA).
15. ABI PRISM 96-well Optical Reaction Plates with Barcode and ABI PRISM Optical Adhesive Covers (Applied Biosystem, Foster City, CA).

2.5. In Vitro Cytokine Detection

1. Ficoll-HyPaque (Amersham Biosciences, Piscataway, NJ).
2. PBS (Invitrogen, Carlsbad CA).
3. Iscove's modified DMEM (IMDM) (Mediatech Inc., Manassas, VA).
4. Nonessential amino acids (NEAA) (Invitrogen, Carlsbad, CA).
5. Glutamine (Invitrogen, Carlsbad, CA).
6. Bleach.
7. Lipofectamine™ RNAiMAX transfection reagent (Invitrogen, Carlsbad, CA).
8. OptiMEM reduced serum media (Invitrogen, Carlsbad, CA).
9. 50 mL Conical tubes (VWR, Westchester, PA).
10. Collected human peripheral blood (Golden West Biological, Temecula, CA).
11. Isolated human peripheral blood mononuclear cells (hPBMC) (Astarte Biologics, Redmond, WA).
12. Human Interferon ELISA kit (PBL Biomedical Laboratories, Piscataway, NJ).
13. Procarta™ custom 10-plex cytokine profiling kit.
14. Luminex 100 IS System (Bio-Rad Life Sciences, Hercules, CA).

2.6. Global Gene Expression Profiling

1. Mammalian cells such as HeLa or HepG2 (ATCC, Manassas, VA).
2. RNAiMAX Transfection reagent (Invitrogen, Carlsbad, CA).
3. 6-Well tissue culture plates (BD Falcon, Franklin Lakes, NJ).
4. Bioanalyzer 2100 (Agilent Technologies, Santa Clara, CA).
5. RNeasy Mini Kit (Qiagen, Valencia, CA).
6. NanoDrop spectrophotometer, ND-1000 (Thermo Scientific, Wilmington, DE).
7. RNA 6000 Nano LabChip® (Agilent Technologies, Santa Clara, CA).
8. 7900 Real-Time PCR System (Applied Biosystems, Foster City, CA).
9. ABI PRISM 96-well Optical Reaction Plates with Barcode and ABI PRISM Optical Adhesive Covers (Applied Biosystem, Foster City, CA).
10. Two-Cycle cDNA Synthesis Kit (Affymetrix, Santa Clara, CA).
11. Human Genome U133 Plus 2.0 GeneChip (Affymetrix, Santa Clara, CA).
12. GeneChip Scanner 3000 7G System (Affymetrix, Santa Clara, CA).

13. Access to microarray data analysis software, R/Bioconductor (http://www.bioconductor.org) or GeneSpring GX 11 (Agilent, Santa Clara, CA).

3. Methods

3.1. Bioinformatic Design of Highly Functional siRNAs

High activity of siRNA is the single most important criterion to ensure a strong and long lasting downregulation of a gene via RNAi while using minimal quantities of siRNA to limit the toxicity and off-target issues. The most favored siRNA construct is the 21 nt RNA duplex containing 2 nt TT or UU overhangs at both 3′-ends. This construct mimics the natural cleavage product of Dicer processing of longer RNA duplexes. Various other constructs have been pursued to increase the potency of siRNA. Dicer substrates are designed to enhance the uptake of siRNA duplexes early in the RNAi pathway, before the formation of the RNA-induced silencing complex (RISC), thus increasing the potency (26). Both short hairpin (shRNA) and 25/27 mer RNA duplexes containing one blunt end and one 2 nt 3′-overhang are Dicer substrates, and are reported to have higher activity than corresponding conventional 21/21 mer siRNA duplexes (27, 28). It is proposed that Dicer-substrates increase the potency by enhancing the incorporation of the product into the RISC after cleavage by the Dicer enzyme. Other siRNA formats include 19 mer blunt-end siRNA (29), fork-siRNA (fsiRNA) (30) and siRNA containing a pre-cleaved passenger strand (sisiRNA) (31). All of the siRNA formats described above target 19 nt segments to the mRNA and can be designed using similar oligo selection criteria.

The design of appropriate siRNAs requires careful inspection of the siRNA target sites for maximum siRNA efficacy and minimum off-target effects. A variety of siRNA selection algorithms have been developed and are freely available for design of siRNAs with high activity (see Note 1). Steps for designing highly potent siRNAs are described as follows:

1. Identify the target gene of interest against which an siRNA is desired.
2. Search and download the mRNA sequence of the target gene, including all splice variants, from a public database (http://www.ncbi.nlm.nih.gov/). To get the mRNA sequence, go to the reference sequence division of the NCBI (RefSeq database) and retrieve all the transcript variants of a gene. Some genes will have one while others will have multiple transcript variants. It is desirable to design siRNA against all transcript variants unless targeting a specific transcript is required by the experiment. Download transcript variants in FASTA format.

3. Align the transcripts to identify the region of shared sequences. The sequence alignment can be accomplished using either desktop software such as Vector NTI or a Web based tool such as ClustalW. Identify the regions of the shared positions for designing siRNAs targeting all transcripts.
4. Species homology: If the experiment demands evaluation of siRNA activity in a multispecies setting, such as in mouse, rat, and primates for pharmacological development, identify mRNA sequences and their splice variants for all of the species needed for siRNA by homology search. Identify the homologous regions with the target gene as described in step 3 above.
5. If there is a preference to design siRNA for a particular region, identify the open reading frame (ORF) and untranslated (5′-UTR and 3′-UTR) regions for the primary target sequence. In general, the ORF is considered best for siRNA design because the 5′-UTR and 3′-UTR are usually associated with proteins in the translational machinery.
6. Identify the SNP positions for the target gene. It is preferable to avoid SNP sites because a single nucleotide mismatch between an siRNA and its target can dramatically reduce knockdown activity, especially when the SNP position falls around the target cleavage site.
7. Parse the primary mRNA sequence to generate 19 nt candidate siRNA sequences. This can be accomplished by custom developed scripts, or Excel function "MID" can be utilized. To accomplish this, enter the target sequence in the cell A1. Populate the column B with the numbers in the increasing order (1, 2, 3, 4, etc.). Enter the MID formula in column C (cells 1, 2, 3, 4, etc.) (=MID(A1,$B1,19)) to populate the 19 nt sequences.
8. Discard all 19 nt sequences that are unsuitable for any of the following reasons:
   (a) They do not target all splice variants.
   (b) They are contained within a species nonhomologous region.
   (c) They are contained within a potential SNP site.
9. Remove the sequences with features known to be detrimental for RNAi activity. This includes sequences with the following features:
   (a) >3 GGG or CCC in a row.
   (b) Potential for self-folding.
   (c) GC content that is <40% or >50%.
   (d) Very high or very low duplex melting temperature.
10. For the remaining candidates, assign a functionality score based on the terminal thermodynamic stability and base preferences

at certain sequence positions. This can be accomplished by calculating the thermodynamic parameters on the nucleotides 1–5 and 15–19. Sequences with higher free energy values for the 3′-end compared to the 5′-end are likely to be more active than sequences with higher free energy for the 5′-end. As a more simplistic approach, sequences containing three or more A or U residues at positions 15–19 could be sufficient for selecting highly active sequences. An A or U at position 19 and G or C at position 1 is highly preferred. Among other base preferences, A/U at position 10 and any other base except G at position 13 is desirable (32–34). In addition to their potentially higher silencing activity, sequences with a low melting temperature for positions 15–19 could also be highly useful for selecting sequences with low off-target potential (6, 35).

11. Rank the sequences based on the functionality score.
    (a) Select the top ranking sequences with high scores, i.e., those with more propensities to guide strand activity.
    (b) Eliminate sequences that fall around the same region on the target sequence.
12. Analyze top candidate sequences for off-target potential using the Blastn function in the NCBI database against the target organism of interest (see Note 2).
    (a) Eliminate sequences with greater match to off-target genes.
    (b) Eliminate sequences with excessive similarity in the 15 nt towards the 3′-end.
13. Analyze the target secondary structure for potentially strong thermodynamically stable regions. Discard the candidate siRNAs targeting stable secondary structure regions of the target mRNA.
14. Select 10–15 siRNA candidate sequences with a high functionality score.
15. Design the passenger and guide strands.
16. Synthesize siRNAs either using an in house facility or a commercial source such as IDT or Dharmacon-Thermo Fisher.

3.2. In Vitro Activity Screen of siRNAs

In silico designed siRNAs against a specific clinical target are first evaluated for inhibition of target mRNAs by biological end-point assays in vitro. An activity screen is carried out by direct measurement of endogenous mRNA knockdown using species and tissue relevant cell line(s). For example, if the mRNA target is specifically expressed in liver, such as *ApoB*, then the activity of the siRNAs is evaluated in cells originating from liver, such as HepG2 and Hep3B. It is preferable to determine the siRNA activity in more than one cells line. If functionality across species is a consideration, then the siRNA activity must be determined in relevant cells lines for each of the species, such as mouse, rat, and primate cells. The potency of the selected siRNAs is established by titrating the siRNAs at concentrations ranging from 0.1 pM to 25 nM to establish the IC_{50} and maximal inhibition of the mRNA by each siRNA.

3.2.1. Assessment of the In Vitro Knockdown Activity by Measuring Levels of the Endogenous Target

1. Seed cells growing in log phase in a 96-well flat-bottom plate in the appropriate growth medium 1 day before transfection. The optimal number of cells per well should be determined for each cell type. For example, 10,000 cells per well in 96-well plates works well for HeLa and HepG2.
2. The following day, the cells should be 70–80% confluent. Thaw the siRNAs and reagents. Before transfection, remove the media and replace with 75 μL/well serum-free OptiMEM, followed by the addition of 25 μL of siRNA–RNAiMAX complex. For the transfection of primary cells or slowly dividing cells, reverse transfection is preferable, whereby the siRNA is first added to each well followed by the seeding of cells.
3. Carry out an activity screen for three concentrations of siRNAs ranging from 0.1 to 10 nM in the desired cell-line(s).
4. Incubate for 4 h.
5. After incubation, add 100 μL of 20% serum-containing media to bring the final concentration of serum in the cell culture to10%.
6. After 24 h, remove the growth media, lyse the cells, and isolate total RNA or isolate mRNAs using a poly-A coated resin (see Notes 3 and 4).
7. Carry out qRT-PCR to measure mRNA levels (see Note 5).
8. Select the 3–5 most active sequences.
   (a) If the first set of siRNAs does not lead to highly active siRNAs, synthesize additional siRNAs from the candidate siRNA list from the bioinformatic design (Subheading 3.1).
9. Determine the IC_{50} of the most active siRNAs to assess the inhibition concentration at which 50% of the mRNA is downregulated.
10. Pick the most potent siRNA as a final candidate for further development.

3.3. The Addition of a 3′-Overhang Modification by UNA to Increase the siRNA Specificity

Synthetic siRNAs can interact with the cytokine pathway and activate innate immunity, and may regulate a large number of genes in an siRNA sequence independent manner. siRNAs can interact with various cell surface, endosomal and cytosolic cytokine sensors and exert nonspecific effects (11–15, 22). With the exception of TLR-7 mediated cytokine activation, which was shown to be dependent on specific RNA sequence motifs in single-stranded RNA (11, 12), the rest of the cytokine pathways activated by siRNA appear to be dependent on the double-stranded nature of the RNA duplex and are independent of the 2 nt 3′-overhangs (13–15, 20). Chemical modification of the siRNA duplex is an attractive approach to mitigate the non-sequence specific effect of siRNAs (21). Vaish et al. demonstrated that introduction of UNA in the 3′-overhangs can have a profound effect in reducing the nonspecific gene regulation by siRNAs (10). Using microarray gene profiling, these authors demonstrated that replacement of the 2 nt 3′-end overhang with UNA's on both guide and passenger strands reduced the off-target gene signature by ~50% compared to the unmodified siRNA control, suggesting that 3′-end modification of siRNA duplexes could be a useful method to reduce siRNA toxicity.

3.4. 5′-End Chemical Modification to Increase the Strand Selection Bias During RISC Assembly

Both strands of siRNAs are capable of effecting gene silencing. Incorporation of the passenger strand is undesirable and promotes sequence specific off-target activity by siRNA. Participation of the passenger strand in RISC assembly also results in relatively lower incorporation of the guide strand in the RISC complex, thus reducing the potency of the siRNA duplex. Bioinformatic selection of siRNA target sites that result in asymmetric design of the siRNA duplex is an attractive approach to limit the passenger strand participation in RISC assembly. However, the design of asymmetric siRNA to promote the guide strand assembly into RISC does not guarantee the complete inactivity of the passenger strand. Modification of siRNA duplexes that block the incorporation of the passenger strand into the RISC can be a useful alternative to achieve this goal. Asymmetric siRNA (asiRNA) designs containing a short passenger strands have demonstrated reduced off-target effects compared to the full-length siRNAs (36, 37). 5′-OMe modification has also demonstrated the reduced activity of the passenger strand (38). An sisiRNA design containing a precleaved passenger strand is particularly appealing for eliminating passenger strand mediated sequence specific off-target effects; however, this format may require careful modification of passenger strand fragments by locked nucleic acids to maintain the siRNA duplex (31). Recently, Vaish et al. have demonstrated that the additions of a single UNA residue at the 5′-end of the passenger strand completely eliminated the passenger strand activity and thus the off-target activity (10). These results are in agreement with those of Kenski et al. that suggested that UNA substitution at the position one of the guide strand prevented phosphorylation, which is critical for RISC assembly (39). In addition, elimination of the passenger strand activity by 5′-UNA modification increased the activity of the guide strand threefold (10). Thus, 5′-end modification by a single UNA residue can help prevent the off-target activity of the passenger strand and improve the potency of siRNA duplexes.

3.5. Seed Region Chemical Modification to Decrease miRNA-Like Off-Target Activity

Elimination of the seed region mediated miRNA-like off-target activity of the guide strand without reducing the siRNA potency is particularly challenging. Because miRNA-like off-target activity is primarily mediated by a 7 nt seed region, an siRNA can potentially silence hundreds of gene (7, 8, 40–43). Chemical modifications in the seed region of the guide strand should be used to reduce the miRNA-like off-target activity while preserving the on-target activity. 2′-OMe modification at position 2 from the 5′-end of the guide strand has been shown to reduce the seed regions mediated off-target activity (44). Deoxynucleotide substitution in the seed region has also been useful in reducing the off-target activity (45). Design of siRNA duplexes with high thermal stability is useful in reducing non-Ago2 mediated off-target activity (46). Selection of siRNAs guide strand seed region with low thermal stability may also be useful in lowering the miRNA-like off-target activity (6, 35). Recently studies from two independent groups have demonstrated that UNA monomers substitution in the seed region may be useful in eliminating the miRNA-like off-target activity of the guide strand without compromising siRNA on-target activity (9, 10). Using an mRNA expression profiling study, Vaish and coworkers further demonstrated that a combined placement of UNA at both the 3′-end of the passenger and the guide strands, at the 5′-end of the passenger strand, and one UNA in the seed region of the guide strand reduced the off-target activity of an siRNA by more than tenfold without reducing the on-target activity (10). Thus, an siRNA construct containing UNAs at the 3′-ends, one UNA at the 5′-end of the passenger strand and one UNA in the seed region (Fig. 1b) could be used as a general strategy for elimination of majority of off-target activity of siRNAs (see Notes 6 and 7).

3.6. Design of Luciferase Reporter Constructs

The off-target activity of the passenger strand and the miRNA-like off-target activity of the siRNA containing UNAs can be assessed using a luciferase assay by co-transfecting siRNAs with a plasmid construct carrying either the passenger strand target sequence or an miRNA-like target sequence cloned in the 3′-UTR of Renilla luciferase. Luciferase assays are carried out in a mammalian cell line. Dual-luciferase plasmid constructs are prepared by the following method:

1. To clone the guide strand target, choose a 150–200 nt long mRNA segment harboring the guide strand complementary sequence in the middle. Design a reverse complement sequence of the mRNA target segment and add a *Xho I* restriction sequence at the 5′-end and a *Not* I restriction sequence at the 3′end to allow cloning into the vector. Also synthesize a double-strand DNA molecule containing the passenger strand target, and a *Xho I* restriction site at the 5′-end and a *Not I* restriction site at the 3′-end. Alternately, a dsDNA containing

the appropriate target sequence can be prepared by polymerase chain reaction (PCR) using target mRNA specific PCR primers, followed by a nested PCR reaction using nested PCR primers containing *Xho I* and *Not I* sequences for cloning. To assess miRNA-like activity, design a target sequence containing only the seed-region complementary sequence plus 2–5 additional nucleotides in the 3′-half of the siRNA sequence. Add *Xho I* and *Not I* restriction sequences at the 5′- and 3′-end for cloning. Synthesize a corresponding dsDNA molecule. Two to four miRNA targets cloned in the 3′-UTR usually produces a good downregulation of the luciferase signal by the siRNA.

2. Digest the psiCHECK II vector with *Xho I* and *Not I* restriction enzymes present downstream of the Renilla sequence.
3. Remove the smaller digested segment by using small molecular weight cutoff filters.
4. Ligate the synthetic dsDNA molecules containing the siRNA or miRNA targets to the digested psiCHECK II plasmid using DNA ligase.
5. Transform the plasmid in suitable competent *E. coli* cells, and select for ampicillin resistance.
6. Select 1–2 colonies and prepare 5 mL cultures. Screen cultures for the presence of the psiCHECK plasmid with inserts by PCR.
7. Grow a 500 mL bacterial culture containing the psiCHECK plasmid with the target sequence.
8. Isolate the plasmid using a Maxiprep kit.
9. Identify the correct target sequence insert by DNA sequencing.

3.6.1. Dual Luciferase Assay

1. Seed HeLa cells at 7,500 cells per well in 100 μL growth media in a 96-well clear flat-bottom black Optiplate-96 one day before transfection. This should result in 70–80% confluency after 24 h growth.
2. The following day, thaw the siRNAs, transfection lipid mixture, and the plasmid containing the target.
3. Mix the siRNA (0.01–25 nM final) and plasmid DNA (75 ng/well) in OptiMEM. Dilute the Lipofectamine 2000 (0.2 μL/well for a 96-well plate) in OptiMEM and incubate at room temperature for 5 min.
4. Combine the siRNA–plasmid DNA solution with the lipofectamine-complex solution in a V-bottom 96-well plate, and incubate for 20 min at room temperature. Cover the plate to prevent evaporation.
5. Include a scrambled siRNA, and irrelevant plasmid DNA or untransfected cells as a control.

6. Add 50 μL of the siRNA-plasmid-lipid cocktail per well and incubate at 37°C for 24 h.
7. Remove 75 μL media and measure the Renilla and firefly luciferase activities using Promega's luciferase detection kit according to the manufacturer's protocol. Alternatively, the assay can be stopped at this stage and plates frozen at −80°C for later use.

Measuring Firefly Luciferase Activity (Based on Kit's Instructions)

1. Prepare the "working" Dual-Glo Luciferase reagent by gently mixing the contents of one bottle of Dual-Glo Luciferase Buffer to one bottle of Dual-Glo Luciferase substrate. Ensure that the substrate is completely dissolved. Store the "working" reagent in aliquots at -20°C.
2. Equilibrate an aliquot of premade "working" Dual-Glo Luciferase reagent to room temperature.
3. Bring the transfection plate to room temperature.
4. Add 75 μL of "working" Dual-Glo Luciferase reagent into each well containing the cells and 75 μL of culture medium, and mix by pipetting without generating bubbles.
5. Gently rock the plate at room temperature for 10 min (no longer than 2 h).
6. Measure the luminescence using a luminometer.

Measuring Renilla Luciferase Activity

1. Dilute the Dual-Glo Stop & Glo Substrate 1:100 into an appropriate volume of Dual-Glo Stop & Glo Buffer in a fresh falcon tube.
2. Add 75 μL of Dual-Glo Stop & Glo Reagent to each well.
3. Mix thoroughly.
4. Seal plate with a foil cover and rock gently at room temperature for 10 min (no longer than 2 h).
5. Measure the luminescence.
6. Calculate the ratio of Renilla/firefly luciferase activity from averages of triplicate reactions. Calculate the percentage of siRNA activity using the formula (1 – siRNA activity/negative control activity).

3.7. Assessment of In Vitro Cytokine Responses

3.7.1. Isolation and Transfection of Human Peripheral Blood Mononuclear Cells

Peripheral blood mononuclear cells (PBMCs) are isolated using a Ficoll-HyPaque density-gradient method that separates lymphocytes from other components in the blood. The blood sample is layered onto a Ficoll-sodium metrizoate gradient of specific density. Following centrifugation, lymphocytes are collected from the plasma–Ficoll interface (see Notes 8 and 9).

1. Dilute 40 mL of human blood with 80 mL PBS (without Ca^{2+}, Mg^{2+}). Set up four 50 mL conical tubes and add 15 mL Ficoll per tube.

2. Gently layer 30 mL blood–PBS mix on top of the Ficoll layer (hold the tube at an angle and layer the blood–PBS mix over the Ficoll from the side of tube to avoid direct mixing of blood with Ficoll).
3. Centrifuge at 600 rcf for 30 min at RT without brakes.
4. Carefully collect the buffy white thin layer below the plasma into a 50 mL conical tube.
5. Add PBS up to 50 mL. Centrifuge at 400 × g for 10 min at RT.
6. After the second wash, resuspend the cells in 30 mL of Iscove's DMEM (IMDM) with 10% FBS, 1× NEAA, 2 mM Glutamine, 100 U/mL penicillin, and 100 μg/mL streptomycin.
7. Count the cells using a hemocytometer and seed isolated PBMCs 1 day prior to the transfection assay in triplicate at a density of 200,000 cells per well in a flat-bottom 96-well plate in 100 μL IMDM.
8. Disinfect all Ficoll tubes with 10% bleach for 30 min prior to disposal.
9. After 24 h, transfect cells by adding 20 μL of siRNA–lipid complexes (outlined below) directly into the 100 μL of growth media. Only inner wells are used for transfections.
10. Dilute 0.25 μL/well of RNAiMAX in 10 μL OptiMEM and separately prepare a 12× concentration of siRNA in 10 μL OptiMEM, for 120 μL final transfection volume. Incubate siRNA and RNAiMAX solutions for 5 min at room temperature prior to mixing together. Incubate the siRNA–RNAiMAX mixture for an additional 20 min at room temperature, followed by addition to each well. After 3 h incubation at 37°C, cells are supplemented with an additional 80 μL/well of 10% FBS/IMDM. Incubate the plates for another 24 h at 37°C.
11. Briefly centrifuge the transfection plate at 700 × *g* at room temperature to pellet any debris and cells. Collect the supernatants into a new V-bottom 96-well plate and store at −80°C until the ELISA is performed.

3.7.2. Detection of Human Interferon-α and Human Interferon-β Using ELISA

siRNAs have been shown to be highly immuno-stimulatory, promoting inflammatory cytokine production including type I interferon. This secondary effect of siRNA can complicate the interpretation of biological assays such as those involving viruses and may hamper the development of siRNAs as therapeutics (14). To assess whether siRNA and corresponding modified siRNAs containing UNAs induces type I interferons in vitro, human peripheral mononuclear cells are transfected with siRNAs and assayed for interferon-α (IFN-α) and interferon-β (IFN-β) levels 24 h post transfection. ELISA is performed using PBL Biomedical Human Interferon-α and Interferon-β kits from R&D Systems according to manufacturer's instructions.

1. Prepare test samples of unknown interferon concentration using dilution buffer as required. Measurements in duplicate are recommended. Refrigerate until use.
2. Construct a high sensitivity standard curve ranging from 12.5 to 500 pg/mL or an extended range standard curve ranging from 156 to 5,000 pg/mL for human IFN-α and a standard curve ranging from 25 to 2,000 pg/mL for human IFN-β by diluting the Interferon standard samples in dilution buffer.
3. Dilute antibody concentrates with dilution buffer. Refer to the lot specific Certificate of Analysis (COA) for the correct amounts of antibody solution to prepare. Keep on ice until use.
4. Dilute HRP conjugate concentrate with HRP conjugate diluent. Keep on ice until use.
5. All incubations should be performed in a closed chamber at room temperature. During all wash steps remove the contents of the plate by inverting and blotting the plate on lint-free absorbent paper; tap the plate dry. All wells should be filled with a minimum of 250 μL of diluted wash buffer.
6. Add 100 μL per well of the samples, interferon standards, and blanks. Cover and incubate for 1 h at room temperature. Empty the contents of the plate and wash the wells once with wash buffer.
7. Add 100 μL of diluted anti-interferon secondary antibody solution to all wells. Cover and incubate for 1 h. Empty the contents of the plate and wash the wells three times with wash buffer.
8. Add 100 μL of diluted HRP conjugated anti-secondary interferon antibody solution to all wells. Cover and incubate for 1 h. During this incubation period, warm the TMB substrate solution to room temperature (22–25°C). Empty the contents of the plate and wash the wells four times with diluted wash buffer.
9. Add 100 μL of the TMB substrate solution to each well. Incubate in the dark for 15 min. Do not use a plate sealer during this incubation.
10. After the 15 min incubation of TMB, add 100 μL of stop solution to each well.
11. Using a microplate reader, determine the absorbance at 450 nm within 5 min after the addition of the stop solution.
12. Determine the interferon titers by plotting the optical densities (OD) using a four-parameter fit for the standard curve. Subtract blank ODs from the standards and sample ODs to eliminate background. The interferon values from the curves can be determined in units/mL as well as pg/mL.

3.7.3. Cytokine Profiling Using a Procarta™ Assay

Procarta's cytokine profile kit uses the xMAP® technology (multi-analyte profiling beads) and is a powerful method to detect multiple proinflammatory cytokines in a 96-well format. The protocol uses a combination of a flow cytometer, fluorescent-dyed microspheres (beads), lasers, and digital signal processing to quantify multiplex cytokine proteins within a single sample.

1. Prepare 1× Wash Buffer by diluting 20 mL of the 10× buffer into 180 mL of deionized water. Prepare serial dilutions of the premixed standard ranging from 2 pg/mL to 20 ng/mL.
2. Pre-wet the filter plate by adding 150 μL of reading buffer to each well. Incubate for 5 min at room temperature and remove the buffer by vacuum filtration.
3. Vortex the premixed antibody beads for 30 s at room temperature. Add 50 μL of antibody beads to each standard and sample well. Remove the buffer using vacuum filtration.
4. Wash beads with 150 μL of 1× Wash Buffer. Thoroughly remove the residual buffer from the bottom of the filter plate by blotting on paper towels.
5. Add 25 μL of standards and samples plus 25 μL of assay buffer to the appropriate wells. Seal the plate and gently rock for 30 min at 150 × *g* at room temperature.
6. Wash the plate three times with Wash Buffer.
7. Add 25 μL/well of the detection antibody. Seal the Filter Plate and rock for another 30 min at 150 × *g* at room temperature. Wash the plate three times.
8. Add 50 μL/well of streptavidin-PE. Seal the plate and gently shake for 30 min at 150 × *g* at room temperature.
9. Remove the buffer by vacuum filtration and wash the plate. Add 120 μL/well of the reading buffer. Shake the plate for an additional 5 min at 150 × *g* at room temperature. Analyze on a Luminex instrument that has been calibrated appropriately.
10. Calculate the fold change in the cytokine levels in the supernatants of the cell cultures treated with siRNA as compared to the untreated control.

3.8. Microarray Analysis to Determine Global Off-Target Events

The artificial luciferase assay system has limited capability to assess the true off-target capability of an siRNA molecule. Assessment of the global off-target potential using microarray analysis incorporating the known transcriptome could determine the extent of the off-target signature of siRNA drug candidates. The microarray analysis can be performed using total RNA isolated from an siRNA treated and a mock transfected cell culture in order to measure the differential expression of genes between the two sample types.

3.8.1. Microarray Protocol and Data Analysis

1. Transfection for each condition should be carried out in triplicate to represent three biological replicates. To determine the impact of modification on the siRNA off-target events, transfect siRNA, modified siRNAs, and a mock-transfection without siRNA in three biological replicates.
2. Seed 500,000 mammalian cells, for example HepG2, in each well of a 6-well plate in 2.5 mL of growth medium 1 day before transfection.
3. The following day, when cells are 70–80% confluent, thaw siRNAs and reagents. Before transfection, the growth media is removed and replaced with 2 mL/well serum-free OptiMEM, followed by the addition of 0.5 mL of siRNA–RNAiMAX complex (see Note 10).
4. After 4 h incubation, remove the transfection media and replace with the growth media. Continue incubation for another 20 h.
5. After 24 h, completely remove the growth media, and isolate total RNA using an RNeasy Mini Kit (Qiagen), following the manufacturer's protocol. Dissolve the RNA in deionized water. If the RNA is prepared at a later time, cells can be frozen on dry ice at this stage without growth medium. Thaw cells on ice before lysis.
6. Confirm the quality of RNA sample using a NanoDrop. The ratio of 260/280 absorbances should be 1.8 to ensure high quality RNA.
7. Determine the concentration of total RNA and equalize each sample to the same final concentration.
8. Confirm the extent of target RNA knockdown in each sample by qRT-PCR to ensure that transfection of siRNAs in each condition was efficient.
9. Prepare the biotin labeled cRNA from total RNA using a Two-Cycle cDNA synthesis kit according to the manufacturer's protocol.
10. Hybridize the labeled cRNA on a microarray chip in a hybridization oven for 16 h at 45°C. Include a hybridization control as a quality control.
11. After incubation, wash and stain the chip according to the protocol recommended by Affymetrix.
12. Scan the chip to generate raw microarray data.
13. Raw data is normalized using the GC-RMA package in R/Bioconductor.
14. Calculate the fold change and *P*-values from the average of three biological samples, and use the average of the mock treated no siRNA control as a base line.
15. Plot the heat maps, volcano plots, and Venn diagrams to determine the statistically significant off-target events.

3.9. Strategies for Chemical Modification to Improve Nuclease Stability

Incorporation of UNA substitutions at strategic positions in the siRNA duplex can be beneficial in reducing nonspecific activity. Further chemical modification of siRNA may still be needed to protect it from nucleases and in vivo activity. In stable cell lines siRNA stabilization seems to provide no advantage over unmodified siRNA (47). Chemical modification of siRNA to enhance serum stability has been successful for in vivo applications (22, 23, 48–50). Nuclease and chemical stability of siRNAs for in vivo application should be optimized empirically by incorporating modifications such as Deoxy, UNA, 2′-OMe, 2′-F, and locked nucleic acid in specific positions that would abrogate off-target activity and immune induction while retaining RNAi activity in vivo (51–54).

3.10. Clinical Significance

With increased understanding of the RNAi mechanism and significant advances made in delivery approaches (55), siRNA technology has been quickly applied to clinical settings (56). While siRNAs have a potential to interfere with any gene target, they are particularly appealing as drugs for neurological disorders requiring allele specific silencing such as dystonia (57) and Huntington's diseases (58, 59). For these examples, siRNAs were used to silence disease specific mRNAs in vitro. Because siRNAs can silence genes with limited homology (60), and UNA modifications destabilize the RNAs duplex (61) and increase the specificity without loss in efficacy (10), UNA technology may be useful for more effective silencing of allele specific mRNA. Using exosome mediated delivery (62), it may be possible to deliver allele specific siRNAs across the blood–brain barrier in clinical settings.

4. Notes

1. siRNA design tools widely accessible via World Wide Web include the following: siDESIGN Center (http://www.dharmacon.com/designcenter/DesignCenterPage.aspx), Block-iT RNAi (https://rnaidesigner.invitrogen.com/rnaiexpress/?), DSIR (http://biodev.extra.cea.fr/DSIR/DSIR.html), OptiRNA (http://optirna.unl.edu/), siRNA selection server (http://jura.wi.mit.edu/bioc/siRNAext/), siRNA Target Finder (http://www.ambion.com/techlib/misc/siRNA_finder.html), siDirect (http://sidirect2.rnai.jp/), RNAxs (http://rna.tbi.univie.ac.at/cgi-bin/RNAxs), City of Hope design center (http://www1.infosci.coh.org/hpcdispatcher/siRNA.aspx#sequence), Oligo Walk (http://rna.urmc.rochester.edu/cgi-bin/server_exe/oligowalk/oligowalk_form.cgi). Each of the above siRNA design tools utilizes various overlapping sets of rules for selection of active siRNAs and has their limitations. siDESIGN Center utilizes duplex asymmetry rules and also

checks off-target potential of sequences by BLAST search. RNAxs utilizes local secondary structure of mRNAs to account for target accessibility.

2. In assessing the off-target potential of siRNAs, adjust the word size to 7 from the default value of 11 to allow detection of shorter regions of sequence similarity. Because both sense and antisense strands of siRNA duplexes can engage into off-targeting, Blastn analysis should be carried out for both strands.
3. Extreme care should be taken in handling unmodified siRNAs and RNA in general to prevent degradation by RNase contamination. RNase AWAY® reagent can be highly beneficial to remove RNase contamination from work surfaces. Use of disposable, sterile plasticware, and sterile pipette tips with aerosol filters are advisable. Always wear gloves while handling reagents, plasticware, and RNA samples to prevent RNase contamination from the surface of the skin. When working with RNA, use proper microbiological aseptic technique.
4. For RNA isolation from cells and animal tissue samples, it is good practice to complete the procedure once started. For example, storing the lysate in RNA lysis solution at −80°C adversely affects the quality of RNA.
5. After dispensing the sample mix into a PCR plate for real-time PCR, remove air bubbles by centrifuging the plate at 600 × g or 1–2 min.
6. Dicer substrates are 25/27 mer siRNA duplexes and depend on the Dicer enzyme for processing to siRNA duplexes of 21/21 nt. Because the design of Dicer substrates relies on first identifying the active 21/21 mer siRNA duplexes, and they share the off-target issues associated with standard siRNAs (28), some of the design principles of siRNAs containing UNAs to eliminate the seed region mediated off-targeting can, in principle, be incorporated into the Dicer siRNA design. The effect of UNA substitution in the seed region of the putative Dicer product should be first evaluated on the processing of the Dicer substrate by the Dicer enzyme.
7. Both Bramsen et al. (9) and Vaish et al. (10) demonstrated that UNA at position 7 of the guide strand was useful in eliminating the miRNA-like off-target activity; however, for each siRNA the optimal position for the substitution of UNA in the seed region should be determined by the on-target vs. off-target activity ratio by substituting a UNA at positions 4–8 in the seed region of the guide strand.
8. Human blood samples should be handling in accordance with the institutional policy and special care should be taken. Human blood should be obtained from trusted sources and should be prescreened for contagious pathogens such as HIV and HCV.

9. The in vitro optimized siRNA with improved drug-like properties should be tested in an animal model to assess potency and cytokine stimulatory potential in context with the delivery vehicle appropriate for in vivo delivery of the siRNA. In vivo cytokine induction should be assessed for the interferon alpha and beta, and proinflammatory cytokines in the animal blood and the target tissue.
10. RNAiMAX should be used for transfecting siRNA in microarray analyses. Mock treated cells with RNAiMAX show very little change in the global gene expression profile when compared to the untreated cells. Use of Lipofectamine 2000 should be avoided in microarray analysis. Lipofectamine is quite toxic to cells and drastically changes the global gene expression profile compared to the untreated cells, thus complicating the detection of changes in gene expression profiles between unmodified and modified siRNAs.

References

1. Fire A, Xu S, Montgomery MK, Kostas SA, Driver SE, Mello CC (1998) Potent and specific genetic interference by double-stranded RNA in Caenorhabditis elegans. Nature 391:806–811
2. Elbashir SM, Harborth J, Lendeckel W, Yalcin A, Weber K, Tuschl T (2001) Duplexes of 21-nucleotide RNAs mediate RNA interference in cultured mammalian cells. Nature 411:494–498
3. Burnett JC, Rossi JJ, Tiemann K (2011) Current progress of siRNA/shRNA therapeutics in clinical trials. Biotechnol J 6:1130–1146
4. Davidson BL, McCray PB Jr (2011) Current prospects for RNA interference-based therapies. Nat Rev Genet 12:329–340
5. Pecot CV, Calin GA, Coleman RL, Lopez-Berestein G, Sood AK (2011) RNA interference in the clinic: challenges and future directions. Nat Rev Cancer 11:59–67
6. Naito Y, Yoshimura J, Morishita S, Ui-Tei K (2009) siDirect 2.0: updated software for designing functional siRNA with reduced seed-dependent off-target effect. BMC Bioinformatics 10:392
7. Jackson AL, Bartz SR, Schelter J, Kobayashi SV, Burchard J, Mao M, Li B, Cavet G, Linsley PS (2003) Expression profiling reveals off-target gene regulation by RNAi. Nat Biotechnol 21:635–637
8. Jackson AL, Burchard J, Schelter J, Chau BN, Cleary M, Lim L, Linsley PS (2006) Widespread siRNA "off-target" transcript silencing mediated by seed region sequence complementarity. RNA 12:1179–1187
9. Bramsen JB, Pakula MM, Hansen TB, Bus C, Langkjaer N, Odadzic D, Smicius R, Wengel SL, Chattopadhyaya J, Engels JW et al (2010) A screen of chemical modifications identifies position-specific modification by UNA to most potently reduce siRNA off-target effects. Nucleic Acids Res 38:5761–5773
10. Vaish N, Chen F, Seth S, Fosnaugh K, Liu Y, Adami R, Brown T, Chen Y, Harvie P, Johns R et al (2011) Improved specificity of gene silencing by siRNAs containing unlocked nucleobase analogs. Nucleic Acids Res 39:1823–1832
11. Hornung V, Guenthner-Biller M, Bourquin C, Ablasser A, Schlee M, Uematsu S, Noronha A, Manoharan M, Akira S, de Fougerolles A et al (2005) Sequence-specific potent induction of IFN-alpha by short interfering RNA in plasmacytoid dendritic cells through TLR7. Nat Med 11:263–270
12. Judge AD, Sood V, Shaw JR, Fang D, McClintock K, Maclachlan I (2005) Sequence-dependent stimulation of the mammalian innate immune response by synthetic siRNA. Nat Biotechnol 23:457–462
13. Reynolds A, Anderson EM, Vermeulen A, Fedorov Y, Robinson K, Leake D, Karpilow J, Marshall WS, Khvorova A (2006) Induction of the interferon response by siRNA is cell type- and duplex length-dependent. RNA 12:988–993
14. Kleinman ME, Yamada K, Takeda A, Chandrasekaran V, Nozaki M, Baffi JZ,

Albuquerque RJ, Yamasaki S, Itaya M, Pan Y et al (2008) Sequence- and target-independent angiogenesis suppression by siRNA via TLR3. Nature 452:591–597

15. Marques JT, Devosse T, Wang D, Zamanian-Daryoush M, Serbinowski P, Hartmann R, Fujita T, Behlke MA, Williams BR (2006) A structural basis for discriminating between self and nonself double-stranded RNAs in mammalian cells. Nat Biotechnol 24:559–565
16. Hornung V, Ellegast J, Kim S, Brzozka K, Jung A, Kato H, Poeck H, Akira S, Conzelmann KK, Schlee M et al (2006) 5′-Triphosphate RNA is the ligand for RIG-I. Science 314:994–997
17. Pichlmair A, Schulz O, Tan CP, Naslund TI, Liljestrom P, Weber F, Reis e Sousa C (2006) RIG-I-mediated antiviral responses to single-stranded RNA bearing 5′-phosphates. Science 314:997–1001
18. Kato H, Takeuchi O, Mikamo-Satoh E, Hirai R, Kawai T, Matsushita K, Hiiragi A, Dermody TS, Fujita T, Akira S (2008) Length-dependent recognition of double-stranded ribonucleic acids by retinoic acid-inducible gene-I and melanoma differentiation-associated gene 5. J Exp Med 205:1601–1610
19. Robbins M, Judge A, Ambegia E, Choi C, Yaworski E, Palmer L, McClintock K, MacLachlan I (2008) Misinterpreting the therapeutic effects of small interfering RNA caused by immune stimulation. Hum Gene Ther 19:991–999
20. Cho WG, Albuquerque RJ, Kleinman ME, Tarallo V, Greco A, Nozaki M, Green MG, Baffi JZ, Ambati BK, De Falco M et al (2009) Small interfering RNA-induced TLR3 activation inhibits blood and lymphatic vessel growth. Proc Natl Acad Sci U S A 106:7137–7142
21. Judge AD, Bola G, Lee AC, MacLachlan I (2006) Design of noninflammatory synthetic siRNA mediating potent gene silencing in vivo. Mol Ther 13:494–505
22. Morrissey DV, Lockridge JA, Shaw L, Blanchard K, Jensen K, Breen W, Hartsough K, Machemer L, Radka S, Jadhav V et al (2005) Potent and persistent in vivo anti-HBV activity of chemically modified siRNAs. Nat Biotechnol 23:1002–1007
23. Frank-Kamenetsky M, Grefhorst A, Anderson NN, Racie TS, Bramlage B, Akinc A, Butler D, Charisse K, Dorkin R, Fan Y et al (2008) Therapeutic RNAi targeting PCSK9 acutely lowers plasma cholesterol in rodents and LDL cholesterol in nonhuman primates. Proc Natl Acad Sci U S A 105:11915–11920
24. Mantei A, Rutz S, Janke M, Kirchhoff D, Jung U, Patzel V, Vogel U, Rudel T, Andreou I, Weber M et al (2008) siRNA stabilization prolongs gene knockdown in primary T lymphocytes. Eur J Immunol 38:2616–2625
25. Langkjær N, Pasternak A, Wengel J (2009) UNA (unlocked nucleic acid): a flexible RNA mimic that allows engineering of nucleic acid duplex stability. Bioorg Med Chem 17:5420–5425
26. Kim DH, Behlke MA, Rose SD, Chang MS, Choi S, Rossi JJ (2005) Synthetic dsRNA Dicer substrates enhance RNAi potency and efficacy. Nat Biotechnol 23:222–226
27. Siolas D, Lerner C, Burchard J, Ge W, Linsley PS, Paddison PJ, Hannon GJ, Cleary MA (2005) Synthetic shRNAs as potent RNAi triggers. Nat Biotechnol 23:227–231
28. Amarzguioui M, Rossi JJ (2008) Principles of Dicer substrate (D-siRNA) design and function. Methods Mol Biol 442:3–10
29. Czauderna F, Fechtner M, Dames S, Ayguen H, Klippel A, Pronk GJ, Giese K, Kaufmann J (2003) Structural variations and stabilising modifications of synthetic siRNAs in mammalian cells. Nucleic Acids Res 31:2705–2716
30. Hohjoh H (2002) RNA interference (RNAi) induction with various types of synthetic oligonucleotide duplexes in cultured human cells. FEBS Lett 521:195–199
31. Bramsen JB, Laursen MB, Damgaard CK, Lena SW, Babu BR, Wengel J, Kjems J (2007) Improved silencing properties using small internally segmented interfering RNAs. Nucleic Acids Res 35:5886–5897
32. Reynolds A, Leake D, Boese Q, Scaringe S, Marshall WS, Khvorova A (2004) Rational siRNA design for RNA interference. Nat Biotechnol 22:326–330, Epub 2004 Feb 2001
33. Jagla B, Aulner N, Kelly PD, Song D, Volchuk A, Zatorski A, Shum D, Mayer T, De Angelis DA, Ouerfelli O et al (2005) Sequence characteristics of functional siRNAs. RNA 11:864–872
34. Matveeva O, Nechipurenko Y, Rossi L, Moore B, Saetrom P, Ogurtsov AY, Atkins JF, Shabalina SA (2007) Comparison of approaches for rational siRNA design leading to a new efficient and transparent method. Nucleic Acids Res 35:e63
35. Ui-Tei K, Naito Y, Nishi K, Juni A, Saigo K (2008) Thermodynamic stability and Watson-Crick base pairing in the seed duplex are major determinants of the efficiency of the siRNA-based off-target effect. Nucleic Acids Res 36:7100–7109
36. Chang CI, Yoo JW, Hong SW, Lee SE, Kang HS, Sun X, Rogoff HA, Ban C, Kim S, Li CJ et al (2009) Asymmetric shorter-duplex siRNA structures trigger efficient gene silencing with

reduced nonspecific effects. Mol Ther 17: 725–732

37. Sun X, Rogoff HA, Li CJ (2008) Asymmetric RNA duplexes mediate RNA interference in mammalian cells. Nat Biotechnol 26: 1379–1382
38. Chen PY, Weinmann L, Gaidatzis D, Pei Y, Zavolan M, Tuschl T, Meister G (2008) Strand-specific 5′-O-methylation of siRNA duplexes controls guide strand selection and targeting specificity. RNA 14:263–274
39. Kenski DM, Cooper AJ, Li JJ, Willingham AT, Haringsma HJ, Young TA, Kuklin NA, Jones JJ, Cancilla MT, McMasters DR et al (2010) Analysis of acyclic nucleoside modifications in siRNAs finds sensitivity at position 1 that is restored by 5′-terminal phosphorylation both in vitro and in vivo. Nucleic Acids Res 38:660–671
40. Lin X, Ruan X, Anderson MG, McDowell JA, Kroeger PE, Fesik SW, Shen Y (2005) siRNA-mediated off-target gene silencing triggered by a 7 nt complementation. Nucleic Acids Res 33:4527–4535
41. Lewis BP, Burge CB, Bartel DP (2005) Conserved seed pairing, often flanked by adenosines, indicates that thousands of human genes are microRNA targets. Cell 120:15–20
42. Birmingham A, Anderson EM, Reynolds A, Ilsley-Tyree D, Leake D, Fedorov Y, Baskerville S, Maksimova E, Robinson K, Karpilow J et al (2006) 3′ UTR seed matches, but not overall identity, are associated with RNAi off-targets. Nat Methods 3:199–204
43. Aleman LM, Doench J, Sharp PA (2007) Comparison of siRNA-induced off-target RNA and protein effects. RNA 13:385–395
44. Jackson AL, Burchard J, Leake D, Reynolds A, Schelter J, Guo J, Johnson JM, Lim L, Karpilow J, Nichols K et al (2006) Position-specific chemical modification of siRNAs reduces "off-target" transcript silencing. RNA 12:1197–1205
45. Ui-Tei K, Naito Y, Zenno S, Nishi K, Yamato K, Takahashi F, Juni A, Saigo K (2008) Functional dissection of siRNA sequence by systematic DNA substitution: modified siRNA with a DNA seed arm is a powerful tool for mammalian gene silencing with significantly reduced off-target effect. Nucleic Acids Res 36:2136–2151
46. Petri S, Dueck A, Lehmann G, Putz N, Rudel S, Kremmer E, Meister G (2011) Increased siRNA duplex stability correlates with reduced off-target and elevated on-target effects. RNA 17:737–749
47. Layzer JM, McCaffrey AP, Tanner AK, Huang Z, Kay MA, Sullenger BA (2004) In vivo activity of nuclease-resistant siRNAs. RNA 10: 766–771
48. Soutschek J, Akinc A, Bramlage B, Charisse K, Constien R, Donoghue M, Elbashir S, Geick A, Hadwiger P, Harborth J et al (2004) Therapeutic silencing of an endogenous gene by systemic administration of modified siRNAs. Nature 432:173–178
49. Morrissey DV, Blanchard K, Shaw L, Jensen K, Lockridge JA, Dickinson B, McSwiggen JA, Vargeese C, Bowman K, Shaffer CS et al (2005) Activity of stabilized short interfering RNA in a mouse model of hepatitis B virus replication. Hepatology 41:1349–1356
50. Zimmermann TS, Lee AC, Akinc A, Bramlage B, Bumcrot D, Fedoruk MN, Harborth J, Heyes JA, Jeffs LB, John M et al (2006) RNAi-mediated gene silencing in non-human primates. Nature 441:111–114
51. Bramsen JB, Kjems J (2011) Chemical modification of small interfering RNA. Methods Mol Biol 721:77–103
52. Chiu YL, Rana TM (2003) siRNA function in RNAi: a chemical modification analysis. RNA 9:1034–1048
53. Czauderna F, Fechtner M, Aygun H, Arnold W, Klippel A, Giese K, Kaufmann J (2003) Functional studies of the PI(3)-kinase signalling pathway employing synthetic and expressed siRNA. Nucleic Acids Res 31:670–682
54. Amarzguioui M, Holen T, Babaie E, Prydz H (2003) Tolerance for mutations and chemical modifications in a siRNA. Nucleic Acids Res 31:589–595
55. Love KT, Mahon KP, Levins CG, Whitehead KA, Querbes W, Dorkin JR, Qin J, Cantley W, Qin LL, Racie T et al (2010) Lipid-like materials for low-dose, in vivo gene silencing. Proc Natl Acad Sci U S A 107: 1864–1869
56. Kim DH, Rossi JJ (2007) Strategies for silencing human disease using RNA interference. Nat Rev Genet 8:173–184
57. Gonzalez-Alegre P, Miller VM, Davidson BL, Paulson HL (2003) Toward therapy for DYT1 dystonia: allele-specific silencing of mutant torsinA. Ann Neurol 53:781–787
58. Pfister EL, Kennington L, Straubhaar J, Wagh S, Liu W, DiFiglia M, Landwehrmeyer B, Vonsattel JP, Zamore PD, Aronin N (2009) Five siRNAs targeting three SNPs may provide therapy for three-quarters of Huntington's disease patients. Curr Biol 19:774–778

59. Hu J, Liu J, Corey DR (2010) Allele-selective inhibition of huntingtin expression by switching to an miRNA-like RNAi mechanism. Chem Biol 17:1183–1188
60. Dahlgren C, Zhang HY, Du Q, Grahn M, Norstedt G, Wahlestedt C, Liang Z (2008) Analysis of siRNA specificity on targets with double-nucleotide mismatches. Nucleic Acids Res 36:e53
61. Pasternak A, Wengel J (2010) Thermodynamics of RNA duplexes modified with unlocked nucleic acid nucleotides. Nucleic Acids Res 38:6697–6706
62. Alvarez-Erviti L, Seow Y, Yin H, Betts C, Lakhal S, Wood MJ (2011) Delivery of siRNA to the mouse brain by systemic injection of targeted exosomes. Nat Biotechnol 29: 341–345

Chapter 7

The Design, Preparation, and Evaluation of Asymmetric Small Interfering RNA for Specific Gene Silencing in Mammalian Cells

Chanil Chang, Sun Woo Hong, Pooja Dua, Soyoun Kim, and Dong-ki Lee

Abstract

RNA interference (RNAi) is a highly efficient endogenous gene silencing mechanism mediated by short double-stranded RNAs termed small interfering RNAs (siRNAs). The current standard siRNA structure, which is used by most researchers to trigger sequence-specific target gene silencing, consists of a double strand region of 19 bp with 2 nt 3′-overhangs at both ends. However, in addition to the desired target gene silencing, this conventional siRNA structure also exhibits several unintended effects that constitute obstacles to the use of siRNA in gene function studies and therapeutics development. Here, we provide protocols for designing and preparing an alternative structure for RNAi trigger, termed asymmetric shorter-duplex RNA (asiRNA). The asiRNA structure has a duplex region shorter than 19 bp and has an asymmetric 3′-overhang structure. Importantly, the asiRNA structure not only triggers efficient target gene silencing comparable to that of the 19 bp standard siRNA structure but also significantly reduces nonspecific effects triggered by 19 bp siRNAs such as sense-strand-mediated off-target silencing and the saturation of RNAi machinery. Procedures are described for verifying that asiRNA activates gene silencing through an Ago2-dependent pathway and for assessing the miRNA pathway competition potency and specific and nonspecific silencing abilities of asiRNAs. We propose that asiRNA, an improved RNAi trigger that can overcome the nonspecific effects evoked by standard siRNA structures, can be developed as a precise and effective tool for both functional genomics and therapeutic applications.

Key words: RNAi, siRNA, asiRNA, Off-target silencing, RNAi machinery saturation

1. Introduction

RNA interference (RNAi) is a powerful, endogenous post-transcriptional gene silencing mechanism which is present in most eukaryotic organisms (1). In mammalian cells, sequence-specific target gene silencing via an RNAi mechanism is executed by exogenously introduced small interfering RNAs (siRNAs) (2).

Debra J. Taxman (ed.), *siRNA Design: Methods and Protocols*, Methods in Molecular Biology, vol. 942, DOI 10.1007/978-1-62703-119-6_7, © Springer Science+Business Media, LLC 2013

An early structure–activity relationship study conducted in *Drosophila melanogaster* embryo lysates identified double-stranded (ds) RNAs with 19 bp duplexes and 2 nt 3′-overhangs at both ends as the most efficient RNAi triggers (3). In addition, the study revealed that siRNAs with different overhang structures, such as 5′ overhangs or blunt ends, and siRNAs with duplex region shorter than 19 bp were inefficient. Henceforth, this 19 bp siRNA structure was accepted as the standard structure to carry out mammalian gene silencing by most researchers in this field. Since the development of this siRNA structure, which could induce specific gene silencing in mammalian cells, RNAi-based gene silencing technology has become one of the most popular methods for gene function studies and therapeutic applications (4–6).

Despite the significant promise of RNAi technology, several studies have reported a number of unexpected adverse effects triggered by siRNA structures. First, siRNA can induce downregulation of nontarget genes, termed as off-target silencing (7, 8). Off-target silencing can be caused by several distinct mechanisms. The antisense (guide) strand of siRNA could silence nontarget mRNAs complementary to the seed sequence (defined as the 2nd to 7th nucleotides from the 5′-end of the antisense strand) via imperfect base pairing. Also, the sense (passenger) strand of siRNA could be incorporated into the RNA-induced silencing complex (RISC) and guide cleavage of fully or partially complementary nontarget mRNA (9). Second, introduction of excess amounts of siRNA into cells can saturate the cellular RNAi machinery and interfere with the function of endogenous microRNAs or other siRNAs (10–13). Finally, although the 19 bp siRNA structure was originally designed to circumvent long dsRNA-mediated antiviral responses in mammalian cells, several studies have reported that nonspecific immune responses can be activated by siRNAs (14).

For the development of safe RNAi therapeutics and precise gene function studies using siRNAs, it is crucial to achieve specific target gene silencing with minimal nonspecific effects. Although a variety of chemical modification strategies have been developed that overcome some of the siRNA associated nonspecific effects (15, 16), only a few studies have reported siRNA backbone modifications to alleviate nonspecific effects.

Herein, we describe methods for designing a novel RNAi-triggering molecular structure termed asymmetric shorter-duplex siRNA (asiRNA) (17). Compared with the standard 19 bp siRNA structure, asiRNA has a shortened sense strand of 15–16 nt in length, which results in a shortened dsRNA region. Also, the 5′-end of the antisense strand is blunt-ended, whereas the 3′-end of the antisense strand has a 3–6 nt overhang, which makes the RNA structure asymmetric in contrast to the symmetric standard 19 bp siRNA (Fig. 1). asiRNAs showed comparable gene silencing activity when compared with 19 bp siRNAs, and like siRNAs, asiRNAs executed target gene silencing via RNAi mechanisms, as confirmed

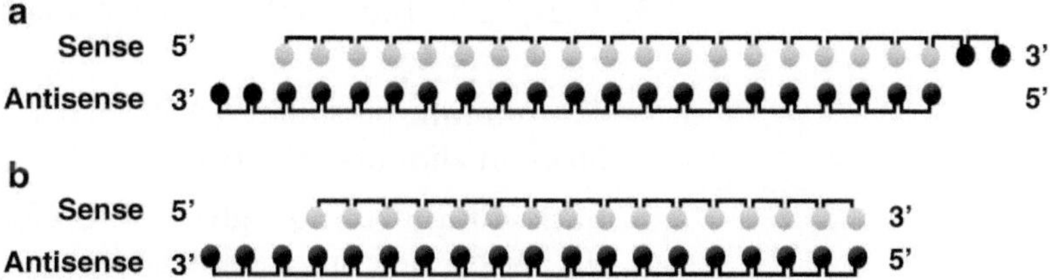

Fig. 1. Structure of conventional siRNA and asiRNA. (**a**) Structure of conventional siRNA. (**b**) Structure of asiRNA.

by Argonaute-2 (Ago2) dependency and the analysis of mRNA cleavage products by 5′-RACE assay (18).

Importantly, we found that asiRNA structure significantly reduces nonspecific effects caused by 19 bp siRNA structures, such as sense-strand-mediated off-target gene silencing and saturation of the cellular RNAi machinery (9, 19). Because the asiRNA structure only has backbone modifications, further chemical modification of asiRNAs can be expected to add additional benefits such as reduction in TLR7/8 activation (20) and increase in stability (21). The increased specificity of asiRNA makes it an excellent candidate tool for next generation gene silencing research and therapeutics development.

Methods are described in this article for optimizing the structure of asiRNA to trigger efficient gene silencing. We also describe methods for verifying the Ago2 dependency and reduced off-target gene silencing of asiRNA-mediated gene silencing.

2. Materials

2.1. Annealing of siRNAs and Asymmetric siRNAs

1. 5× siRNA buffer (Thermo): 100 mM KCl, 30 mM HEPES (pH = 7.5), and 1 mM $MgCl_2$.
2. RNase free water and buffers. Add 1 ml diethylpyrocarbonate (DEPC) to 1,000 ml of water or buffers to be treated and stir it overnight with a magnetic stirrer at room temperature. Autoclave for 15 min to inactivate the DEPC. Store in aliquots at −20°C.
3. The antisense strand and sense strand RNAs for siRNAs and asymmetric siRNAs are chemically synthesized and purified with a column (Bioneer). Make stock solutions of each (100 μM) using RNase-free water and store in aliquots at −20°C.

2.2. Verification of siRNA Annealing

1. 40% acrylamide/bis-acrylamide solution (19:1, Sigma). This is a neurotoxin when unpolymerized. Store at 4°C.
2. 5× Tris Borate EDTA (TBE) buffer: 1.1 M Tris, 900 mM Borate, and 25 mM EDTA (pH = 8.3).

3. *N,N,N,N'*-Tetramethyl-ethylenediamine (TEMED, Sigma). Store at 4°C.
4. Ammonium persulfate (APS): 10% (W/W) solution in water. Store in aliquots at –20°C.
5. 6× loading dye: 30% glycerol, 0.25% bromophenol blue, and 0.25% xylene cyanol solution in water.
6. Staining solution: 0.1% EtBr in 1× TBE buffer.

2.3. Cell Culture and Transfection of asiRNA

1. HeLa cells and T98G cells (ATCC).
2. DMEM (Invitrogen) supplemented with 10% fetal bovine serum (FBS, Invitrogen), 1× Pen-Strep (Invitrogen). Store at 4°C.
3. Trypsin-EDTA solution (1×) (Sigma). Store at 4°C.
4. Opti-MEM (Invitrogen) and Lipofectamine2000 (Invitrogen). Store at 4°C.

2.4. RNA Extraction

1. Tri-reagent (Ambion).
2. Chloroform, 70% ethanol, and isopropanol.
3. DEPC-treated water.

2.5. Reverse Transcription and Quantitative RT-PCR

1. High Capacity cDNA Reverse Trancription Kit (Applied Biosystems) containing 10× RT Buffer, 10× Random Primers, 25× dNTP Mix (100 mM), MultiScribe Reverse Transcriptase (50 U/μl) and RNase Inhibitor.
2. The relevant forward and reverse PCR primers (Bioneer). First prepare 100 μM stock solutions of each primer in water, and then prepare a mix of 2 μM each of the forward and reverse primers. Store in aliquots at –20°C.
3. Fast SYBR Green Master Mix (Applied biosystems). Store at –20°C.

2.6. Western Blotting for Ago2

1. siRNA oligos directed against Ago2 (siAgo2), prepared and annealed as described in Subheading 3.1.1.
2. 1× Phosphate Buffered Saline (PBS): 0.0038 M NaH_2PO_4, 0.0162 M Na_2HPO_4, pH = 7.4.
3. RIPA buffer: 150 mM NaCl, 20 mM Tris–HCl (pH = 7.5), 0.5% sodium dodecyl sulfate, 0.1% sodium deoxycholate, 0.02% sodium azide, 1 mM EDTA, and protease inhibitors.
4. BCA protein assay kit (Pierce).
5. 30% acrlyamide/bis-acrylamide solution (37.5:1, Sigma).
6. 10% SDS.
7. 4× Tris–HCl (1.5 M Tris–HCl, pH = 8.8).
8. *N,N,N,N'*-Tetramethyl-ethylenediamine (TEMED, Sigma). Store at 4°C.

9. 10% APS (W/W) in water. Store in aliquots at −20°C.
10. 1× running buffer: 25 mM Tris–HCl (pH = 8.8), 250 mM glycine, and 0.1% SDS.
11. 2× sample buffer: 125 mM Tris–HCl (pH = 6.8), 10% glycerol, 1% b-mercaptoethanol, 0.02% bromophenol blue, and 2% SDS.
12. Protein molecular weight markers (TaKaRa).
13. Cellulose nitrate membrane (Whatman).
14. hAgo2 antibody (Abcam) and peroxidase-conjugated goat polyclonal secondary antibody to rabbit IgG (Abcam). Store in aliquots at −20°C.
15. 5% skim milk solution in 1× Tris-buffered saline (TBS): 20 mM Tris–HCl, (pH = 7.4) and 150 mM NaCl.
16. 0.1% Tween-20 solution in TBS.
17. Enhanced chemiluminescence (ECL) detection system (Amersham Biosciences, NJ). X-ray films (Kodak, Rochester, NY).

2.7. 5′ RACE Assay

1. Extracted total RNA.
2. Gene Racer RNA oligo (Invitrogen). Store in aliquots at −20°C.
3. T4 RNA ligase (5 U/μl) (NEB),10× T4 Ligase reaction buffer (NEB), 10 mM ATP (NEB), RNaseOut (40 U/μl) (Invitrogen), phenol:chloroform (1:1 mixture), 10 mg/ml glycogen, 3 M sodium acetate, 100% ethanol, 70% ethanol, and nuclease free water.
4. SuperScript III First-Strand Synthesis System for RT-PCR (Invitrogen) which contains SuperScript III RT, DEPC-treated water, 10× RT buffer, 0.1 M DTT, 10 mM dNTP mix, oligo dT (50 uM), 25 mM $MgCl_2$, and RNaseOUT.
5. The relevant PCR primers and nested PCR primers (Invitrogen).
   (a) GeneRacer 5′ Primer 5′-CGACTGGAGCACGAGGACA CTGA-3′
   (b) GeneRacer 5′ Nested Primer 5′-GGACACTGACATGGA CTGAAGGAGTA-3′
   (c) Gene specific 3′ primer designed to have 50–70% GC content to obtain a high annealing temperature (>72°C) and 23–28 nucleotides in length to increase specificity of binding.
   (d) Gene specific 3′ Nested Primer designed to have annealing temperature similar to the annealing temperature for the GeneRacer 5′ Nested Primer. Nested 3′ primers should be

far enough from the original gene specific primer so that you can distinguish the products of original and nested PCR by size.

Prepare 10 μM stock solutions in water for each primer and store in aliquots at −20°C.

6. Taq polymerase (NEB), 10× Taq polymerase buffer (NEB), $MgSO_4$ (50 mM), and dNTPs (10 mM each) (NEB). Store at −20°C.
7. 1% agarose gel (Invitrogen).
8. 50× TAE (Boston BioProducts).
9. Running buffer (1× TAE): dilute 50× TAE in water. Store at room temperature.
10. T&A cloning kit (RBC).

2.8. Dual Luciferase Assay

1. Plasmids
   (a) DNA oligonucleotides corresponding to the antisense target or sense target sequences cloned into the SpeI and HindIII sites of the pMIR Report-luciferase vector (Ambion). Prepare a 200 ng/μl dilution in water. Store in aliquots at −20°C.
   (b) 2 ng/μl of Renilla luciferase plasmid (pRL-SV40) (Promega) in water. Store in aliquots at −20°C.
2. Dual luciferase reporter assay kit (Promega) which contains 5× Passive Lysis Buffer, Luciferase Assay Buffer II, Luciferase assay Substrate (lyophilized product), Stop&Glo buffer, and Stop&Glo Substrate (50×). Prepare 1× stock solutions of Luciferase assay Substrate in Luciferase Assay Buffer II and Stop&Glo Substrate in Stop&Glo Substrate and store in aliquots at −20°C.

2.9. Assay for RNAi Machinery Saturation

1. CREB3 siRNA (100 μM stock) (Bioneer). Store in aliquots at −20°C.
2. CREB3 realtime PCR primers (2 μM each forward and reverse primers).
3. pMIR-Report-luciferase (miR21 target sequence; TAGCTT ATCAGACTGATGTTGA) and pRL-SV40 (Promega), 200 ng/μl stocks in water. Store in aliquots at −20°C.
4. AntagomiR antagonist against miR21 (Anti-miR21) (Ambion).

2.10. DNA Microarray

1. DNA chip: Nimblegen Human Gene Expression 12x135K Array (Roche NimbleGen, Inc., Madison, WI).
2. Double strand cDNA synthesis: SuperScript Double-Stranded cDNA Synthesis Kit (Invitrogen).
3. Sample labeling: NimbleGen One-Color DNA Labeling Kit (Roche NimbleGen, Inc., Madison, WI).

4. Hybridization and washing: NimbleGen Hybridization Kit, NimblGen Wash Buffer Kit (Roche NimbleGen, Inc., Madison, WI).
5. Equipment: NimbleGen Hybridization System 4 (Roche NimbleGen, Inc., Madison, WI), InnoScan900 (Innopsys, France).
6. Software: Mapix (Innopsys, France), NimbleScan, (Roche NimbleGen, Inc., Madison, WI), Sylamer (http://www.ebi.ac.uk/enright/sylamer/).

3. Methods

siRNA structural variants against four different target genes were designed to have either conventional 19 bp siRNA structure or asymmetric siRNA structures harboring 19 nt antisense strand with 13–17 nt long sense strands. An activity test comparing each structural variant was performed in HeLa and T98G cells by quantitative real-time PCR. The results indicated that the length of the sense strand could be reduced to 16 nt without significant loss of activity (Fig. 2) (17) Methods for preparing the asiRNA and for performing RT-PCR to test its activity are provided in Subheading 3.1.

To confirm that asiRNAs also execute gene silencing via conventional RNAi mechanisms, target gene silencing by asiRNA was tested in cells treated with siRNA targeting Ago2. As expected for RNAi mechanism-based gene silencing, the target gene silencing activity of asiRNA was reduced upon Ago2 knockdown (12). Methods for verifying conventional RNAi mechanisms by co-transfection of siRNA against Ago2 (siAgo2) are provided in Subheading 3.2. Next, target mRNA cleavage sites by both siRNA and asiRNA were analyzed by 5′-RACE assay. Sequencing analysis of 5′-RACE products showed that both siRNA and asiRNA cleaved the same site within the target mRNA (18). Methods for 5′-RACE are provided in Subheading 3.3.

The use of an asiRNA structure can reduce nonspecific effects triggered by siRNAs. First, sense-strand-mediated off-target silencing activity of asiRNA was compared with siRNA using a dual luciferase assay. The result shows that the asiRNA structure can significantly reduce the sense-strand-mediated off-target activity while maintaining antisense strand-based on-target activity (17). Methods for assessing the antisense-strand-mediated silencing and sense-strand-mediated off-target activity by using a reporter system are provided in Subheading 3.4. The saturation of the endogenous RNAi machinery by exogenously introduced siRNAs can also be ameliorated by using asiRNA structures. A competition assay with other siRNAs and a luciferase reporter assay to measure the

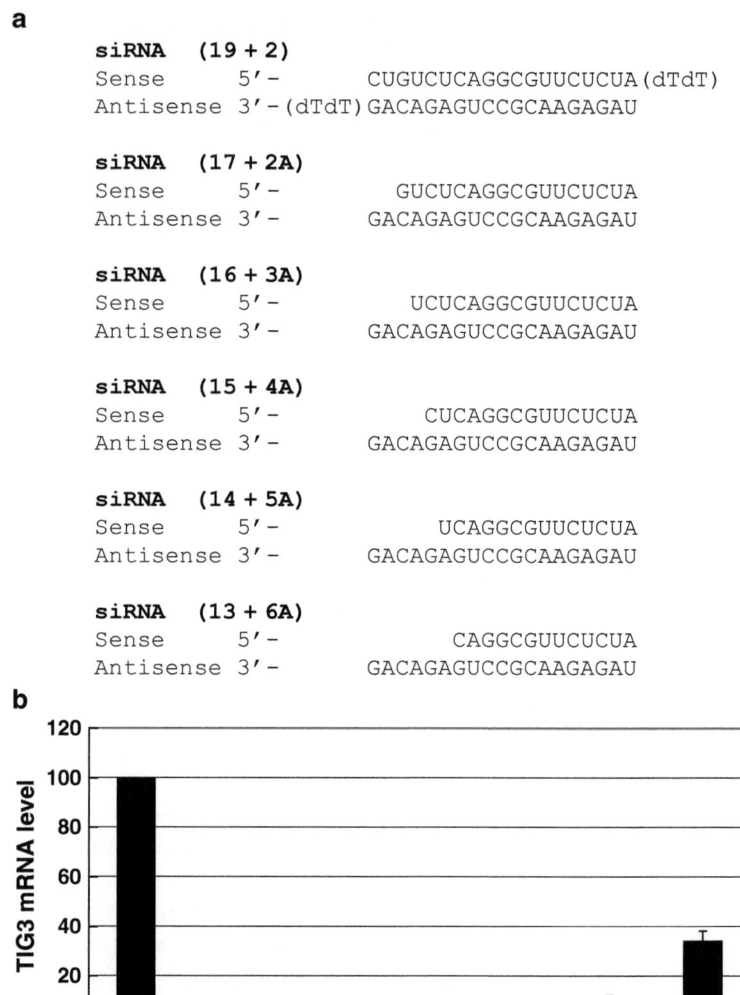

Fig. 2. Structures and activities of asymmetric siRNAs targeting TIG3 (17). (**a**) Structures of siRNA and asiRNAs that target TIG3 mRNA. (**b**) Expression of TIG3 mRNA as a measurement of silencing activity for the structure in *panel A* following transfection into HeLa cells.

endogenous miRNA activity both show that asiRNA have less or no inhibition of endogenous RNAi machinery (17). Methods for performing these assays are provided in Subheading 3.5. Genome-wide analysis through DNA microarray also revealed that the asiRNA structure can alleviate off-target effects mediated by the seed sequence of the siRNA sense strand (22). Methods for assessment of siRNA and asiRNA off-targeting effects by microarray analysis are provided in Subheading 3.6.

3.1. asiRNA Annealing, Transfection, and Assessment of RNA Silencing Activity by Quantitative RT-PCR

3.1.1. asiRNA Preparation and Annealing

1. Design asiRNA with 19–21 nt antisense and 17–13 nt sense strand so that the asiRNA has a duplex region shorter than 19 bp and the antisense strand has an overhang of variable lengths at the 3′ end (Fig. 1). The antisense sequence of siRNA could be directly used for asiRNA design. Optimized asiRNA structure harbors a 16 bp duplex region and 3 nt of 3′-antisense overhang. (Examples of asiRNA structures are shown in Fig. 2a).
2. To prepare siRNA or asiRNA, mix 20 μl of sense strand (100 μM), 20 μl of antisense strand (100 μM), 120 μl of water, and 40 μl of 5× siRNA buffer, and anneal the strands by heating at 95°C for 3 min and then incubating at 37°C for 1 h.
3. Prepare 5 mL of 10% nondenaturing polyacrylamide gel by mixing 1 mL of 5× TBE buffer, 1.25 mL of 40% acrylamide/bis solution, 2.75 mL of water, 50 μl of 10% APS solution, and 5 μl of TEMED. Wait for 30 min so that the gel is completely polymerized.
4. Assemble the gel unit and wash the wells with 1× TBE buffer. Add 0.5 μl of 6× loading dye to 2.5 μl of siRNA, sense strand, or antisense strand and run on a 10% nondenaturing polyacrylamide gel at 100 V at room temperature for 1 h, until the second dye (bromophenol blue) runs to the middle of the gel.
5. Stain the gel with 0.1% EtBr solution in 1× TBE buffer, wash with 1× TBE buffer and confirm the formation of double-stranded RNA by comparing the size of the siRNA with sense and antisense strands on the UV trans-illuminator.

3.1.2. asiRNA Transfection

1. One day before the transfection, grow HeLa and T98G cells in a 12-well plate to allow about 30–50% confluency in 24 h.
2. Just prior to transfection, exchange the medium with 800 μl of pre-warmed complete growth medium without antibiotics and incubate cells at 37°C.
3. To prepare the Lipofectamine2000 mixture, mix 2 μl of lipofectamine2000 and 98 μl of Opti-MEM. Then incubate at room temperature for 10 min.
4. To prepare the siRNA or asiRNA mixture, dilute annealed siRNA or asiRNA in a total volume of 100 μl Opti-MEM solution.
5. Mix the diluted Lipofectamine2000 and siRNA/asiRNA from steps 3 and 4, and vortex vigorously. Incubate the mixture at room temperature for 20 min so that the siRNA and Lipofectamine2000 form a complex.
6. Add the siRNA/lipofectamine2000 mixture solution to the cells and incubate for 24 h.

3.1.3. RNA Extraction

1. Discard the medium and lyse the cells by adding 500 μl of Tri-reagent to each well of a 12-well plate. Incubate the mixture for 10 min at room temperature.
2. Transfer 500 μl of cell/Tri-reagent solution to an Eppendorf tube, and add 100 μl of chloroform. Close the lid and vortex vigorously for 15 s. Incubate at room temperature for 10 min.
3. Centrifuge samples at 13,400 × *g* for 20 min at 4°C. Following centrifugation, transfer the upper aqueous phase carefully to a fresh tube.
4. Measure the volume of the aqueous phase and add 1 volume of isopropanol. Vortex vigorously for 10 s and incubate at −20°C for 20 min.
5. Centrifuge at 13,400 × *g* for 15 min at 4°C.
6. Remove the supernatant and wash the pellet with 500 μl of 75% ethanol. Centrifuge at 13,400 × *g* for 5 min.
7. Remove the supernatant completely and dry the pellet for 5 min at room temperature.
8. Dissolve the pellet in 12 μl of DEPC-treated water and measure the concentration of the RNA.

3.1.4. Reverse Transcription

1. Reverse-transcribe single stranded cDNA from the RNA template isolated as described in Subheading 3.1.3 using the High Capacity cDNA Reverse Transcription Kit described in Subheading 2.5. Prepare a 2× RT master mix by combining:
   (a) 2 μl of 10× RT buffer
   (b) 0.8 μl of 25× dNTP mix (100 mM)
   (c) 2 μl of 10× Random Primers
   (d) 1 μl of Reverse transcriptase
   (e) 1.0 μl of RNase inhibitor
   (f) 4.2 μl of Nuclease-free water.
2. Dilute 1 μg of RNA in a total volume of 10 μl RNase free water.
3. Add 10 μl of 2× RT master mix from step 1 to the RNA solution from step 2.
4. Load the reactions onto a thermocycler and run the program given below.
   (a) 25°C for 10 min
   (b) 37°C for 120 min
   (c) 80°C for 5 min
5. Store reactions at −20°C.

3.1.5. Real-Time PCR

1. Prepare a Real-time PCR mix by combining:
   (a) 1 μl of cDNA template prepared in Subheading 3.1.4.
   (b) 10 μl of 2× SYBR Green Master mix
   (c) 3 μl of primer mix (2 μM each forward and reverse primers)
   (d) 6 μl of water
2. Load the reactions into a thermocycler and run the program below.
   (a) 95°C 10 min
   (b) 95°C 20 s
   (c) 60°C 60 s (b–c 40 cycles)
3. Normalize the expression levels of each gene to GAPDH and graph the results to assess the relative expression levels. An example of results for quantitative RT-PCR to assess knockdown efficacies is shown in Fig. 2b.

3.2. Assessment of the Requirement of Ago2 for asiRNA-Mediated Gene Silencing

To assess the requirement for Ago2, Ago2 siRNA is transfected, and knockdown of Ago2 is verified by western blotting. Subsequently, the siRNA for the gene of interest is transfected into Ago2 knockdown cells, and its ability to block the Ago2 knockdown is measured.

1. Seed HeLa cells 1 day prior to the transfection in a 100 mm dish.
2. Exchange the growth medium before the transfection. Prepare an siAgo2/Lipofectamine2000 mix as described in Subheading 3.1.2.
3. Add the siAgo2/Lipofectamine2000 mix to the cells and incubate at 37°C.
4. 48 h after transfection, discard the medium, and wash the cells with PBS.
5. Lyse the cells using RIPA buffer and measure the protein concentration using a BCA protein assay kit.
6. Prepare a separating gel by mixing 10 ml of 30% acrylamide/bis-acrylamide solution, 200 μl of 10% SDS, 12.5 ml of water, 7.5 ml of 4× Tris–HCl (pH = 8.8), 150 μl 10% APS, and 15 μl TEMED. Let it polymerize for 30 min.
7. Prepare a 4% stacking gel mix (1 ml of acrylamide/bis-acrylamide, 100 μl 10%SDS, 2.5 ml of 0.5 M Tris–HCl (pH = 6.8), 75 μl of 10% APS, and 15 μl TEMED). Pipette the stacking solution onto the gel and insert the comb. Wait for another 20 min for the polymerization of the stacking gel.
8. Assemble the gel unit and wash the wells with 1× running buffer. In the meantime, add 2× sample buffer to protein samples and denature at 95°C for 5 min.

9. Load samples and protein marker onto the wells.
10. Run the samples at 60 V for 30 min and thereafter at 100 V until the loading dye reaches the bottom of the gel.
11. Assemble a western transfer apparatus and transfer proteins to cellulose nitrate membrane. Block the membrane for 1 h in TBS buffer containing 5% milk powder.
12. After overnight incubation at 4°C with an hAgo2 antibody, wash the membranes in TBS containing 0.1% Tween-20.
13. Incubate the membrane with peroxidase-coupled secondary antibody, wash the membranes in TBS containing 0.1% Tween-20, and develop the membrane using the ECL detection system.
14. Expose to X-ray film and verify that Ago2 has been knocked down by >50%.
15. To test whether asiRNA-mediated gene silencing is Ago2 dependent, seed the HeLa cells a day prior to the transfection in a 12-well plate. 24 h after siAgo2 transfection, exchange the medium and transfect asiRNA or siRNA into the siAgo2 treated cells.
16. 24 h after transfection, compare the silencing activity of asiRNA in siAgo2 transfected cells and control cells using real-time PCR as described in Subheadings 3.1.3–3.1.5.

3.3. Mapping of the Target mRNA Cleavage Site by 5′ RACE Assay

1. To map the site of asiRNA-mediated cleavage of the target, transfect asiRNAs or siRNAs into HeLa cells. (See Subheading 3.1.2.)
2. 12 h after transfection, isolate RNA as described in Subheading 3.1.3.
3. Combine 2 μg of RNA from siRNA transfected cells and 0.25 μg of GeneRacer RNA oligo in a total volume of 7 μl nuclease-free water. To relax the RNA secondary structure, incubate at 65°C for 5 min. Place on ice for 5 min and centrifuge briefly.
4. Ligate the Gene Racer RNA oligo to the cleaved mRNA using the ligation reaction mixture given below:
   (a) 10× T4 ligase buffer 1 μl
   (b) 10 mM ATP 1 μl
   (c) RNaseOut (40 U/μl) 1 μl
   (d) T4 RNA ligase (5 U/μl) 1 μl

   and incubate at 37°C for 1 h.
5. After incubation, add 90 μl of nuclease free water and 100 μl of phenol:chloroform. Vortex vigorously for 30 s.
6. Centrifuge at 13,400 × *g* for 5 min at room temperature and transfer the aqueous phase to a new Eppendorf tube.

7. To precipitate RNA, add 2 μl of 10 mg/ml glycogen and 10 μl of 3 M sodium acetate (pH = 5.2), and vortex briefly. Add 220 μl of ethanol, and vortex for 5 s.
8. Incubate at −80°C for 20 min, and centrifuge at 13,400 × *g* for 20 min at 4°C.
9. Discard supernatant, and wash with 500 μl of 70% ethanol. Centrifuge at 12,000 × *g* for 5 min.
10. Discard supernatant, and dry pellet at room temperature for 2 min.
11. Resuspend the pellet in 10 μl of nuclease free water.
12. Reverse-transcribe single stranded cDNA from the RNA template using the SuperScript III RT kit as described below.
    (a) Combine total RNA and oligo dT in a total volume of 10 μl water.
    (b) Incubate at 90°C for 5 min, then place on ice for 5 min.
    (c) Prepare reverse transcription mix by combining 10× RT buffer (2 μl), 25 mM $MgCl_2$ (4 μl), 0.1 M DTT (2 μl), RNaseOUT (40 U/μl) (1 μl), SuperScript III RT (200 U/μl) (1 μl). Add 10 μl of mix to each RNA/primer mixture and mix gently. Incubate 50 min at 50°C
    (d) Terminate the reactions by heating at 85°C for 5 min. Chill on ice. The cDNA can be stored at −20°C or used for PCR.
13. To amplify cDNA ends, prepare a mixture as described below
    (a) GeneRacer 5′ primer 3 μl (10 μM)
    (b) Gene Specific 3′ primer 1 μl (10 μM)
    (c) Template cDNA from step 12 1 μl
    (d) 10× Taq polymerase buffer 5 μl
    (e) dNTP mix solution 1 μl (10 mM)
    (f) Taq DNA polymerase 0.5 μl (5 U/μl)
    (g) $MgSO_4$ 2 μl (50 mM)
    (h) Water 36.5 μl
14. Design PCR cycles as described below and load reactions
    (a) 94°C 2 min
    (b) 94°C 30 s
    (c) 72°C 1 min (b–c: 5 cycles)
    (d) 94°C 30 s
    (e) 70°C 1 min (d–e: 5 cycles)
    (f) 94°C 30 s
    (g) 60–68°C (depends upon primer sequence) 30 s
    (h) 72°C 30 s (f–h: 20 cycles)
    (i) 72°C 5 min

15. To increase specificity and sensitivity, perform nested PCR by preparing the mixture described below
    (a) GeneRacer 5′ Nested 1 μl (10 μM)
    (b) Nested Gene Specific 3′ primer 1 μl (10 μM)
    (c) For the template, use 1 μl of PCR product from step 14
    (d) 10× Taq polymerase buffer 5 μl
    (e) dNTP mix solution 1 μl (10 mM)
    (f) Taq DNA polymerase 0.5 μl (5 U/μl)
    (g) Water 40.5 μl
16. Design PCR cycles as described below and load reactions
    (a) 94°C 2 min
    (b) 94°C 30 s
    (c) 60–68°C (depends upon primer sequence) 30 s
    (d) 72°C 30 s (b–c: 25–30 cycles)
    (e) 72°C 5 min
17. Analyze 10 μl of Nested PCR product on a 1% agarose gel in 1× TAE running buffer.
18. Sequence the PCR products following cloning into a plasmid. Sample results are shown in Fig. 3.

3.4. Assessment of the Antisense-Mediated Silencing and the Sense-Strand-Mediated Off-Target Silencing Activity of asiRNA Using a Reporter Assay System

1. One day prior to transfection, grow HeLa cells in a 12-well plate and allow them to grow to 70% confluency on the day of transfection.
2. Prepare an asiRNA/Luciferase reporter plasmid DNA mixture as described below
    (a) 1 μl of asiRNA (10 μM)
    (b) 200 ng of a pMIR-Report-luciferase plasmid harboring the sense strand target or antisense strand target sequences
    (c) 2 ng of Renilla luciferase plasmid
    (d) Add water to a total volume of 100 μl
3. Prepare lipofectamine2000 mix as described in Subheading 3.1.2, and combine with the solution from step 2.
4. Exchange the cell culture medium with 800 μl of complete medium, and add the mixture from step 3.
5. 24 h after transfection, discard the medium, and lyse cells by adding 200 μl of 5× Passive Lysis Buffer. Place the plate on an orbital shaker for 15 min.
6. To measure the firefly luciferase activity, which represents expression of sense or antisense target, transfer 50 μl of supernatant from step 5 to an Eppendorf tube and add 100 μl of 1× Luciferase assay substrate solution. Measure the luciferase activity.

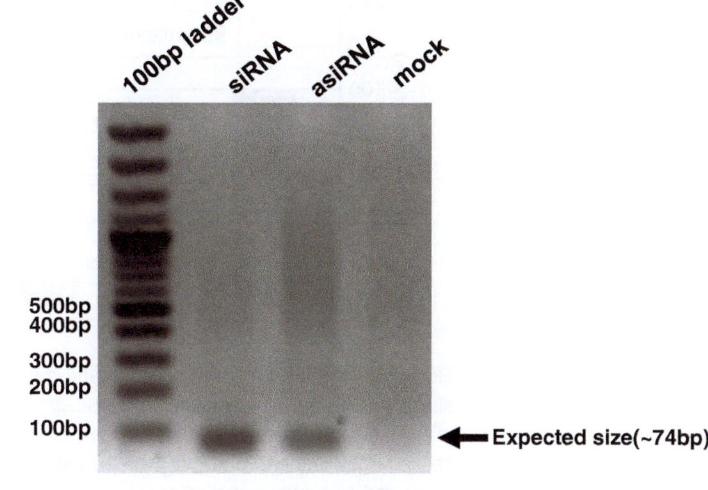

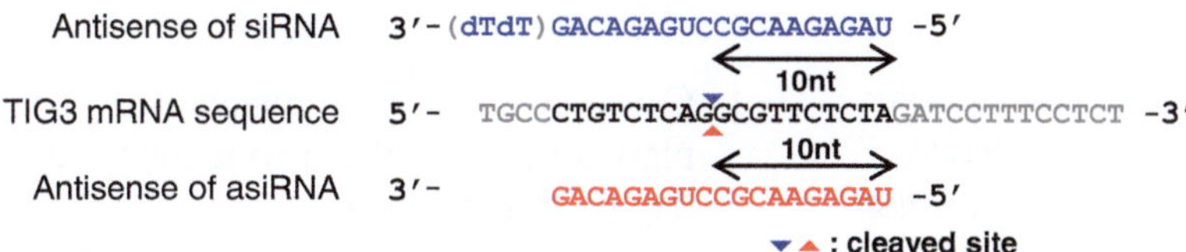

Fig. 3. asiRNA and siRNA cleave identical site within target mRNAs (17). 5′ RACE PCR products from siTIG3-transfected cells. TIG3 mRNA cleavage sites were analyzed by 5′-RACE assay and sequencing. The antisense strand of the siRNA is shown in *blue* and the antisense strand of asiRNA is shown in *red*. Cleavage sites are marked with arrowheads.

7. To measure the Renilla luciferase activity, which is the transfection control, add 100 μl of 1× Stop&Glo substrate solution and measure the luciferase activity.
8. Normalize the activity of firefly luciferase by dividing the Renilla luciferase activity for each reaction. Sample results are shown in Fig. 4.

3.5. Assessment of RNAi Machinery Saturation by Exogenously Introduced asiRNA and siRNA

3.5.1. Comparison of the Potency of asiRNA and siRNA to Compete Against Exogenously Introduced siCREB3

1. One day prior to transfection, seed HeLa cells in a 12-well plate and allow them to grow to 50% confluency on the day of transfection.
2. Co-transfect siCREB3 (final concentration 1 nM) with asiRNA (10 nM) or siRNA (10 nM) using lipofectamine2000 as described in Subheading 3.1.2.
3. 24 h after transfection, analyze the expression of CREB3 mRNA using realtime PCR as described in Subheading 3.1.

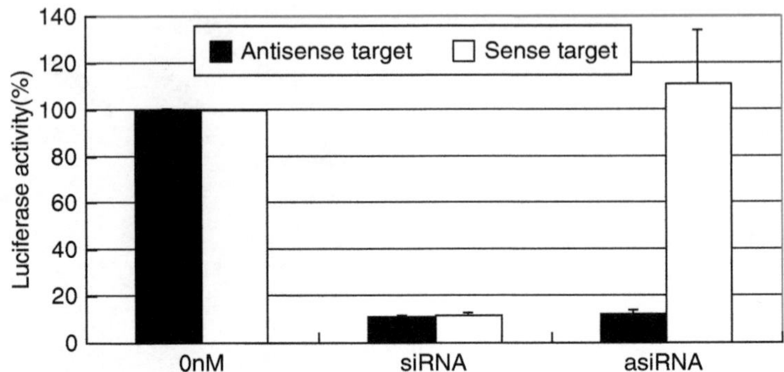

Fig. 4. asiRNA shows reduced sense-strand-mediated off-target gene silencing (17). Gene silencing activities of sense and antisense strands of asiRNA. HeLa cells were transfected with a luciferase reporter plasmid that carried either a Survivin antisense target or Survivin sense target, without (0 nM) or with 10 nM of asiRNA or siRNA. Luciferase activity was measured 48 h after transfection.

3.5.2. Comparison of the Potency of asiRNA and siRNA to Compete Against Endogenous miRNA

1. One day prior to transfection, grow HeLa cells in a 12-well plate and allow them to grow to about 50% confluency on the day of transfection.
2. Co-transfect 200 ng of pMIR-Report-luciferase plasmid harboring an miR21 target sequence and 2 ng of pRL-SV40 Renilla luciferase plasmid with asiRNA (10 nM), siRNA (10 nM), or Anti-miR21 (10 nM) using lipofectamine2000 as described in Subheading 3.1.2. The Anti-miR21 serves as a positive control for miR21 competition.
3. 24 h after transfection, analyze miR21 activity using a dual luciferase assay as described in Subheading 3.4.

3.6. Genome-Wide Analysis of siRNA Versus asiRNA Specificity Through DNA Microarray

3.6.1. Analysis of Genome-Wide Transcription Profiles

1. Transfect siRNA or asiRNA into HeLa cells in a 6-well plate. Mock transfection such as transfection reagent only is generally used for a control. Avoid including additional siRNAs as a control that may also have their own off-target effects and interfere with subsequent analysis.
2. 24 h after transfection, extract total RNA, and conduct a DNA microarray experiment. Perform the DNA microarray experiment following the detailed DNA microarray protocols provided by the DNA microarray platform supplier.
3. From the normalized expression data, calculate the ratio of gene expression for siRNA or asiRNA to control samples, and depict MA-plots to examine the genome-wide transcript level change upon siRNA or asiRNA transfection. The MA-plot represents the distribution of the intensity ratio ("M") plotted

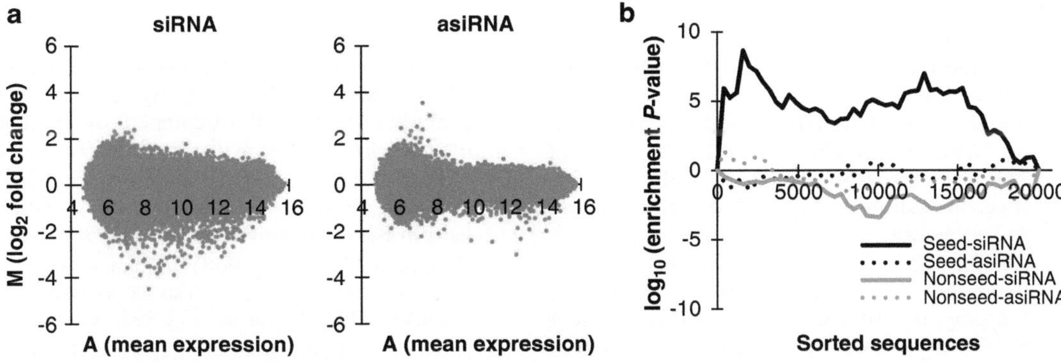

Fig. 5. Genome-wide analysis showed reduced off-target silencing by asiRNA (22). (**a**) MA scatter plots illustrating genome-wide expression patterns. "M" denotes the log2-transformed expression ratios (siRNA- or asiRNA-treated versus control) and "A" denotes the average of the log2-transformed expression level. (**b**) Sylamer analysis for visualizing off-target effects. From the DNA microarray data, a gene list ranked according to fold change (from most downregulated genes to most upregulated genes) was generated and loaded into Sylamer software. Sylamer software was used to compute the enrichment of 3′ UTRs targeted by the "seed sequence" or "non-seed sequence" of the sense strand. The solid and dotted lines represent siRNA and asiRNA, respectively. The *x*-axes represent the sorted gene lists from most downregulated (*left*) to most upregulated (*right*). The *y*-axes show the hypergeometric significance of each word at each leading bin. Positive values indicate enrichment (–log 10(*P*-value)) and negative values indicate depletion (log 10(*P*-value)).

by the average intensity ("A"). In this plot, the Y axis contains the log-ratio intensity of one array to the control array, and the X axis contains the average log-intensity of both arrays. Plotting of MA-plots allows the comparison of global gene expression change after siRNA or asiRNA transfection. Sample results are shown in Fig. 5a.

3.6.2. Analysis of Sense-Strand Seed Sequence Mediated Off-Target Effects by DNA Microarray

1. Install the Sylamer program from the authors' Web site (http://www.ebi.ac.uk/enright/sylamer/) (23).
2. According to the manual, execute the Sylamer. Load the gene list and sequences, and, by using "–w" option, input sequences targeted by the sense-strand seed sequence.
3. After the Sylamer run, compare sense-strand seed mediated off-target effects. Sample results are shown in Fig. 5b.

Acknowledgments

This work was supported by the Global Research Laboratory program by the Ministry of Education and Science and Technology in Korea (grant 2008–00582).

References

1. Hannon GJ (2002) RNA interference. Nature 418:244–251
2. Elbashir SM, Harborth J, Lendeckel W, Yalcin A, Weber K, Tuschl T (2001) Duplexes of 21-nucleotide RNAs mediate RNA interference in cultured mammalian cells. Nature 411:494–498
3. Elbashir SM, Martinez J, Patkaniowska A, Lendeckel W, Tuschl T (2001) Functional anatomy of siRNAs for mediating efficient RNAi in Drosophila melanogaster embryo lysate. EMBO J 20:6877–6888
4. de Fougerolles A, Vornlocher HP, Maraganore J, Lieberman J (2007) Interfering with disease: a progress report on siRNA-based therapeutics. Nat Rev Drug Discov 6:443–453
5. Lares MR, Rossi JJ, Ouellet DL (2010) RNAi and small interfering RNAs in human disease therapeutic applications. Trends Biotechnol 28:570–579
6. Scherer L, Rossi JJ, Weinberg MS (2007) Progress and prospects: RNA-based therapies for treatment of HIV infection. Gene Ther 14:1057–1064
7. Jackson AL, Linsley PS (2004) Noise amidst the silence: off-target effects of siRNAs? Trends Genet 20:521–524
8. Jackson AL et al (2003) Expression profiling reveals off-target gene regulation by RNAi. Nat Biotechnol 21:635–637
9. Clark PR, Pober JS, Kluger MS (2008) Knockdown of TNFR1 by the sense strand of an ICAM-1 siRNA: dissection of an off-target effect. Nucleic Acids Res 36:1081–1097
10. Yoo JW, Kim S, Lee DK (2007) Competition potency of siRNA is specified by the 5′-half sequence of the guide strand. Biochem Biophys Res Commun 367:78–83
11. Koller E et al (2006) Competition for RISC binding predicts in vitro potency of siRNA. Nucleic Acids Res 34:4467–4476
12. Vickers TA, Lima WF, Nichols JG, Crooke ST (2007) Reduced levels of Ago2 expression result in increased siRNA competition in mammalian cells. Nucleic Acids Res 35:6598–6610
13. Grimm D et al (2006) Fatality in mice due to oversaturation of cellular microRNA/short hairpin RNA pathways. Nature 441:537–541
14. Kleinman ME et al (2008) Sequence- and target-independent angiogenesis suppression by siRNA via TLR3. Nature 452:591–597
15. Amarzguioui M, Holen T, Babaie E, Prydz H (2003) Tolerance for mutations and chemical modifications in a siRNA. Nucleic Acids Res 31:589–595
16. Chiu YL, Rana TM (2003) siRNA function in RNAi: a chemical modification analysis. RNA 9:1034–1048
17. Chang CI et al (2009) Asymmetric shorter-duplex siRNA structures trigger efficient gene silencing with reduced nonspecific effects. Mol Ther 17:725–732
18. Soutschek J et al (2004) Therapeutic silencing of an endogenous gene by systemic administration of modified siRNAs. Nature 432:173–178
19. Barik S (2006) RNAi in moderation. Nat Biotechnol 24:796–797
20. Marques JT, Williams BR (2005) Activation of the mammalian immune system by siRNAs. Nat Biotechnol 23:1399–1405
21. Choung S, Kim YJ, Kim S, Park HO, Choi YC (2006) Chemical modification of siRNAs to improve serum stability without loss of efficacy. Biochem Biophys Res Commun 342:919–927
22. Jo SG, Hong SW, Yoo JW, Lee CH, Kim S, Lee DK (2011) Selection and optimization of asymmetric siRNA targeting the human c-MET gene. Mol Cells 32:543–548
23. van Dongen S, Abreu-Goodger C, Enright AJ (2008) Detecting microRNA binding and siRNA off-target effects from expression data. Nat Methods 5:1023–1025

Chapter 8

Design of Nuclease-Resistant Fork-Like Small Interfering RNA (fsiRNA)

Elena L. Chernolovskaya and Marina A. Zenkova

Abstract

Small interfering RNAs (siRNAs) are potent inducers of RNA interference—the conservative cellular process of posttranscriptional gene silencing. The silencing activity of siRNAs depends on the thermodynamic asymmetry of the siRNA duplex. Here, we describe the design of chemically modified fork-like siRNA (fsiRNAs) containing mismatches at the 3′-end region of the sense strand and 2′-O-methyl modifications in nuclease-sensitive sites, capable of silencing of thermodynamically unfavorable targets.

Key words: siRNA, 2′-O-methyl analogs, Fork-like siRNA, Nucleotide substitution, Mismatches, Thermodynamic asymmetry

1. Introduction

RNA interference (RNAi) is a conserved posttranscriptional mechanism of sequence-specific gene silencing triggered by long double-stranded RNAs (dsRNAs). The key step of RNAi is the formation of an activated RNA-induced silencing complex (RISC) (1–4). Initially, siRNA interacts with a Dicer/TRBP heterodimer in a specific manner governed by duplex thermoasymmetry (5). The assembly of this complex then recruits Ago2, which replaces the heterodimer and subsequently determines which strand will stay in the complex and guide the target recognition. The antisense strand of the siRNA complementary to the mRNA of the target gene must be incorporated into the active RISC to achieve the silencing effect. Consequently, siRNAs with less thermodynamically stable sequences at the 5′-part of antisense strand are more active in gene silencing (6, 7). Several computational algorithms for the selection of active siRNA sequences based on their thermodynamic properties have been proposed (8–10); however, the nucleotide sequence of the mRNA-target may not comply with the

Debra J. Taxman (ed.), *siRNA Design: Methods and Protocols*, Methods in Molecular Biology, vol. 942,
DOI 10.1007/978-1-62703-119-6_8,

requirements for "good" siRNA. This problem can arise when the target is a chimeric or point mutated gene or a gene belonging to a highly homologous family of genes. In order to circumvent this problem, the unfavorable thermodynamic asymmetry of an siRNA duplex can be corrected by introducing duplex destabilizing nucleotide substitutions at the 3′-end of the sense strand. The resulting so-called "fork-like siRNAs" (fsiRNAs) were shown to be able to effectively silence the expression of target genes (11, 12).

The presence of mismatches and long dangling ends in fsiRNAs leads to increased sensitivity to nuclease degradation and reduces the duration of the silencing effect. Numerous studies have shown that resistance of siRNA to nucleases can be improved by introducing modifications involving the 2′-position of the ribose ring, the 5′-end phosphate, or internucleotide phosphates (13–17). Several analogs of ribonucleotide, including boranophosphates (18), 2′-fluoro (2′-F) (19), and 2′-deoxy-2′-fluoro-beta-D-arabino (20), have been applied for the generation of siRNA with enhanced serum stability with no or very little influence on the silencing activity. Because heavy modification of siRNAs often decreases or even blocks their interfering activity (21–24), it is important to define the optimal scheme of modification providing the balance between the activity and the stability of siRNAs.

Recently, we proposed a methodology for the rational design of nuclease-resistant siRNAs (25). This methodology includes mapping of nuclease-sensitive sites in siRNA and protection of these sites with 2′-O-methyl modifications. These selective modifications improve both the stability of siRNA in the presence of serum and the duration of the silencing effect with minimal loss of silencing efficiency. Subsequently, this methodology was applied to the design of nuclease resistant fsiRNAs (26). We found that fsiRNA duplexes with 2′-O-methyl modifications in nuclease sensitive sites suppress the expression of the target gene more effectively than conventional siRNAs. Moreover, this selective modification greatly improved the stability of fsiRNAs in the presence of serum and provided long-lasting silencing effect (12 days or even more) (26). Here, we provide detailed protocols for implementing this methodology for the design of nuclease-resistant fsiRNAs. This includes methods for designing the fsiRNA, for mapping the nuclease-sensitive sites to identify nucleotides for selective modification, and for testing the gene silencing efficacy of the modified fsiRNA.

2. Materials

All solutions should be prepared using ultrapure water (for example, purified on MilliQ) and molecular biology grade reagents.

2.1. Oligoribonucleotides (See Notes 1–5)

Modified 21 nt oligoribonucleotides can be obtained commercially (for example, from Thermo Scientific http://www.dharmacon.com or Ambion http://www.ambion.com/). Alternatively, synthesize oligoribonucletides in your lab using an automatic DNA/RNA synthesizer (for example, ASM-800 Biosset, Russia or 3400 DNA Synthesizer Applied Biosystems, USA) with ribo- and 2′-O-methylribo β-cyanoethyl phosphoramidites (Glen Research, USA). Use a protocol optimized for your instrument for the programing of the synthesizer. After standard deprotection, purify oligoribonucleotides by denaturing polyacrylamide gel electrophoresis as described for labeled oligoribonucleotides below and isolate as sodium salts.

2.2. Buffers (See Notes 6 and 7)

1. Annealing Buffer: 30 mM HEPES-KOH, pH 7.5, 100 mM potassium acetate, 2 mM magnesium acetate. Store at +4°C or at −20°C for long-term storage.
2. Imidazole Buffer: 2 M imidazole, pH 7.0, 1 mM EDTA, 250 μg/ml total tRNA from *Escherichia coli*.
3. T1 Buffer: 6 M urea, 25 mM sodium citrate, pH 4.5, 1 mM EDTA, 100 μg/ml total tRNA from *E. coli*.
4. 3′-Labeling Buffer: 50 mM HEPES-KOH, pH 7.5, 10 mM $MgCl_2$, 2 mM DTT, 10 μg/ml BSA.
5. Elution Buffer: 0.3 M sodium acetate, pH 5.5.
6. T4 PNK Buffer: 50 mM Tris–HCl, pH 7.6, 10 mM $MgCl_2$, 1 mM DTT, 0.1 mM spermidine.
7. RL Buffer: 50 mM HEPES-KOH, pH 7.5, 10 mM $MgCl_2$, 2 mM DTT, 10 μg/ml BSA.
8. TBE 10× Buffer: 0.5 M Tris–HCl, pH 8.3, 0.5 M boric acid, 10 mM EDTA.
9. Loading Buffer D: 8 M urea, 0.025% bromphenol blue, 0.025% xylene cyanol.
10. Loading Buffer N: 15% Ficoll-400, 0.05% bromphenol blue, 0.05% xylene cyanol.
11. Stains-All solution: 0.1% Stains-All (Sigma, USA) in 50% deionized formamide/water.
12. Water saturated phenol: equilibrate distilled phenol twice with ultrapure water, aspirate water (upper layer), and store frozen in aliquots. Do not use phenol if it turns pale yellow or pale pink.

2.3. Polyacrylamide Gels (See Note 8)

1. Denaturing Sequencing Polyacrylamide Gel: 20% acrylamide, 1% *N,N′*-methylene-bis-acrylamide, 8 M urea, TBE 1× Buffer.
2. Denaturing Polyacrylamide Gel for oligoribonucleotide isolation: 15% acrylamide, 0.5% *N,N′*-methylene-bis-acrylamide, 8 M urea, TBE 1× Buffer.

3. Native Polyacrylamide Gel: 15% acrylamide, 0.5% *N,N'*-methylene-bis-acrylamide, TBE 1×Buffer.
4. Ammonium persulfate: 10% solution in water, freshly prepared.
5. *N, N, N, N'*-tetramethyl-ethylenediamine (TEMED) (Sigma, USA). Store at 4°C.

2.4. Cell Culture

1. Complete Growth medium: Use a growth medium appropriate for your cell line (as indicated in the documentation supplied by the cell line bank, for example American Tissue Culture Collection (ATCC), USA or Institute of Cytology Russian Academy of Sciences, Russia). For HEK 293 cells we use DMEM (Dulbecco's Modified Eagle's Medium, Sigma, USA) supplemented with 10% FBS (Fetal Bovine Serum, for example from HyClone, USA) and antibiotic/antimycotic mixture (Sigma, USA) (100 units/ml penicillin, 0.1 μg/ml streptomycin and 0.25 μg/ml amphotericin). The medium components can be obtained as sterile solutions ready to use. Alternatively, solutions can be prepared from dry powders and filtered through a sterile 22 nm filter in sterile bottles under a tissue culture hood.
2. Serum and antibiotic free growth medium: the same as in the latter section but without FBS and antibiotic/antimycotic mixture.
3. Trypsin solution: 0.25% trypsin in modified Hank's balanced salt solution with 1 mM EDTA.
4. Phosphate buffered saline (PBS): 1.7 mM KH_2PO_4, 5.2 mM Na_2HPO_4, pH 7.4, 150 mM NaCl. This solution could be prepared from ready-to-use tablets (for example, MP Biomedicals, USA) by dissolving the tablets in the volume of ultrapure water indicated by the manufacturer and sterilizing the solution by autoclaving.
5. Opti-MEM (Invitrogen, USA).
6. Lipofectamine 2000 (Invitrogen, USA).

2.5. Enzymes

1. T4 polynucleotide kinase (for example, Fermentas, Lithuania).
2. T4 RNA ligase.
3. RNase T1.

2.6. Radiolabeled Compounds

1. [γ^{32}P]-ATP.
2. 5′- [^{32}P] -cytidine-3′,5′-di-phosphate.

2.7. General Reagents and Solutions

1. 0.1 mM ATP: 0.1 mM ATP in 0.1 M Tris–HCl, pH 8.0.
2. 75% and 96% ethanol.

3. DMSO.
4. 2% lithium perchlorate in acetone.
5. Acetone.
6. A eukaryotic expression plasmid containing the sequence of the target gene fused to the sequence of GFP, for example, pGFP/MDR1 (26) for testing anti-MDR1 siRNA.

3. Methods

3.1. Sequence Selection (See Notes 1, 3–5)

The following section provides an overall guideline for preparing and assessing the fsiRNA for serum stability and functional efficacy. Details for the steps within this section are provided in Subheadings 3.2–3.9 below.

1. Find the accession number and sequence of your mRNA target in NCBI GenBank at http://www.ncbi.nlm.nih.gov/nucleotide/. Use siRNA sequence design tools for the selection of potential siRNA sequences, for example, the siRNA Selection Program at Whitehead http://jura.wi.mit.edu/bioc/siRNAext/home.php, siDirect http://sidirect2.rnai.jp/, OligoWalk http://rna.urmc.rochester.edu/cgi-bin/server_exe/oligowalk/oligowalk_form.cgi, or Block-iT™ RNAi Designer https://rnaidesigner.invitrogen.com/rnaiexpress/.
2. Replace the pyrimidine nucleotides at the 5′-end of $^{5\prime}$CpA$^{3\prime}$, $^{5\prime}$UpA^{3}, and $^{5\prime}$UpG^{3} motives with their 2′-O-methyl analogs. Substitute four nucleotides at positions 16–19 of the sense strand by four other nucleotides that are noncomplementary to the opposite strand.
3. Order or synthesize oligoribonucleotides (see Subheading 2.1).
4. Anneal the fsiRNAs as describe below in Subheading 3.2.
5. Analyze the stability of the fsiRNA in the presence of 10% FBS as describe below in the Subheading 3.3. If fsiRNA is not stable enough for the purpose of your experiment, map the location of nuclease sensitive sites as described in Subheadings 3.4–3.8.
6. Replace nucleotides located at the 5′-side of the nuclease sensitive bond with a 2′-O-methyl- or 2′-F-analog of ribonucleotide and resynthesize the affected oligoribonucleotides.
7. Finally, test the biological activity of fsiRNA in the model system as described in Subheading 3.9.

3.2. siRNA Annealing (See Notes 9 and 10)

1. Prepare a mixture of the antisense and the sense strands of siRNA (100 μM each) in Annealing Buffer.

2. Heat the mixture at 85°C for 3 min, and then cool down slowly to room temperature (for example, leave in a switched-off thermostat for 1 h).
3. Analyze siRNA duplexes by electrophoresis in a 15% native PAAG, followed by staining of the gel with Stains-All solution.
4. Store siRNA frozen in aliquots at –20°C, and thaw when needed on ice.

3.3. Analysis of siRNA Stability in the Presence of 10% FBS (See Note 11)

1. Add 0.3 nmole of siRNA dissolved in Annealing Buffer to 140 μl of complete growth medium with 10% FBS, and incubate at 37°C.
2. After 15, 30 min, 1, 2, 4, and 8 h of incubation (or a similar timecourse of your preference) take 20 μl aliquots; put them immediately into 1.5 ml tubes with an equal volume of water saturated phenol and mix vigorously.
3. Transfer the water phases to the new tubes and precipitate by the addition of 0.1 volume of 3 M sodium acetate, pH 5.5 and 2.5 volume of cold ethanol. Leave at –20°C for at least 2 h.
4. Collect the precipitate by centrifugation at 12,000×*g* for 5 min, aspirate the ethanol, rinse the precipitate with cold 75% ethanol, and dry at 37°C.
5. Dissolve in Loading Buffer N and analyze by electrophoresis in 15% PAAG under native conditions with subsequent staining of the gel with Stains-All solution (see an example of the gel image in Fig. 1).

Sequence	Name
5'-AUCAUCCAUGGGGCUGGACUU-3' 3'-GGUAGUAGGUACCCCGACCUG-5'	siE
5'-AUCAUCCAUGGGGCUUACGUU-3' 3'-GGUAGUAGGUACCCCGACCUG-5'	fsiE-4
5'-AUCAUCCAUGGGGCUGGACUU-3' 3'-GGUAGUAGGUACCCCGACCUG-5'	siEm
5'-AUCAUCCAUGGGGCUUACGUU-3' 3'-GGUAGUAGGUACCCCGACCUG-5'	fsiE-4m

C, U - 2'-O-methyl- C and U

■ - «mismatches»

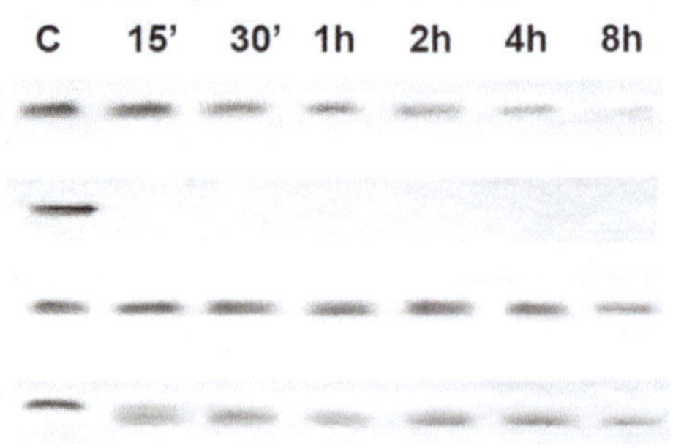

Fig. 1. Degradation of siRNAs and fsiRNAs in DMEM supplemented with 10% FBS. The duplexes shown (*left*) were incubated in DMEM, 10% FBS for the times indicated, separated on a 15% native PAGE and visualized by Stains-All staining (*right*). (Reproduced with modifications from ref. 26).

3.4. 5′-End Labeling of Oligoribonucleotides (See Notes 12–13)

1. Assemble a 10 μl reaction mixture, containing 0.1 mCi [γ^{32}P]-ATP, 0.01 OD260 units of oligoribonucleotide and 10 units of T4 polynucleotide kinase in T4 PNK Buffer, and incubate for 1 h at 37°C or overnight at 4°C.
2. At the end of the incubation, add an equal volume of Loading Buffer D, and separate the 5′-[^{32}P]-oligoribonucleotide from unincorporated ATP by electrophoresis in a 15% PAAG under denaturation conditions.
3. Excise a slice of the gel containing labeled oligoribonucleotide, put it in a disposable plastic tube of appropriate size (typically, a 1.5 ml Eppendorf tube) and add enough Elution Buffer to cover the slice (approx. 300–400 μl).
4. Incubate for 2 h at room temperature with constant inversion. Collect the buffer and repeat the elution with a fresh portion of the buffer.
5. Combine the two solutions containing eluted material and precipitate the radiolabeled oligoribonucleotide by the addition of 2.5 volumes of cold ethanol. Incubate at −20°C for at least 2 h (or leave at −20°C overnight).
6. Collect the precipitate by centrifugation at 12,000×*g* for 5 min, aspirate the ethanol, rinse the precipitate with cold 75% ethanol, and dry at 37°C.
7. Dissolve the oligoribonucleotide in MilliQ water and store in aliquots at −20°C.

3.5. 3′-End Labeling of Oligoribonucleotides

1. Assemble a 20 μl reaction mixture containing 0.01 OD260 units of oligoribonucleotide, 0.1 mM ATP, 10% DMSO, 0.1 mCi 5′- [^{32}P]-cytidine-3′,5′-di-phosphate, and 20 units T4 RNA ligase in RL Buffer. Incubate 18 h at 4°C.
2. At the end of the incubation, add an equal volume of Loading Buffer D, and isolate the radiolabeled product by electrophoresis and gel isolation as described above in the Subheading 3.4.

3.6. Mapping of Nuclease Sensitive Sites (See Note 14)

1. Add 0.6 pmole [^{32}P]-siRNA (2 mCi/nmole) to 200 μl DMEM, 10% FBS, and incubate at 37°C. Samples containing 3′- end and 5′-end labeled siRNA can be processed in parallel or sequentially.
2. Remove 20 μl aliquots after 3, 15, 30 min, 1, 2, and 4 h of incubation, mix with an equal volume of Loading Buffer D, and freeze in liquid nitrogen. Store samples at −20°C.
3. Prepare imidazole ladder and T1 ladder markers as described below (Subheadings 3.7 and 3.8, respectively) to run together with experimental samples for band definition.
4. Pre-run the gel before loading to warm it up to 60°C and keep it warm during the run.

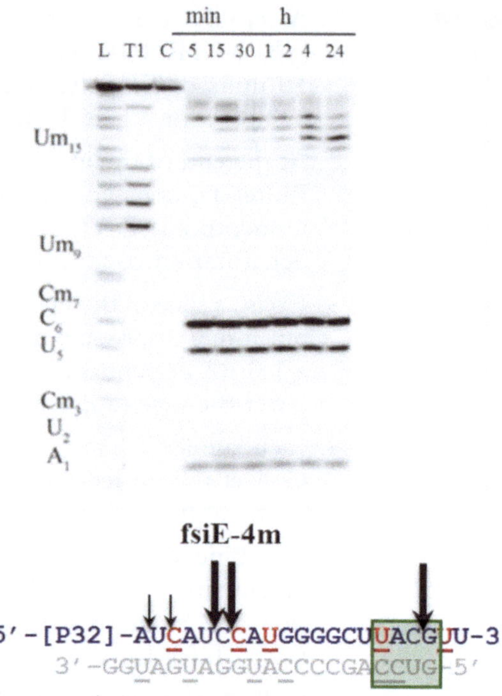

Fig. 2. Mapping of nuclease sensitive sites in fsiE-4 m containing a 5′-[32P]-labeled sense strand. *Top*: An autoradiograph is shown for a 20% polyacrylamide/8 M urea gel. Lanes L and T1, imidazole and RNase T1 ladders, respectively; lanes C, siRNA incubated in DMEM without serum. siRNA (3 nM) was incubated in DMEM supplemented with 10% FBS at 37°C for different times. The incubation times are shown at the *top*. The positions of migration for the truncated nuclease-digested oligonucleotides are designated on the *left*. Modified analogs are designated with an "m" preceding the position number. *Bottom*: The sequence of the 5′-[32P]-labeled oligonucleotide sense strand is shown. *Thin* and *bold arrows* indicate weak and strong cleavage, respectively, in the sense strand of the siRNA. (Reproduced with modifications from ref. 26).

5. Before loading, incubate all samples at 65°C for 5 min, place them on ice briefly, and then load them on a 20% denaturing PAAG of 30–40 cm length running at 40 V/cm.
6. After separation, dry the gel on a gel dryer and expose it to X-ray film, or visualize it using a phosphorimager. An example image of a nuclease mapping experiment is provided in Fig. 2.

3.7. Preparation of an Imidazole Ladder

1. Incubate a reaction mixture (20 μM) containing 6 nM [32P]-oligoribonucleotide (2 μCi/nmole) in Imidazole Buffer at 90°C for 10–15 min.
2. Precipitate the RNA by the addition of 100 μl of 2% lithium perchlorate in acetone.
3. Collect the precipitate by centrifugation (12,000×*g*, 10 min, 4°C), rinse with acetone, air-dry, and dissolve in Loading Buffer D; store frozen at −20°C.

3.8. Preparation of a T1 Ladder by Partial Digestion of Oligoribonucleotides with RNase

1. Incubate a reaction mixture (9 μl) containing 1 nM [^{32}P]-oligoribonucleotide (2 μCi/nmole) in T1 Buffer at 50°C for 10 min.
2. Add 4 units of RNase T1 and incubate for an additional 10 min.
3. After completion, add to the reaction mixture 1/10 volume 10× TBE and store frozen at −20°C.

3.9. Gene Silencing Assay Using a GFP-Tagged Target (See Notes 15–21)

1. One day before transfection, split HEK 293 cells into 12-well plates at a density 3.5×10^5 cells per well in 1.2 ml of Complete Growth Medium with 10% FBS.
2. On the day of transfection replace the medium with the Serum and Antibiotic Free Growth Medium.
3. For each well, mix 0.4 μg of reporter plasmid (for anti-MDR1 fsiRNA we used pEGFP/MDR1 (26)) and 15–150 pmol siRNA. Add Opti-MEM to a total volume of 150 μl.
4. Mix 4 μl of Lipofectamine 2000 with 146 μl of Opti-MEM and incubate 5 min.
5. Combine the two mixtures and gently mix. Incubate at room temperature for 20 min.
6. Add the resulting mixture to the cells dropwise, distributing equally all over the well.
7. Incubate the cells in a 5% CO_2 incubator at 37°C 5% CO_2/95% air.
8. After 4 h of incubation, add 1/3 volume of Complete Growth Medium with 30% FBS and incubate for an additional 20 h.
9. Remove the medium, rinse the cells carefully with PBS twice and add warm trypsin solution for 30 s–1 min. Monitor the dissociation of the cells under an inverted microscope.
10. Add complete growth medium with 10% FBS to stop the action of the trypsin.
11. Collect the cells by centrifugation at $1{,}000 \times g$ for 5 min, suspend them in 250 μl PBS, and repeat the centrifugation and resuspension steps.
12. Add an equal volume of freshly prepared 4% formaldehyde in PBS.
13. Assay the expression level of the EGFP reporter by flow cytometry. Use a 488 nm laser and a 530 ± 30 nm emission filter for EGFP. For each sample, analyze at least 10,000–30,000 cells. To obtained statistically relevant data, collect data from at least three independent experiments, and calculate the average extent of the inhibition of gene expression, the standard deviations and the p-value of the difference between the experimental and control samples.

4. Notes

1. If the choice of the siRNA target is not limited to a specific region of mutation within the mRNA or the site of junction of two genes, it is better to design an active siRNA with canonical duplex structure using an siRNA design tool. We recommend selecting among sequences (a) characterized by relatively high scores in more than one program and (b) with maximum difference in the thermostability of the duplexes formed by nucleotides 1–4 and 16–19 from the 5′-end of the antisense strand. In the case of a point mutated target, fix the position of the mutation inside the "seed" region of the siRNA (2–8 nucleotides from the 5′-terminus of the antisense strand), and then, if required, correct the thermoasymmetry of the duplex by substitutions in the 3′-end region of the sense strand. In the case of a chimeric target, position the siRNA target so that the junction is in the central region of the siRNA duplex and check the thermoasymmetry. Please note that 2′-O-methyl modifications increase the thermostability of the duplex, and their location and number should be considered. The same methodology for selective modification of nuclease sensitive sites could be used for the design of nuclease-resistant canonical siRNA.
2. The oligoribonucleotides (the sense and the antisense strands of siRNA duplexes) have to be electrophoretically pure (approx. 96–99%). The preparations should not contain high molecular weight contamination. The purity of oligoribonucleotides can be also confirmed by mass-spectrometry, reverse-phase high performance liquid chromatography, capillary electrophoresis, etc. After the annealing procedure, verify the formation of siRNA duplexes to avoid problems during gene-silencing experiments. Analyze the formation of siRNA duplexes by gel-electrophoresis in 15% polyacrylamide/0.5% *N,N′*-methylene-bis-acylamide gels under native conditions as described in Subheading 3.2.
3. According to our experience assessing siRNA susceptibility to RNA cleavage by mapping of nuclease sensitive sites, all cytosines and uridines located within $^{5'}$CpA$^{3'}$, $^{5'}$UpA$^{3'}$, and $^{5'}$UpG$^{3'}$ motives have to be protected with 2′-O-methyl modifications. Sometimes it is necessary to introduce 2′-O-methyl modifications in other motives, for example $^{5'}$CpC$^{3'}$ and $^{5'}$UpU$^{3'}$, but these sites have to be identified experimentally.
4. The average number of mismatches that optimally increases the activity of low and moderately active siRNA is 4, and we recommend that initial fsiRNA designs have this number of substitutions. In the case of extremely low or extremely high thermostability of an siRNA duplex, the optimal number of mismatches could be varied from 2 to 5.

5. Please avoid the nuclease sensitive motives $^{5'}$CpA$^{3'}$, $^{5'}$UpA$^{3'}$, and $^{5'}$UpG$^{3'}$ within 3′-overhangs of siRNA duplexes.
6. Because RNA and oligoribonucleotides are very sensitive to traces of ribonucleases, we advise that you take the following extra precautions in the buffer preparation. First, wear gloves when preparing all solutions. Second, it is preferable to prepare stock solutions of 0.5 M HEPES, 1 M Tris, 250 mM sodium citrate, and 2 M solutions of mono- and divalent salts. Sterilize all the stock solutions in an autoclave, filter buffers through a 0.22 μM pore nitrocellulose filter, and store them at 4°C for up to 6 months. For buffers that are to be used for manipulation with oligoribonucleotides, we recommend preparing small portions (up to 50 ml of 5× or 10× concentrated stock). Sterilize the buffers after pH adjustment by filtering through a 0.22 μM pore nitrocellulose filter, add total tRNA from *E. coli* or BSA to achieve the desired concentration, aliquot the buffers into 1 ml portions and store at −20°C until use. When an aliquot of the buffer is opened, use it within a week, and then discard.
7. Please note that you will need to prepare stock solutions of 10 mg/ml total tRNA from *E. coli* and 10 mg/ml of BSA. These solutions are used without sterilization to minimize loss of reagents. Use autoclaved MilliQ water for their preparation, and store these solutions in 100 μl aliquots at −20°C until use.
8. Wear a mask and gloves when weighing acrylamide. Unpolymerized acrylamide is a neurotoxin and extreme care should be exercised to avoid skin contact or inhalation. To avoid exposure to acrylamide, cover the acrylamide with Parafilm after weighing, and transport it to a fume hood. Transfer the acrylamide to a glass cylinder inside the fume hood, add the weighed *N,N′*-methylene-bis-acylamide, urea, and water, and mix on a stir plate placed inside the hood. It is better to prepare a stock solution of 20% acrylamide/1% *N,N′*-methylene-bis-acrylamide/8 M urea. When the solution is ready, add 15 g per liter of Mixed resin AG 501-X8 (anion and cation exchange resin) and stir for at least 1 h until the conductivity of the acrylamide solution is close to that of distilled water. Deionization of acrylamide/urea solution is needed to achieve further resolution. Filter the acrylamide/urea solution through a 0.45 μM pore size nitrocellulose filter, and store it in the dark at 4°C. This solution can be used for 1 month. Before use, remove the required amount, equilibrate to room temperature, and add 1/20 V of TBE 20× buffer and other ingredients for polymerization.

   To prepare 15% acrylamide/0.5% *N,N′*-methylene-bisacrylamide/8 M urea solution for oligoribonucleotide isolation use the same protocol, changing only the acrylamide and *N,N′*-methylene-bis-acrylamide concentrations.

To prepare a native 15% acrylamide/0.5% *N,N'*-methylene-bisacrylamide gel for analysis of siRNA nuclease resistance prepare a stock solution of 15% acrylamide/0.5% *N,N'*-methylene-bisacrylamide using the same protocol. Please note that no urea is added to the solution in this case, but deionization, filtering, and storage conditions are the same.

9. Be careful upon handling of dry pellets of oligoribonucleotides. Usually the oligonucleotide pellet after ethanol precipitation is visible at the bottom of the tube as a fine white powder. Open the tube with the pellet slowly; otherwise the pellet can fly out of the tube upon opening. If the oligonucleotide pellet is overly dry (seen as a white spot in the tube) it takes time to dissolve.

10. Extra precautions are needed when working with RNA to avoid contamination with ribonucleases. Use autoclaved tubes and tips and wear gloves when handling oligoribonucleotides, buffers, etc. Never work with RNA in a room in which ribonucleases are routinely used: for example, ribonuclease is added during the initial steps of plasmid isolation. It is also best to avoid rooms where work is done with *E. coli*, yeast, or other live cultures. Use a reserved set of pipetmen for RNase T1 ladder preparation. Immediately throw out all tubes in which RNase T1 was diluted after use, and clean lab bench, pipets and gloves with ethanol solution after using even a diluted solution of RNase T1. If necessary, use 0.1% diethylpyrocarbonate solution to remove ribonucleases from pipets. Note that diethylpyrocarbonate is a neurotoxin, and care should be exercised to avoid skin contact and inhalation: work under a fume hood.

11. Blood serum contains ribonucleases that maintain their cleavage activity in the presence of Loading Buffer, after freezing and even in the wells of a gel. Stopping the reaction by phenol extraction is required to get proper kinetics data for siRNA degradation in the presence of serum. Do not attempt to stop the reaction by freezing the samples. Take care to prevent the formation of water condensation on the tube caps during incubation because this will result in dramatic changes in the concentration of the components due to the small volume. For this purpose we recommend using a water bath or the upper shelf of a heated air box rather than a heating block.

12. All experiments with radioactive compounds ([γ^{32}P]-ATP or [5′-^{32}P]-cytosine 3′,5′-diphosphate) should be done according to the safety requirements at your institution. Do not throw away tubes, tips, or polyacrylamide gels in which any radioactive material was used into a general trash can. They must be placed in a designated container for radioactive waste. After elution from the gel and ethanol precipitation, dissolve [5′-^{32}P]- or [3′-^{32}P]-labeled oligoribonucleotides in autoclaved MilliQ water, and store the tubes in appropriate polypropylene

or lead containers with tightly adjusted covers at –20°C until use. Never store labeled oligoribonucleotides as a dry pellet: radiolysis can take place.

13. For isolation of 5′-[^{32}P]- or 3′-[^{32}P]-labeled oligoribonucleotides use a gel of at least 20 cm length and 0.4 mm thickness. 21-mer oligoribonucleotides run on a 15% polyacrylamide /0.5% *N,N′*-methylene-bis-acylamide /8 M urea gel between bromophenol and xylene cyanol and unincorporated γ-[^{32}P]-ATP, 5′-[^{32}P]-cytosine 3′,5′-diphosphate migrates much faster than the first dye. Do not allow unincorporated label to run off the gel to avoid contamination of the gel electrophoresis apparatus with radioactivity. It is optimal to allow the first dye to migrate 7–8 cm.
14. Blood serum contains phosphatases that catalyze the dephosphorylation of labeled oligoribonucletides, resulting in the formation of radioactive pyrophosphate, which can be easily lost during precipitation of the products of oligoribonucleotides degradation. We recommend avoiding precipitation to keep the loading balance proportional on the gel. The addition of an equal volume of Loading Buffer to the samples is necessary for good separation of the products on the sequencing gel. Samples loaded on the gel still contain ribonucleases from the serum which maintain their activity, and may even cause accelerated cleavage at the pH in the gel. For this reason, the data obtained in this experiment can be used only for the mapping of nuclease sensitive sites, but not for determining the kinetics.
15. siRNA experiments should include a number of controls to ensure the validity of the data. The editors of Nature Cell Biology have recommended several controls (27). These controls include a negative control siRNA with the same nucleotide composition as your siRNA but lacking significant sequence homology to the genome. To design a negative control siRNA, scramble the nucleotide sequence of the gene-specific siRNA, and conduct a search to make sure it lacks homology to any other gene. Another recommended control is the use of an additional siRNA sequences targeting the same mRNA.
16. For experiments with cell cultures, use only annotated cell lines obtained from an established cell culture collection. Prepare a sufficient number of vials of frozen cells with a low passage number. Do not use cells longer then 1.5–2 month (depending on the cell line), after which time the cells should be discarded and a new vial thawed from liquid nitrogen. Do not return cells to storage after passaging for long time. Do not allow cells to overgrow: it is better to discard the cells after extensive overgrowth and to take a fresh vial of the cells. Maintain the cells in an exponential growth phase before transfection.

The condition of the cells is critical for the efficiency of the transfection.

17. For manipulations with the cells, avoid long-term exposure of small volumes of the culture medium to the air because it can lose CO_2 and change pH. Do not allow cells in wells to stay without medium longer then absolutely required. Note that prolonged incubation of cells with trypsin solution can reduce their viability. After treatment with trypsin, disaggregate cells carefully by pipetting to avoid the presence of cell clumps. Cell monolayers should be distributed evenly in the wells, and the degree of confluence should be in accordance with the recommendations of the manufacture of the transfection reagent.

18. We use Lipofectamine 2000 (Invitrogen, USA) for the transfection of fsiRNA into HEK293 cell. In this case, only Opti-MEM should be used as diluent for the fsiRNA and the Lipofectamine 2000 during the formation of fsiRNA complexes with the transfection reagent. Do not substitute Opti-MEM with DMEM or PBS, as this will result in poor transfection efficiency. For some cell lines, Lipofectamine 2000 may be toxic; in this case, another transfection reagent should be selected. For example, for the transfection of SK-N-MC, we use Oligofectamine (Invitrogen, USA). For each cell line, the conditions of transfection (amount of transfection reagent, cell culture medium, and serum free or serum-containing medium) should be optimized individually. If transfection reagents do not give satisfactory results, electroporation could be used for the delivery of fsiRNA into cells.

19. The design of the Silencing assay depends on your target. If you do not have a convenient cellular model, co-transfection of a plasmid encoding the target mRNA (or its fragment) fused to the coding region of a fluorescent protein could be used (as described in Subheading 3.9). If you are planning to use a cell line endogenously expressing the target mRNA, you should ensure that the silencing of the target gene is not lethal to the cells because in this case, the negative selection would not allow you to evaluate the silencing activity of fsiRNAs. For example, this can happen when testing anti-MDR1 fsiRNA in drug-resistant cell lines growing in the presence of a cytostatic agent.

20. The length of time between fsiRNA transfection and the analysis depends on your target. If you use co-transfection systems containing CMV or another strong promoter, the time should be 24–48 h (sufficient for the synthesis of the fluorescent protein). If your target gene is expressed endogenously in the model cells, an assay time should be selected in the range of 48–72 h when monitoring the mRNA level by RT-PCR. For Western blot analysis, the assay time should be selected taking into account the half-life of the target protein.

21. Avoid cell clumping during preparation of samples for FACS analysis. We recommend adding a fixing solution with double concentration of formaldehyde (4%) to a suspension of the cells in PBS, rather than adding the fixing solution directly to the cell pellet. If your cells tend to form clumps, pass them through a 40–70 μm filter or mesh to eliminate cell clumps (for example, BD Falcon strainer or strainer cap-test tube).

Acknowledgement

This work was supported by the Russian Academy of Science under the programs "Molecular and Cell Biology" grant No. 21.1; "Science to Medicine" grant No. 37; Russian Foundation for Basic Research grants Nos. 11-04-01017-a and 11-04-12095-ofi-m-2011; Ministry of Science and Education of the Russian Federation grant No. 14.740.11.1058 and Siberian Branch of Russian Academy of Sciences grant No. 41.

References

1. Hammond SM, Bernstein E, Beach D, Hannon GJ (2000) An RNA-directed nuclease mediates post-transcriptional gene silencing in Drosophila cells. Nature 404:293–296
2. Martinez J, Patkaniowska A, Urlaub H, Luhrmann NR, Tuschl T (2002) Single-stranded antisense siRNAs guide target RNA cleavage in RNAi. Cell 110:563–574
3. Gregory RI, Chendrimada TP, Cooch N, Shiekhattar R (2005) Human RISC couples microRNA biogenesis and posttranscriptional gene silencing. Cell 123:631–640
4. Matranga C, Tomari Y, Shin C, Bartel DP, Zamore PD (2005) Passenger-strand cleavage facilitates assembly of siRNA into Ago2-containing RNAi enzyme complexes. Cell 123:607–620
5. Aronin N (2006) Target selectivity in mRNA silencing. Gene Ther 13:509–516
6. Khvorova A, Reynolds A, Jayasena SD (2003) Functional siRNAs and miRNAs exhibit strand bias. Cell 115:209–216
7. Schwarz DS, Hutvagner G, Du T, Xu Z, Aronin N, Zamore PD (2003) Asymmetry in the assembly of the RNAi enzyme complex. Cell 115:199–208
8. Amarzguioui M, Prydz H (2004) An algorithm for selection of functional siRNA sequences. Biochem Biophys Res Commun 316: 1050–1058
9. Reynolds A, Leake D, Boese Q, Scaringe S, Marshall WS, Khvorova A (2004) Rational siRNA design for RNA interference. Nat Biotechnol 22:326–330
10. Patzel V (2007) In silico selection of active siRNA. Drug Discov Today 12:139–148
11. Hohjoh H (2004) Enhancement of RNAi activity by improved siRNA duplexes. FEBS Lett 557:193–198
12. Ohnishi Y, Tokunaga K, Hohjoh H (2005) Influence of assembly of siRNA elements into RNA-induced silencing complex by fork-siRNA duplex carrying nucleotide mismatches at the 3′- or 5′-end of the sense-stranded siRNA element. Biochem Biophys Res Commun 329: 516–521
13. Manoharan M (2004) RNA interference and chemically modified small interfering RNAs. Curr Opin Chem Biol 8:570–579
14. Corey DR (2007) Chemical modification: the key to clinical application of RNA interference? J Clin Invest 117:3615–3622
15. De Paula D, Bentley MV, Mahato RI (2007) Hydrophobization and bioconjugation for enhanced siRNA delivery and targeting. RNA 13:431–456
16. Watts JK, Deleavey GF, Damha MJ (2008) Chemically modified siRNA: tools and applications. Drug Discov Today 13: 842–855
17. Chernolovskaya EL, Zenkova MA (2010) Chemical modification of siRNA. Curr Opin Mol Ther 12(2):158–67

18. Hall AH, Wan J, Shaughnessy EE, Ramsay Shaw B, Alexander KA (2004) RNA interference using boranophosphate siRNAs: structure-activity relationships. Nucleic Acids Res 32:5991–6000
19. Layzer JM, McCaffrey AP, Tanner AK, Huang Z, Kay MA, Sullenger BA (2004) In vivo activity of nuclease-resistant siRNAs. RNA 10: 766–771
20. Dowler T, Bergeron D, Tedeschi AL, Paquet L, Ferrari N, Damha MJ (2006) Improvements in siRNA properties mediated by 2′-deoxy-2′-fluoro-beta-D-arabinonucleic acid (FANA). Nucleic Acids Res 34:1669–1675
21. Chiu YL, Rana TM (2003) siRNA function in RNAi: a chemical modification analysis. RNA 9:1034–1048
22. Czauderna F, Fechtner M, Dames S, Aygun H, Klippel A, Pronk GJ, Giese K, Kaufmann J (2003) Structural variations and stabilising modifications of synthetic siRNAs in mammalian cells. Nucleic Acids Res 31:2705–2716
23. Choung S, Kim YJ, Kim S, Park HO, Choi YC (2006) Chemical modification of siRNAs to improve serum stability without loss of efficacy. Biochem Biophys Res Commun 342:919–927
24. Bramsen JB, Laursen MB, Nielsen AF, Hansen TB, Bus C, Langkjaer N, Babu BR, Hojland T, Abramov M, Van Aerschot A, Odadzic D, Smicius R, Haas J, Andree C, Barman J, Wenska M, Srivastava P, Zhou C, Honcharenko D, Hess S, Muller E, Bobkov GV, Mikhailov SN, Fava E, Meyer TF, Chattopadhyaya J, Zerial M, Engels JW, Herdewijn P, Wengel J, Kjems J (2009) A large-scale chemical modification screen identifies design rules to generate siRNAs with high activity, high stability and low toxicity. Nucleic Acids Res 37:2867–2881
25. Volkov AA, Kruglova NS, Meschaninova MI, Venyaminova AG, Zenkova MA, Vlassov VV, Chernolovskaya EL (2009) Selective protection of nuclease-sensitive sites in siRNA prolongs silencing effect. Oligonucleotides 19:191–202
26. Petrova Kruglova NS, Meschaninova MI, Venyaminova AG, Zenkova MA, Vlassov VV, Chernolovskaya EL (2010) 2′-O-methyl-modified anti-MDR1 fork-siRNA duplexes exhibiting high nuclease resistance and prolonged silencing activity. Oligonucleotides 20: 297–308
27. Editors of Nature Cell Biology (2003) Whither RNAi? Nat Cell Biol 5:489–490

Chapter 9

Designing Dual-Targeting siRNA Duplexes Having Two Active Strands that Combine siRNA and MicroRNA-Like Targeting

Pål Sætrom

Abstract

Short interfering RNAs (siRNAs) have become valued tools for knocking down specific genes. As such, siRNAs are routinely used to study gene function and are also being explored as therapeutic agents. Traditionally, siRNAs are designed to target one specific gene, but this chapter describes a procedure for designing dual-targeting siRNAs where the two strands in the siRNA duplex are both active and down-regulate different target genes through both siRNA and miRNA-like effects. The procedure can be used to create siRNAs that robustly target pairs of genes.

Key words: siRNA, Design, RNA interference, Bioinformatics

1. Introduction

Short 21–22 nt duplex RNAs, such as microRNAs (miRNAs) and short interfering RNAs (siRNAs), are the triggers in RNA interference, as the RNA-induced silencing complex (RISC) binds one of the strands in the siRNA duplex and uses this strand as a guide to recognize and down-regulate target messenger RNAs (mRNAs). Being artificial exogenous molecules, siRNAs are designed to bind and knockdown one specific target mRNA. Typically, the siRNA guide strand has perfect complementarity to this intended target and limited complementarity to other mRNAs, but siRNAs can also act as miRNAs and bind short regions in 3′ untranslated regions (UTRs) and thereby down-regulate these targeted mRNAs (1, 2). Moreover, the other strand in the siRNA duplex—the passenger strand—can also be incorporated into RISC and target and regulate mRNAs. These additional targets are collectively referred to as the siRNA's off-targets and can be a strong confounding factor in siRNA functional screens (3).

Debra J. Taxman (ed.), *siRNA Design: Methods and Protocols*, Methods in Molecular Biology, vol. 942,
DOI 10.1007/978-1-62703-119-6_9, © Springer Science+Business Media, LLC 2013

Here, I describe an siRNA design procedure that harnesses the siRNAs' off-targeting potential into dual-targeting siRNAs that robustly inhibit two different mRNA targets (4). Specifically, the procedure creates siRNA duplexes that have two functional strands such that each strand recognizes distinct target sites in distinct mRNAs. Each strand in the dual-targeting siRNAs has perfect complementarity to its intended target site, but the dual-targeting siRNA duplex itself can contain up to six mismatches. Three additional key aspects of the dual-targeting siRNA design are (1) that the duplex ends have similar thermodynamic stability such that RISC does not preferentially incorporate one strand over the other (5, 6), (2) that each duplex strand has the same sequence characteristics as highly effective regular siRNAs (7–9), and (3) that the duplex strands have miRNA-like target sites in the target mRNA's 3'UTRs. These three aspects separate this dual-targeting design approach from earlier attempts (10). Using this design approach, we created dual-targeting siRNAs that successfully targeted and knocked down for six of six distinct target gene pairs (4).

2. Materials

2.1. Sequence Data

1. Target gene sequence: Use a Web-browser to access the University of California Santa Cruz (UCSC) Table Browser Web site (11) (http://genome.ucsc.edu/cgi-bin/hgTables). Download from the site the target gene's complete mRNA sequence and 3′ UTR sequence and store the sequences in two separate FASTA files (see Note 1).

2.2. Computer Programs

1. RNAfold (12): Use a Web-browser to access the RNAfold Web server (http://rna.tbi.univie.ac.at/cgi-bin/RNAfold.cgi). Paste or type the hairpin sequence in the designated text field and press "Proceed" to compute the hairpin minimum free energy (see Note 2).
2. GPboost (8): Use a Web-browser to access the GPboost Web server (http://demo1.interagon.com/sirna/). Input the 19mer target site sequence into the "Nucleotide" text field and press "Submit."

3. Methods

Designing dual-targeting siRNAs consists of first finding partially complementary dual-targeting siRNA duplex candidates and then, for each duplex candidate (1) calculating the duplex end stability

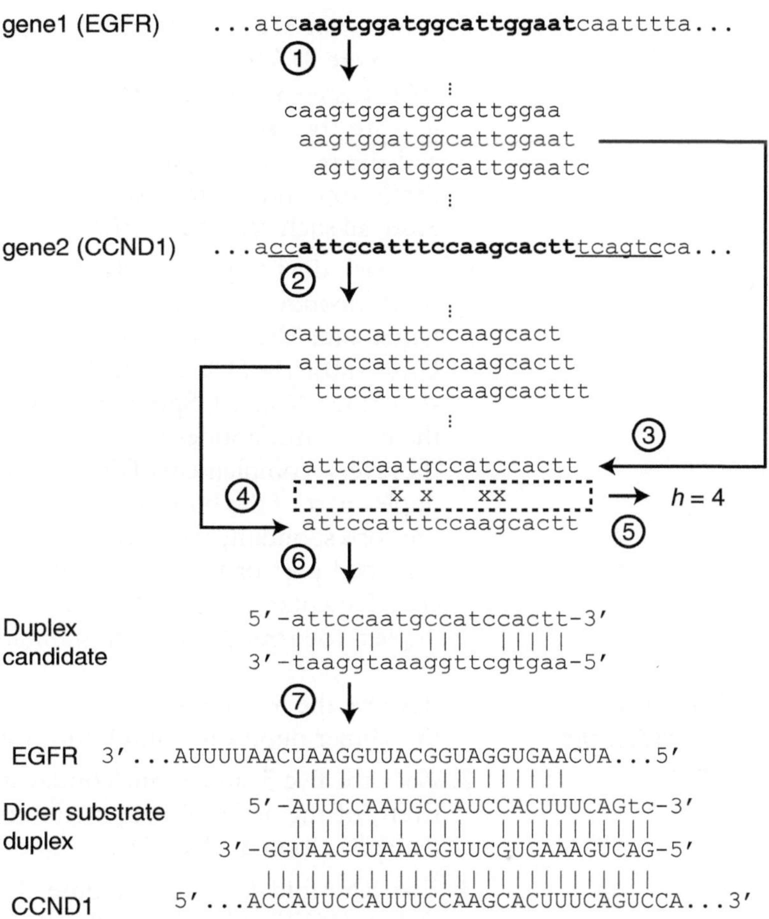

Fig. 1. Steps in finding partially complementary dual-targeting siRNA candidates. The two input sequences (EGFR and CCND1 in this example) are broken down into candidate 19mer sequences (steps 1–2), the Hamming distances between all pairs of candidate 19mer sequences from the two input sequences are computed to identify candidate duplexes (steps 3–6), and each candidate duplex is extended to create a Dicer substrate duplex (step 7).

difference, (2) predicting both duplex strands' siRNA efficacies, (3) predicting both duplex strands' miRNA efficacies, and (4) scoring and prioritizing the duplex candidates.

3.1. Finding Partially Complementary Dual-Targeting siRNA Candidates

1. Retrieve the mRNA sequence for the first target gene, *gene1*, and split the sequence into all 19mer subsequences (Fig. 1, step 1).
2. Retrieve the mRNA sequence for the second target gene, *gene2*, and split the sequence into all 19mer subsequences (Fig. 1, step 2).
3. For each 19mer subsequence *g1,i* in *gene1*, create its reverse complement (Fig. 1, step 3). Then, for each 19mer subsequence

g2,j in *gene2* (Fig. 1, step 4), compute the Hamming distance (see Note 3) h between *g2,j* and the reverse complement of *g1,i* (Fig. 1, step 5; see Note 4). If $h \leq 5$ (see Note 5), then *g1,i* and *g2,j* are the target sites for a dual-targeting siRNA candidate and the reverse complements of *g1,i* and *g2,j* form the candidate's top and bottom strands, respectively (Fig. 1, step 6). Store all such candidates *g1,i* and *g2,j* for further processing.

4. For each dual-targeting siRNA candidate, retrieve the two and six nucleotides 5′ and 3′, respectively, of the bottom strand's target site. Use these two and six nucleotides to construct a Dicer-substrate (13, 14) dual-targeting siRNA duplex (Fig. 1, step 7; see Note 6). Specifically, use the reverse complement of the two 5′ nucleotides for the bottom strand's 3′ overhang and the reverse complement of the six 3′ nucleotides for the extended duplex part of the bottom strand. Use the six 3′ nucleotides for the corresponding extended part of the top strand such that the extended part of the Dicer-substrate siRNA forms a perfectly complementary duplex. Replace the two 3′ RNA residues in the duplex' top strand with corresponding DNA residues.

3.2. Calculating the Duplex End Stability Difference

1. Retrieve the five 5′ nucleotides and the five 3′ nucleotides from the 19mer duplex top and bottom strands, respectively.
2. Place the five 5′ and 3′ nucleotides at the 5′ and 3′ end, respectively, of the artificial hairpin cassette 5′-CCLLLLLGG-3′ (L denotes a loop nucleotide).
3. Use RNAfold (12) to compute the hairpin's minimum free energy (MFE). Use this MFE value ΔG_T as an estimate of the duplex end stability of the top strand's 5′ end.
4. Retrieve the five 5′ nucleotides and the five 3′ nucleotides from the 19mer duplex bottom and top strands, respectively, and repeat steps 2 and 3 above by using these nucleotide sequences. Use the resulting MFE value ΔG_B as an estimate of the duplex end stability of the bottom strand's 5′ end.
5. Compute the top strand's difference in duplex end stability $\Delta\Delta G_T = \Delta G_T - \Delta G_B$.
6. Compute the bottom strand's difference in duplex end stability $\Delta\Delta G_B = \Delta G_B - \Delta G_T$.

3.3. Predicting Both Duplex Strands' siRNA Efficacy

1. Input the top strand's 19mer target sequence into the GPboost siRNA efficacy predictor (8). The resulting score S_T is the top strand's predicted siRNA efficacy.
2. Repeat step 1 for the bottom strand's 19mer target sequence the get the bottom strand's siRNA predicted efficacy S_B.
3. Alternatively, use Reynolds and colleagues' algorithm (7) to score efficacies R_T and R_B of the top and bottom strands,

respectively. Use these scores to compute the corresponding strands' predicted siRNA efficacies as $S_T = (R_T - 5)/10$ and $S_B = (R_B - 5)/10$.

3.4. Predicting Both Duplex Strands' miRNA Efficacy

1. Retrieve the eight nucleotides from the top strand's 5′ end, remove the first (5′) nucleotide, and reverse-complement the resulting seven nucleotides to get the sequence for the top strand's 7mer seed site *s7T* (see Note 7).
2. Remove the first (5′) nucleotide from the top strand's 7mer seed site sequence *s7T* to get the top strand's 6mer seed site sequence *s6T*.
3. Add an Adenine (A) nucleotide to the end of the 6mer and 7mer seed site sequences to get the 6mer-A1 and 7mer-A1 seed site sequences *s6AT* and *s7AT*, respectively.
4. Retrieve the 3′ UTR sequence for the top strand's target gene, *gene1*, and find the positions of all occurrences of the top strand's four seed site sequences *s6T*, *s6AT*, *s7T*, and *s7AT* (see Note 8).
5. Remove from the list of seed site positions seed sites that overlap with a stronger seed site. Use the following hierarchy to determine seed site strength: $s6T < s6AT < s7T < s7AT$.
6. Use the resulting seed site positions to compute the distance between each consecutive seed site.
7. Create a set of seed modules $SM = \{M_1, \dots, M_k\}$ by grouping into a module all neighboring seed sites where the distance between two consecutive seed sites is at most 35 nucleotides.
8. For each seed module M_i, consisting of l seed sites and $l-1$ distances $Di = \{d_1, \dots, dl_{-1}\}$, use the following function to compute the score f for the seed site module: $f(Mi, Di) = 1 + \sum j\ g(dj)$, where $g(dj) = \{0 \text{ if } dj \leq 13;\ 1 \text{ if } dj \geq 17; \text{ and } (dj - 13)/4 \text{ otherwise}\}$.
9. Compute the final miRNA score M_T by taking the score for the highest scoring seed module and adding the score for each of the other seed modules, multiplied by a weight factor h that depends on the distance d to the neighboring seed module; that is, $h(d) = \{0.25 \text{ if } d \geq 70;\ 1.75 - 0.75 \times d/35 \text{ otherwise}\}$.
10. Repeat steps 1–9 for the bottom strand to compute the bottom strand's miRNA score M_B.

3.5. Scoring Duplex Candidates

1. Use the following function to compute the strand score for the duplex candidate's top strand: $S_T = r(\Delta\Delta G_T) + S_T + M_T$, where $r(\Delta\Delta G_T) = \{-1 \text{ if } \Delta\Delta G_T < -1;\ 1 \text{ if } \Delta\Delta G_T > 1;\ 0 \text{ otherwise}\}$.
2. Use the same as in step 1 to compute the strand score for the duplex candidate's bottom strand; that is, $S_B = r(\Delta\Delta G_B) + S_B + M_B$.

3. Use the following function to compute the dual-targeting siRNA candidate's duplex score S: $S = (S_T + S_B)/2 - stdev(S_T, S_B)$, where $stdev(S_T, S_B)$ is the sample standard deviation of S_T and S_B and is equal to $\sqrt{((S_T + S_B)^2/2 - 2 \times S_T \times S_B)}$.
4. Repeat steps 1–3 for each dual-targeting siRNA candidate and order the dual-targeting siRNA candidates by their duplex score. The highest-scoring duplex candidate is the best dual-targeting siRNA candidate (see Note 9).

4. Notes

1. You can retrieve target gene sequences from other sources than UCSC, such as the NCBI nucleotide database (http://www.ncbi.nlm.nih.gov/nucleotide/) or Ensembl (http://www.ensembl.org/) or store the target gene sequence in other file formats than FASTA. What is important for this design protocol is to have the target gene's mRNA and 3′ UTR as separate sequences, and the UCSC Table Browser provides a user-friendly interface for retrieving both.
2. We use a locally executable version of RNAfold instead of the RNAfold Web server because this locally installed version can easily be integrated into a design pipeline. The source code and instructions for installing RNAfold can be downloaded from http://rna.tbi.univie.ac.at/.
3. We use the Hamming distance metric instead of the more commonly used Levenshtein (edit) distance metric. The reason is that the Levenshtein metric considers insertions and deletions in addition to substitutions, and in our experience, including these two operations tend to give very asymmetric candidate duplexes, which may be suboptimal for Dicer processing.
4. We use custom search hardware (15) to compute these Hamming distances. Specifically, for a given 19mer subsequence *g1,i* in *gene1*, we run a single search on the search hardware to find all 19mers in *gene2* that have a Hamming distance less than 6 (the threshold) to the reverse complement of *g1,i*. Alternative but slower methods to compute Hamming distances include the nrgrep program with the option "–k 5s" ((16); http://www.dcc.uchile.cl/~gnavarro/software/nrgrep.tar.gz) or the Python code in Table 1.
5. We use an initial Hamming distance threshold of 5 as this threshold usually gives a reasonable number of candidates for further

Table 1
Python function for computing the Hamming distance between two strings *g1* and *g2*

```
from itertools import izip_longest
def hammingDistance(g1, g2):
return sum(n1 != n2 for n1, n2 in izip_longest(g1, g2,
  fillvalue='X'))
```

processing; overall, this threshold has a 55% chance of yielding dual-targeting siRNA candidates (4). If we do not find any dual-targeting candidates for this Hamming distance threshold, we increase the threshold to 6 and repeat the design process. We do not use higher thresholds, as Hamming distances of 7 or above will likely compromise the duplex' stability.

6. Expectedly, Dicer cleaves Dicer-substrate double-stranded RNAs 21 nucleotides from the duplex end with the 2 nucleotide 3′ overhang (left side in Fig. 1, step 7). In some cases, however, Dicer's actual cleavage site differs slightly from the expected—especially for duplexes that contain bulges or internal loops, which can result in Dicer creating both longer and shorter duplexes. As the 5′ end is more important for siRNA and miRNA targeting than is the 3′ end, such differences have little effect on the top strand. In contrast, a bottom strand that is 1–2 nucleotides longer than expected may have reduced efficacy if the 1–2 additional nucleotides do not base-pair with the strand's intended target site. By using the target site for the bottom strand as template when extending the duplex candidate into a Dicer-substrate siRNA, we ensure that the bottom strand always has perfect complementarity to its intended target irrespective of the actual Dicer cleavage site (Fig. 1, step 7).
7. These seven nucleotides form the strand's seed region and their reverse complement forms the strand's 3′ UTR seed sites.
8. We use the Python code in Table 2 to find the positions of all such seed sites.
9. This complete design procedure is available from the following Web server: http://demo1.interagon.com/DualTargeting/.

Table 2
Python function for finding the positions of all occurrences of the seed sequence *seed* in the 3′ UTR *utr*

```
def findSeedPositions(seed, utr):
positions = []
hitpos = utr.find(seed, 0)
while hitpos != -1:
positions.append(hitpos)
hitpos = utr.find(seed, hitpos + 1)
return positions
```

Acknowledgement

This work was supported by the Norwegian Functional Genomics Program of the Norwegian Research Council.

References

1. Birmingham A, Anderson EM, Reynolds A, Ilsley-Tyree D, Leake D, Fedorov Y, Baskerville S, Maksimova E, Robinson K, Karpilow J, Marshall WS, Khvorova A (2006) 3' UTR seed matches, but not overall identity, are associated with RNAi off-targets. Nat Methods 3:199–204
2. Jackson AL, Burchard J, Schelter J, Chau BN, Cleary M, Lim L, Linsley PS (2006) Widespread siRNA "off-target" transcript silencing mediated by seed region sequence complementarity. RNA 12:1179–1187
3. Lin X, Ruan X, Anderson MG, McDowell JA, Kroeger PE, Fesik SW, Shen Y (2005) siRNA-mediated off-target gene silencing triggered by a 7 nt complementation. Nucleic Acids Res 33:4527–4535
4. Tiemann K, Hohn B, Ehsani A, Forman SJ, Rossi JJ, Saetrom P (2010) Dual-targeting siRNAs. RNA 16:1275–1284
5. Khvorova A, Reynolds A, Jayasena SD (2003) Functional siRNAs and miRNAs exhibit strand bias. Cell 115:209–216
6. Schwarz DS, Hutvagner G, Du T, Xu Z, Aronin N, Zamore PD (2003) Asymmetry in the assembly of the RNAi enzyme complex. Cell 115:199–208
7. Reynolds A, Leake D, Boese Q, Scaringe S, Marshall WS, Khvorova A (2004) Rational siRNA design for RNA interference. Nat Biotechnol 22:326–330
8. Saetrom P (2004) Predicting the efficacy of short oligonucleotides in antisense and RNAi experiments with boosted genetic programming. Bioinformatics 20:3055–3063
9. Saetrom P, Snove O Jr (2004) A comparison of siRNA efficacy predictors. Biochem Biophys Res Commun 321:247–253
10. Hossbach M, Gruber J, Osborn M, Weber K, Tuschl T (2006) Gene silencing with siRNA duplexes composed of target-mRNA-complementary and partially palindromic or partially complementary single-stranded siRNAs. RNA Biol 3:82–89
11. Karolchik D, Hinrichs AS, Furey TS, Roskin KM, Sugnet CW, Haussler D, Kent WJ (2004)

The UCSC table browser data retrieval tool. Nucleic Acids Res 32:D493–496

12. Hofacker IL (2003) Vienna RNA secondary structure server. Nucleic Acids Res 31:3429–3431
13. Kim DH, Behlke MA, Rose SD, Chang MS, Choi S, Rossi JJ (2005) Synthetic dsRNA Dicer substrates enhance RNAi potency and efficacy. Nat Biotechnol 23:222–226
14. Rose SD, Kim DH, Amarzguioui M, Heidel JD, Collingwood MA, Davis ME, Rossi JJ, Behlke MA (2005) Functional polarity is introduced by Dicer processing of short substrate RNAs. Nucleic Acids Res 33: 4140–4156
15. Halaas A, Svingen B, Nedland M, Saetrom P, Snove O, Birkeland OR (2004) A recursive MISD architecture for pattern matching. Ieee T Vlsi Syst 12:727–734
16. Navarro G (2001) NR-grep: a fast and flexible pattern-matching tool. Software Pract Exper 31:1265–1312

Chapter 10

Strategies for Designing and Validating Immunostimulatory siRNAs

Michael P. Gantier

Abstract

Specific clinical applications of RNA interference (RNAi) can benefit from a concurrent activation of the immune system. This is the case for small interfering RNAs (siRNAs) with antitumor or antiviral activities. This chapter provides a brief overview of the strategies reported to date to design siRNAs with gene silencing and immune activating properties, as well as methods for the validation of immunostimulatory activities.

Key words: Innate immunity, RNA interference, siRNA, RIG-I, TLR7, TLR8, TLR9, Bifunctional siRNA

1. Introduction

Blood immune cells express a set of receptors that aid in the detection of pathogens and the mounting of an appropriate immune response. For instance, Toll-like receptors (TLRs) detect foreign DNA, RNA or bacterial/fungal components, together with modified endogenous molecules (e.g., oxidized low density lipoprotein (LDL) and fibrillar amyloid-β peptides). It is now well accepted that siRNAs have the potential to recruit immune receptors specialized in RNA detection, such as TLR3 and TLR7 (1–3). This relates to the route of siRNA delivery and their inherent propensity to be cleared by blood phagocytes. Double-stranded RNA-induced immune activation results in the production of a range of cytokines normally associated with antiviral activity (including type I interferons (IFNs) such as IFN-α and tumor necrosis factor alpha (TNF-α)), which promotes flu-like symptoms (4). This is generally perceived as an unwanted nonspecific effect of in vivo siRNA administration and can be prevented through the use of chemical modifications of the

Debra J. Taxman (ed.), *siRNA Design: Methods and Protocols*, Methods in Molecular Biology, vol. 942,
DOI 10.1007/978-1-62703-119-6_10, © Springer Science+Business Media, LLC 2013

siRNA oligonucleotides, such as the use of 2′-O-methyl ribonucleotides (5, 6). Nonetheless, for specific applications, siRNA-driven immune activation can be harnessed to the gene-targeting capabilities of the siRNA and result in enhanced activity. For example, this is the case for antitumoral and antiviral siRNAs, where the dual effect of gene-targeting and immune activation can potentiate tumor or viral restriction (7–11).

In this chapter, we review the current design strategies to increase siRNA-driven immune activation, together with methods allowing for the assessment of immune activation by bifunctional siRNAs. Three distinct strategies have currently been applied to generate bifunctional siRNAs, with gene targeting and immunostimulatory capabilities. Each approach relies on the activation of a different innate immune receptor, be it TLR7/8, TLR9 or retinoic acid-inducible gene I (RIG-I). While TLR7/8 and TLR9 exclusively detect endosomal RNAs/DNAs in immune cells, RIG-I detects cytoplasmic RNAs in both immune and nonimmune cells (12). Recruitment of these receptors therefore depends on the type of strategy adopted to deliver the siRNA. For example, certain lipids favor cytoplasmic versus endosomal uptake (13).

1.1. siRNAs Recruiting TLR7/8 in Human Immune Cells

1.1.1. TLR7/8 Sensing of Small RNAs

RNAi studies in vivo rapidly identified that siRNAs could induce a strong innate immune response when administered systemically in mammals with cationic lipids or cationic polymers (1, 2, 14), which favor endosomal uptake by phagocytes (13, 15). This response relies strictly on TLR7 in the mouse, and specific uridine immunostimulatory motifs in the siRNA duplex (1, 2). While such immunostimulatory motifs usually also elicit activation of an immune response in human blood, endosomal small RNA sensing in humans is more complex and involves both TLR7 and TLR8 (13, 16, 17). This results in the detection of a greater variety of small RNAs (16, 18). In addition to specific uridine motifs, we and others have recently proposed that the affinity between the two complementary strands of an siRNA could influence recruitment of TLR7/8 in human immune cells (19, 20). Making use of this observation, we characterized an siRNA scaffold allowing for increased TLR7/8 activation while retaining full RNAi efficacy (19).

1.1.2. Rational Design of siRNAs Recruiting TLR7/8

Because sequence-dependent siRNA TLR7/8 sensing is strongly reliant on uridine bases (16, 17, 21), one approach to design bifunctional siRNAs is to screen for siRNAs containing uridine-rich motifs such as UGUGU or GUCCUUCA (1, 2). However the presence of such uridine-rich motifs in one of the two strands of the siRNA does not necessarily predict immunostimulation (22). For instance, we observed that incorporation of UUGGUU or UUUUUU in the 5′-end of the targeting strand of an asymmetric Dicer-substrate siRNA did not promote immune stimulation (19).

In addition, because RNAi efficacy can vary greatly between siRNAs, identification of sequences that perform best in both immunostimulation and silencing is particularly challenging.

An alternate approach is to introduce a specific mismatch in the passenger strand of the siRNA duplex to create an immunostimulatory motif. This was recently applied to increase the immunostimulatory properties of an siRNA targeting H5N1 influenza in chicken cells (9), but could also be applicable to mammalian studies. Interestingly, when such a mismatch is introduced between bases 9 and 12 from the 5′-end of the passenger strand, it can also result in increased RNAi potency (23). Preservation of the targeting strand is however essential, as single mismatches with the target can ablate silencing efficacy of the duplex (24). Nevertheless, the ability to introduce such an "immunostimulatory" mismatch is strongly dependent on the siRNA sequence considered.

We have recently reported a novel siRNA scaffold, which circumvents these restrictions and confers TLR8-driven immunostimulation upon any given siRNA sequence while preserving RNAi efficacy (19). This siRNA design relies on the asymmetric Dicer-substrate siRNA scaffold (D-siRNA), which consists of a 25-nt passenger strand and a 27-nt targeting strand (see Fig. 1) and can have increased RNAi potency (25). Critically, when applied to conventional 19+2 siRNAs, we found that this design resulted in loss of silencing efficacy (19). This design should therefore be restricted to the D-siRNA scaffold.

Design rules:

1. Following identification of the 19-mer siRNA target site for optimal RNAi potency, the passenger strand (sense strand) of the D-siRNA scaffold is designed by selecting the next six bases of the target mRNA (in 5′–3′ orientation) (Fig. 1a).
2. The complementary targeting strand is inferred from this sequence and extended at the 3′ end with a 2-nt overhang, which can be complementary to the target. This 2-nt overhang should preferentially be synthesized with RNA moieties for increased RNAi efficacy (25).
3. The sense strand is subsequently mutated to contain a UUUU motif from bases 9–12 (from the 5′-end of the sense strand) (see Note 1). Bases 24 and 25 should preferentially be synthesized with DNA moieties for increased RNAi efficacy (25).

Increased immunostimulatory activity of such siRNAs results from the converging contribution of the poly-uridine motif, together with the structural distortion that lowers the affinity of the two strands. Importantly, our data indicate that these activities are prevalently attributed to TLR8 sensing (19). The use of uridine-modified D-siRNAs is restricted to species where TLR8 is involved in RNA sensing, thereby excluding mice (17, 26). Finally, TLR8

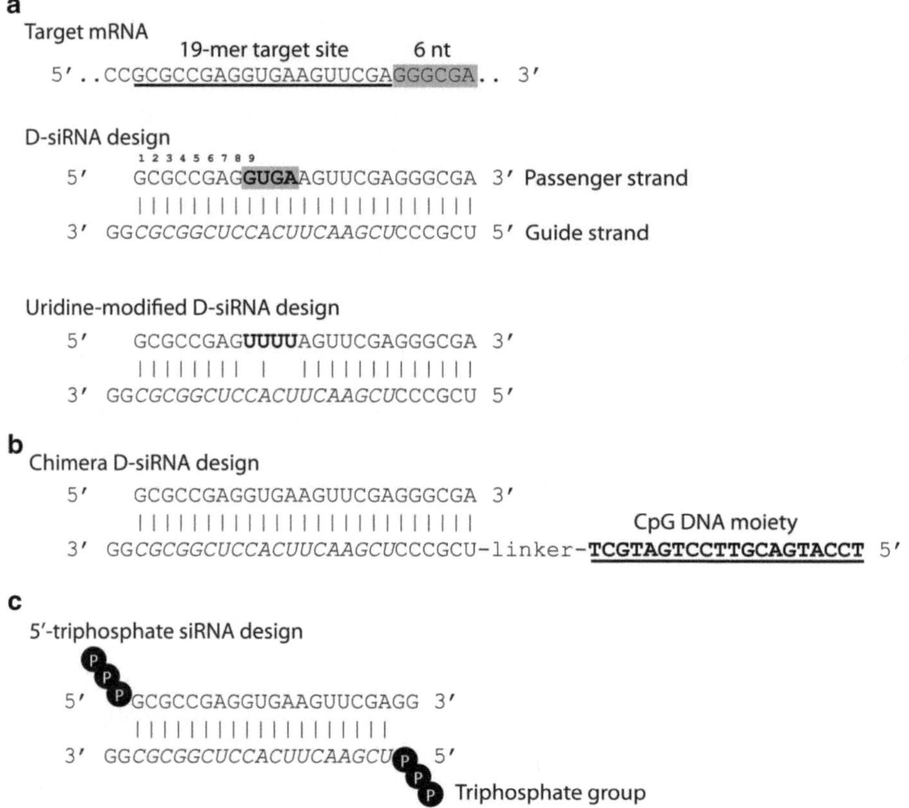

Fig. 1. Design of immunostimulatory siRNAs. All the designs presented here rely on the enhanced Green Fluorescent Protein (EGFP) siRNA sequence previously reported (19). (**a**) Design of D-siRNAs with a uridine bulge, increasing TLR8 activation. The uridine bulge is replacing bases 9–12 (in this case "GUGA") of the D-siRNA passenger strand. (**b**) D-siRNA–CpG chimera design (based on ref. 7). A CpG DNA moiety (phosphorothioate-modified) is fused to the guide strand of the D-siRNA by a linker. In this example, the CpG sequence used is the type B CpG 1668, which is mouse-specific. (**c**) siRNA with 5′-triphosphate moieties for enhanced RIG-I activation. Such triphosphate moieties are added during bacteriophage in vitro transcription of each siRNA strand, or using chemical synthesis.

recruitment by uridine-modified D-siRNAs following systemic injection intrinsically relies on their endosomal uptake by phagocytes. Immunostimulation is therefore contingent upon the siRNA delivery approach used.

1.2. siRNAs Recruiting TLR9

TLR9 is specialized in the detection of unmethylated DNA. In addition, TLR9 can be artificially activated by synthetic oligonucleotides with cytidine–phosphate–guanosine-containing motifs (CpG). Similarly to that of TLR7/8, TLR9 sensing is restricted to the endosomal compartment of immune cells. In a recent study, Kortylewski et al. reported the use of a D-siRNA fused to a CpG oligonucleotide moiety using 6 units of C3 carbon chain linker $(CH_2)_3$, which retained RNAi efficacy and promoted TLR9 activation in mouse immune cells (7, 27) (Fig. 1b).

Critically, such chimera siRNAs can be delivered spontaneously to the endosome of phagocytes, both in vitro and in vivo, without the need for liposomal transfection reagent ((7); Gantier MP, unpublished). In addition, these bifunctional chimera siRNAs promoted specific silencing in phagocytes following TLR9-mediated endosomal maturation (7).

Given that certain CpG oligonucleotides can trigger TLR9 activation in both human and mouse, this approach presents the great advantage over that of uridine-modified D-siRNAs of being applicable to mouse models of disease. Nevertheless, TLR9 differential expression in immune cell subtypes between species—and possibly functional outcome—is an important parameter to be taken into account when designing preclinical studies.

1.3. siRNAs Recruiting RIG-I

RIG-I is an intracellular sensor of viral RNA (4, 28). In recent years, it has been shown that RIG-I specifically detects viral RNAs on the basis of 5′-triphosphate moieties added by viral polymerases (29, 30). When combined with double-stranded structures, presence of the 5′-triphosphate moiety activates RIG-I-dependent immune responses (28, 31). Consequently, siRNAs that possess 5′-triphosphates are potent activators of RIG-I while retaining RNAi capacity (8, 11, 31) (Fig. 10.1c). While this approach was shown to promote synergistic antitumoral effects against melanoma in animal model experiments (8) or to have increased antiviral effects (10, 11), it relied on in vitro synthesized siRNAs, using bacteriophage RNA polymerase. Such enzymatic synthesis of siRNAs, although applicable for preclinical studies, presents challenges for their purity (due to the synthesis of aberrant side-products), which is a critical issue for their large-scale usage in the clinic (31). In addition, alternative chemical syntheses of synthetic 5′-triphosphate siRNAs, although previously reported by Schlee et al. (31), remains extremely challenging and a likely hurdle for large-scale industrial production.

2. Materials

2.1. Cell Culture

1. THP-1: ATCC reference TIB-202.
2. Cell medium: Royal Park Memorial Institute medium (RPMI) 1640 plus L-glutamine medium (Life Technologies) complemented with 1× antibiotic/antimycotic (Life Technologies) and 10% sterile fetal bovine serum (FBS; Life Technologies) (referred to as complete RPMI).
3. Dulbecco's Phosphate-Buffered Saline (PBS).
4. TrypLE™ Express Stable Trypsin (Life Technologies).

5. Sterile tissue culture treated microtest™ 96-well plates (Falcon).
6. Stimulant of human TLR8: 1 mg/mL CL075 (InvivoGen).
7. *N*-[1-(2,3-Dioleoyloxy)propyl)-*N*,*N*,*N*-trimethylammonium methylsulfate (DOTAP) (Roche).
8. Stimulant of human TLR9: 500 μg/mL ODN 2216 (InvivoGen).
9. 1 mg/mL Phorbol-12-myristate-13-acetate (PMA; Calbiochem) in 100% dimethyl sulfoxide (DMSO).
10. 10^6 U/mL Recombinant human IFN-γ (Chemicon).
11. siRNA duplex buffer: 100 mM potassium acetate, 30 mM HEPES, pH 7.5, in UltraPure™ DNase/RNase-Free Distilled Water—Life Technologies.
12. siRNAs are synthesized as single-stranded RNAs (ssRNAs) and purified using high-performance liquid chromatography (HPLC) by Integrated DNA Technologies (IDT). The ssRNAs are resuspended into filter-sterilized siRNA duplex buffer to a concentration of 80 μM. Each siRNA duplex results from the annealing of two complementary ssRNAs at 92°C for 2 min, which are then left for 30 min at room temperature before being aliquoted and frozen at −80°C (at a final concentration of 40 μM).

2.2. TNF-α and IFN-α Enzyme-Linked ImmunoSorbent A ssay

1. TNF-α OptEIA Enzyme-Linked ImmunoSorbent Assay (ELISA) set (BD Biosciences).
2. TNF-α coating buffer: 0.084 mg $NaHCO_3$ with 0.036 mg Na_2CO_3 in 10 mL of double-distilled H_2O (ddH_2O), freshly made up.
3. Capture monoclonal mouse antibody to human IFN-α, clone MMHA-11 (PBL Interferon Source).
4. Detection rabbit polyclonal antibody to human IFN-α (PBL Interferon Source).
5. 200 μg/0.5 mL goat antirabbit IgG horseradish peroxidase (HRP) (SC-2004) from Santa Cruz.
6. PBS 10×: NaCl 8% (w/v), KCl 0.2% (w/v), Na_2HPO_4 1.22% (w/v), KH_2PO_4 0.2% (w/v) in ddH_2O—pH 7.4.
7. PBS-tween (PBST): 1× PBS diluted in H_2O complemented with 0.05% tween 20.
8. Pharmingen Assay Diluent (BD Biosciences Pharmingen).
9. Three million IU (11.1 μg/0.5 mL) Human recombinant IFN-α-2a (Roferon-a, Roche Pharmaceuticals)—injection solution.
10. F96 maxisorp plates (Nunc).
11. Tetramethyl benzidine substrate (TMB, Sigma-Aldrich).

12. Sulfuric acid 2N.
13. Plate reader with 450 nm absorbance filter.

3. Methods

Two protocols are described here, allowing for the in vitro identification of bifunctional siRNAs with increased immunostimulatory activities through TLR7/8 and 9. Importantly, because they rely on targeted endosomal delivery of the siRNAs, these two assays are limited to bifunctional siRNAs recruiting TLR7/8 and TLR9, and are not applicable to 5′-triphosphate siRNAs activating RIG-I (13). Both protocols rely on similar procedures of cell transfection and cytokine analyses, and are therefore grouped together.

3.1. Bifunctional siRNAs Recruiting Human TLR7 and/or TLR9

The first protocol described in this section relies on the use of human peripheral blood mononuclear cells (PBMCs), which consist of a mixed population of immune cells with varying expression levels of TLRs, and varying abilities to produce selective cytokines. In this assay, we measure the production of human IFN-α by PBMCs following endosomal delivery of siRNAs. This reflects the activation of plasmacytoid dendritic cells (pDCs), which predominantly express TLR7 and 9, and are the main producers of IFN-α under TLR7/9 recruitment (13, 32). This assay is referred to as *the PBMC assay.*

3.2. Bifunctional siRNAs Recruiting Human TLR7/8

The second protocol described relies on the use of differentiated human monocytic THP-1 cells. We have previously reported that this cell line could be used to identify small RNAs recruiting TLR7 and TLR8 (16). In this assay, we use the production of TNF-α as a readout of immune activation by the cells. Although not as sensitive as the analysis of human TNF-α produced by primary human PBMCs (33), this approach allows for the more convenient screening of bifunctional siRNAs in a very common monocytic cell line. This assay is referred to as *the THP-1 assay.*

3.3. Preparation of the Cells

3.3.1. For the PMBC Assay

Following purification of human PBMCs as previously reported (33), seed an average of 130,000–200,000 cells in 150 μL of complete RPMI medium in each well of a 96-well plate. Rest the cells for a minimum of 1 h at 37°C in 5% CO_2 prior to stimulation.

3.3.2. For the THP-1 Assay

1. THP-1 cells are grown in tissue-treated plasticware, and passaged two to three times a week in RPMI complete. While a fraction of the cells will differentiate and become adherent, the majority of the cells should remain in suspension. The cells should be split when large clumps are visible, by transferring

one-third of the volume grown into a new flask. It is critical to ensure that the clumps separate into individual cells during this operation, by pipetting vigorously several times (10–20) with a 10 mL pipette.

2. The remainder of the cells can be used for an experiment. We usually use ~80,000 cells per well of a 96-well plate. Following a hemocytometer-based cell count, calculate the volume of cells required to plate the number of wells desired (using 150 μL per well). If necessary, supplement with fresh complete RPMI medium. Importantly however, the volume of fresh medium added should not exceed the volume of conditioned medium containing the cells (see Note 2).
3. Add PMA to a final concentration of 20 ng/mL and aliquot the cells into the wells. Leave the cells to differentiate overnight at 37°C in 5% CO_2 (~16–18 h).
4. Following differentiation, the cells will now be adhering to the bottom of the wells. Carefully aspirate the medium from each well, making sure not to touch the adhering cells (see Note 3).
5. Replace medium with 150 μL of fresh complete RPMI supplemented with 100 U/mL of recombinant human IFN-γ. Incubate the cells for another 6 h. This promotes increased expression of TLR7 and TLR8—with a preference for TLR8—which restores sequence-dependent RNA sensing by the THP-1 cells (16).
6. Aspirate the medium from each well and replace with 150 μL of fresh complete RPMI. The cells are now ready for TLR stimulation.

3.4. TLR Stimulation of the Cells

We always perform each treatment in biological triplicate. This is particularly important given that the output of the assay relies on cytokine production, which often varies between replicates of the same condition. The amount/volumes indicated here are for treatments in biological triplicate. Treatment of the cells is the same for PBMCs and THP-1 cells.

1. In sterile microcentrifuge tubes, aliquot 63.8 μL of RPMI that has not been complemented with antibiotic/antimycotic and FBS (referred to as pure RPMI). Dilute 11.2 μL of 40 μM siRNA into each tube (resulting in 75 μL per tube).
2. In a separate tube, mix 21 μL DOTAP with 54 μL pure RPMI (a mastermix conserving this ratio can be made). Mix the tube by gentle tapping and then incubate at room temperature for 5 min (see Note 4).
3. Add 75 μL of DOTAP/RPMI mix to each diluted siRNA, mix gently, and then incubate the tubes for a further 10 min at room temperature.

4. Add 50 μL of the DOTAP–siRNA mixture to each well of plated cells (three wells per condition) to give a final volume of 200 μL and a final siRNA–DOTAP concentration of 750 nM. Positive controls for TLR7/8/9 stimulation should also be added (see Note 5). Incubate the plate overnight at 37°C for 14–18 h.
5. The following morning, inspect the cells using inverted microscopy (see Note 6). Collect 100 μL of supernatant and dilute 1:2 with OPti-EA buffer for PBMCs (there is no need to dilute for THP-1 cells). Freeze the supernatants at −80°C and keep until cytokine analysis by ELISA.

3.5. Cytokine Production Analysis by ELISA

1. The day before the assay (or a few days before), coat a maxisorp 96-well plate with 100 μL of diluted capture antibody and leave sealed with tape at 4°C:
   (a) For the IFN-α ELISA, use 1:1,500 dilution of the monoclonal antihuman IFN-α in fresh 1× PBS.
   (b) For the TNF-α ELISA, use 1:500 dilution of the capture antibody in TNF-α coating buffer.
2. The morning of the assay, rinse the plate three times with PBST and block for 1 h at room temperature with 130 μL Assay Diluent per well, with rocking.
3. Following blocking, wash the plate three times with PBST.
   (a) Prepare the human IFN-α standard curve using a two-step dilution of human recombinant IFN-α-2a (Roferon-a). First, dilute 1 μL of Roferon-a in 49 μL of Assay Diluent (see Note 7). Vortex well and dilute further 0.9 μL of this into 800 μL of Assay Diluent. This gives a concentration of ~500 pg/mL (used as the top standard), further diluted in 1:2 series dilutions to 7.8125 pg/mL (7 points).
   (b) Prepare the TNF-α standard curve following the Analysis Certificate leaflet from the kit, to give a concentration range from 1,000 pg/mL to 15.6 pg/mL (7 points).
4. Add 75–100 μL of diluted/neat supernatant or standard to each well of the ELISA plate, and incubate for 2 h at room temperature, with rocking.
5. Wash the plate four times with PBST and prepare the diluted capture antibody.
   (a) For the IFN-α ELISA, dilute the detection antibody (rabbit polyclonal antibody to human IFN-α) to 1:1,750, together with the goat α-rabbit HRP antibody diluted to 1:1,000 in Assay Diluent. Incubate for 10 min before adding 100 μL per well and further incubate for 1 h at room temperature, with rocking.

(b) For the TNF-α ELISA, dilute both the detection antibody and streptavidin–horseradish peroxidase (SAv–HRP) to 1:500 in Assay Diluent. Incubate for 10 min before adding 100 μL per well and further incubate for 1 h at room temperature, with rocking.

6. Following 5–7 PBST washes, perform the enzymatic assay. Add 100 μL of prewarmed TMB (at 25–37°C) per well and stop the reaction with 50 μL sulfuric acid (see Note 8). Read the absorbance in a plate reader within 30 min at 450 nm (correction using absorbance at 570 nm can be applied) (see Note 9).

4. Notes

1. Our data rely exclusively on the UUUU motif. However, the literature indicates that other motifs can be used to increase immunostimulation (1, 2, 9, 22). Our data suggest that the structural distortion introduced by the mismatches between the UUUU motif and the siRNA targeting strand contributes significantly to the increased immune activation (19). While other motifs could therefore be substituted for the UUUU, such motifs should be selected by trying to keep the minimum of complementarity with the targeting strand.
2. When passaging 1:3 confluent THP-1 cells in 15 mL in a T-75 flask, we usually find that the remainder of the cells (i.e., ~10 mL) can be directly plated with PMA into 60 wells of a 96-well plate (150 μL per well), resulting in the appropriate cell confluency per well (~60–80,000 cells). It is critical to PMA-differentiate the cells in the conditioned medium in which the cells have been grown (this can be supplemented with up to 50% fresh medium, if necessary), to ensure the correct responsiveness of the cells to the TLR agonists.
3. To minimize the loss of cells while aspirating the supernants and ensure the proper removal of PMA/IFN-γ-supplemented media, the plate can be tilted on an angle.
4. In this protocol, we rely on a DOTAP–RNA ratio of 1.87 μL DOTAP per μL of siRNA at 40 μM for both THP-1 and PBMCs. This ratio can be increased further in the PBMC assay to 2.65 μL DOTAP per μL of siRNA at 40 μM to obtain maximal immune activation.
5. Positive and negative controls should be included for each experiment. CL75 (at a final concentration of 1–2 μg/mL) is a strong activator of human TLR8 and TNF-α in the THP-1 assay (16). For the PBMC assay and IFN-α induction, we use the TLR9 agonist ODN2216 (at a final concentration of 3 μM). TLR7-specific agonists can also be used.

6. Following overnight stimulation with DOTAP–siRNA complexes, phagocytic activation should be visible and reflected by the formation of an increased number of cell clumps. While it is not always possible to draw significant conclusions from such clumps as for immunostimulation, their formation confirms the proper cellular uptake of DOTAP–siRNA complexes.
7. Other recombinant IFN-α-2a can be used—however, Roferon is widely available from hospital pharmacies. We use batches that have passed their expiry date, but these remain stable for several years when kept at 4°C.
8. The reaction should be stopped when a blue coloration for each standard of the standard curve is visible or when the intensity of the blue coloration of the samples is much more intense than at the highest point of the standard curve. As long as the substrate has not been fully processed, the coloration is proportional to the amount of enzyme. Thus, even though the coloration of the positive controls is more than that of the top standard, meaningful results can be drawn from their absorbance as long as the substrate was not extinguished. Following sulfuric acid addition, we often find that Optical Densities up to a value of 3.0 yield significant results.
9. In these experiments, we use a final concentration of 750 nM. Although very high, these concentrations might not be sufficient to activate TLR7 and 8 in vitro but could still induce immune activation in vivo.

Acknowledgments

The author thanks Bryan Williams, Cameron Stewart, Soroush Sarvestani, and Frances Cribbin, for their useful comments and their help with the redaction of this review. The author was supported by funding from the Australian NHMRC (1006590 and 1022144) and the Victorian Government's Operational Infrastructure Support Program.

References

1. Hornung V, Guenthner-Biller M, Bourquin C, Ablasser A, Schlee M, Uematsu S, Noronha A, Manoharan M, Akira S, de Fougerolles A, Endres S, Hartmann G (2005) Sequence-specific potent induction of IFN-alpha by short interfering RNA in plasmacytoid dendritic cells through TLR7. Nat Med 11:263–270
2. Judge AD, Sood V, Shaw JR, Fang D, McClintock K, MacLachlan I (2005) Sequence-dependent stimulation of the mammalian innate immune response by synthetic siRNA. Nat Biotechnol 23:457–462
3. Kleinman ME, Yamada K, Takeda A, Chandrasekaran V, Nozaki M, Baffi JZ, Albuquerque RJ, Yamasaki S, Itaya M, Pan Y, Appukuttan B, Gibbs D, Yang Z, Kariko K, Ambati BK, Wilgus TA, DiPietro LA, Sakurai E, Zhang K, Smith JR, Taylor EW, Ambati J

(2008) Sequence- and target-independent angiogenesis suppression by siRNA via TLR3. Nature 452:591–597

4. Gantier MP, Williams BR (2007) The response of mammalian cells to double-stranded RNA. Cytokine Growth Factor Rev 18:363–371
5. Judge AD, Bola G, Lee AC, MacLachlan I (2006) Design of noninflammatory synthetic siRNA mediating potent gene silencing in vivo. Mol Ther 13:494–505
6. Zamanian-Daryoush M, Marques JT, Gantier MP, Behlke MA, John M, Rayman P, Finke J, Williams BR (2008) Determinants of cytokine induction by small interfering RNA in human peripheral blood mononuclear cells. J Interferon Cytokine Res 28:221–233
7. Kortylewski M, Swiderski P, Herrmann A, Wang L, Kowolik C, Kujawski M, Lee H, Scuto A, Liu Y, Yang C, Deng J, Soifer HS, Raubitschek A, Forman S, Rossi JJ, Pardoll DM, Jove R, Yu H (2009) In vivo delivery of siRNA to immune cells by conjugation to a TLR9 agonist enhances antitumor immune responses. Nat Biotechnol 27:925–932
8. Poeck H, Besch R, Maihoefer C, Renn M, Tormo D, Morskaya SS, Kirschnek S, Gaffal E, Landsberg J, Hellmuth J, Schmidt A, Anz D, Bscheider M, Schwerd T, Berking C, Bourquin C, Kalinke U, Kremmer E, Kato H, Akira S, Meyers R, Hacker G, Neuenhahn M, Busch D, Ruland J, Rothenfusser S, Prinz M, Hornung V, Endres S, Tuting T, Hartmann G (2008) 5′-Triphosphate-siRNA: turning gene silencing and Rig-I activation against melanoma. Nat Med 14:1256–1263
9. Stewart CR, Karpala AJ, Lowther S, Lowenthal JW, Bean AG (2011) Immunostimulatory motifs enhance antiviral siRNAs targeting highly pathogenic avian influenza H5N1. PLoS One 6:e21552
10. Han Q, Zhang C, Zhang J, Tian Z (2011) Reverse of HBV-induced immune tolerance by an immunostimulatory 3p-HBx-siRNA in a retinoic acid inducible gene I (RIG-I) -dependent manner. Hepatology 54(4):1179–1189
11. Ebert G, Poeck H, Lucifora J, Baschuk N, Esser K, Esposito I, Hartmann G, Protzer U (2011) 5′ Triphosphorylated small interfering RNAs control replication of hepatitis B virus and induce an interferon response in human liver cells and mice. Gastroenterology 141(2):696–706
12. Melchjorsen J, Jensen SB, Malmgaard L, Rasmussen SB, Weber F, Bowie AG, Matikainen S, Paludan SR (2005) Activation of innate defense against a paramyxovirus is mediated by RIG-I and TLR7 and TLR8 in a cell-type-specific manner. J Virol 79:12944–12951
13. Ablasser A, Poeck H, Anz D, Berger M, Schlee M, Kim S, Bourquin C, Goutagny N, Jiang Z, Fitzgerald KA, Rothenfusser S, Endres S, Hartmann G, Hornung V (2009) Selection of molecular structure and delivery of RNA oligonucleotides to activate TLR7 versus TLR8 and to induce high amounts of IL-12p70 in primary human monocytes. J Immunol 182:6824–6833
14. Sioud M, Sorensen DR (2003) Cationic liposome-mediated delivery of siRNAs in adult mice. Biochem Biophys Res Commun 312:1220–1225
15. Schlee M, Hornung V, Hartmann G (2006) siRNA and isRNA: two edges of one sword. Mol Ther 14:463–470
16. Gantier MP, Tong S, Behlke MA, Xu D, Phipps S, Foster PS, Williams BR (2008) TLR7 is involved in sequence-specific sensing of single-stranded RNAs in human macrophages. J Immunol 180:2117–2124
17. Heil F, Hemmi H, Hochrein H, Ampenberger F, Kirschning C, Akira S, Lipford G, Wagner H, Bauer S (2004) Species-specific recognition of single-stranded RNA via toll-like receptor 7 and 8. Science 303:1526–1529
18. Forsbach A, Nemorin JG, Montino C, Muller C, Samulowitz U, Vicari AP, Jurk M, Mutwiri GK, Krieg AM, Lipford GB, Vollmer J (2008) Identification of RNA sequence motifs stimulating sequence-specific TLR8-dependent immune responses. J Immunol 180:3729–3738
19. Gantier MP, Tong S, Behlke MA, Irving AT, Lappas M, Nilsson UW, Latz E, McMillan NA, Williams BR (2010) Rational design of immunostimulatory siRNAs. Mol Ther 18: 785–795
20. Goodchild A, Nopper N, King A, Doan T, Tanudji M, Arndt GM, Poidinger M, Rivory LP, Passioura T (2009) Sequence determinants of innate immune activation by short interfering RNAs. BMC Immunol 10:40
21. Sioud M (2006) Single-stranded small interfering RNA are more immunostimulatory than their double-stranded counterparts: a central role for 2′-hydroxyl uridines in immune responses. Eur J Immunol 36:1222–1230
22. Jurk M, Chikh G, Schulte B, Kritzler A, Richardt-Pargmann D, Lampron C, Luu R, Krieg AM, Vicari AP, Vollmer J (2011) Immunostimulatory potential of silencing RNAs can be mediated by a non-uridine-rich toll-like receptor 7 motif. Nucleic Acid Ther 21:201–214
23. Addepalli H, Meena, Peng CG, Wang G, Fan Y, Charisse K, Jayaprakash KN, Rajeev KG, Pandey RK, Lavine G, Zhang L, Jahn-Hofmann K, Hadwiger P, Manoharan M, Maier MA (2010) Modulation of thermal stability can

enhance the potency of siRNA. Nucleic Acids Res 38:7320–7331

24. Hamada M, Ohtsuka T, Kawaida R, Koizumi M, Morita K, Furukawa H, Imanishi T, Miyagishi M, Taira K (2002) Effects on RNA interference in gene expression (RNAi) in cultured mammalian cells of mismatches and the introduction of chemical modifications at the 3′-ends of siRNAs. Antisense Nucleic Acid Drug Dev 12:301–309
25. Rose SD, Kim DH, Amarzguioui M, Heidel JD, Collingwood MA, Davis ME, Rossi JJ, Behlke MA (2005) Functional polarity is introduced by Dicer processing of short substrate RNAs. Nucleic Acids Res 33:4140–4156
26. Gantier MP, Irving AT, Kaparakis-Liaskos M, Xu D, Evans VA, Cameron PU, Bourne JA, Ferrero RL, John M, Behlke MA, Williams BR (2010) Genetic modulation of TLR8 response following bacterial phagocytosis. Hum Mutat 31:1069–1079
27. Gantier MP, Williams BR (2009) siRNA delivery not toll-free. Nat Biotechnol 27:911–912
28. Schlee M, Hartmann G (2010) The chase for the RIG-I ligand – recent advances. Mol Ther 18:1254–1262
29. Hornung V, Ellegast J, Kim S, Brzozka K, Jung A, Kato H, Poeck H, Akira S, Conzelmann KK, Schlee M, Endres S, Hartmann G (2006) 5′-Triphosphate RNA is the ligand for RIG-I. Science 314:994–997
30. Pichlmair A, Schulz O, Tan CP, Naslund TI, Liljestrom P, Weber F, Reis e Sousa C (2006) RIG-I-mediated antiviral responses to single-stranded RNA bearing 5′-phosphates. Science 314:997–1001
31. Schlee M, Roth A, Hornung V, Hagmann CA, Wimmenauer V, Barchet W, Coch C, Janke M, Mihailovic A, Wardle G, Juranek S, Kato H, Kawai T, Poeck H, Fitzgerald KA, Takeuchi O, Akira S, Tuschl T, Latz E, Ludwig J, Hartmann G (2009) Recognition of 5′ triphosphate by RIG-I helicase requires short blunt double-stranded RNA as contained in panhandle of negative-strand virus. Immunity 31:25–34
32. Hornung V, Rothenfusser S, Britsch S, Krug A, Jahrsdorfer B, Giese T, Endres S, Hartmann G (2002) Quantitative expression of toll-like receptor 1–10 mRNA in cellular subsets of human peripheral blood mononuclear cells and sensitivity to CpG oligodeoxynucleotides. J Immunol 168:4531–4537
33. Gantier MP, Williams BR (2010) Monitoring innate immune recruitment by siRNAs in mammalian cells. Methods Mol Biol 623:21–33

Chapter 11

Designing Efficient and Specific Endoribonuclease-Prepared siRNAs

Vineeth Surendranath, Mirko Theis, Bianca H. Habermann, and Frank Buchholz

Abstract

RNA interference (RNAi) has grown to be one of the main techniques for loss-of-function studies, leading to the elucidation of biological function of genes in various cellular systems and model organisms. While for many invertebrates such as *Drosophila melanogaster* (*D. melanogaster*) and *Caenorhabditis elegans* (*C. elegans*) long double-stranded RNA (dsRNA) can directly be used to induce a RNAi response, chemically synthesized small interfering RNAs (siRNAs) are typically employed in mammalian cells to avoid an interferon-like response triggered by long dsRNA (Reynolds et al., RNA 12:988–993, 2006). However, siRNAs are expensive and beset with unintentional gene targeting effects (off-targets) confounding the analysis of results from such studies. We, and others, have developed an alternative technology for RNAi in mammalian cells, termed endoribonuclease-prepared siRNA (esiRNA), which is based on the enzymatic generation of siRNA pools by digestion of long dsRNAs with recombinant RNase III *in vitro* (Yang et al., Proc Natl Acad Sci USA 99: 9942–9947, 2002; Myers et al., Nat Biotechnol 21:324–328; 2003). This technology has proven to be cost-efficient and reliable. Furthermore, several studies have demonstrated that complex pools of siRNAs, as inherent in esiRNAs, which target one transcript reduce off-target effects (Myers et al., J RNAi Gene Silencing 2:181, 2006; Kittler et al., Nat Methods 4:337–344, 2007). Within this chapter we describe design criteria for the generation of target-optimized esiRNAs.

Key words: RNA interference, esiRNA, siRNA, Off-target effects, siRNA pool, siRNA efficiency

1. Introduction

RNA interference (RNAi) is a cellular process that takes part in the control of gene expression. The molecular machinery of RNAi is required for both the production and control of micro-RNAs (miRNAs), as well as small interfering RNAs (siRNAs). It is believed that RNAi has several functions in an organism or a cell, including the defense against parasitic genes from viruses or transposons (6), and the control of developmental processes through

Debra J. Taxman (ed.), *siRNA Design: Methods and Protocols*, Methods in Molecular Biology, vol. 942,
DOI 10.1007/978-1-62703-119-6_11, © Springer Science+Business Media, LLC 2013

posttranscriptional gene regulation (7). In the laboratory, RNAi has advanced to a powerful and widely used method to silence genes in cells or whole organisms (8).

Several enzymatic functions are part of the RNAi pathway. The endoribonuclease Dicer digests long double-stranded RNA (dsRNA) or miRNA precursors into small pieces of about 20–25 nucleotides, leading to the double-stranded form of siRNAs and miRNAs (9, 10). These RNAs activate the RISC (RNA induced silencing complex), which unwinds the si- or miRNAs, and incorporates one of the strands, the so-called guide strand (11). The RISC complex uses the enclosed guide strand to identify target mRNAs, which leads to their translational repression or degradation (11). Argonaute, the catalytically active RNase in the RISC complex, is involved in the selection of the RNA strand that is incorporated into the complex. It seems that the strand with the lower thermodynamic stability at its 5′end is preferentially chosen, suggesting that unwinding and strand incorporation are linked to each other (11). This step is a crucial factor in target specificity, as the RISC complex can only regulate the expression of target mRNAs complementary to the guide strand.

Most eukaryotic organisms, including mammals, express the RNA machinery to regulate their genes in a variety of biological processes, and this fact has been widely exploited to perform functional screens in invertebrate model organisms and plants. In these model systems, long dsRNA can be used for the silencing of genes. However, long dsRNA, when applied to most mammalian cells led to unspecific responses and cell death due to triggering an interferon-like response (1). In contrast, with the discovery that long dsRNA is processed *in vivo* and that RNAi is indeed mediated by short 21-mers (12), RNAi also became available for mammalian cells (13). This work has paved the way for making RNAi one of the most widely and successfully used techniques for functional genomics studies in mammalian cells. Yet it seemed that the usage of synthetic siRNAs also led to silencing of unintended targets (14), making results of gene knockdowns sometimes difficult to interpret. Strikingly, many of the affected unintended targets did not share similarity over the entire length of the siRNA, but rather resembled the miRNA-based silencing mechanism based on seed matches in the 3′UTR (15). Hence, the unspecificity of siRNAs probably results from eliciting a miRNA-like response (15). Since this discovery, much effort has been focused on improving the specificity of RNAi experiments in mammalian cells. Some progress has been reported by carefully selecting siRNA sequences (16) and/or by modifying certain bases in the RNA molecules (17). We and others could show that the off-target effect of the siRNAs could be greatly reduced by using complex mixtures of siRNAs targeting a gene rather than a single siRNA (4, 5) (Fig. 1).

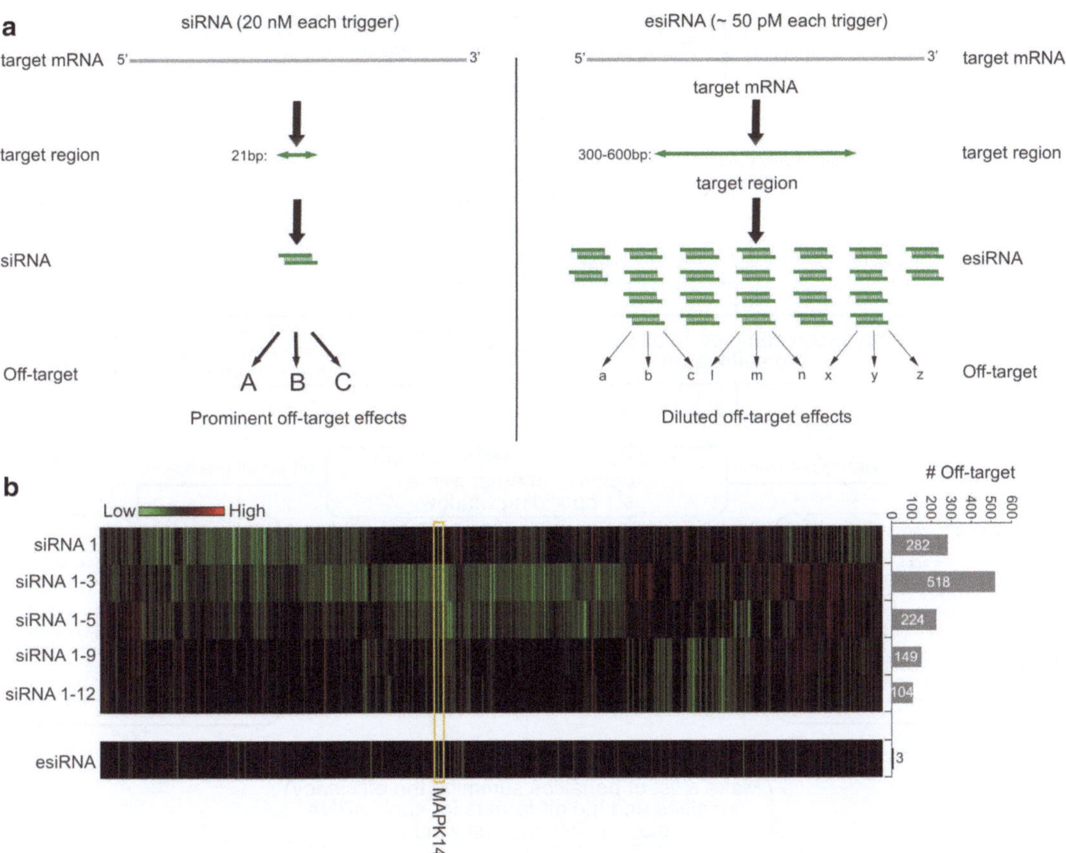

Fig. 1. Complex pools of siRNAs reduce off-target effects. (**a**) Individual siRNAs typically target a region of 21 base pairs in length in the target transcript sequence. In contrast, esiRNAs are complex pools of different siRNAs covering a region of 300–600 base pairs in length. Note that every siRNA has its typical sequence-dependent prominent off-target signature (*indicated by capital letters*). Because the various siRNAs comprising an esiRNA pool are each at reduced concentration compared to an individual siRNA off-target effects are diluted out (*indicated by small letters*), whereas every silencing trigger contributes to on-target silencing. (**b**) Expression array analysis of the changes in transcript levels after target gene (MAPK14) depletion by RNAi. Transcripts that were significantly altered ($p<0.01$; *green*: downregulated; *red*: upregulated) are shown in clusters. A quantification of the off-target effects is shown to the right. Increasing numbers of pooled siRNAs (1–12 individual siRNAs) reduce the number of off-target events. As a comparison, the transcript changes after esiRNA transfection is shown on the *bottom*. Note the clean signature produced by the complex siRNA pool inherent to the esiRNAs.

Even before the increased specificity of siRNA mixtures was revealed, research laboratories explored alternatives to chemically synthesized siRNAs for usage in mammalian cells. A sensible approach was to take the step of dsRNA digestion out of the cell into the test tube and transfect cells with the resulting mixture of siRNAs (2, 18). Endoribonuclease-prepared siRNAs (esiRNAs) have since been successfully used to perform genome-scale knockdown screens (19–27). While the first screens were carried out using esiRNAs based on cDNA clones (28), later screens were based on target-optimized esiRNAs through in silico selection of regions most suitable for gene silencing (5).

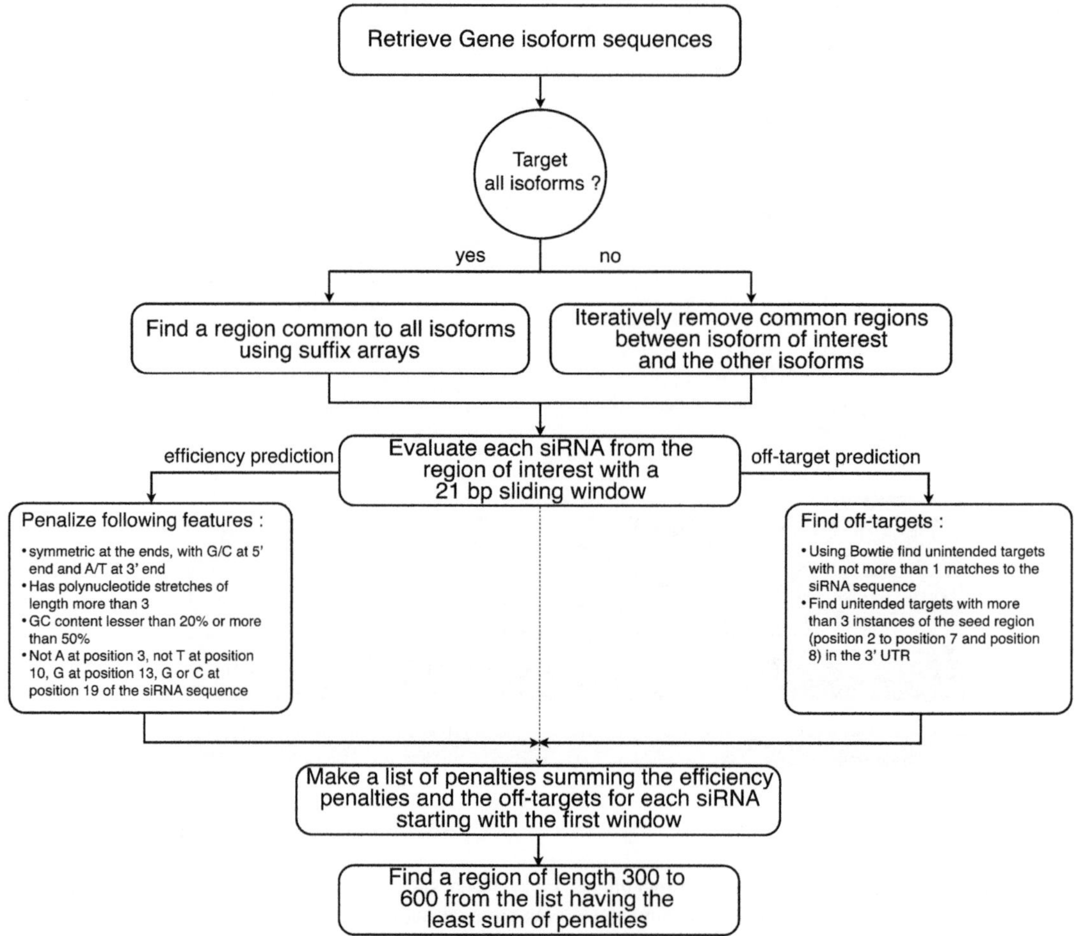

Fig. 2. Workflow for the design of esiRNAs corresponding to Subheading 3.

Mechanistic insights into the RNA interference pathway (29, 30) and systematic studies of silencing efficiency and specificity (16, 31) have taught us about sequence features that should be considered when selecting an efficient siRNA. As stated earlier, the strand bias of the RISC complex for instance favors the incorporation of the strand with a thermodynamically less stable 5′ end. Based on experimental and computational work, preferences for several positions in a siRNA have been defined (13, 16, 32–38), and systematic analysis of observed off-targets has led to rules governing the specificity of a siRNA (15, 39). When designing an esiRNA, the same principles apply, aiming for a mixture of optimized siRNAs with high potency and specificity. In this chapter, we introduce a step-by-step protocol (Fig. 2) to design specific and efficient esiRNAs for gene knockdown studies.

2. Tools and Resources Needed for the Design of esiRNAs

The design of esiRNAs involves two primary steps in succession. The first is the identification of the entity to be knocked down and the subsequent sequence region from which to design the esiRNA. The second is to predict, *in silico*, the silencing potential (the efficiency) and the off-target effect (the specificity) of the esiRNA.

2.1. Resources for Obtaining the Sequence Region of Interest

mRNA sequences can be obtained from NCBI (http://www.ncbi.nlm.nih.gov) or ENSEMBL (http://www.ensembl.org) for most model organisms. When designing esiRNAs for use in *D. melanogaster* or *C. elegans*, sequences are preferably obtained from FlyBase (http://www.flybase.org) or WormBase (http://www.wormbase.org).

2.2. Tools for Evaluating Efficiency and Specificity

1. An interpreted programming language to encode the sequence features to be evaluated for efficiency calculations such as Python (http://www.python.org) or Perl (http://www.perl.org).
2. Bowtie (http://bowtie-bio.sourceforge.net/index.shtml) to compare the sequences of the constituent siRNAs against the reference transcriptome.
3. A bioinformatics programming library designed to handle biological sequence data, such as BioPython or BioPerl. These, and other frameworks corresponding to the programming language of choice, are listed at http://www.open-bio.org.

3. Instructions for Designing esiRNAs (Summarized in Fig. 2)

3.1. Choosing a Region of Sequence Against Which an esiRNA Is To Be Designed

Because an esiRNA is designed against a specific sequence region, first, this region has to be determined. For efficient knockdown, the region against which the esiRNA is designed should be of a length ranging between 300 and 600 base pairs. This is to ensure efficient generation of dsRNAs and to target a region that does not code for highly conserved domains and has a high density of efficient constituent siRNAs. Furthermore, a minimum length of 300 base pairs ensures sufficient complexity of the resulting esiRNA pool for maximum specificity. As many genes are differentially spliced, the first step is to determine whether an esiRNA should target all possible splice variants or only a specific isoform.

In both cases, it is necessary to find a sequence common to all of the splice variants of a gene. In the former case, the esiRNA will be designed against this common region, while in the latter, the design process will consider sequence regions that are exclusive to the mRNA sequence to be targeted.

3.1.1. Steps for Finding the Region of Interest

1. Look for the gene of interest at http://www.ncbi.nlm.nih.gov/gene/advanced. Additionally, when querying NCBI at this page, in the Search Builder, choose the organism field, and specify the organism of interest. It is also possible to use ENSEMBL (http://www.ensembl.org) to find sequences of interest, leveraging the BioMart utility available on the Web site. For the purposes of this protocol, we will adhere to NCBI. All of these steps can be easily applied to sequences retrieved from elsewhere as well. For designing esiRNA for a set of genes, sequences can be retrieved automatically (see Note 1).
2. On the relevant gene page, scroll down to the *mRNA and protein(s)* subsection of the *NCBI Reference Sequences (RefSeq)* section. Retrieve the FASTA sequences for each of the transcripts listed by choosing the *FASTA (text)* option from the *Display Settings* menu on the transcript page. If the gene has only one splice form, then this entire sequence will be used to design an esiRNA.
3. Use the procedure described under "Identification of the longest common substring (LCS)" in the Supplementary Methods section of (5) to determine a sequence stretch common to all of the transcripts. This procedure extends the suffix arrays idea of Manber and Meyers (40) by way of creating suffixes of multiple strings. An implementation of the procedure is available from the authors upon request.
4. If all of the splice forms of a gene are to be targeted, the region found in step 3 can be used for the esiRNA design. If a specific mRNA is to be targeted, then iteratively find the LCS between the mRNA of interest and all the other splice variants. Remove these regions from the specific target mRNA, and continue with the remaining sequence regions.

3.2. In Silico Procedure for Predicting the Efficiency of Constituent siRNAs

Over the course of the last decade, with the pervasive use of siRNAs as tools for loss-of-function studies, many parameters derived from experimental data have been evaluated with mixed results (13, 16, 32). These parameters address questions of thermodynamic properties, stability, and positional nucleotide preferences. It is prudent to use a set of parameters recurring across studies as predictors of siRNA efficiency.

With the sequence region determined as in Subheading 3.1, use a sliding window of 21 base pairs from the start of the sequence to enumerate the constituent siRNAs. Each of these siRNAs is to be checked for the sequence features described below. Assign a weight factor to each of the parameters under consideration; if a particular parameter does not have the desired value, this weight factor is added to the theoretical efficiency penalty (for suggested weight factors, see ref. 41). The objective of this procedure is to deduce a region whose sum of penalties, for the constituent siRNAs, is lowest.

3.2.1. Desirable Features of a siRNA for Efficient Knockdown

1. The siRNA should be asymmetric, with an A/T at its 5′ end and a G/C at the 3′ end (33).
2. There should be no polynucleotide stretches of length more than 3 in the siRNA sequence(34, 35).
3. The GC content of the siRNA should be between 20% and 50% (16).
4. Positional nucleotide preferences: A at position 3 of the siRNA sequence, T at position 10, no G at position 13 and A or T at position 19 (16, 36–38).

3.3. In Silico Procedure for Predicting Off-Targets of esiRNAs

siRNAs with near-perfect sequence similarity to an unintended target tend to down regulate that mRNA, a phenomenon known to induce off-targeting. Off-target effects produce confounding findings when used for loss of function studies (14). Subsequent studies interrogating off-target effects have found that siRNAs seem to behave like microRNAs, in that a hexamer or heptamer beginning at the second position in the siRNA sequence having matches in the 3′ UTR region of an mRNA tends to down regulate the mRNA (15). When predicting the specificity of the constituent siRNAs of an esiRNA, both of these modes of target down regulation have to be considered. Use the steps described below to analyze the 21 bp siRNA pool produced in Subheading 3.2.

3.3.1. Steps for Finding Unintended mRNAs Containing a Near-Perfect Match to an siRNA Sequence

While it has been a standard procedure to use BLAST to find short sequences with near-perfect matches, Bowtie (42), a program designed for analysis of Second Generation Sequencing data, is much faster, and its results are easier to process.

1. Download the RefSeq mRNA sequences for the species of interest from NCBI as a FASTA file, available at ftp://ftp.ncbi.nlm.nih.gov/refseq.
2. Use the bowtie-build program to build an index of the downloaded RefSeq FASTA file.
3. Run bowtie with the options -v 1 --norc --all iterating through the constituent siRNA sequences; the -v option specifies the number of mismatches allowed, the --norc tells bowtie to map only in the forward orientation.
4. For each constituent siRNA sequence, remove the match/es corresponding to the intended target/s, and count the remaining mRNAs that have near-perfect complementarity to the siRNA sequence. A region with high numbers of near-perfect matches should not be considered for further esiRNA production.

3.3.2. Steps for Finding Unintended mRNAs Which Have 3′ UTR Matches to the siRNA Seed Region

1. Download the GenBank file corresponding to the species of interest from NCBI, available at ftp://ftp.ncbi.nlm.nih.gov/genbank.
2. Locate the 3′UTR of the transcripts. For this, use for instance the BioPython framework (or one of the other frameworks corresponding to the programming language of choice—listed at http://www.open-bio.org) to parse the downloaded GenBank file to determine the end of the CDS.
3. For each constituent siRNA sequence, search the 3′ UTR sequences corresponding to mRNAs other than the intended mRNA target/s for matches with the hexamer and heptamer starting at position 2 in the siRNA sequence; count the number of occurrences of the hexamer and heptamer in the 3′ UTR sequences. If esiRNAs are to be designed frequently, it would be advisable to create a repository of seeds (see Note 2).
4. If there are three or more matches to the constituent siRNA sequence's seed region, then flag the mRNA corresponding to the 3′ UTR in which the matches are found as an off-target. Birmingham and colleagues (15) concluded from their analysis of experimental data that three or more matches to the seed region have 100% predictive power for determining miRNA-like targets.

3.4. Combining Efficiency and Off-Target Predictions for Evaluating and Selecting an Optimal Region for Endoribonuclease-Based Digestion

The penalties assigned in Subheadings 3.2 and 3.3 corresponding to the violation of efficiency conditions and the off-targeting pervasiveness have to be combined to pick a region that has the minimal penalty across its constituent siRNAs, such that this fragment has the highest density of efficient and specific silencers.

1. Initialize a 1-dimensional vector with size equal to the length of the sequence region determined in Subheading 3.1.
2. For each index of the 1 dimensional vector, find all the siRNAs constituting an esiRNA of a range of lengths from a minimum of 300 to a maximum of 600 starting at that same index in the sequence region determined in Subheading 3.1. *Note that each element of the 1 dimensional vector will now be a set of sets of siRNAs.*
3. For each set of siRNAs in every element of the vector, find the average efficiency penalty and the average off-target count corresponding to the siRNAs as determined in Subheadings 3.2 and 3.3.
4. For every element of the vector, corresponding to each esiRNA in that element, compute a rank based on the average efficiency penalty and the average off-target count. Compute the average rank for each esiRNA in that element, and assign

to the element the average efficiency penalty and average off-target count corresponding to the esiRNA with the least average rank.

5. Repeat the ranking procedure described in step 4, but now applying it to the entire vector.
6. Find the element of the vector with the least average rank. The position of the minimal element yields the starting position of the most optimal stretch in the sequence region that can be endoribonuclease digested resulting in a pool of silencing triggers.

Given that these steps are quite elaborate, to design esiRNAs frequently or for large sets of genes, it would be prudent to automate the procedure (see Note 3). On the other hand, for knockdown of a single gene, it would be easier to use a Web service that already implements the analysis described in this chapter (see Note 4).

4. Notes

1. For designing esiRNAs for a set of genes, it is advisable to retrieve the sequences using the E-utilities framework from NCBI (http://www.ncbi.nlm.nih.gov/books/NBK25500) or the ID filter in BioMart at ENSEMBL (http://www.ensembl.org/biomart/martview).
2. If the design of esiRNAs is a frequent task, or for a large set of genes, whilst checking for seed matches in the 3′ UTR, it would be more efficient to preprocess the 3′ UTRs. This can be done by creating an index of all the seeds (both hexamers and heptamers) from the entire set of 3′ UTRs and querying this index.
3. For regularly designing esiRNAs, an object oriented framework implementing the design steps listed in this chapter would be ideal. Such a framework would afford a platform to easily incorporate changes in design principles accruing from continuing research into RNAi mechanisms.
4. To design single esiRNAs, a Web application such as DEQOR (http://deqor.mpi-cbg.de) can be used. An input sequence is scored based on the steps described in this chapter, and a graphical display (Fig. 3) eases the selection of a region from which to design an esiRNA.

Deqor

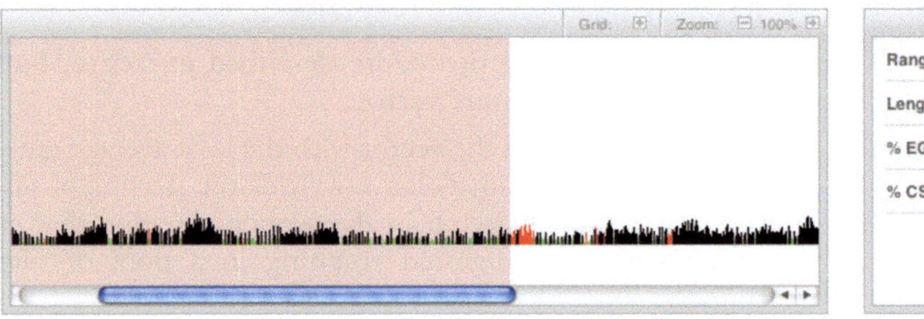

Selection Sequence

Windows: 105 - 479
GCGGCCGCCGCGATGCAGAAATACGAGAAACTGGAAAAGATTGGGGAAGGCACCTACGGAACTGTGTTCAAGGCCAAAAACCGGGAGACTCAT
GAGATCGTGGCTCTGAAACGGGTGAGGCTGGATGACGATGATGAGGGTGTGCCGAGTTCCGCCCTCCGGGAGATCTGCCTACTCAAGGAGCTG
AAGCACAAGAACATCGTCAGGCTTCATGACGTCCTGCACAGCGACAAGAAGCTGACTTTGGTTTTTGAATTCTGTGACCAGGACCTGAAGAAGTA
TTTTGACAGTTGCAATGGTGACCTCGATCCTGAGATTGTAAAGTCATTCCTCTTCCAGCTACTAAAAGGGCTGGGATTCTGTCATAGCCGCAATGT
GCTACACAGGGACCTGAA

Off-Targets

#	Cross Silencing Score	Transcripts
	Off-Targets successfully received.	
133	9	▸ Show all 1
134	11	▸ Show all 1
212	26.5714285714	▸ Show all 2

Fig. 3. Screenshot of efficiency and specificity analysis of a coding sequence. The *top panel* indicates a graphical representation of individual siRNAs color coded for their efficiency and specificity: *green bars* represent siRNAs which are predicted to be very efficient with no predicted off-targets, *black bars* represent those which are not efficient and with no off-target and the *red bars* represent siRNAs with off-targets. The *middle panel* shows the sequence of the region that can be highlighted in the *top-panel*, while the *bottom panel* lists individual siRNAs and their predicted off-targets.

References

1. Reynolds A et al (2006) Induction of the interferon response by siRNA is cell type- and duplex length-dependent. RNA 12: 988–993
2. Yang D et al (2002) Short RNA duplexes produced by hydrolysis with Escherichia coli RNase III mediate effective RNA interference in mammalian cells. Proc Natl Acad Sci USA 99: 9942–9947
3. Myers JW, Jones JT, Meyer T, Ferrell JE (2003) Recombinant Dicer efficiently converts large dsRNAs into siRNAs suitable for gene silencing. Nat Biotechnol 21:324–328
4. Myers JW et al (2006) Minimizing off-target effects by using diced siRNAs for RNA interference. J RNAi Gene Silencing 2:181
5. Kittler R et al (2007) Genome-wide resources of endoribonuclease-prepared short interfering RNAs for specific loss-of-function studies. Nat Methods 4:337–344
6. Obbard DJ et al (2009) The evolution of RNAi as a defence against viruses and transposable

elements. Philos Trans R Soc Lond B Biol Sci 364:99

7. Carrington JC, Ambros V (2003) Role of microRNAs in plant and animal development. Science 301:336–338
8. Carpenter AE, Sabatini DM (2004) Systematic genome-wide screens of gene function. Nat Rev Genet 5:11–22
9. Bernstein E, Caudy AA, Hammond SM, Hannon GJ (2001) Role for a bidentate ribonuclease in the initiation step of RNA interference. Nature 409:363–366
10. Hutvagner G (2001) A cellular function for the RNA-interference enzyme dicer in the maturation of the let-7 small temporal RNA. Science 293:834–838
11. Rana TM (2007) Illuminating the silence: understanding the structure and function of small RNAs. Nat Rev Mol Cell Biol 8:23–36
12. Elbashir SM, Lendeckel W, Tuschl T (2001) RNA interference is mediated by 21- and 22-nucleotide RNAs. Genes Dev 15:188–200
13. Elbashir SM et al (2001) Duplexes of 21-nucleotide RNAs mediate RNA interference in cultured mammalian cells. Nature 411:494–498
14. Jackson AL et al (2003) Expression profiling reveals off-target gene regulation by RNAi. Nat Biotechnol 21:635–637
15. Birmingham A et al (2006) 3′ UTR seed matches, but not overall identity, are associated with RNAi off-targets. Nat Methods 3: 199–204
16. Reynolds A et al (2004) Rational siRNA design for RNA interference. Nat Biotechnol 22: 326–330
17. Chiu Y-L, Rana TM (2003) siRNA function in RNAi: a chemical modification analysis. RNA 9:1034–1048
18. Selinger CI, Day CJ, Morrison NA (2005) Optimized transfection of diced siRNA into mature primary human osteoclasts: inhibition of cathepsin K mediated bone resorption by siRNA. J Cell Biochem 96:996–1002
19. Kittler R et al (2007) Genome-scale RNAi profiling of cell division in human tissue culture cells. Nat Cell Biol 9:1401–1412
20. Collinet C et al (2010) Systems survey of endocytosis by multiparametric image analysis. Nature 464:243–249
21. Fazzio TG, Huff JT, Panning B (2008) An RNAi screen of chromatin proteins identifies Tip60-p400 as a regulator of embryonic stem cell identity. Cell 134:162–174
22. Galvez T et al (2007) siRNA screen of the human signaling proteome identifies the PtdIns(3,4,5)P3-mTOR signaling pathway as a primary regulator of transferrin uptake. Genome Biol 8:R142
23. Krastev DB et al (2011) A systematic RNAi synthetic interaction screen reveals a link between p53 and snoRNP assembly. Nat Cell Biol 13:809–818
24. Słabicki M et al (2010) A genome-scale DNA repair RNAi screen identifies SPG48 as a novel gene associated with hereditary spastic paraplegia. PLoS Biol 8:e1000408
25. Theis M et al (2009) Comparative profiling identifies C13orf3 as a component of the Ska complex required for mammalian cell division. EMBO J 28:1453–1465
26. Leushacke M et al (2011) An RNA interference phenotypic screen identifies a role for FGF signals in colon cancer progression. PLoS One 6:e23381
27. Ding L et al (2009) A genome-scale RNAi screen for Oct4 modulators defines a role of the Paf1 complex for embryonic stem cell identity. Cell Stem Cell 4:403–415
28. Kittler R et al (2004) An endoribonuclease-prepared siRNA screen in human cells identifies genes essential for cell division. Nature 432: 1036–1040
29. Sontheimer EJ (2005) Assembly and function of RNA silencing complexes. Nat Rev Mol Cell Biol 6:127–138
30. Tijsterman M, Plasterk RHA (2004) Dicers at RISC; the mechanism of RNAi. Cell 117:1–3
31. Mittal V (2004) Improving the efficiency of RNA interference in mammals. Nat Rev Genet 5:355–365
32. Matveeva O et al (2007) Comparison of approaches for rational siRNA design leading to a new efficient and transparent method. Nucleic Acids Res 35:e63
33. Khvorova A, Reynolds A, Jayasena SD (2003) Functional siRNAs and miRNAs exhibit strand bias. Cell 115:209–216
34. Hardin CC, Watson T, Corregan M, Bailey C (1992) Cation-dependent transition between the quadruplex and Watson-Crick hairpin forms of d(CGCG3GCG). Biochemistry 31:833–841
35. Geiduschek EP, Kassavetis GA (2001) The RNA polymerase III transcription apparatus. J Mol Biol 310:1–26
36. Jagla B et al (2005) Sequence characteristics of functional siRNAs. RNA 11:864–872
37. Shabalina SA, Spiridonov AN, Ogurtsov AY (2006) Computational models with thermodynamic and composition features improve siRNA design. BMC Bioinformatics 7:65
38. Amarzguioui M, Prydz H (2004) An algorithm for selection of functional siRNA sequences. Biochem Biophys Res Commun 316:1050–1058

39. Anderson EM et al (2008) Experimental validation of the importance of seed complement frequency to siRNA specificity. RNA 14: 853–861
40. Manber U, Myers G (1990) In: Proceedings of the first annual ACM-SIAM symposium on discrete algorithms, SODA '90. Society for Industrial and Applied Mathematics, pp 319–327
41. Henschel A, Buchholz F, Habermann B (2004) DEQOR: a web-based tool for the design and quality control of siRNAs. Nucleic Acids Res 32:W113–W120
42. Langmead B, Trapnell C, Pop M, Salzberg SL (2009) Ultrafast and memory-efficient alignment of short DNA sequences to the human genome. Genome Biol 10:R25

Chapter 12

Short Hairpin RNA-Mediated Gene Silencing

Luke S. Lambeth and Craig A. Smith

Abstract

Since the first application of RNA interference (RNAi) in mammalian cells, the expression of short hairpin RNAs (shRNAs) for targeted gene silencing has become a benchmark technology. Using plasmid and viral vectoring systems, the transcription of shRNA precursors that are effectively processed by the RNAi pathway can lead to potent gene knockdown. The past decade has seen continual advancement and improvement to the various strategies that can be used for shRNA delivery, and the use of shRNAs for clinical applications is well underway. Driving these developments has been the many benefits afforded by shRNA technologies, including the stable integration of expression constructs for long-term expression, infection of difficult-to-target cell lines and tissues using viral vectors, and the temporal control of shRNA transcription by inducible promoters. The use of different effector molecule formats, promoters, and vector types, has meant that experiments can be tailored to target specific cell types and minimize cellular toxicities. Through the application of combinatorial RNAi (co-RNAi), multiple shRNA delivery strategies can improve gene knockdown, permit multiple transcripts to be targeted simultaneously, and curtail the emergence of viral escape mutants. This chapter reviews the history, cellular processing, and various applications of shRNAs in mammalian systems, including options for effector molecule design, vector and promoter types, and methods for multiple shRNA delivery.

Key words: RNA interference, Short hairpin RNA, microRNA, Combinatorial RNAi

1. Introduction

Immediately after the first application of synthetic small interfering RNAs (siRNAs) for gene silencing in mammalian cells, DNA vectors that could transcribe short RNA molecules capable of entering the RNA interference (RNAi) pathway were developed (1–4). These molecules, known as short hairpin RNAs (shRNAs), spontaneously form hairpin structures that are recognized by the cellular RNAi machinery and are processed to form active siRNAs. It was clear from early experiments that shRNAs maintained the highly efficacious nature of their synthetic siRNA counterparts, demonstrated by effective reporter and endogenous gene knockdown

Debra J. Taxman (ed.), *siRNA Design: Methods and Protocols*, Methods in Molecular Biology, vol. 942,
DOI 10.1007/978-1-62703-119-6_12, © Springer Science+Business Media, LLC 2013

(1–4). Moreover, in mammalian cells, shRNAs (like siRNAs) are mostly able to avoid the unwanted nonspecific toxic cellular responses to longer dsRNAs mediated by type I interferons. The transcription of shRNAs therefore offers not only equally effective performance compared to synthetic siRNAs but also the potential for a range of molecule optimizations and enormous flexibility in delivery options.

Despite the undeniable utility of siRNA-mediated gene knockdown for both research and therapeutic purposes, the transcription of shRNAs as siRNA precursors does offer several key advantages. One major pitfall of siRNA-mediated RNAi is that target repression is transient and usually only lasts for 3–5 days in cell culture. Although this may be sufficient for many applications, in some situations this is not satisfactory. The stable integration of shRNA expression vectors through the use of antibiotic selection markers can allow long-term knockdown of stable proteins or suppression of viral replication (5). The use of viral vectors can potentially alleviate problems with difficult-to-transfect cell lines, and can also be more useful for in vivo applications. shRNA transcription can be temporally controlled by the use of drug-inducible promoters and shRNA-expressing cells can be traced via the expression of reporter genes, which also allows for sorting of shRNA-enriched populations. Vector construction can be less expensive than chemical synthesis of siRNA, and particular tissue types can be targeted through the use of tissue-specific promoter sequences. There is also evidence that shRNAs might offer more effective silencing compared to siRNAs (6, 7) and that they induce fewer nonspecific gene expression changes (8).

The past decade has seen considerable development in the art of shRNA delivery. Nearly every imaginable factor that governs shRNA efficacy has been scrutinized and tested, resulting in wide-ranging options for shRNA delivery depending on the exact requirements of the application. Variations in shRNA design parameters such as molecule length can have marked impact on molecule efficacy. shRNA sequences can be modeled after endogenous microRNAs (miRNAs) by inserting the shRNA sequence into an miRNA backbone sequence. This allows them to be processed in a more biologically natural manner, thereby potentially increasing efficacy whilst minimizing unwanted side effects. By identifying and comparing various promoter sequences and types, a vast range of optimized shRNA expression systems have been developed. Various viral vectors of different types have been engineered for use in different cell lines, tissues, embryos, and whole animals. The delivery of multiple RNAi effectors simultaneously, known as combinatorial RNAi (co-RNAi), can offer considerable advantages over the use of single molecule knockdown strategies. For example, targeting multiple locations on the same transcript can induce enhanced suppression, the targeting of different genes can allow for more powerful pathway analysis, and the simultaneous targeting of multiple highly mutable viral transcripts can offer extended protection against infection. Taken together, these developments have ensured that the use of shRNAs for gene knockdown studies is now not only a standard laboratory technique both in vitro and in vivo but also a tool with enormous therapeutic potential.

2. Cellular Processing of shRNAs

It is no coincidence that the anatomy of an expressed shRNA strongly resembles that of an miRNA a naturally occurring class of ubiquitous gene suppressors that were discovered at about the same time siRNAs were first used (9, 10). Indeed shRNAs and miRNAs are both processed by essentially the same RNAi machinery (Fig. 1). A fundamental structural difference between these two molecules is that miRNAs typically have internal mismatches, creating so-called bulges in the secondary structure, whereas shRNAs usually have complete internal base pairing and no bulging (Fig. 1). miRNA precursors (known as pri-miRNA) are first processed in the nucleus by the RNAse III enzyme Drosha and the double-stranded RNA binding domain protein DGCR8 to form a pre-miRNA (11). Both shRNAs and pre-miRNAs are then exported from the nucleus by Exportin 5 (Exp5) via a Ran-GTP-dependent mechanism. Once in the cytoplasm, both molecules are processed into about 20–25 nt siRNAs by the RNAse III enzyme Dicer. Dicer was originally isolated from Drosophila cell extracts as it was capable of producing fragments of 22 nt and is conserved in worms, flies, fungi, plants, and mammals. Dicer's enzymatic activity involves its ATP-dependent translocation along dsRNAs prior to cleavage, the efficiency of which has been shown to be directly proportional to the length of the target (12). Dicer has four distinct domains; an amino-terminal helicase domain, dual RNAse III motifs, a dsRNA binding domain, and a PAZ domain (13, 14). This enzyme shows specificity for dsRNA, usually cleaving them to leave 3′ overhangs of two or three nucleotides with 5′-phosphate and 3′-hydroxyl termini (15). This very specific action of Dicer permits the rapid and predictable processing of expressed shRNA precursors into active siRNAs, which can then effectively initiate targeted RNAi.

After the cleavage of shRNA into siRNAs and pre-miRNAs into miRNAs by Dicer, the second important stage of RNAi-mediated mRNA degradation begins. This involves the formation of the RNA-protein structure known as the RNA-induced silencing complex (RISC), which was originally identified based on its requirement for siRNA binding for gene silencing in cultured *Drosophila* cells (16). Binding of the siRNA or miRNA to RISC results in the ATP-dependent activation of this complex (17). Upon the subsequent unwinding of the siRNA, the RISC complex

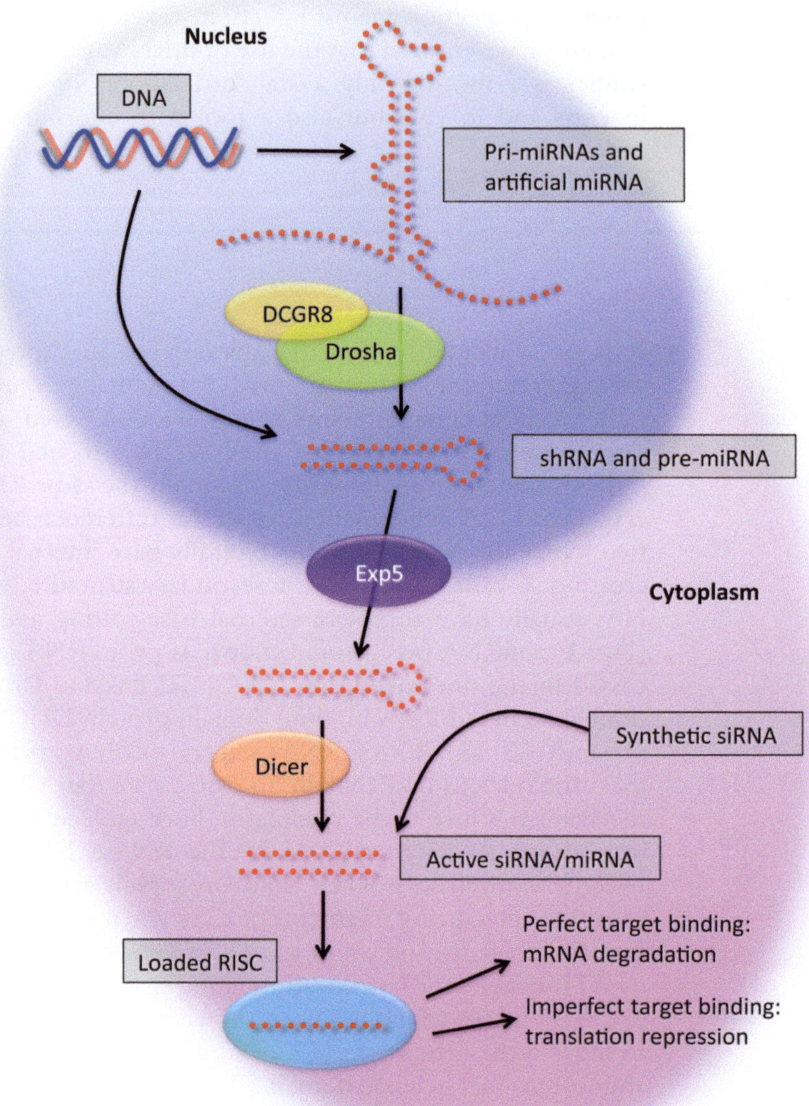

Fig. 1. Cellular processing of shRNAs and miRNAs. Endogenous miRNAs, artificial miRNAs and shRNAs are all expressed within the nucleus where they spontaneously form as hairpin RNAs. Pre-miRNAs and artificial miRNAs are then excised by Drosha together with its dsRNA-binding partner DGCR8, and together with shRNAs are transported from the nucleus to the cytoplasm by Exportin 5 (Exp5). Pre-miRNAs and shRNAs are then cleaved by Dicer into active shRNAs and miRNAs, at which point transfected synthetic siRNAs also join the RNAi pathway. Binding of the siRNA or miRNA to RISC results in the ATP-dependent activation of this complex, and after subsequent duplex unwinding, RISC facilitates in the binding of target homologous mRNAs. Perfect binding sequences result in cleavage and degradation of target mRNAs, whereas imperfect binding, commonly seen with miRNAs, results in translational repression.

facilitates binding of the target homologous mRNA, after which cleavage occurs at a single site in the center of the duplex region 10 nt from the 5′-end of the siRNA (15). RISC catalyzes hydrolysis of the target RNA phosphodiester linkage, similar to the reaction that occurs when Dicer generates siRNA from dsRNAs (18, 19). Unlike the base-perfect pairing of siRNA to cognate mRNAs, the pairing of processed miRNAs to target RNAs by RISC often contains mismatches and leads to translational repression of target RNAs (20).

3. shRNA Structures and Factors Effecting Activity

Approaches for the expression of shRNA have advanced rapidly since the initial discovery of RNAi in mammalian cells. Although many subtle variations exist, the anatomy of an expressed shRNA molecule has remained relatively constant and features a number of characteristic components that are essential for function (Fig. 2a).

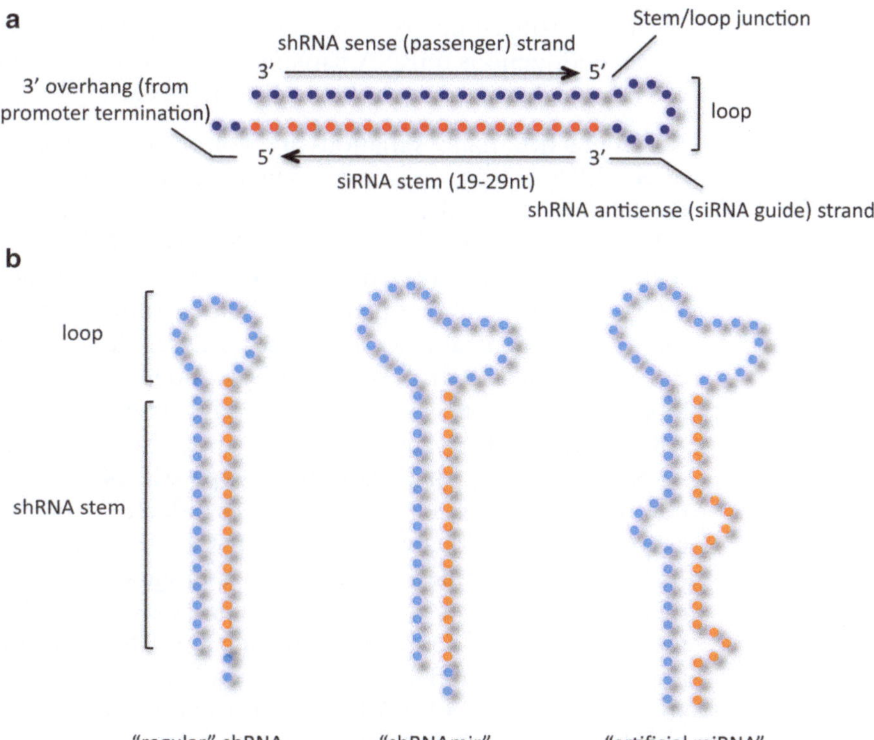

Fig. 2. The structure of a short hairpin RNA. (**a**) Schematic representation of the typical structure of an expressed shRNA. The siRNA stem sequence is shown in *red* and is usually from 19 to 29 bp in length. The loop sequence connects the 3′ end of the upper siRNA strand (shRNA sense strand) to the 5′ end of the lower siRNA (shRNA antisense strand). The shRNA sense strand is identical to the target sequence and gives rise to the siRNA "passenger" strand, whereas the shRNA antisense strand is complementary to the target mRNA and gives rise to the siRNA "guide" strand. (**b**) Schematic representation of the common shRNA subtypes; the "regular" type shRNA forms a perfect stem–loop, the "shRNAmir" also forms a perfect-loop style shRNA but features an miRNA loop sequence; "artificial miRNA" is an shRNA sequence that mimics an endogenous miRNA due to its miRNA flanking sequence, its length, and the inclusion of any mismatches present in the original miRNA it was modeled on.

In its simplest form, the fundamental structure of an shRNA comprises a stem region of paired sense and antisense strands, which are separated by a stretch of unpaired nucleotides known as the loop. A multitude of detailed studies on the various factors that govern shRNA efficacy and vector design have, however, resulted in the development of evermore-effective expression systems. Numerous factors influencing activity have been identified, including the loop sequence, thermodynamic properties of the hairpin, secondary structure, and the surrounding sequences. As a result of these studies, a number of different shRNA subtypes exist, and these can be categorized into three main groups (although there is no strict nomenclature so names for these vary considerably) (Fig. 2b):

1. The original "regular" type shRNA is a perfect stem–loop style molecule.
2. The "shRNAmir" is perfect stem shRNA with an miRNA loop sequence derived from an endogenous miRNA.
3. The "artificial miRNA" is an shRNA sequence surrounded by an endogenous miRNA flanking sequence and can include the mismatches present in the original miRNA.

In practice, many variations and combinations of each of these three shRNA subtypes exist, each potentially offering different benefits. In particular, although closely resembling the regular shRNA subtype, the use of short shRNAs (sshRNAs) has been one such recently described modification. These molecules contain 19 bp or shorter stem sequences and differ from longer shRNAs in that they are generally not subjected to Dicer cleavage and as such can be designed so that the antisense strand is either 5′ or 3′ to the loop (L- or R-type sshRNA respectively). It has been found that ssRNAs can induce more effective knockdown than equivalent siRNA sequences, particularly the L-type sshRNAs (21). Another new strategy takes advantage of two different types of shRNA expression strategies by the development of bifunctional shRNA (bi-shRNA) molecules. bi-shRNAs are designed so that they can thermodynamically accommodate the processing of both strands of the shRNA simultaneously; one being an miRNA-like effector that is processed through cleavage-independent RISC incorporation, and the other is an siRNA-like effector that is processed in a cleavage-dependent fashion. The bifunctional construct is therefore designed to concurrently repress the translation of a target mRNA (cleavage independent) through mRNA sequestration and degradation, as well as to promote mRNA degradation (cleavage dependent) through RNase H activity. Enhanced activity of bi-shRNAs compared to regular shRNAs and siRNAs has been demonstrated by effectively suppressing the expression of the cancer-related gene Stathmin1 (STMN1), both in vitro (22) and in vivo (23).

Although a large statistical analysis of factors that determine shRNA efficacy has not yet been published, it appears that in general the same rules that determine siRNA activity can also be applied to the selection of shRNA sequences. It has been demonstrated, however, that published algorithms for siRNA design can be improved upon by considering the internal thermodynamic properties of individual molecules. In particular, it has been found that the average stability of the six central bases of effective shRNAs was lower than that of ineffective sequences, despite all scoring well by the original design algorithm (24). Li and colleagues have found that functional shRNAs exhibit a preference for GC nucleotides at position 11 of the shRNA, as well as a preference for AU but not GC at position 9, and accordingly developed an advanced shRNA selection computer program that incorporates these constraints (25).

The effect of stem length on siRNA activity has also been the subject of a number of reports, with somewhat contrasting findings. Earlier reports showed that increasing the duplex length to 27–29 bp can enhance siRNA and shRNA activity, presumably through enhanced Dicer processing (6, 26, 27). Other reports have specifically analyzed the effect of shRNA stem length on gene silencing activity. It was found that increasing the stem length of a 19 bp shRNA to 28 bp improved initially poor activity, but an already active 19 bp was not further enhanced by a similar extension (28). Another report that tested 22–29 bp shRNAs found that the most active stem length was 25–29 bp (3), whilst findings from a separate study showed that when shRNAs from 19 to 25 bp were tested, 21 bp was the most active (29). In a comprehensive study, Mcintyre and colleagues tested a large number of anti-human immunodeficiency virus (HIV) shRNAs with varying stem lengths and found that there was no fixed correlation between shRNA length and suppressive ability, but instead concluded that the placement of the siRNA core sequence within the molecule had a greater influence on activity (30).

Although in most cases shRNAs are transcribed from DNA vectors, the use of synthetic shRNAs has not only helped define the parameters that determine efficacy and aided our understanding of shRNA processing, but these molecules also show promise for therapeutic applications as well. Studies have looked at the effect of stem length on resultant Dicer processing and gene knockdown (6, 26, 31), and a range of shRNA types have been synthesized, including sshRNAs (21) and artificial miRNAs (32). One potential advantage of synthetic shRNAs over DNA-transcribed versions can be achieved through chemical modifications to enhanced activity and stability. In particular, it has been shown that 2′-O-methyl (2′-OMe) substitution of sshRNAs abrogated the innate immune response and improved the serum stability of the hairpins (33).

A number of different arbitrary loop sequences to separate each shRNA strand have been successfully used (1, 3, 4, 34), although none of these appear to offer any particular benefit. The most commonly used sequence (5′-UUCAAGAGA-3′) was originally tested by Brummelkamp et al. (2002) and has been used in more than 60% of reported studies (30, 35). It has recently been suggested that the use of these unrelated loop sequences could be improved upon by using naturally occurring sequences. To generate molecules referred to here as shRNAmirs, the loop that features in human miRNA miR-23 (36) has been utilized in a number of instances resulting in enhanced gene suppression (37, 38). Likewise, the loop of the well characterized miRNA miR-30 has also been used to enhance the activity of shRNAs targeting HIV (39), a reporter gene (40), and Influenza A (41).

The insertion of shRNA sequences within naturally occurring miRNA precursor sequences represents a potentially favorable strategy as effector molecules can be processed and exported by the same cellular pathways as endogenous miRNAs. Furthermore, it has be shown that these artificial miRNAs can induce gene silencing equally (42, 43), or even more efficiently (44, 45) as equivalent shRNA sequences, although two comprehensive studies that compared miR-30 embedded shRNAs to regular shRNAs found that shRNAs are generally the more potent of the two (25, 46). Potency, however, is not the main driving force behind the increasing popular use of artificial miRNA in preference to the more traditional perfect stem–loop style shRNAs. Evidence is mounting that shRNA expression in vivo using highly active promoters such as U6 can trigger toxic responses (47–49) (see Subheading 6). It turns out that in some cases at least, these detrimental effects can be largely avoided by the use of artificial miRNAs instead of shRNAs. The improved safety of this method has been demonstrated by significant decreases in neurotoxicity in mouse striatum whilst maintaining effective gene silencing (50).

4. Promoters for shRNA Transcription

Although the various properties of individual shRNA effector molecules are well known to be critical for function, the appropriate choice of promoter for shRNA transcription is also vital for effective vector delivered RNAi. It was shown even in the early days of shRNA-mediated RNAi that promoter choice was an important aspect for effective RNAi, and that more effective silencing could be achieved by comparing various sequences (37, 51). Since then, the testing and validation of various different promoter types and sequences has provided researchers with numerous options for shRNA expression, allowing for increased flexibility and for a more tailored approach to experimental design. For example, the use of a U6 promoter ensures a high level of constitutive shRNA transcription regardless of cell type, whereas tissue-specific polymerase II (pol II) promoters can allow for tissue-restricted gene knockdown (52–54).

In eukaryotes, the task of transcribing nuclear genes is shared between three RNA polymerases. Polymerase I (pol I) synthesizes large ribosomal RNA (rRNA), pol II synthesizes mRNA, some small nuclear RNA (snRNA) (reviewed in ref. (55)) and miRNA (56), and polymerase III (pol III) synthesizes 5S rRNA, transfer RNA (tRNA), 7SL RNA, U6 snRNA, and a number of other small stable RNAs (57). Early experiments described the transcription of shRNAs by the human U6 snRNA promoter, a member of the pol III promoter family. It was clear from the outset that this promoter was efficient, as demonstrated by the robust knockdown of reporter (3) and endogenous genes in mouse cells (2). As a consequence of these and other groundbreaking studies, the use of these types of promoters has been the most frequently used approach for shRNA transcription. In particular, the pol III promoters U6 and H1 have been by far the most commonly used for shRNA expression, although other members including 7SK (58), BC1 (59), the $tRNA^{val}$ (37), $tRNA^{lys3}$ (60), and the modified promoter tRNA promoter MTD (51) have also been effectively applied.

4.1. RNA Polymerase III Promoters

RNA pol III promoters are classified into three different categories (type 1, 2 or 3) based upon the composition of the promoter element sequences and their position relative to the transcriptional start site (61) (Fig. 3a). The type 1, 2, and 3 RNA pol III promoters are

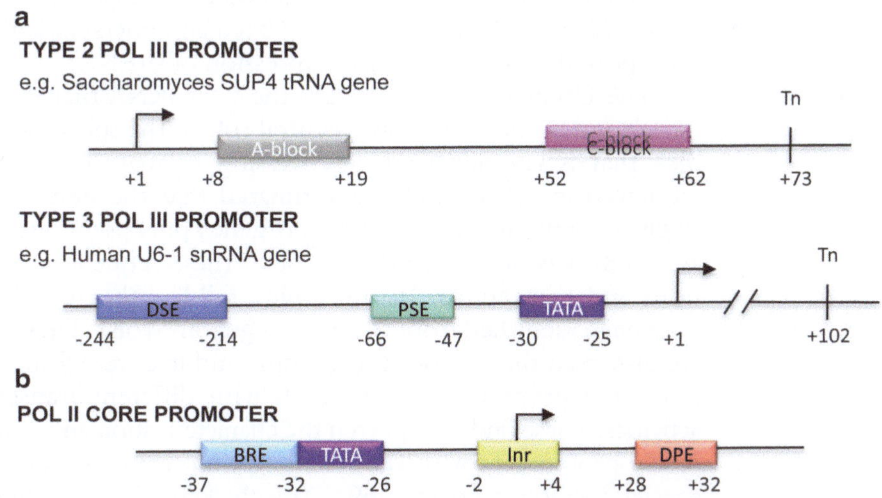

Fig. 3. Promoters used for shRNA transcription. (**a**) Schematic representation of examples of the two classes of pol III promoters commonly used for shRNA transcription. Tn refers to termination sequence, and +1 refers to the transcriptional start site. *Top*: The type 2 promoters consist of two highly conserved sequence blocks, A and B, located within the transcribed region (86). *Bottom*: The type 3 promoter elements are located within ~250 nt of the 5′ regions flanking the transcript and contain a core region comprised by a proximal sequence element (PSE) and a TATA-like element, and the enhancer region referred to as the distal sequence element (DSE) is located further upstream (reviewed in ref. 65). (**b**) Schematic representation of the core pol II promoter apparatus. These elements are found in only a subset of core promoters and any particular promoter may contain some, all, or none of these motifs. The initiator (Inr), TFIIB recognition element (BRE), and downstream core promoter element (DPE) are shown in this example (reviewed in ref. 153).

characterized by very different promoter structures. However, the majority of these promoters share the unusual feature of the requirement for sequence elements downstream of the transcribed region (after the transcription start site). Vertebrate U6 and 7SK snRNA genes were the first examples found to contain promoters of the type 3 RNA pol III subtype (62–64). These promoters represent a minority of the pol III promoters and drive transcription purely by the use of *cis*-acting elements located within ~250 nt of the 5′ regions flanking the transcript (Fig. 3a). These promoter element sequence motifs are located within two areas known as the enhancer and core regions. The core region comprises a proximal sequence element (PSE) and a TATA-like element, and the enhancer region referred to as the distal sequence element (DSE) is located further upstream (reviewed in ref. 65).

Together with the H1 promoter, the U6 and 7SK promoters are self-contained and possess numerous advantageous features that have resulted in their extremely frequent and widespread use for shRNA expression. Type 3 pol III promoters naturally direct the synthesis of small, highly abundant noncoding RNA transcripts, have a compact and relatively simple organization and are located entirely upstream of the transcribed region (reviewed in refs. 57, 65). Transcription is terminated within a stretch of 4–5 thymidine residues yielding small RNA transcripts with a defined 3′-overhang. Since this feature has been shown to be critical for siRNA activity (15) and considering that these promoters are ideal for synthesis of small and defined RNAs, they have been an understandable choice for repeated application in the expression of shRNA.

The U6 promoter transcribes the U6 snRNA that is 107 nt in length and is known to play a central role in the spliceosome complex that is responsible for processing premature RNA species (reviewed in ref. 66). It was estimated that the human genome might contain as many as 200 U6 genes plus pseudogenes (67), although it is now clear that most of these sequences have deletions and truncations. In total, nine full-length U6 snRNA loci have been identified from the human genome (68). These nine loci are dispersed throughout the genome and five were found to have associated promoter regions, each with different transcriptional activities. These findings spurred the characterization and comparison of U6 promoters for shRNA transcription from a variety of other species including bovine (69, 70), chicken (71, 72), fugu (73), zebrafish (74), sheep (75), and pig (76).

In contrast to the U6 family of promoters, only one 7SK and one H1 gene have been identified from the human genome. The organization and size of the 7SK promoter closely resembles that of the U6 family of promoters, whereas despite sharing the same basic structure with these promoters, the H1 promoter features a more compact organization and is located within 100 bp of the 5′-flanking sequences (77). It has been shown that the human 7SK

(78) and bovine 7SK (70) were more effective for shRNA-mediated gene suppression when compared to selected U6 promoters; however, similar studies using chicken (79) and porcine (80) sequences found that knockdown from 7SK was similar to that of U6. Although the effectiveness of the human H1 promoter for shRNA transcription has been clearly proven (1), and that this sequence can work at similar levels to the U6 promoter (5), some reports suggest that its activity might be somewhat reduced in comparison (51, 81, 82). One study also characterized and applied the Y1 and Y3 small cytoplasmic RNAs (scRNAs) promoters for shRNA transcription, which are also members of the type 3 family (83), and found that they displayed similar activity to a U6 promoter (76).

Despite the proven effectiveness of the various type 3 pol III promoters for shRNA transcription, potential drawbacks and limitations are also associated with their use. One avenue to circumvent toxicities seen in vivo using strong promoters to express shRNAs is through the use of similar type 3 pol III promoters that display lower levels of transcription, as decreased cellular concentrations of shRNA might avoid saturation of the RNAi machinery (see Subheading 6). Grimm and colleagues, who reported liver damage in mice expressing high levels of shRNA (47), found in a later study that the use of weaker H1 and 7SK promoters in mice reduced cytotoxicity, implying that the lower shRNA levels remained below the saturation threshold of the hepatic RNAi machinery and thus averted toxicity (84).

Another potential limitation of type 3 pol III promoters such as U6 is their lack of capacity to transcribe longer transcripts, which might render them unsuitable for expression of some pri-miRNA, long hairpin RNAs (lhRNAs), and polycistronic miRNA sequences. One study observed that H1 or U6 promoters can express the longer transcript reporter genes including firefly luciferase, green fluorescent protein, and red fluorescent protein, however, concluded that this was likely due a cryptic pol II activity (85). Type 2 pol III promoters are capable of transcribing much longer transcripts than the type 3 promoters, and numerous reports have described their use for shRNA expression. The tRNA promoter is a typical type 2 promoter, consisting of two highly conserved sequence blocks, A and B, located within the transcribed region (86) (Fig. 3a). Transcription from these promoters leads (87) to the production of long RNAs that can also be engineered to encode downstream shRNA precursor sequences that are processed to form active molecules for RNAi (79). Numerous studies have successfully employed this approach using a number of different tRNA promoters resulting in gene knockdown at comparable levels to U6 promoters (37, 51, 60, 88–90).

4.2. RNA Polymerase II Promoters

Considering that RNA pol II promoters are frequently used for endogenous miRNA expression (56), strategies that utilize these sequences may more closely mimic the natural expression of short RNA and may therefore be an efficient strategy for RNAi (Fig. 3b). As such, effective gene suppression mediated by pol II transcribed shRNAs has been demonstrated using the human cytomegalovirus immediate-early (CMV) promoter (43, 91, 92), the human ubiquitin C promoter (1), the human U1 promoter (93), and the adenovirus E1b promoter (94). In addition, an innovative approach that utilized an HIV-1 variant vector to encode a 300 bp-long hairpin RNA (lhRNA) sequence under the control of the viral LTR has also been described (95).

However, while the use of pol II-driven expression of shRNAs in mammalian cells has been effective, construct design has been complicated (43) and a low efficacy of these molecules has been reported (3, 51, 91). Unlike the extremely defined initiation and termination signals that ensure the predictable transcription products of pol III promoters, pol II generates longer and more complex sequences, requiring a poly adenylation signal which results in the addition of a poly(A) tail at the 3′ end of the RNA. Whilst this may not be ideal for the transcription of the very short defined structures of shRNAs, it is ideal for mono-and polycistronic primary miRNA sequences embedded with shRNA precursors. Indeed this approach has been used very effectively to transcribe modified endogenous polycistronic miRNA clusters to achieve targeted gene suppression. For example, using lentiviral vectors, the CMV promoter was used to transcribe multiple anti-HIV-1 shRNAs (96), and the CMV-enhancer chicken beta-actin promoter (CAGGS) was used to transcribe anti-influenza shRNAs (97).

Arguably the single greatest advantage of pol II promoters, however, is the potential for tissue-specific delivery of shRNAs to cells of interest. Pol II promoters have a long history of application in achieving tissue-specific mRNA delivery in model animals and have been particularly useful in transgenic mice. It is therefore not surprising that those using RNAi technologies have also employed such promoters to target transcripts in restricted cell types. Neuronal cell-specific reporter gene RNAi has been achieved using the enolase promoter (NSE) (53), and a survivin promoter-driven RNAi system efficiently and specifically downregulates eIF4E expression in human breast carcinoma cells but not in normal human mammary epithelial cell (98). In an effort to increase the therapeutic potential by improving the safety of liver expressed shRNAs in mice, the liver-specific ApoE/hAAT promoter effectively reduced target expression without any signs of liver damage (99).

4.3. Inducible shRNA Expression

A number of inducible promoter systems have been developed that allow temporal control over shRNA transcription, thereby further increasing the control over gene knockdown studies. The most commonly used method for shRNA transcription is the tetracycline-inducible system, Tet-On and Tet-Off. In this system transcription is reversibly and quantitatively turned on or off by the addition of the antibiotic tetracycline (Tc), or its more stable derivative doxycycline (Dox) (100). In the Tet-Off system, a tetracycline-controlled transactivator protein (tTA) regulates the transcription of a tetracycline-responsive promoter element (TRE), which comprises Tet operator (tetO) sequence concatemers fused to a promoter. In the absence of Tc or Dox, TRE is bound by tTA to activate transcription. Conversely, when Tc or Dox is added, tTA is unable to bind to TRE and expression remains inactive. In the Tet-On system, the reverse tetracycline-controlled transactivator (rtTA) fusion protein (comprised of the TetR repressor and the VP16 transactivation domain) can only recognize the tetO in the presence of the Dox. Therefore, transcription is stimulated by rtTA only in the presence of Dox. van de Wetering and colleagues were quick to adopt the Tet-On approach to modify the H1 promoter for inducible transcription of shRNAs (101), and the use of the U6 promoter for inducible shRNA expression also soon followed (102). This system can also be effectively used in vivo for temporally controlled shRNA expression. Using modified lentiviral vectors encoding inducible pol II and pol III promoters, the addition of Dox into the drinking water of mice (103) and rats (104) leads to tightly regulated shRNA expression and targeted gene suppression.

Other inducible and reversible transcription approaches have been adapted for use in shRNA experiments and potentially offer some benefits over the tetracycline-based systems. The insect steroid molting hormone ecdysone has a lipophilic nature, which facilitates efficient penetration, and can also be used as an inducer of gene expression. This response is mediated through the functional ecdysone receptor, which is a heterodimer of the ecdysone receptor (EcR) and the product of the ultraspiracle gene USP, and is able to regulate an optimized ecdysone responsive promoter in mammalian cells (105). This system can also be used to tightly induce shRNA transcription delivered retrovirally through the use of a modified U6 promoter (106), and using a modified pol II promoter, ecdysone can also induce the expression of shRNAs expressed in a miR-30 context (107).

The Cre-Lox recombination system involves the deletion of loxP-flanked chromosomal DNA sequences by the enzyme Cre-recombinase (108) and is a technique that is used extensively for studies in transgenic mice. This system has also been successfully adapted to allow induction of shRNA transcription by the insertion of LoxP-flanked stuffer sequences in the promoter region. This results in promoter inactivation, and reactivation is achieved by expression of a Cre-recombinase, which removes the intervening sequence and reactivates promoter activity. Using the SIN lentiviral vector encoding a modified U6 promoter, this approach for shRNA induction was demonstrated in vitro and in mice, and a similar approach was used to abolish an already active promoter (102).

5. Vectors Used for shRNA Delivery

The easiest and most straightforward approach for an shRNA experiment is via the use of plasmid vectors. Like synthetic siRNAs, their use in vitro is extremely widespread and relies on the one basic requirement that the cell line used responds well to transfection. A plethora of RNAi plasmid options exist and can be selected based on the particular experimental requirement, including commercially available shRNA cloning vectors and pre-validated shRNA libraries. Although shRNA plasmid vectors can be used for transient gene knockdown, the use of mammalian selection markers allows for the stable integration and hence long-term suppression of target genes.

The first use of shRNAs in vivo was by hydrodynamic injection of plasmid vectors expressing anti-hepatitis C (HCV) shRNAs into the tail veins of mice (109), and then subsequently using the same approach to inhibit hepatitis B replication (110). Although there is no denying the exceptional impact of these experiments, the application of this approach is mostly limited to RNAi studies in the liver and is therefore not very useful for other organs. Many of the limitations of plasmid-delivered shRNAs can be overcome by the use of viral-vectoring systems, including the infection of hard-to-transfect cell lines and the widespread infection in vivo. Viral vectors have been developed that belong to two main classes; RNA viruses (such as retro- and lentiviruses) that can integrate into the host cell chromosomes, and nonintegrating DNA viruses (such as the adenoviruses, adeno-associated viruses (AAV), and herpesviruses), which generally persist as episomes.

Retro- and lentiviruses contain a reverse transcriptase that allows them to integrate into the host genomes as a part of their normal life cycle. Recombinant viral vectors of these types can thereby permit continual expression of the inserted shRNA. In most cases, non-replicating viruses are used for transgene expression due to biological safety reasons, which is achieved by introducing minimal components of the viral genome in three or four separate plasmid vectors. This allows for a single round of viral infection, and then after stable integration of the recombinant viral genome the shRNA is continually expressed without production of infectious virus. Shortly after the first report of plasmid vector delivered RNAi, the self-inactivating pMSCV retroviral vector was used to stably deliver shRNAs in human tumor cells to knockdown the mutant K-RAS^{v12} oncogene (111), and in HeLa cells, human nuclear Dbf2-related (NDR) kinase was targeted and suppressed (112). The use of a replication-competent retrovirus for shRNA delivery soon followed (113), which made use of the modified Rous sarcoma virus-derived vector RCANBP (114).

Unlike the modified replication incompetent retroviral vectors that do not infect nondividing cells, lentiviruses can transduce actively dividing cells, in both resting and differentiated states. Lentiviruses are a subcategory of the retrovirus family, the HIV subtype-1 (HIV-1) being the most common example. Using the self-inactivation (SIN) lentiviral vector (115), the first demonstration of the use of these viruses for RNAi was shown in HEK293 cells (116). A short time later, Rubinson and colleagues developed a lentiviral system for the delivery of shRNAs into mammalian cells, stem cells and transgenic mice to achieve stable gene silencing (117). Since then, the use of lentiviruses such as the HIV-based vectors for RNAi-mediated gene therapy has been pushing ahead and offers considerable promise in the treatment of various conditions. The first successful clinical trial using lentiviral vectors for a gene therapy of hematopoietic stem cells was reported (118), and considering that lentiviruses are efficient transducers of the major target cells for an HIV-AIDS therapy, CD34-positive blood stem cells and CD4-positive T cells, strategies to target this virus are continually advancing (119).

Similar to the retro- and lentiviruses, AAV were also quickly adapted for RNAi experiments shortly after the first application of shRNAs (120). The AAV contain a very small (<5 kb) single-stranded DNA genome and offer numerous features that make them attractive vectors for RNAi; they can rapidly infect both dividing and nondividing cells, they are not associated with any known disease, they do not induce host immune responses, and they can integrate specifically into a host genome or persist episomally without any known side effects (121).

In addition to the viruses that are most commonly used for shRNA delivery, a number of other virus-types, including herpesviruses and baculoviruses, have also been adapted for this task. A herpes simplex virus (HSV) amplicon was engineered to express shRNAs targeting epidermal growth factor receptor (122), and a replication defective HSV vector was used for shRNA transcription in the hippocampus of mice (123). The replication competent vaccine herpesvirus of Turkey (HVT) was engineered to expresses shRNAs from a U6 promoter to target Marek's disease virus replication in chickens (40). Baculoviruses also offer potential advantages over human-derived gene therapy vectors, as they naturally cannot replicate in human cells but can transduce many vertebrate cell types with high efficiency. The baculovirus *Autographa californica* nuclear polyhedrosis virus was used to express shRNAs targeting reporter and endogenous genes in numerous cell lines (124).

6. Potential Toxicity of shRNAs In Vivo

In a seminal study published in 2006, Grimm and colleagues observed dose-dependent liver injuries and fatalities in the majority of AAV/shRNA infected mice (47). The authors also observed a reduction in miR-122 function in the livers of AAV/shRNA-infected mice, opening up the possibility that the observed morbidity was due to oversaturation of the RNAi machinery. More recently, the cellular toxicity of shRNAs in the nervous systems of animals injected with AAV vectors has also been described. In particular, high-level expression of shRNAs in the dorsal root ganglia of rats (48) and in mouse striatum (49) resulted in weight loss, neuronal degeneration, and animal demise. Efforts to circumvent these toxic effects have proven effective and have involved relatively simple and easily employed alternatives. McBride and colleagues unexpectedly found that two active shRNAs induced significant neurotoxicity in mouse striatum, including a control shRNA that contained mismatches. When the same sequences were expressed as artificial miRNAs, molecular and neuropathological readouts of neurotoxicity were significantly reduced without compromising targeted gene silencing, thereby providing a potentially more appropriate approach for shRNA expression in the brain (50).

Studies into the differences in processing of siRNA and shRNAs compared to miRNAs has helped shed light on the reasons for the reduced toxicity of artificial miRNA. One study found that both synthetic siRNAs and expressed shRNAs compete against each other and with the endogenous miRNAs for transport and incorporation into RISC (125). However, competition was abolished when the same siRNA sequences were expressed from an miRNA backbone, further supporting the notion that artificial miRNAs may represent a potentially safer option for shRNA delivery. The role of Exportin 5 in observed shRNA induced toxicities has been explored by overexpression studies. It appears that although shRNA efficacy can be increased and competition between RNAi molecules reduced, these studies also suggest that other factors must also be involved (47, 125). Indeed, on closer examination of the pathway constituents involved in RNAi saturation, it was found that four members of human Argonaute (Ago), particularly Ago-2, were the primary rate-limiting determinants of both in vitro and in vivo RNAi efficacy, toxicity, and persistence (84).

In addition to the use of artificial miRNAs to avoid cellular toxicities, it has also been found that by simply using weaker promoters for shRNA transcription these effects can also be alleviated. Evidence of the improved safety of this method has been provided through use of polymerase II expressed molecules in the HBV transgenic mouse model (99). Grimm and colleagues also showed that the use of weaker H1 and 7SK promoters in mice actually achieved greater long-term knockdown and reduced cytotoxicity compared to the same shRNA sequence transcribed by U6 (84).

7. Combinatorial-RNAi

Building upon the early experimental success of expressed shRNAs, the delivery of multiple RNAi effectors, known as combinatorial RNAi (co-RNAi), can offer considerable advantages over the use of single molecule knockdown strategies (126). Co-RNAi is particularly important for evolving targets that require long-term and multifaceted treatment such as highly mutable RNA viruses like HIV and HCV. In addition, the prospect of increased levels of gene suppression and multiple-gene silencing is of importance for other transcripts that are not particularly susceptible to spontaneous mutation such as host genes and the complex genomes of large DNA viruses such as herpesviruses.

Like the many other viruses that have been tested, HIV-1 is highly susceptible to siRNA-mediated degradation. It has been found, however, that HIV-1 can very quickly evade the inhibitory effects of shRNA treatment by the emergence of viral quasispecies harboring nonsynonymous mutations in the siRNA binding site (127). On closer examination, it seems that mutations not only arise within the siRNA binding sequence but also in regions outside the targeted sequence that give rise to changes to the RNA secondary structure (119). In an attempt to avoid viral escape, it has been found that the simultaneous expression of up to four separate shRNAs can effectively suppress HIV-1 replication for periods of at least 75 days (5). In addition, several others have also used a number of different multi-shRNA strategies to target both HIV-1 (95, 96, 128–131) and HCV (132), which not only demonstrates the potential power of co-RNAi but also opens up potential new therapies to help combat viral diseases.

The expression of multiple shRNAs from a single construct encoding several separate promoter/shRNA cassettes offers the potential for relatively straightforward vector construction, as previously validated RNAi cassettes can be simply assembled as tandem repeats (Fig. 4). Recent studies have included the use of shRNA cassettes in combinations of two (133), three (132, 134, 135), four (5, 136, 137), six (138, 139), and in one study a cloning strategy for the production of up to seven was described but not validated (140). Results consistently show that such approaches provide an additive effect for single- and multiple-gene knockdown as demonstrated on a variety of host and viral gene targets. In one report, individual shRNAs were transcribed at much lower levels

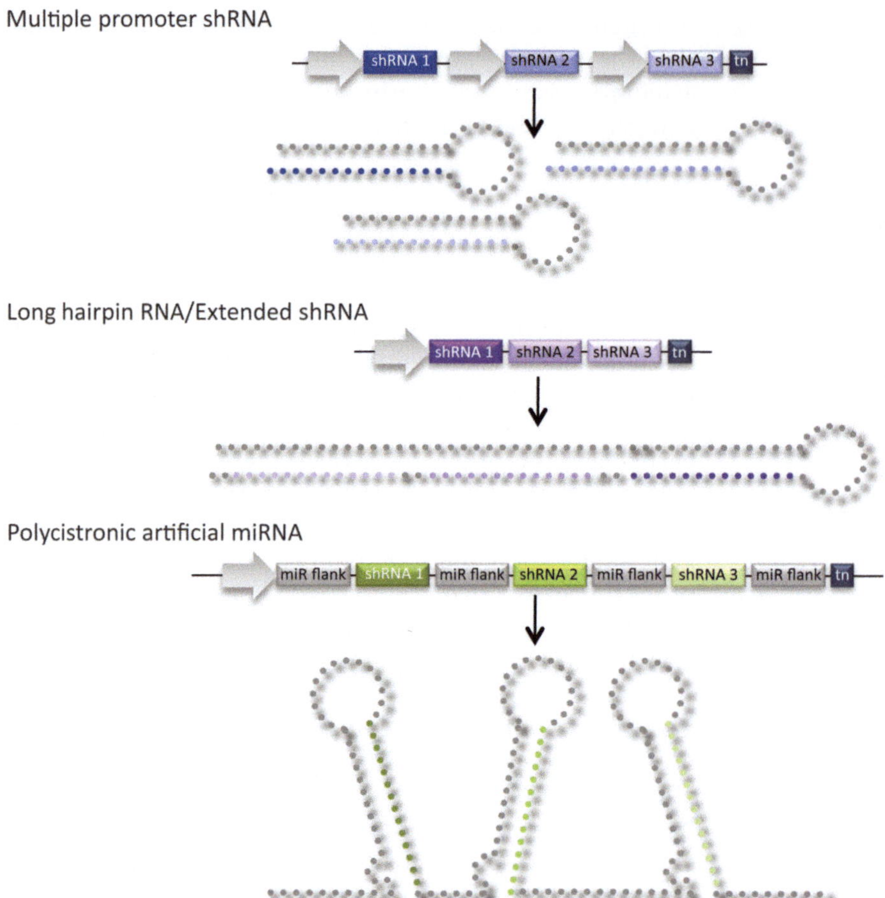

Fig. 4. Schematic representation of the strategies used for combinatorial RNAi. Each method of co-RNAi results in the transcription of dsRNA precursor molecules that encode the siRNA stem (*colored*), a loop sequence, and a passenger strand, which are processed into active siRNAs. The *top panel* shows a multiple promoter/shRNA construct that encodes three promoters to express three separate shRNAs simultaneously. The *middle panel* shows a long hairpin RNA (lhRNA) or extended-shRNA (e-shRNA) construct that uses a single promoter to express an extended shRNA that encodes three separate siRNA stems. The *bottom panel* shows an artificial polycistronic miRNA construct that uses a single promoter to express an miRNA cluster that features three introduced shRNA sequences that mimic naturally expressed miRNAs.

when expressed from a 4-cassette construct compared to single copy vectors (5), but in a separate study, vectors encoding up to five separate promoter/shRNA cassettes offered similar levels of gene knockdown compared to single shRNAs (141). One potential hazard of this strategy was identified by the unwanted recombination and loss of identical promoter sequences cloned into lentiviral vectors, an event that was then avoided by using different promoter sequences (5).

In addition to the transcription of the very short and defined shRNA and miRNA sequences used in other co-RNAi strategies, a number of approaches to express longer transcripts that can be processed to form multiple active siRNAs have been developed.

The use of lhRNA and extended shRNAs (e-shRNAs) represents a likely progression from single site targeting as long dsRNA are naturally processed as part of the RNAi pathway, and such molecules have been shown not to induce interferon mediated responses (142–144) (Fig. 4). In general, an lhRNA is transcribed by a pol II promoter and is a relatively long RNA sequence that does not contain any characterized siRNAs, but should be processed by the RNAi machinery into active siRNA molecules. On the other hand, e-shRNAs are usually expressed from pol III promoters and encode multiple already characterized siRNAs that have been carefully stitched next to one another. A number of recent studies have successfully utilized e-shRNA approaches to target and suppress HIV replication (95, 128–131, 143, 145), and as a result, the parameters that determine efficient processing have been well defined. In particular, the effect of varying siRNA stem length and positioning, spacing between siRNAs stems, and the relative abundance of processed molecules have been tested (129, 130). Despite these advances, siRNAs have been produced from both e-shRNA and lhRNA precursors in a gradient with the most abundant and active being at the end distal from the loop, resulting in reduced silencing for the second, third, and fourth siRNAs (128–131). It was also found that compared to other methods of co-RNAi, the same shRNA sequences were much less predictable when expressed as e-shRNAs and generally exhibited reduced silencing activity (141). Nonetheless, these examples clearly show that these approaches can be very successfully used to initiate potent gene knockdown, particularly in the case of HIV.

As described in Subheading 3, the expression of shRNAs embedded into an miRNA context can offer numerous advantages, none more important than potentially reducing the cellular toxicity seen with shRNAs (Fig. 4). Strategies for expressing multiple shRNAs using naturally occurring miRNA structures are therefore logical, especially considering that individual miRNAs often appear in clusters located together throughout the genome. Strategies have been described that link together multiple miRNA-based shRNAs, either as concatemers of the same surrounding sequence, or as modified naturally occurring miRNA clusters. Two separate studies showed that shRNA sequences embedded between the 5′ and 3′ flanking sequences of miR-30 could be stitched together in repeats of up to three and used to induce enhanced and multiple gene knockdown (146, 147). Similar methods were also used to express two shRNAs by modifying the noncoding RNA BIC transcript that encodes miR-155 (42).

To more closely mimic miRNA transcripts as they occur biologically, the potential for multiple shRNA expression by modifying naturally occurring polycistronic miRNA clusters has also been explored. The conserved chicken miRNA operon encoding miR-106a, 18b, 20b, 19b-2, 92-2, and 363 located on chromosome 4 was modified to encode two shRNA sequences that were capable of suppressing two separate genes simultaneously in chicken embryos (148). Several studies have also used similar approaches to modify human polycistronic miRNA operon sequences. The human mir-17-92 cluster on chromosome 13, which encodes a 1 kb pri-miRNA polycistronic transcript that produces seven mature miRNAs, was modified to encode up to four embedded shRNA sequences, leading to effective HIV inhibition (96). Similar studies have used the chicken miR-126 cluster to express shRNAs targeting influenza (97), the human mir106b-mir93-mir2548 cluster to express anti-HIV molecules (149), and five of the miRNAs from the miRNA-17-92 cluster were modified to target different regions of the HCV genome (150).

It appears that although each of the different methods can result in efficient gene suppression, these studies often involve the detailed optimization of the chosen system and experimental findings are not reported in the context of the other delivery options. Two recent studies attempted to shed light on this inconsistency by comparing various co-RNAi methods directly to one another. Both essentially concluded that the use of multiple promoter/shRNA cassettes encoded in a single vector offered the most useful approach for immediate application in gene therapy (20, 141, 151, 152). This was attributed to the predictable and effective nature of these constructs, the simplicity of vector construction, and the ability to use existing shRNA sequences without further optimizations. Other methods such as the miRNA cluster embedded approaches were less reliable, and lhRNAs encoding multiple shRNA were much less efficacious. It should be mentioned, however, that given the proven potentially toxic nature of pol III-delivered shRNAs (47) and other unknown effects on endogenous miRNA processing, the use of RNAi effectors that more closely resemble naturally occurring molecules such as miRNAs might represent a more balanced strategy for co-RNAi.

8. Summary

The expression of shRNAs for targeted gene silencing has now become a standard technology. Over the past decade we have witnessed the continual expansion, advancement, and improvement of the various strategies that can be used for shRNA delivery and the use of shRNAs for clinical applications is well underway. These developments have also uncovered many potential pitfalls and hazards associated with the application of vector-delivered RNAi, particularly cellular toxicity resulting from the saturation of key factors of the RNAi pathways. However, alongside the discovery of such potential setbacks, research into alternative methods and approaches is providing viable alternatives that can circumvent adverse affects. For example, it now seems that the use of weaker promoter sequences and/or artificial miRNA sequences can alleviate potential toxicities associated with highly expressed shRNAs. With the enormous depth and variety of techniques and platforms for the delivery of shRNAs, individual experiments can be tailored to best suit particular situations. From the availability of various different effector molecule formats, to options in promoter type and the ability for multiple shRNA delivery, approaches can be custom designed to target specific cell types, minimize the emergence of escape mutants, and minimize cellular toxicities. Since its discovery, the Nobel Prize winning cellular process of RNAi has held indisputable promise for both research and therapeutic applications. The substantial advancements of shRNA technologies over the past decade has not only seen their use become a standard laboratory technique but has seen their use move increasingly into clinical applications, a trend that is likely to continue.

References

1. Brummelkamp TR, Bernards R, Agami R (2002) A system for stable expression of short interfering RNAs in mammalian cells. Science 296:550–553. doi:10.1126/science.1068999
2. Yu J-Y, DeRuiter SL, Turner DL (2002) RNA interference by expression of short-interfering RNAs and hairpin RNAs in mammalian cells. Proc Natl Acad Sci U S A 99:6047–6052. doi:10.1073/pnas.092143499
3. Paddison PJ, Caudy AA, Bernstein E, Hannon GJ, Conklin DS (2002) Short hairpin RNAs (shRNAs) induce sequence-specific silencing in mammalian cells. Genes Dev 16:948–958. doi:10.1101/gad.981002
4. Paul CP, Good PD, Winer I, Engelke DR (2002) Effective expression of small interfering RNA in human cells. Nat Biotechnol 20:505–508. doi:10.1038/nbt0502-505
5. Ter Brake O, t'Hooft K, Liu YP, Centlivre M, von Eije KJ, Berkhout B (2008) Lentiviral vector design for multiple shRNA expression and durable HIV-1 inhibition. Mol Ther 16:557–564. doi:10.1038/sj.mt.6300382
6. Siolas D, Lerner C, Burchard J, Ge W, Linsley PS, Paddison PJ, Hannon GJ, Cleary MA (2005) Synthetic shRNAs as potent RNAi triggers. Nat Biotechnol 23:227–231. doi:10.1038/nbt1052
7. McAnuff MA, Rettig GR, Rice KG (2007) Potency of siRNA versus shRNA mediated knockdown in vivo. J Pharm Sci 96:2922–2930. doi:10.1002/jps.20968
8. Klinghoffer RA, Magnus J, Schelter J, Mehaffey M, Coleman C, Cleary MA (2010) Reduced seed region-based off-target activity with lentivirus-mediated RNAi. RNA 16:879–884. doi:10.1261/rna.1977810
9. Lee RC, Ambros V (2001) An extensive class of small RNAs in *Caenorhabditis elegans*. Science 294:862–864. doi:10.1126/science.1065329
10. Lagos-Quintana M, Rauhut R, Yalcin A, Meyer J, Lendeckel W, Tuschl T (2002) Identification of tissue-specific microRNAs from mouse. Curr Biol 12:735–739
11. Cullen BR (2004) Transcription and processing of human microRNA precursors. Mol Cell 16:861–865. doi:10.1016/j.molcel.2004.12.002
12. Bernstein E, Caudy AA, Hammond SM, Hannon GJ (2001) Role for a bidentate ribonuclease in the initiation step of RNA interference. Nature 409:363–366. doi:10.1038/35053110
13. Catalanotto C, Azzalin G, Macino G, Cogoni C (2000) Gene silencing in worms and fungi. Nature 404:245. doi:10.1038/35005169
14. Tabara H, Sarkissian M, Kelly WG, Fleenor J, Grishok A, Timmons L, Fire A, Mello CC (1999) The rde-1 gene, RNA interference, and transposon silencing in *C. elegans*. Cell 99:123–132
15. Elbashir SM, Lendeckel W, Tuschl T (2001) RNA interference is mediated by 21- and 22-nucleotide RNAs. Genes Dev 15:188–200
16. Hammond SM, Bernstein E, Beach D, Hannon GJ (2000) An RNA-directed nuclease mediates post-transcriptional gene silencing

in Drosophila cells. Nature 404:293–296. doi:10.1038/35005107

17. Nykänen A, Haley B, Zamore PD (2001) ATP requirements and small interfering RNA structure in the RNA interference pathway. Cell 107:309–321
18. Martinez J, Tuschl T (2004) RISC is a 5′ phosphomonoester-producing RNA endonuclease. Genes Dev 18:975–980. doi:10.1101/gad.1187904
19. Schwarz DS, Tomari Y, Zamore PD (2004) The RNA-induced silencing complex is a Mg2+-dependent endonuclease. Curr Biol 14:787–791. doi:10.1016/j.cub.2004.03.008
20. Doench JG, Petersen CP, Sharp PA (2003) siRNAs can function as miRNAs. Genes Dev 17:438–442. doi:10.1101/gad.1064703
21. Ge Q, Ilves H, Dallas A, Kumar P, Shorenstein J, Kazakov SA, Johnston BH (2010) Minimal-length short hairpin RNAs: the relationship of structure and RNAi activity. RNA 16:106–117. doi:10.1261/rna.1894510
22. Rao DD, Maples PB, Senzer N, Kumar P, Wang Z, Pappen BO, Yu Y, Haddock C, Jay C, Phadke AP et al (2010) Enhanced target gene knockdown by a bifunctional shRNA: a novel approach of RNA interference. Cancer Gene Ther 17:780–791. doi:10.1038/cgt.2010.35
23. Phadke AP, Jay CM, Wang Z, Chen S, Liu S, Haddock C, Kumar P, Pappen BO, Rao DD, Templeton NS et al (2011) In vivo safety and antitumor efficacy of bifunctional small hairpin RNAs specific for the human Stathmin 1 oncoprotein. DNA Cell Biol 30:715–726. doi:10.1089/dna.2011.1240
24. Taxman DJ, Livingstone LR, Zhang J, Conti BJ, Iocca HA, Williams KL, Lich JD, Ting JP-Y, Reed W (2006) Criteria for effective design, construction, and gene knockdown by shRNA vectors. BMC Biotechnol 6:7. doi:10.1186/1472-6750-6-7
25. Li L, Lin X, Khvorova A, Fesik SW, Shen Y (2007) Defining the optimal parameters for hairpin-based knockdown constructs. RNA 13:1765–1774. doi:10.1261/rna.599107
26. Kim D-H, Behlke MA, Rose SD, Chang M-S, Choi S, Rossi JJ (2005) Synthetic dsRNA Dicer substrates enhance RNAi potency and efficacy. Nat Biotechnol 23:222–226. doi:10.1038/nbt1051
27. Rose SD, Kim D-H, Amarzguioui M, Heidel JD, Collingwood MA, Davis ME, Rossi JJ, Behlke MA (2005) Functional polarity is introduced by Dicer processing of short substrate RNAs. Nucleic Acids Res 33:4140–4156. doi:10.1093/nar/gki732
28. Yu J-Y, Taylor J, DeRuiter SL, Vojtek AB, Turner DL (2003) Simultaneous inhibition of GSK3alpha and GSK3beta using hairpin siRNA expression vectors. Mol Ther 7:228–236
29. Miyagishi M, Sumimoto H, Miyoshi H, Kawakami Y, Taira K (2004) Optimization of an siRNA-expression system with an improved hairpin and its significant suppressive effects in mammalian cells. J Gene Med 6:715–723. doi:10.1002/jgm.556
30. Mcintyre GJ, Yu Y-H, Lomas M, Fanning GC (2011) The effects of stem length and core placement on shRNA activity. BMC Mol Biol 12:34. doi:10.1186/1471-2199-12-34
31. Vlassov AV, Korba B, Farrar K, Mukerjee S, Seyhan AA, Ilves H, Kaspar RL, Leake D, Kazakov SA, Johnston BH (2007) shRNAs targeting hepatitis C: effects of sequence and structural features, and comparison with siRNA. Oligonucleotides 17:223–236. doi:10.1089/oli.2006.0069
32. Terasawa K, Shimizu K, Tsujimoto G (2011) Synthetic pre-miRNA-based shRNA as potent RNAi triggers. J Nucleic Acids 2011:131579. doi:10.4061/2011/131579
33. Ge Q, Dallas A, Ilves H, Shorenstein J, Behlke MA, Johnston BH (2010) Effects of chemical modification on the potency, serum stability, and immunostimulatory properties of short shRNAs. RNA 16:118–130. doi:10.1261/rna.1901810
34. Lee NS, Dohjima T, Bauer G, Li H, Li M-J, Ehsani A, Salvaterra P, Rossi J (2002) Expression of small interfering RNAs targeted against HIV-1 rev transcripts in human cells. Nat Biotechnol 20:500–505. doi:10.1038/nbt0502-500
35. Mcintyre GJ, Fanning GC (2006) Design and cloning strategies for constructing shRNA expression vectors. BMC Biotechnol 6:1. doi:10.1186/1472-6750-6-1
36. Lagos-Quintana M, Rauhut R, Lendeckel W, Tuschl T (2001) Identification of novel genes coding for small expressed RNAs. Science 294:853–858. doi:10.1126/science.1064921
37. Kawasaki H, Taira K (2003) Short hairpin type of dsRNAs that are controlled by tRNA(Val) promoter significantly induce RNAi-mediated gene silencing in the cytoplasm of human cells. Nucleic Acids Res 31:700–707
38. Carmona S, Ely A, Crowther C, Moolla N, Salazar FH, Marion PL, Ferry N, Weinberg MS, Arbuthnot P (2006) Effective inhibition

of HBV replication in vivo by anti-HBx short hairpin RNAs. Mol Ther 13:411–421. doi:10.1016/j.ymthe.2005.10.013

39. Boden D (2004) Enhanced gene silencing of HIV-1 specific siRNA using microRNA designed hairpins. Nucleic Acids Res 32:1154–1158. doi:10.1093/nar/gkh278
40. Lambeth LS, Zhao Y, Smith LP, Kgosana L, Nair V (2009) Targeting Marek's disease virus by RNA interference delivered from a herpesvirus vaccine. Vaccine 27:298–306. doi:10.1016/j.vaccine.2008.10.023
41. Hinton TM, Wise TG, Cottee PA, Doran TJ (2008) Native microRNA loop sequences can improve short hairpin RNA processing for virus gene silencing in animal cells. J RNAi Gene Silencing 4:295–301
42. Chung K-H, Hart CC, Al-Bassam S, Avery A, Taylor J, Patel PD, Vojtek AB, Turner DL (2006) Polycistronic RNA polymerase II expression vectors for RNA interference based on BIC/miR-155. Nucleic Acids Res 34:e53. doi:10.1093/nar/gkl143
43. Zhou H, Xia X-G, Xu Z (2005) An RNA polymerase II construct synthesizes short-hairpin RNA with a quantitative indicator and mediates highly efficient RNAi. Nucleic Acids Res 33:e62. doi:10.1093/nar/gni061
44. Silva JM, Li MZ, Chang K, Ge W, Golding MC, Rickles RJ, Siolas D, Hu G, Paddison PJ, Schlabach MR et al (2005) Second-generation shRNA libraries covering the mouse and human genomes. Nat Genet 37:1281–1288. doi:10.1038/ng1650
45. Dickins RA, Hemann MT, Zilfou JT, Simpson DR, Ibarra I, Hannon GJ, Lowe SW (2005) Probing tumor phenotypes using stable and regulated synthetic microRNA precursors. Nat Genet 37:1289–1295. doi:10.1038/ng1651
46. Boudreau RL, Monteys AM, Davidson BL (2008) Minimizing variables among hairpin-based RNAi vectors reveals the potency of shRNAs. RNA 14:1834–1844. doi:10.1261/rna.1062908
47. Grimm D, Streetz KL, Jopling CL, Storm TA, Pandey K, Davis CR, Marion P, Salazar F, Kay MA (2006) Fatality in mice due to oversaturation of cellular microRNA/short hairpin RNA pathways. Nature 441:537–541. doi:10.1038/nature04791
48. Ehlert EM, Eggers R, Niclou SP, Verhaagen J (2010) Cellular toxicity following application of adeno-associated viral vector-mediated RNA interference in the nervous system. BMC Neurosci 11:20. doi:10.1186/1471-2202-11-20
49. Martin JN, Wolken N, Brown T, Dauer WT, Ehrlich ME, Gonzalez-Alegre P (2011) Lethal toxicity caused by expression of shRNA in the mouse striatum: implications for therapeutic design. Gene Ther 18:666–673. doi:10.1038/gt.2011.10
50. McBride JL, Boudreau RL, Harper SQ, Staber PD, Monteys AM, Martins I, Gilmore BL, Burstein H, Peluso RW, Polisky B et al (2008) Artificial miRNAs mitigate shRNA-mediated toxicity in the brain: implications for the therapeutic development of RNAi. Proc Natl Acad Sci U S A 105:5868–5873. doi:10.1073/pnas.0801775105
51. Boden D, Pusch O, Lee F, Tucker L, Shank PR, Ramratnam B (2003) Promoter choice affects the potency of HIV-1 specific RNA interference. Nucleic Acids Res 31:5033–5038
52. Rao MK, Wilkinson MF (2006) Tissue-specific and cell type-specific RNA interference in vivo. Nat Protoc 1:1494–1501. doi:10.1038/nprot.2006.260
53. Nielsen TT, Marion IV, Hasholt L, Lundberg C (2009) Neuron-specific RNA interference using lentiviral vectors. J Gene Med 11:559–569. doi:10.1002/jgm.1333
54. Dong K, Wang R, Wang X, Lin F, Shen J-J, Gao P, Zhang H-Z (2009) Tumor-specific RNAi targeting eIF4E suppresses tumor growth, induces apoptosis and enhances cisplatin cytotoxicity in human breast carcinoma cells. Breast Cancer Res Treat 113:443–456. doi:10.1007/s10549-008-9956-x
55. Hernandez N (2001) Small nuclear RNA genes: a model system to study fundamental mechanisms of transcription. J Biol Chem 276:26733–26736. doi:10.1074/jbc.R100032200
56. Lee Y, Kim M, Han J, Yeom K-H, Lee S, Baek SH, Kim VN (2004) MicroRNA genes are transcribed by RNA polymerase II. EMBO J 23:4051–4060. doi:10.1038/sj.emboj.7600385
57. Paule MR, White RJ (2000) Survey and summary: transcription by RNA polymerases I and III. Nucleic Acids Res 28:1283–1298
58. Czauderna F, Santel A, Hinz M, Fechtner M, Durieux B, Fisch G, Leenders F, Arnold W, Giese K, Klippel A et al (2003) Inducible shRNA expression for application in a prostate cancer mouse model. Nucleic Acids Res 31:e127
59. Kobayashi S, Higuchi T, Anzai K (2005) Application of the BC1 RNA gene promoter for short hairpin RNA expression in cultured neuronal cells. Biochem Biophys Res Commun 334:1305–1309. doi:10.1016/j.bbrc.2005.07.033
60. Scherer LJ, Frank R, Rossi JJ (2007) Optimization and characterization of

tRNA-shRNA expression constructs. Nucleic Acids Res 35:2620–2628. doi:10.1093/nar/gkm103

61. Gunnery S, Ma Y, Mathews MB (1999) Termination sequence requirements vary among genes transcribed by RNA polymerase III. J Mol Biol 286:745–757. doi:10.1006/jmbi.1998.2518
62. Murphy S, Tripodi M, Melli M (1986) A sequence upstream from the coding region is required for the transcription of the 7SK RNA genes. Nucleic Acids Res 14:9243–9260
63. Das G, Henning D, Wright D, Reddy R (1988) Upstream regulatory elements are necessary and sufficient for transcription of a U6 RNA gene by RNA polymerase III. EMBO J 7:503–512
64. Kunkel GR, Pederson T (1988) Upstream elements required for efficient transcription of a human U6 RNA gene resemble those of U1 and U2 genes even though a different polymerase is used. Genes Dev 2:196–204
65. Schramm L, Hernandez N (2002) Recruitment of RNA polymerase III to its target promoters. Genes Dev 16:2593–2620. doi:10.1101/gad.1018902
66. Valadkhan S (2005) snRNAs as the catalysts of pre-mRNA splicing. Curr Opin Chem Biol 9:603–608. doi:10.1016/j.cbpa.2005.10.008
67. Hayashi K (1981) Organization of sequences related to U6 RNA in the human genome. Nucleic Acids Res 9:3379–3388
68. Domitrovich AM, Kunkel GR (2003) Multiple, dispersed human U6 small nuclear RNA genes with varied transcriptional efficiencies. Nucleic Acids Res 31:2344–2352
69. Lambeth LS, Moore RJ, Muralitharan M, Dalrymple BP, Mcwilliam S, Doran TJ (2005) Characterisation and application of a bovine U6 promoter for expression of short hairpin RNAs. BMC Biotechnol 5:13. doi:10.1186/1472-6750-5-13
70. Lambeth LS, Wise TG, Moore RJ, Muralitharan MS, Doran TJ (2006) Comparison of bovine RNA polymerase III promoters for short hairpin RNA expression. Anim Genet 37:369–372. doi:10.1111/j.1365-2052.2006.01468.x
71. Kudo T, Sutou S (2005) Usage of putative chicken U6 promoters for vector-based RNA interference. J Reprod Dev 51:411–417
72. Wise TG, Schafer DJ, Lambeth LS, Tyack SG, Bruce MP, Moore RJ, Doran TJ (2007) Characterization and comparison of chicken U6 promoters for the expression of short hairpin RNAs. Anim Biotechnol 18:153–162. doi:10.1080/10495390600867515
73. Zenke K, Kim KH (2008) Novel fugu U6 promoter driven shRNA expression vector for efficient vector based RNAi in fish cell lines. Biochem Biophys Res Commun 371:480–483. doi:10.1016/j.bbrc.2008.04.116
74. Boonanuntanasarn S, Panyim S, Yoshizaki G (2009) Usage of putative zebrafish U6 promoters to express shRNA in Nile tilapia and shrimp cell extracts. Transgenic Res 18:323–325. doi:10.1007/s11248-009-9249-0
75. Hu S, Ni W, Hazi W, Zhang H, Zhang N, Meng R, Chen C (2011) Cloning and functional analysis of sheep U6 promoters. Anim Biotechnol 22:170–174. doi:10.1080/10495398.2011.580669
76. Chuang C-K, Lee K-H, Fan C-T, Su Y-S (2009) Porcine type III RNA polymerase III promoters for short hairpin RNA expression. Anim Biotechnol 20:34–39. doi:10.1080/10495390802603064
77. Myslinski E, Amé JC, Krol A, Carbon P (2001) An unusually compact external promoter for RNA polymerase III transcription of the human H1RNA gene. Nucleic Acids Res 29:2502–2509
78. Koper-Emde D, Herrmann L, Sandrock B, Benecke B-J (2011) RNA interference by small hairpin RNAs synthesised under control of the human 7S K RNA promoter. Biol Chem 385:791–794. doi:10.1515/BC.2004.103
79. Bannister SC, Wise TG, Cahill DM, Doran TJ (2007) Comparison of chicken 7SK and U6 RNA polymerase III promoters for short hairpin RNA expression. BMC Biotechnol 7:79. doi:10.1186/1472-6750-7-79
80. Cummins D, Doran TJ, Tyack S, Purcell D, Hammond J (2008) Identification and characterisation of the porcine 7SK RNA polymerase III promoter for short hairpin RNA expression. J RNAi Gene Silencing 4: 289–294
81. Mäkinen PI, Koponen JK, Kärkkäinen A-M, Malm TM, Pulkkinen KH, Koistinaho J, Turunen MP, Ylä-Herttuala S (2006) Stable RNA interference: comparison of U6 and H1 promoters in endothelial cells and in mouse brain. J Gene Med 8:433–441. doi:10.1002/jgm.860
82. An DS, Qin FX-F, Auyeung VC, Mao SH, Kung SKP, Baltimore D, Chen ISY (2006) Optimization and functional effects of stable short hairpin RNA expression in primary human lymphocytes via lentiviral vectors. Mol Ther 14:494–504. doi:10.1016/j.ymthe.2006.05.015
83. Farris AD, Gross JK, Hanas JS, Harley JB (1996) Genes for murine Y1 and Y3 Ro RNAs have class 3 RNA polymerase III promoter structures and are unlinked on mouse chromosome 6. Gene 174:35–42

84. Grimm D, Wang L, Lee JS, Schürmann N, Gu S, Börner K, Storm TA, Kay MA (2010) Argonaute proteins are key determinants of RNAi efficacy, toxicity, and persistence in the adult mouse liver. J Clin Invest 120:3106–3119. doi:10.1172/JCI43565

85. Rumi M, Ishihara S, Aziz M, Kazumori H, Ishimura N, Yuki T, Kadota C, Kadowaki Y, Kinoshita Y (2006) RNA polymerase II mediated transcription from the polymerase III promoters in short hairpin RNA expression vector. Biochem Biophys Res Commun 339:540–547. doi:10.1016/j.bbrc.2005.11.037

86. Galli G, Hofstetter H, Birnstiel ML (1981) Two conserved sequence blocks within eukaryotic tRNA genes are major promoter elements. Nature 294:626–631

87. Xia X-G, Zhou H, Samper E, Melov S, Xu Z (2006) Pol II-expressed shRNA knocks down Sod2 gene expression and causes phenotypes of the gene knockout in mice. PLoS Genet 2:e10. doi:10.1371/journal.pgen.0020010

88. Takahashi Y, Yamaoka K, Nishikawa M, Takakura Y (2009) Quantitative and temporal analysis of gene silencing in tumor cells induced by small interfering RNA or short hairpin RNA expressed from plasmid vectors. J Pharm Sci 98:74–80. doi:10.1002/jps.21398

89. Dyer V, Ely A, Bloom K, Weinberg M, Arbuthnot P (2010) tRNA Lys3 promoter cassettes that efficiently express RNAi-activating antihepatitis B virus short hairpin RNAs. Biochem Biophys Res Commun 398:640–646. doi:10.1016/j.bbrc.2010.06.122

90. Weiwei M, Zhenhua X, Feng L, Hang N, Yuyang J (2009) A significant increase of RNAi efficiency in human cells by the CMV enhancer with a tRNAlys promoter. J Biomed Biotechnol 2009:514287. doi:10.1155/2009/514287

91. Xia H, Mao Q, Paulson HL, Davidson BL (2002) siRNA-mediated gene silencing in vitro and in vivo. Nat Biotechnol 20:1006–1010. doi:10.1038/nbt739

92. Zeng Y, Wagner EJ, Cullen BR (2002) Both natural and designed micro RNAs can inhibit the expression of cognate mRNAs when expressed in human cells. Mol Cell 9:1327–1333

93. Denti MA, Rosa A, Sthandier O, De Angelis FG, Bozzoni I (2004) A new vector, based on the PolII promoter of the U1 snRNA gene, for the expression of siRNAs in mammalian cells. Mol Ther 10:191–199

94. Huang M, Jia F-J, Yan Y-C, Guo L-H, Li Y-P (2006) Transactivated minimal E1b promoter is capable of driving the expression of short hairpin RNA. J Virol Methods 134:48–54. doi:10.1016/j.jviromet.2005.11.016

95. Konstantinova P, De Vries W, Haasnoot J, Ter Brake O, De Haan P, Berkhout B (2006) Inhibition of human immunodeficiency virus type 1 by RNA interference using long-hairpin RNA. Gene Ther 13:1403–1413. doi:10.1038/sj.gt.3302786

96. Liu YP, Haasnoot J, Ter Brake O, Berkhout B, Konstantinova P (2008) Inhibition of HIV-1 by multiple siRNAs expressed from a single microRNA polycistron. Nucleic Acids Res 36:2811–2824. doi:10.1093/nar/gkn109

97. Chen SC-Y, Stern P, Guo Z, Chen J (2011) Expression of multiple artificial microRNAs from a chicken miRNA126-based lentiviral vector. PLoS One 6:e22437. doi:10.1371/journal.pone.0022437

98. Dong K, Wang R, Wang X, Lin F, Shen J-J, Gao P, Zhang H-Z (2008) Tumor-specific RNAi targeting eIF4E suppresses tumor growth, induces apoptosis and enhances cisplatin cytotoxicity in human breast carcinoma cells. Breast Cancer Res Treat 113:443–456. doi:10.1007/s10549-008-9956-x

99. Giering JC, Grimm D, Storm TA, Kay MA (2008) Expression of shRNA from a tissue-specific pol II promoter is an effective and safe RNAi therapeutic. Mol Ther 16:1630–1636. doi:10.1038/mt.2008.144

100. Zhu Z, Zheng T, Lee CG, Homer RJ, Elias JA (2002) Tetracycline-controlled transcriptional regulation systems: advances and application in transgenic animal modeling. Semin Cell Dev Biol 13:121–128

101. van de Wetering M, Oving I, Muncan V, Pon Fong MT, Brantjes H, van Leenen D, Holstege FCP, Brummelkamp TR, Agami R, Clevers H (2003) Specific inhibition of gene expression using a stably integrated, inducible small-interfering-RNA vector. EMBO Rep 4:609–615. doi:10.1038/sj.embor.embor865

102. Ventura A, Meissner A, Dillon CP, McManus M, Sharp PA, Van Parijs L, Jaenisch R, Jacks T (2004) Cre-lox-regulated conditional RNA interference from transgenes. Proc Natl Acad Sci U S A 101:10380–10385. doi:10.1073/pnas.0403954101

103. Szulc J, Wiznerowicz M, Sauvain M-O, Trono D, Aebischer P (2006) A versatile tool for conditional gene expression and knockdown. Nat Meth 3:109–116. doi:10.1038/nmeth846

104. Herold MJ, van den Brandt J, Seibler J, Reichardt HM (2008) Inducible and reversible gene silencing by stable integration of an shRNA-encoding lentivirus in transgenic rats. Proc Natl Acad Sci U S A 105:18507–18512. doi:10.1073/pnas.0806213105

105. No D, Yao TP, Evans RM (1996) Ecdysone-inducible gene expression in mammalian cells and transgenic mice. Proc Natl Acad Sci U S A 93:3346–3351
106. Gupta S, Schoer RA, Egan JE, Hannon GJ, Mittal V (2004) Inducible, reversible, and stable RNA interference in mammalian cells. Proc Natl Acad Sci U S A 101:1927–1932. doi:10.1073/pnas.0306111101
107. Rangasamy D, Tremethick DJ, Greaves IK (2008) Gene knockdown by ecdysone-based inducible RNAi in stable mammalian cell lines. Nat Protoc 3:79–88. doi:10.1038/nprot.2007.456
108. Orban PC, Chui D, Marth JD (1992) Tissue- and site-specific DNA recombination in transgenic mice. Proc Natl Acad Sci U S A 89: 6861–6865
109. McCaffrey AP, Meuse L, Pham T-TT, Conklin DS, Hannon GJ, Kay MA (2002) RNA interference in adult mice. Nature 418:38–39. doi:10.1038/418038a
110. McCaffrey AP, Nakai H, Pandey K, Huang Z, Salazar FH, Xu H, Wieland SF, Marion PL, Kay MA (2003) Inhibition of hepatitis B virus in mice by RNA interference. Nat Biotechnol 21:639–644. doi:10.1038/nbt824
111. Brummelkamp TR, Bernards R, Agami R (2002) Stable suppression of tumorigenicity by virus-mediated RNA interference. Cancer Cell 2:243–247
112. Devroe E, Silver PA (2002) Retrovirus-delivered siRNA. BMC Biotechnol 2:15
113. Bromberg-White JL, Webb CP, Patacsil VS, Miranti CK, Williams BO, Holmen SL (2004) Delivery of short hairpin RNA sequences by using a replication-competent avian retroviral vector. J Virol 78:4914–4916
114. Hughes SH, Greenhouse JJ, Petropoulos CJ, Sutrave P (1987) Adaptor plasmids simplify the insertion of foreign DNA into helper-independent retroviral vectors. J Virol 61:3004–3012
115. Déglon N, Tseng JL, Bensadoun JC, Zurn AD, Arsenijevic Y, Pereira de Almeida L, Zufferey R, Trono D, Aebischer P (2000) Self-inactivating lentiviral vectors with enhanced transgene expression as potential gene transfer system in Parkinson's disease. Hum Gene Ther 11:179–190. doi:10.1089/10430340050016256
116. Abbas-Terki T, Blanco-Bose W, Déglon N, Pralong W, Aebischer P (2002) Lentiviral-mediated RNA interference. Hum Gene Ther 13:2197–2201. doi:10.1089/104303402320987888
117. Rubinson DA, Dillon CP, Kwiatkowski AV, Sievers C, Yang L, Kopinja J, Rooney DL, Zhang M, Ihrig MM, McManus MT et al (2003) A lentivirus-based system to functionally silence genes in primary mammalian cells, stem cells and transgenic mice by RNA interference. Nat Genet 33:401–406. doi:10.1038/ng1117
118. Cartier N, Hacein-Bey-Abina S, Bartholomae CC, Veres G, Schmidt M, Kutschera I, Vidaud M, Abel U, Dal-Cortivo L, Caccavelli L et al (2009) Hematopoietic stem cell gene therapy with a lentiviral vector in X-linked adrenoleukodystrophy. Science 326:818–823. doi:10.1126/science.1171242
119. Berkhout B (2009) Toward a durable anti-HIV gene therapy based on RNA interference. Ann N Y Acad Sci 1175:3–14. doi:10.1111/j.1749-6632.2009.04972.x
120. Tomar RS, Matta H, Chaudhary PM (2003) Use of adeno-associated viral vector for delivery of small interfering RNA. Oncogene 22:5712–5715. doi:10.1038/sj.onc.1206733
121. Büning H, Perabo L, Coutelle O, Quadt Humme S, Hallek M (2008) Recent developments in adeno-associated virus vector technology. J Gene Med 10:717–733. doi:10.1002/jgm.1205
122. Saydam O, Glauser DL, Heid I, Turkeri G, Hilbe M, Jacobs AH, Ackermann M, Fraefel C (2005) Herpes simplex virus 1 amplicon vector-mediated siRNA targeting epidermal growth factor receptor inhibits growth of human glioma cells in vivo. Mol Ther 12:803–812. doi:10.1016/j.ymthe.2005.07.534
123. Hong C-S, Goins WF, Goss JR, Burton EA, Glorioso JC (2006) Herpes simplex virus RNAi and neprilysin gene transfer vectors reduce accumulation of Alzheimer's disease-related amyloid-beta peptide in vivo. Gene Ther 13:1068–1079. doi:10.1038/sj.gt.3302719
124. Nicholson LJ, Philippe M, Paine AJ, Mann DA, Dolphin CT (2005) RNA interference mediated in human primary cells via recombinant baculoviral vectors. Mol Ther 11:638–644. doi:10.1016/j.ymthe.2004.12.010
125. Castanotto D, Sakurai K, Lingeman R, Li H, Shively L, Aagaard L, Soifer H, Gatignol A, Riggs A, Rossi JJ (2007) Combinatorial delivery of small interfering RNAs reduces RNAi efficacy by selective incorporation into RISC. Nucleic Acids Res 35:5154–5164. doi:10.1093/nar/gkm543
126. Grimm D, Kay MA (2007) Combinatorial RNAi: a winning strategy for the race against

evolving targets? Mol Ther 15:878–888. doi:10.1038/sj.mt.6300116

127. Boden D, Pusch O, Lee F, Tucker L, Ramratnam B (2003) Human immunodeficiency virus type 1 escape from RNA interference. J Virol 77:11531–11535

128. Liu YP, von Eije KJ, Schopman NCT, Westerink J-T, Ter Brake O, Haasnoot J, Berkhout B (2009) Combinatorial RNAi against HIV-1 using extended short hairpin RNAs. Mol Ther 17:1712–1723. doi:10.1038/mt.2009.176

129. Liu YP, Haasnoot J, Berkhout B (2007) Design of extended short hairpin RNAs for HIV-1 inhibition. Nucleic Acids Res 35:5683–5693. doi:10.1093/nar/gkm596

130. Saayman S, Barichievy S, Capovilla A, Morris KV, Arbuthnot P, Weinberg MS (2008) The efficacy of generating three independent anti-HIV-1 siRNAs from a single U6 RNA Pol III-expressed long hairpin RNA. PLoS One 3:e2602. doi:10.1371/journal.pone.0002602

131. Sano M, Li H, Nakanishi M, Rossi JJ (2008) Expression of long anti-HIV-1 hairpin RNAs for the generation of multiple siRNAs: advantages and limitations. Mol Ther 16:170–177. doi:10.1038/sj.mt.6300298

132. Henry SD, van der Wegen P, Metselaar HJ, Tilanus HW, Scholte BJ, van der Laan LJW (2006) Simultaneous targeting of HCV replication and viral binding with a single lentiviral vector containing multiple RNA interference expression cassettes. Mol Ther 14:485–493. doi:10.1016/j.ymthe.2006.04.012

133. Anderson J, Akkina R (2005) HIV-1 resistance conferred by siRNA cosuppression of CXCR4 and CCR5 coreceptors by a bispecific lentiviral vector. AIDS Res Ther 2:1. doi:10.1186/1742-6405-2-1

134. Hinton TM, Doran TJ (2008) Inhibition of chicken anaemia virus replication using multiple short-hairpin RNAs. Antiviral Res 80:143–149. doi:10.1016/j.antiviral.2008.05.009

135. Ter Brake O, Konstantinova P, Ceylan M, Berkhout B (2006) Silencing of HIV-1 with RNA interference: a multiple shRNA approach. Mol Ther 14:883–892. doi:0.1016/j.ymthe.2006.07.007

136. Song J, Giang A, Lu Y, Pang S, Chiu R (2008) Multiple shRNA expressing vector enhances efficiency of gene silencing. BMB Rep 41:358–362

137. Gou D, Weng T, Wang Y, Wang Z, Zhang H, Gao L, Chen Z, Wang P, Liu L (2007) A novel approach for the construction of multiple shRNA expression vectors. J Gene Med 9:751–763. doi:10.1002/jgm.1080

138. Cheng TL, Teng CF, Tsai WH, Yeh CW, Wu MP, Hsu HC, Hung CF, Chang WT (2009) Multitarget therapy of malignant cancers by the head-to-tail tandem array multiple shRNAs expression system. Cancer Gene Ther 16:516–531. doi:10.1038/cgt.2008.102

139. Gonzalez S, Castanotto D, Li H, Olivares S, Jensen MC, Forman SJ, Rossi JJ, Cooper LJN (2005) Amplification of RNAi-targeting HLA mRNAs. Mol Ther 11:811–818. doi:10.1016/j.ymthe.2004.12.023

140. Mcintyre GJ, Groneman JL, Tran A, Applegate TL (2008) An infinitely expandable cloning strategy plus repeat-proof PCR for working with multiple shRNA. PLoS One 3:e3827. doi:10.1371/journal.pone.0003827.t001

141. Lambeth LS, Van Hateren NJ, Wilson SA, Nair V (2010) A direct comparison of strategies for combinatorial RNA interference. BMC Mol Biol 11:77. doi:10.1186/1471-2199-11-77

142. Akashi H, Miyagishi M, Yokota T, Watanabe T, Hino T, Nishina K, Kohara M, Taira K (2005) Escape from the interferon response associated with RNA interference using vectors that encode long modified hairpin-RNA. Mol Biosyst 1:382–390. doi:10.1039/b510159j

143. Barichievy S, Saayman S, Von Eije KJ, Morris KV, Arbuthnot P, Weinberg MS (2007) The inhibitory efficacy of RNA POL III-expressed long hairpin RNAs targeted to untranslated regions of the HIV-1 5′ long terminal repeat. Oligonucleotides 17:419–431. doi:10.1089/oli.2007.0095

144. Weinberg MS, Ely A, Barichievy S, Crowther C, Mufamadi S, Carmona S, Arbuthnot P (2007) Specific inhibition of HBV replication in vitro and in vivo with expressed long hairpin RNA. Mol Ther 15:534–541. doi:10.1038/sj.mt.6300077

145. Konstantinova P, Ter Brake O, Haasnoot J, De Haan P, Berkhout B (2007) Trans-inhibition of HIV-1 by a long hairpin RNA expressed within the viral genome. Retrovirology 4:15. doi:10.1186/1742-4690-4-15

146. Sun D, Melegari M, Sridhar S, Rogler CE, Zhu L (2006) Multi-miRNA hairpin method that improves gene knockdown efficiency and provides linked multi-gene knockdown. Biotechniques 41:59–63

147. Zhu X, Santat LA, Chang MS, Liu J, Zavzavadjian JR, Wall EA, Kivork C, Simon MI, Fraser ID (2007) A versatile approach to multiple gene RNA interference using microRNA-based short hairpin RNAs. BMC Mol Biol 8:98. doi:10.1186/1471-2199-8-98

148. Das RM, Van Hateren NJ, Howell GR, Farrell ER, Bangs FK, Porteous VC, Manning EM, McGrew MJ, Ohyama K, Sacco MA et al (2006) A robust system for RNA interference in the chicken using a modified microRNA operon. Dev Biol 294:554–563. doi:10.1016/j.ydbio.2006.02.020

149. Aagaard LA, Zhang J, von Eije KJ, Li H, Saetrom P, Amarzguioui M, Rossi JJ (2008) Engineering1 and optimization of the miR-106b cluster for ectopic expression of multiplexed anti-HIV RNAs. Gene Ther 15:1536–1549. doi:10.1038/gt.2008.147

150. Yang X, Haurigot V, Zhou S, Luo G, Couto LB (2010) Inhibition of hepatitis C virus replication using adeno-associated virus vector delivery of an exogenous anti-hepatitis C virus microRNA cluster. Hepatology 52:1877–1887. doi:10.1002/hep. 23908

151. Bartel DP, Chen C-Z (2004) Micromanagers of gene expression: the potentially widespread influence of metazoan microRNAs. Nat Rev Genet 5:396–400. doi:10.1038/nrg1328

152. Mcintyre GJ, Arndt AJ, Gillespie KM, Mak WM, Fanning GC (2011) A comparison of multiple shRNA expression methods for combinatorial RNAi. Genet Vaccines Ther 9:9. doi:10.1186/1479-0556-9-9

153. Butler JEF, Kadonaga JT (2002) The RNA polymerase II core promoter: a key component in the regulation of gene expression. Genes Dev 16:2583–2592. doi:10.1101/gad.1026202

Chapter 13

Design of Lentivirally Expressed siRNAs

Ying Poi Liu and Ben Berkhout

Abstract

RNA interference (RNAi) has been widely used as a tool for gene knockdown in fundamental research and for the development of new RNA-based therapeutics. The RNAi pathway is typically induced by expression of ~22 base pair (bp) small interfering RNAs (siRNAs), which can be transfected into cells. For long-term gene silencing, short hairpin RNA (shRNA), or artificial microRNA (amiRNA) expression constructs have been developed that produce these RNAi inducers inside the cell. Currently, these types of constructs are broadly applied to knock down any gene of interest. Besides mono RNAi strategies that involve single shRNAs or amiRNAs, combinatorial RNAi approaches have been developed that allow the simultaneous expression of multiple siRNAs or amiRNAs by using polycistrons, extended shRNAs (e-shRNAs), or long hairpin RNAs (lhRNAs). Here, we provide practical information for the construction of single shRNA or amiRNA vectors, but also multi-shRNA/amiRNA constructs. Furthermore, we summarize the advantages and limitations of the most commonly used viral vectors for the expression of RNAi inducers.

Key words: RNAi, siRNA, amiRNA, shRNA, Gene therapy, Viral vector, Lentiviral vector, HIV-1, Titer, Transduction

1. Introduction

RNA interference is an evolutionarily conserved posttranscriptional gene silencing mechanism in plants, insects, fungi, and nematodes that induces strong and sequence-specific inhibition of complementary mRNAs (1–8). The major functions of RNAi are to regulate cellular gene expression, to defend the cell against viruses and to suppress transposon activity (7–10). The key players of the RNAi mechanism are small noncoding double-stranded RNAs of ~22 base pairs (bp) in length (11, 12). The main classes of small noncoding RNAs are the small interfering RNAs (siRNAs) and the microRNAs (miRNAs). The siRNAs are processed from double-stranded RNA that is derived from viral replication intermediates, transposable elements, inverted-repeat containing transgenes, aberrant

Debra J. Taxman (ed.), *siRNA Design: Methods and Protocols*, Methods in Molecular Biology, vol. 942,
DOI 10.1007/978-1-62703-119-6_13,

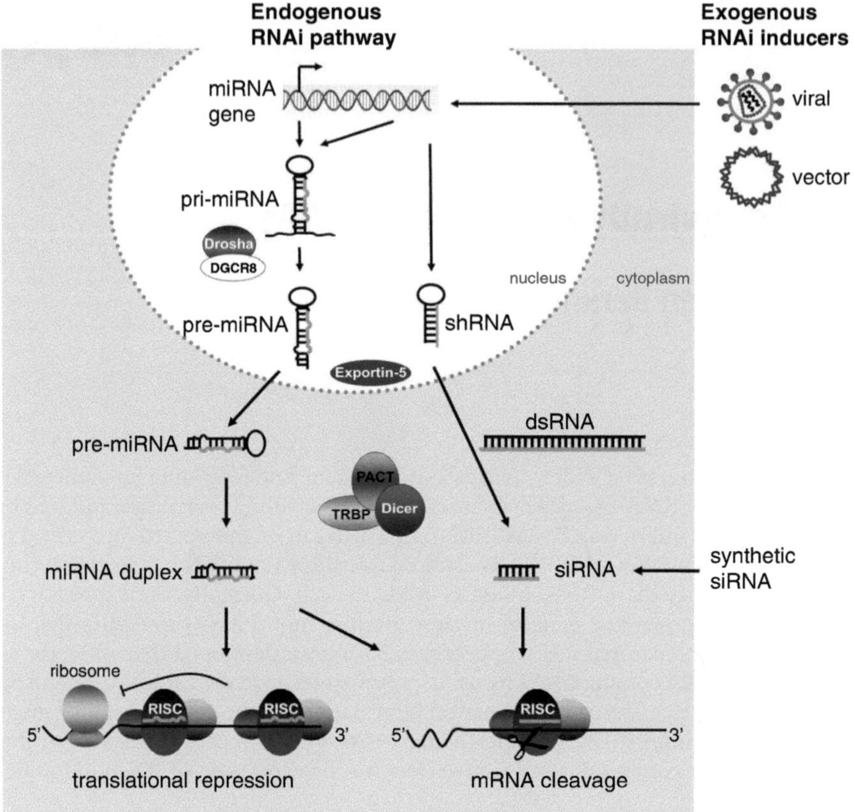

Fig. 1. Schematic of the RNAi pathway. RNAi can be triggered by endogenous and exogenous RNAi inducers (*left* and *right branches*). Endogenous triggers are the ~22 bp noncoding small RNAs known as miRNAs. The miRNA gene encodes a long primary miRNA (pri-miRNA) transcript that is processed in the nucleus by the Drosha-DGCR complex into a precursor miRNA (pre-miRNA). This pre-miRNA typically folds into a hairpin structure containing mismatches and bulges in the stem. After nuclear export by Exportin-5, the Dicer/TRBP/PACT complex cleaves off the loop of the pre-miRNA to generate the miRNA duplex. One strand of this duplex, the mature miRNA, directs RISC to imperfectly or nearly perfectly complementary mRNA targets, where RISC induces translational repression or mRNA cleavage, respectively. Another trigger for RNAi is the presence of double-stranded RNA (dsRNA), which can be processed into small interfering RNAs (siRNAs) by the Dicer/TRBP/PACT complex. The siRNAs are loaded into RISC and the guide strand of the siRNA directs RISC to cleave a perfectly complementary mRNA target. Exogenous RNAi triggers include short hairpin RNAs (shRNAs) that can be expressed from regular plasmids or viral vectors and ready for use siRNAs. Like dsRNA, the shRNA is processed by the Dicer/TRBP/PACT complex into siRNAs that initiate RNAi-mediated cleavage of the mRNA target.

transcription products, or complementary transcripts (1–5). The dsRNA is processed by the cytoplasmic Dicer/TRBP/PACT endonuclease into ~22 bp siRNAs with 2-nt 3′ overhangs and a 5′ phosphate group (Fig. 1, right branch) (13–16). The siRNA duplex is subsequently loaded into the RNA-induced silencing complex (RISC), where it associates with a specific Argonaute (Ago) family protein. One strand of the siRNA, the so-called passenger strand is degraded and discarded (17–19), while the guide strand is retained to instruct RISC to cleave a perfectly complementary mRNA at position 10 or 11 within the base-paired duplex (20).

A growing number of reports have emerged showing that RNAi can also play a role in antiviral defense mechanisms in mammals (21–24). However, the key role of RNAi in mammalian cells is to regulate cellular gene expression at the posttranscriptional level via miRNA genes that express primary miRNA (pri-miRNA) transcripts (Fig. 1, left branch). The pri-miRNA is processed by the Drosha-DGCR8 complex to liberate a ~70 nt precursor miRNA (pre-miRNA) with 2-nt 3′ overhang, which typically folds into a hairpin structure containing mismatches and bulges in the stem (25). The nuclear export protein Exportin-5 recognizes the 2-nt 3′ overhang of the pre-miRNA and transports the hairpin to the cytoplasm (26–28), where it is further processed by the Dicer/TRBP/PACT endonuclease complex into an imperfect ~22 bp miRNA duplex (29). The single-stranded mature miRNA associates with the Ago protein in the RISC complex and directs the complex towards mRNA targets, which are usually located in the 3′ untranslated region (3′UTR) (30, 31). Partial base pairing commonly results in translational repression, whereas perfect or near-perfect base pairing leads to mRNA cleavage (32–34). The specificity of the miRNA for its target mRNA is primarily determined by the "seed sequence" (nucleotides 2–8 from the 5′ end of the miRNA) and the presence of multiple target sites within the mRNA (32, 33, 35, 36). The targeted mRNAs are subsequently transported to cytoplasmic mRNA-processing compartments known as P-bodies (37, 38).

Since its discovery, RNAi-based technology has evolved as a powerful tool to silence any gene of interest for research or therapeutic purposes (24, 39). Initially, siRNAs were used that can be transiently transfected into cells and thus result in short-term gene silencing. Soon thereafter, vectors encoding siRNA precursors known as short hairpin RNAs (shRNAs) were developed. The shRNA is modeled after a pre-miRNA with a stem of 19–29 bp, a small loop and a 3′ UU overhang that resembles the product of Drosha cleavage (Figs. 1, right and 2a) (40–42). Expression of shRNA constructs is mostly driven by RNA polymerase III promoters such as those of the U6 RNA, H1 RNA, or tRNA genes because they allow high-level shRNA expression with well-defined initiation and termination sites (4–6 consecutive U residues in the transcript) in all cell types (40, 41, 43).

The shRNA design has been optimized by including pri-miRNA and pre-miRNA features, including bulges and mismatches in the hairpin stem, flanking sequences, and loop domains (Fig. 2b). A perfectly complementary guide strand is inserted at the location of the mature miRNA in the pri-miRNA backbone. An RNA polymerase II promoter is often used to transcribe artificial miRNAs (amiRNAs) because most miRNA genes have such a promoter (44). A major advantage of RNA polymerase II promoters is that they allow regulatable and tissue-specific gene expression (45, 46).

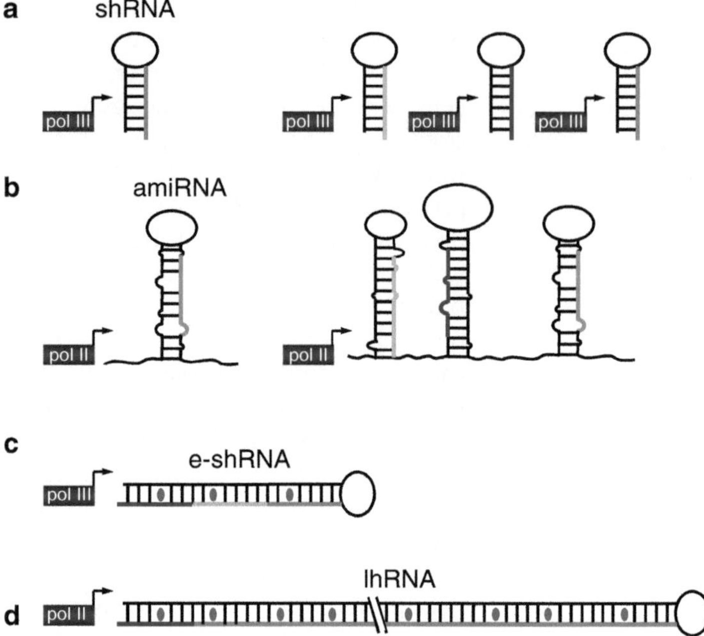

Fig. 2. Vector-based RNAi: mono and combinatorial RNAi approaches. (**a**) *Left*: Mono-RNAi inducer consisting of a single shRNA expressed from an RNA polymerase III promoter (shRNA cassette). *Right*: Combinatorial RNAi-vector containing multiple shRNA cassettes. (**b**) *Left*: Mono-RNAi inducer expressing a single amiRNA. *Right*: Combinatorial RNAi-construct expressing an amiRNA polycistron driven by an RNA polymerase II promoter. (**c**) Combinatorial RNAi can be induced via extended shRNAs (e-shRNAs) that are composed of multiple active siRNAs. *Gray bullets* in the hairpin stem indicate G–U base pairs that can be introduced to facilitate cloning and sequencing. (**d**) Expression of long hairpin RNAs (lhRNAs) theoretically yield many siRNAs that target consecutive mRNA sequences.

Furthermore, this promoter allows the expression of an extended transcript that encodes multiple miRNAs (Fig. 2b, right). The shRNA and miRNA expression cassette can be delivered via a regular expression plasmid or a viral vector (Fig. 1, right).

Besides these mono RNAi-vectors (shRNA or amiRNA) (Fig. 2a, b), combinatorial RNAi-vectors have been developed, e.g., to silence multiple oncogenes, to obtain intensified knockdown of a single target, or to prevent viral escape by simultaneous targeting of multiple viral RNA sequences. In this chapter, we describe methods for preparing the following structures: multiple shRNAs (Fig. 2a, right), an amiRNA polycistron (Fig. 2b, right), extended shRNAs (e-shRNAs) (Fig. 2c), and long hairpin RNAs (lhRNAs) (Fig. 2d). The multiple shRNA approach is based on the expression of multiple shRNAs from cassettes with independent promoters in a single construct (Fig. 2a, right). In the amiRNA polycistron approach, multiple amiRNAs are expressed in a single transcript from a single RNA polymerase II promoter, which closely resembles the natural situation where several miRNAs can be expressed in a coordinated manner from a single transcription unit (Fig. 2b, right). The e-shRNAs are expressed from an RNA polymerase III promoter and encode multiple active siRNAs that are carefully stacked on top of each other (47–51) (Fig. 2c). We previously showed that HIV-1 replication is significantly delayed by the use of an e-shRNA that expresses three siRNAs (48). The main disadvantage of this strategy is that the siRNAs need to be accurately stacked to ensure proper Dicer processing. In addition, the siRNAs are produced in a gradient from the base (maximal) towards the top (minimal) of the hairpin, which is likely due to diminished Dicer processing. Thus, the most effective siRNA should ideally be located at the base of the e-shRNA. Another method to induce combinatorial RNAi is to express lhRNAs from which numerous siRNAs targeting contiguous mRNA targets should be produced (Fig. 2d). Similar to the e-shRNA, these hairpins likely suffer from a decreased siRNA production from the base towards the top of the lhRNA (48). Furthermore, the lhRNA design does not encode well-characterized siRNA units and these transcripts appear unstable in mammalian cells. Notably, in contrast to transfected dsRNA molecules of larger than 30 bp, we previously showed that the intracellular expression of e-shRNAs and lhRNAs does not induce the interferon response (48).

Viral vectors are attractive vehicles for the delivery of desired transgenes to specific target cells. To date, many viral vectors that have distinct characteristics have been developed for gene therapy strategies (52–73). Therefore, depending on the clinical goal and target tissue, one vector may be more suitable than others. We will summarize the characteristics of the most frequently used vector systems for inducing RNAi that are based on the adeno-associated virus, adenovirus, retrovirus, and lentivirus (Table 1). In our laboratory, we mainly use the lentiviral vector to deliver anti-HIV-1 shRNAs, amiRNAs, or e-shRNAs to target cells, either T cells or haematopoietic stem cells. We provide practical information for the production of lentiviral vectors (Figs. 4 and 5) and their titration. As the lentiviral vector is partially composed of HIV-1 sequences, the antiviral RNAi inducers can potentially target these HIV-1 sequences in the vector, which may result in decreased vector titers (vector targeting). We describe a modified lentiviral vector production protocol for such vector-targeting shRNAs or e-shRNAs.

2. Materials

2.1. shRNA Vector Construction

1. pSUPER vector containing the human H1 promoter (OligoEngine, Seattle, WA, USA) (40) (see Note 1).
2. DNA oligonucleotides encoding the shRNA sequence (Subheading 3.1).
3. Annealing buffer: 100 mM NaCl, 50 mM HEPES, pH 7.4.

Table 1
Characteristics of different viral vector types

Vector	Virus group	Virion size (nm)	Packaging capacity (kb)	Advantages	Limitations	References RNAi applications
Adeno-associated virus (AAV)	Parvovirus (ssDNA)	20–30	≤5	Broad target cell specificity, low immunogenicity	Low probability of integration, risk of insertional oncogenesis, small packaging capacity	(74, 75)
Adenovirus (Ad)	Adenovirus (dsDNA)	80–120	≤37	Broad target cell specificity, large packaging capacity	Strong immunogenicity	(76, 77)
Retrovirus	Retrovirus (RNA)	100	≤8	Long-term gene expression due to chromosomal integration, low immunogenicity	Risk of insertional oncogenesis, infects only dividing cells	(78–80)
Lentivirus	Retrovirus (RNA)	100	≤9	Long-term gene expression due to chromosomal integration, low immunogenicity, broad target cell specificity	Risk of insertional oncogenesis, transcriptional silencing	(81, 82)

4. T4 DNA ligase (Life Technologies [LT] Invitrogen, Carlsbad, CA, USA).
5. Ligase buffer (LT Invitrogen, Carlsbad, CA, USA).
6. Restriction enzymes: BglII and HindIII.
7. T7 primer (TAATACGACTCACTATAGGG) and M13 rp primer (CAGGAAACAGCTATGACC).
8. DreamTaq Green PCR Master mix (Fermentas, St. Leon-Rot, Germany).
9. GT116 *Escherichia coli* cells (Δdcm, sbcCD) that are specifically designed for cloning and propagation of shRNA-expressing plasmids with hairpin structures (Invivogen, San Diego, CA, USA).
10. Ampicillin (100 mg/ml).
11. Gene Pulser II, cell electroporator (Bio-Rad, Hercules, CA, USA).
12. 2 mm electroporation cuvettes (Eurogentec, San Diego, CA, USA).
13. Luria Broth (LB) medium.
14. Eppendorf tubes (1.5 ml polypropylene tubes).
15. LB agar plates.
16. SmartLadder (Eurogentec, San Diego, CA, USA).
17. 5 M Betaine (Sigma, St. Louis, MO, USA).
18. BigDye terminator v3.1 cycle sequencing kit (Applied Biosystems Inc., Foster City, CA, USA).
19. Lentiviral vector plasmids derived from the construct pRRL-cpptpgkgfppreSsin (83), which we renamed JS1.
20. Plasmid DNA purification kit (Macherey-Nagel GmBH Co. & KG, Düen, Germany).
21. PCR clean-up Gel extraction kit (Macherey-Nagel GmBH Co. & KG, Düren, Germany).

2.2. Artificial miRNA Vector Construction

1. pcDNA6.2-GW/EmGFP-miR plasmid (LT Invitrogen, Carlsbad, CA, USA) (84).
2. Human cellular genomic DNA to amplify the wild-type miRNA as a template to construct the amiRNA.
3. Oligonucleotides F1 (CAGGTCGACGGATCCTATTT CCTTCAAATGAATG), R1 (ATGGCAGGAAGAAGCGG AGGTGCTACAGAAGCTGTC), F2 (ATGGCGGGAGG AAGCGGTTGGTACTGCTAGCTGTAGAA), R2 (GACCT CGAGTGCGGCCAGATCTAAGCTGGAGTTCTA CAGCTA) and oligonucleotides for generating the antiviral miRNA (forward oligo: CTTCTGTAGCACCTCCGCTT

CTTCCTGCCATGTAGTGTTTAGTTATCTAATGGCGGGAGGAAGCGGTTGGTACTGCTAGC; reverse oligo: GCTAGCAGTACCAACCGCTTCCTCCCGCCATTAGATAACTAAACACTACATGGCAGGAAGAAGCGGAGGTGCTACAGAAG).

4. Restriction enzymes: BglII, BamHI, EcoRV, NruI, and XhoI.
5. The same reagents as described in Subheading 2.1, items 4–7 and 10–21.

2.3. Extended shRNA (e-shRNA) Vector Construction

1. The same reagents as described in Subheading 2.1.

2.4. Long Hairpin RNA (lhRNA) Vector Construction

1. The same reagents as described in Subheading 2.1.

2.5. Lentiviral Vector Production

1. Dulbecco's modified Eagle's medium (DMEM) (LT Invitrogen; brand: GIBCO®) supplemented with 10% fetal calf serum (FCS; Thermo Scientific HyClone, South Logan, UT, USA), penicillin (100 U/ml), streptomycin (100 μg/ml), and minimal essential medium nonessential amino acids (DMEM/10% FCS).
2. Human Embryonic Kidney (HEK) 293T cell line (ATCC nr CRL-11268).
3. Dulbecco's Phosphate Buffered Saline (D-PBS) solution, pH 7.4 (LT Invitrogen; brand: GIBCO®).
4. 0.05% Trypsin with EDTA 4Na solution (LT Invitrogen; brand: GIBCO®).
5. Polystyrene 6-well cell culture plates.
6. Lentiviral vectors encoding RNAi inducers, generated by excising the shRNA, e-shRNA, or amiRNA cassette from the expression plasmid and inserting them in the multiple cloning site of the lentiviral vector JS1, as described in Subheadings 3.1–3.4.
7. Lentiviral vector packaging constructs: pSYNGP (85) for the expression of a human codon-optimized gag-pol sequence without RRE; pRSV-Rev (86) for the expression of Rev, which interacts with the RRE on the vector genome and/or wild-type HIV-1 gag-pol transcripts; and pVSV-G (86) for pseudotyping the vector with the vesicular stomatitis (VSV)-G envelope protein (Fig. 4).
8. RNAi pathway competitor/inhibitor constructs, used to improve lentiviral vector production in the case of RNAi-related production problems against the lentiviral vector:
   (a) A pSUPER construct that contains five shRNA cassettes. Each shRNA cassette targets a different HIV-1 region and

is expressed from an independent H1 promoter, p5xshRNA (87), to provide excess shRNAs.

(b) An siRNA against the human Dicer endonuclease (41) to knock down Dicer function and to saturate the RISC (5′-TCAACCAGCCACTGCTGGA-3′).

(c) pshDrosha, an shRNA expression construct against the Drosha enzyme to knock down the Drosha level (a kind gift from Bryan Cullen, Duke University).

9. 70 μm nylon cell strainers (BD Biosciences, BD Falcon, Bedford, MA, USA).
10. Opti-MEM Reduced Serum Medium (LT Invitrogen; brand: GIBCO®).
11. Opti-MEM supplemented with penicillin (30 U/ml), streptomycin (30 μg/ml), and $CaCl_2$ (100 μg/ml), hereafter called Opti-MEM$^+$ medium.
12. Lipofectamine-2000 reagent (LT Invitrogen, Carlsbad, CA, USA).
13. 0.45 μm cellulose acetate filters (Whatman, Clifton, NJ, USA).
14. Amicon Ultra-15 centrifugal filter units (Millipore, Billerica, MA, USA).
15. Cryovials (Greiner Bio One, Kremsmuenster, Austria).
16. Capsid p24-ELISA. Commercial ELISA kits are available; we perform the CA-p24 ELISA as described previously (88).

2.6. Titration of Lentiviral Vectors

1. SupT1 T cell line (ATCC nr CRL-1942).
2. Advanced Roswell Park Memorial Institute (RPMI) medium (LT Invitrogen) supplemented with 1% fetal calf serum (FCS), 2 mM L-glutamine, 40 U/ml penicillin, and 40 μg/ml streptomycin.
3. Polystyrene 96-well flat bottom cell culture plates.
4. OptiMEM® I Reduced-Serum Medium (LT Invitrogen; brand: GIBCO®).
5. FACS buffer: D-PBS supplemented with 2% FCS.
6. BD FACSCanto II flow cytometer (BD Biosciences, San Jose, CA, USA).

3. Methods

In this chapter we provide the protocols for generating mono RNAi-inducing vectors, including shRNA and amiRNA constructs (Fig. 2a, b). In some cases, an intensified RNAi or a combinatorial RNAi approach is required to obtain the desired effect, which requires a multi-shRNA/amiRNA vector. Different methods exist to obtain combinatorial RNAi. We describe methods for the construction of multiple shRNAs, an amiRNA polycistron, e-shRNAs, and lhRNAs (Fig. 2). Generally the shRNA or amiRNA vectors are first cloned into a nonviral expression vector to test the potency of the hairpins. The efficacy can be tested by transfecting the sh/amiRNA construct into cells that express the endogenous target or by co-transfecting the RNAi inducer with a luciferase reporter construct that expresses the RNAi target sequence. Then, the RNAi gene cassette can be transferred to a lentiviral vector (or any other viral vector of interest). If a combinatorial RNAi construct is required, further cloning can be performed in a nonviral expression vector, until the desired multi-RNAi cassette is generated, which can subsequently be tested and thereafter transferred to a lentiviral vector. Here, we provide the standard procedure of lentiviral vector production, but we also describe an optimized protocol for the production of lentiviral vectors encoding anti-HIV-1 e-shRNAs that target sequences in the vector RNA genome.

3.1. shRNA Vector and Multi-shRNA Vector Construction

Oligonucleotides encoding different shRNAs are designed as follows: compatible sense and antisense DNA oligonucleotides are custom ordered with appropriate overhangs containing BamHI and HindIII restriction sites (Fig. 3a). The annealed oligonucleotides contain the BamHI and HindIII restriction sites which can be inserted into the BglII and HindIII digested pSUPER vector. We use the shRNA design that includes perfect complementary 19-nucleotide sense and antisense strands connected with the 9-nucleotide pSUPER loop design (UUCAAGAGA) driven from an H1 promoter (40) (Fig. 3a). We previously designed optimized loops that can be used to increase RNAi activity (89).

Vectors that contain other human promoters, such as the U6 or 7SK polymerase III promoter, and the human U1 RNA polymerase II promoter can also be considered to drive shRNA expression (*see* Subheading 2.1) (*see* **Note 1**).

1. Dissolve the DNA oligonucleotides in sterile, nuclease free milliQ water to a final concentration of 3 mg/ml.
2. Anneal the forward and reverse DNA oligonucleotides encoding the shRNA sequence by combining 1 μl of each oligonucleotide (3 mg/ml) in 48 μl annealing buffer. The annealed double-stranded oligonucleotides have BamHI- and HindIII-compatible ends for vector insertion (Fig. 3a). This shRNA design is commonly used (40), however other shRNA designs have also been shown to efficiently induce RNAi (41).
3. Heat the sample at 94°C in a beaker of hot water for 5 min and cool down to room temperature by placing the beaker containing the samples on the bench (see Note 2). The annealed oligonucleotides can be used immediately in a ligation reaction or stored at −20°C until further use.

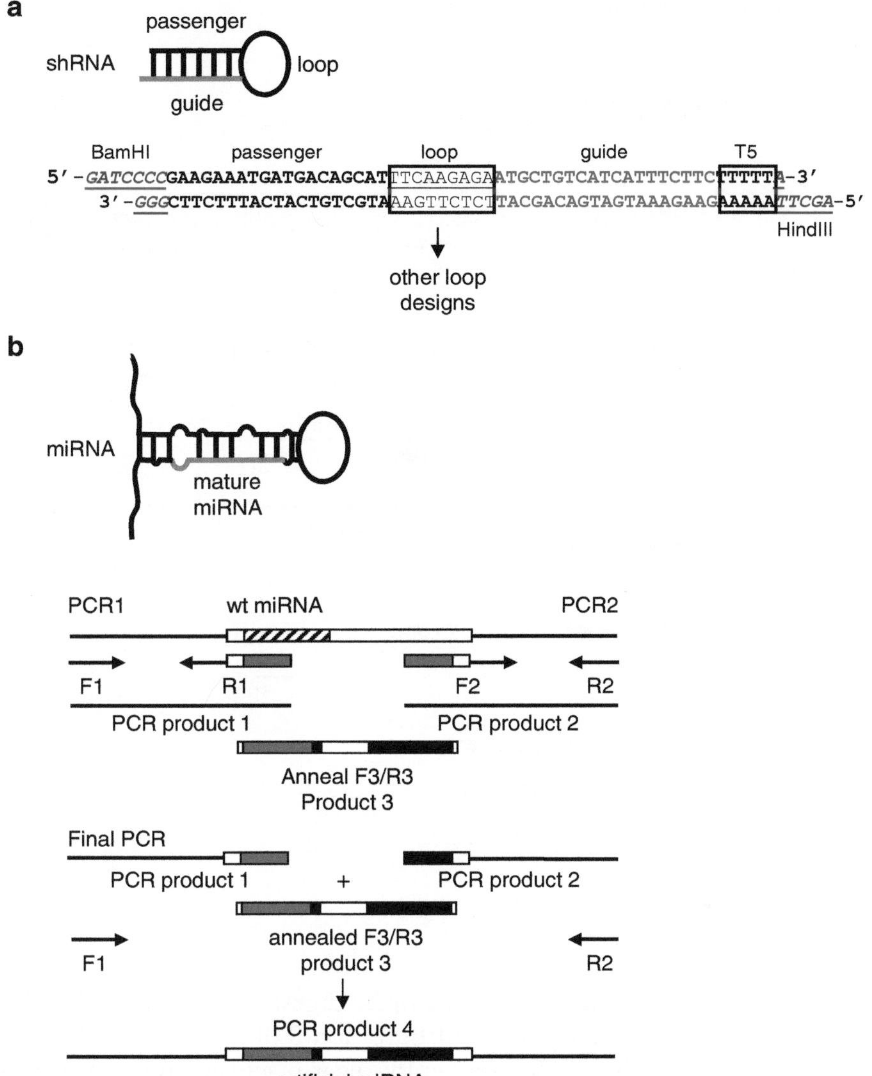

Fig. 3. Generation of shRNA, amiRNA, e-shRNA, or lhRNA expression cassettes. (**a**) *Top*: Schematic of an shRNA with the passenger, loop, guide strand (*gray*), and UU 3′ overhang; *bottom*: the annealed oligonucleotides to clone the shRNA transcript in an expression vector. (**b**) *Top*: Schematic of a pri-miRNA transcript with typical flanking sequences. The mature miRNA strand is shown in *grey*. *Bottom*: The wild-type (wt) miRNA is amplified from cellular genomic DNA. The *white box with diagonal stripes* indicates the mature miRNA sequence, while the *white boxes* indicate the remaining sequences of the pri-miRNA. The following steps are utilized to amplify the amiRNA: (1) PCR 1: the 5′ part of the pri-miRNA is amplified using a forward primer F1 encoding a BamHI site and a reverse primer R1 that partially encodes the HIV-1 sequence (*grey*) to produce PCR product 1; (2) PCR 2: similarly the 3′ part of the pri-miRNA is amplified with forward primer F2 that partially encodes HIV-1 sequences and a reverse primer R2 encoding BglII and XhoI sites to produce PCR product 2; (3) two complementary oligonucleotides F3 and R3, which encode the stem-loop of the amiRNA are annealed to produce annealed F3/R3 product 3; (4) final PCR: the PCR products 1 and 2 and the annealed oligo F3/R3 together are used as template to amplify the full-length artificial pri-miRNA with the outer forward F1 and reverse R2 primers to produce PCR product 4. (**c**) *Top*: Schematic of an e-shRNA transcript with a UU 3′ overhang. *Gray bullets* indicate G–U base pairs in the hairpin stem. *Bottom*: Construction of an e-shRNA transcript requires multiple oligonucleotides that can anneal together to form the forward and reverse strands. The appropriate restriction sites are created upon annealing for cloning into the expression vector. The *black boxed nucleotides* indicate the positions with A to G or C to T mutations in the hairpin stem. The loop sequences are *boxed in gray*. (**d**) *Top*: Schematic of an lhRNA transcript. *Bottom*: Construction of the lhRNA requires three steps. (1) The guide strand of the lhRNA is PCR amplified using a forward primer F1 including the loop sequences and a reverse primer R1 matching the 3′ end of the hairpin (PCR product 1); (2) two oligonucleotides that encode the passenger strand, the loop sequence, and some additional nucleotides of the guide strand are annealed together (product 2); (3) products 1 and 2 are fused together in a final PCR reaction with the FlhRNA primer and R1 primer that encode the BamHI and HindIII site, respectively.

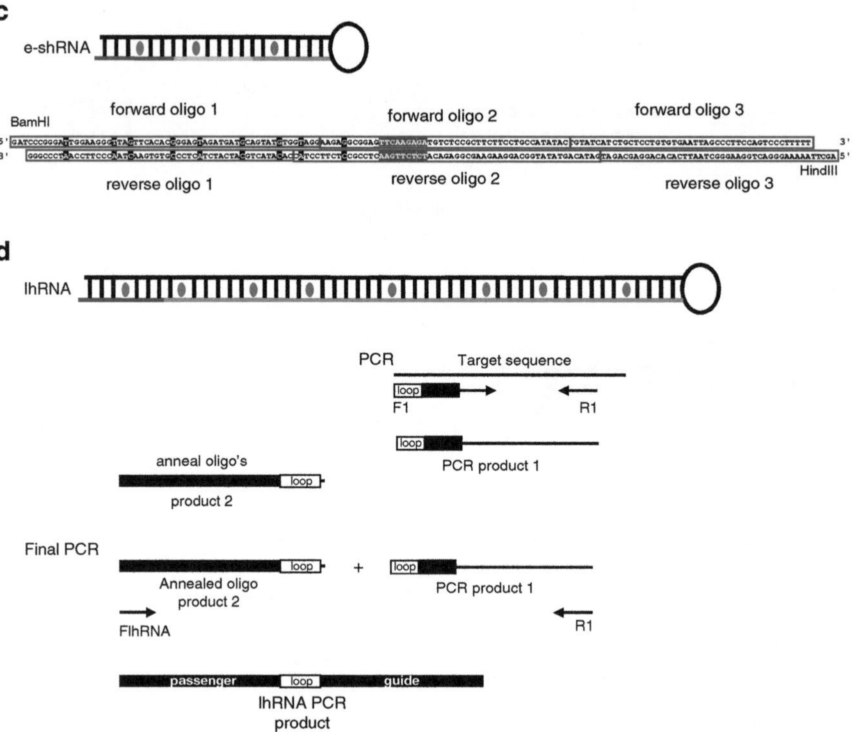

Fig. 3. (continued)

4. Linearize the vector by digesting 5 μg of pSUPER DNA with 3 μl of 10 U/μl BglII and 3 μl of 10 U/μl HindIII for 2 h at 37°C. Subsequently, heat-inactivate the restriction enzymes for 20 min at 65°C. Purify the digested vector from a 1% agarose gel (see Note 3). Note that the insert contains the BamHI and HindIII sites that are compatible with the BglII and HindIII sites of the vector.

   Ligate 1 μl of the annealed oligonucleotides into the BglII/HindIII-digested pSUPER vector backbone (~40 ng) overnight at 16°C. Prior to transformation, ligation mixes should be digested with BglII to reduce colonies derived from vector self-ligation. The BglII site is destroyed upon successful cloning of the annealed oligonucleotides into the vector, thus only the empty vectors contain an intact BglII site and will therefore be eliminated. Add 0.5 μl (10 U/μl) BglII to the ligations and incubate for 30 min at 37°C.

5. Use 1 μl of the ligation mix to transform 50 μl of *E. coli* GT116 cells by electroporation (25 μF, 200 Ω, and 2.5 kV).

6. Add 1 ml of LB medium to the cuvettes and place them in a 37°C incubator for 45 min with shaking for recovery.

7. Transfer the cells to Eppendorf tubes, centrifuge for 5 min at 1,485 ×*g* and remove ~800 μl of the supernatant. Resuspend the cells in the remaining LB medium and plate different volumes of cell suspension on two LB agar plates to increase the chance of obtaining individual colonies.
8. Perform colony PCR to screen for positive clones that contain the shRNA cassette. Pick colonies with tips and dissolve each colony in 50 μl of LB medium. Use 2.5 μl of this suspension as the template for the PCR. Use T7 and M13 rp primers and regular PCR reagents for the PCR reaction and a standard PCR program. Include a negative and positive control in the PCR, e.g., the empty vector and an shRNA construct that gives a PCR product of similar size.
9. To verify the sequence of the shRNA constructs use the BigDye terminator cycle sequencing kit. To sequence palindrome-containing shRNA constructs, we use a modified protocol in which Betaine is added to a final concentration of 1 M and the sample is denatured at 98°C instead of 96°C (see Note 4).
10. To combine multiple identical or distinct shRNA expression cassettes (consisting of the promoter and the shRNA), standard cloning techniques can be used to insert them into a single expression vector or into the lentiviral vector JS1.

3.2. Artificial miRNA Vector or amiRNA Polycistron Construction

Construction of an amiRNA cassette consists of four steps as depicted in the schematic of Fig. 3b (84). This procedure requires a wild-type human miRNA as a template, amplified using human cellular genomic DNA.

1. First, PCR amplify the 5′ flank of the pri-miRNA with a forward primer (F1) with a BamHI site and a reverse primer (R1) containing HIV-1 sequence at its 3′ end using a cellular miRNA as the template. Subsequently, purify the PCR fragment (PCR product 1) from the agarose gel after electrophoresis. The primers are always designed to have at least 18 nt overlapping sequence with the template.
2. Similarly, PCR amplify the 3′ flank of the pri-miRNA with a forward primer (F2) encoding HIV-1 sequences and a reverse primer (R2) with BglII and XhoI sites and purify the PCR product (PCR product 2) from the agarose gel after electrophoresis.
3. Anneal two complementary oligonucleotides F3 and R3 that create the stem-loop structure of the antiviral miRNA as described in Subheading 3.1, steps 1–3 (product 3). The antiviral miRNA is designed such that it resembles as much as possible the natural miRNA (e.g., mismatches, bulges, and loop), by modifying the sequence of the passenger strand of the miRNA. The guide strand of the miRNA is designed to have 100% complementarity with the target to induce RNAi-mediated cleavage of the mRNA.

4. The PCR products 1 and 2 and the annealed oligonucleotides product 3 contain stretches of sequence overlap that allow fusion by overlap extension PCR. To perform the fusion PCR, use 10 ng of each PCR product 1 and 2 and 10 ng of the annealed oligonucleotides product 3 as combined templates to amplify the full-length pri-miRNA product using the outer forward (F1) and reverse (R2) primers.
5. Purify the artificial pri-miRNA product from an agarose gel after electrophoresis, digest with BamHI and XhoI, and ligate the digested fragment into the BamHI and XhoI digested pcDNA6.2-GW/EmGFP-miR vector.
6. Transform 1 μl of the ligation mix into 50 μl of *E. coli* GT116 cells by electroporation (25 μF, 200 Ω, and 2.5 kV) and plate different volumes or dilutions of cell suspension on two LB agar plates to ensure the formation of individual colonies.
7. For colony PCR, inoculate each colony in 50 μl of LB medium and use 2.5 μl of this suspension as a template for the PCR reaction. Include a negative and positive control for the PCR, e.g., the empty vector and a construct that yields a PCR product of similar size.
8. Sequence the amiRNA construct using the standard sequencing reaction and program using the BigDye terminator cycle sequencing kit. If this is not successful, use the hairpin sequencing protocol described in Subheading 3.1, step 9.
9. If the sequence is correct, a starter culture can be inoculated for the bacterial clone and grown for 6 h at 37°C under vigorous shaking. An aliquot of the starter culture can then be used to inoculate a larger overnight culture. Plasmid DNA purification is performed on the overnight cultures.
10. To construct an amiRNA polycistron, digest an miRNA construct with BglII and XhoI and purify the linearized vector. This digestion will open up the vector at the 3′ position of the miRNA hairpin. Digest the miRNA hairpin that will be concatenated to the first miRNA hairpin with BamHI and XhoI and purify this fragment. Ligate the BamHI/XhoI digested fragment into the BglII/XhoI digested vector. By repeating this procedure, multiple hairpins can be chained, resulting in vectors that encode the desired number of antiviral miRNA hairpins. Note that BamHI and BglII have compatible sites and that after ligation the BglII site is destroyed (see Note 5).
11. For cloning of the amiRNA expression cassette into the lentiviral vector, remove the EmGFP encoding sequence by digestion with DraI and religation of the vector (see Note 6). Digest the amiRNA expression cassette (consisting of the amiRNA and the CMV promoter) with NruI and XhoI and ligate into the EcoRV and XhoI digested JS1 vector. Note that NruI and EcoRV generate blunt ends and thus produce compatible sites.

3.3. Extended shRNA (e-shRNA) Vector Construction

The construction procedure of the e-shRNA is similar to that of the shRNA, with the exception that multiple oligonucleotides are used that form the forward primer and multiple oligonucleotides that form the reverse primer (Fig. 3c) (see Note 7). An example is provided in Fig. 3c where six oligonucleotides are needed to form the e-shRNA with the BamHI and HindIII restriction sites. The annealed oligonucleotides can be inserted into the BglII and HindIII digested pSUPER vector. Destabilizing G–U base pairs can be introduced in the hairpin stem of the e-shRNA by modifying the passenger strand (introduction of A to G or C to T mutations), thus maintaining the guide strand sequence (Fig. 3c) (see Note 8).

1. Dissolve DNA oligonucleotides in sterile milliQ water to a final concentration of 3 mg/ml.
2. Add 1 μl of each oligonucleotide (3 mg/ml) in annealing buffer in a total volume of 50 μl for annealing of the forward and the reverse oligonucleotide (see Subheading 3.1, step 3).
3. Clone the annealed oligonucleotides into the pSUPER vector as described in Subheading 3.1, steps 4–8.
4. Sequence the e-shRNA constructs as described in Subheading 3.1, step 9.
5. Digest the e-shRNA cassette (e-shRNA and the H1 promoter) from the pSUPER construct and insert into the multiple cloning site of the lentiviral vector JS1 using standard cloning techniques.

3.4. Long Hairpin RNA (lhRNA) Vector Construction

The construction of lhRNAs requires three steps.

1. PCR amplify the guide strand with a forward primer that includes the loop sequences of the pSUPER system and a reverse primer matching the 3′ end of the hairpin (product 1) (Fig. 3d).
2. The 5′ part of the expression cassette is generated by annealing oligonucleotides that encode the passenger strand, the loop sequence, and some additional nucleotides of the antisense strand (product 2) (see Subheading 3.1, steps 2 and 3).
3. The passenger and guide strand of the lhRNA is fused by an overlap extension PCR with primers (FlhRNA and R1) that have terminal BamHI and HindIII restriction sites. The final product is purified from agarose gel and can be used for subsequent cloning into the BglII and HindIII sites of the pSUPER vector.

3.5. Lentiviral Vector Production (See Fig. 4)

1. On day 1, trypsinize HEK 293T cells in the morning and resuspend the cells in DMEM/10% FCS without antibiotics (see Note 9). Pass the cells through a cell strainer to avoid clumping of cells. Count the cells and seed 6.0×10^5 cells per transfection in a 6-well plate in 2 ml of culture medium without antibiotics.

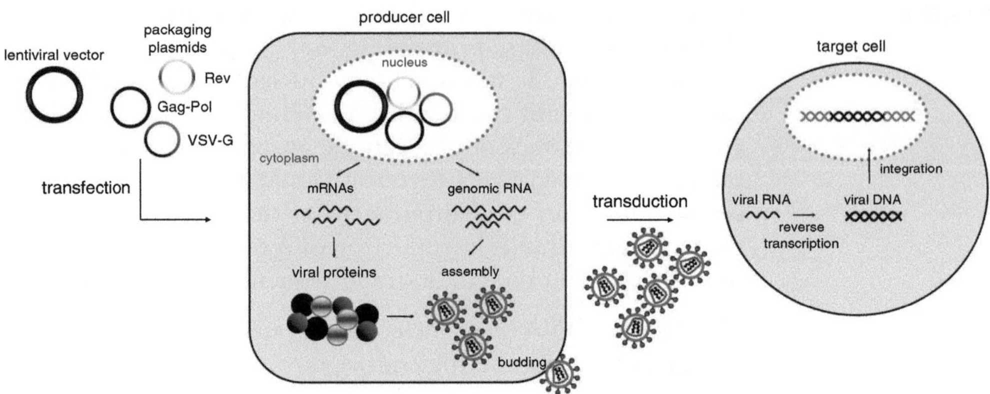

Fig. 4. Procedure of lentiviral vector production and transduction of target cells. HEK 293T producer cells are co-transfected with the lentiviral vector plasmid and three packaging plasmids (constructs expressing gag-pol, rev, and VSV-G envelope protein). Lentiviral vector particles will be assembled and secreted in the supernatant. The vector particles can be used for transduction of target cells, which results in viral DNA integration into the host cell genome.

2. On day 2, transfect the cells in the afternoon using Lipofectamine. Prepare the DNA–Lipofectamine complexes in OptiMEM® I medium for each transfection as follows:
3. Dilute 0.95 μg of lentiviral vector construct, 0.6 μg of pSYNGP, 0.25 μg of pRSV-Rev, and 0.33 μg of pVSV-G in 250 μl OptiMEM® I medium (see Note 10 and 11).
4. When the lentiviral vector encodes an anti-HIV-1 shRNA or e-shRNA that targets the genomic RNA (vector targeting), add 2.9 μg of competitor or inhibitor plasmids (see Subheading 2.5, item 8) or 100 nM of siRNA against Dicer to the co-transfection (see Fig. 5, Notes 12 and 13). These components will saturate the RNAi pathway, such that the vector targeting siRNAs cannot be incorporated into RISC, and prevent targeting of the vector genome. Thus, the vector RNA genome will not be subjected to RISC attack and can be packaged and released as a transducing unit.
5. Mix the Lipofectamine gently before use and dilute for each well 10 μl Lipofectamine in 250 μl Opti-MEM® I medium (see Note 10) and incubate for 5 min at room temperature.
6. Combine the DNA and Lipofectamine solutions and mix gently. Incubate for 20 min at room temperature to allow DNA-Lipofectamine complexes to be formed.
7. Add the DNA-Lipofectamine-Opti-MEM® I mix dropwise to the cells.
8. In the morning of day 3, replace the transfection medium with 1.2 ml of Opti-MEM® I+ medium (see Note 14).

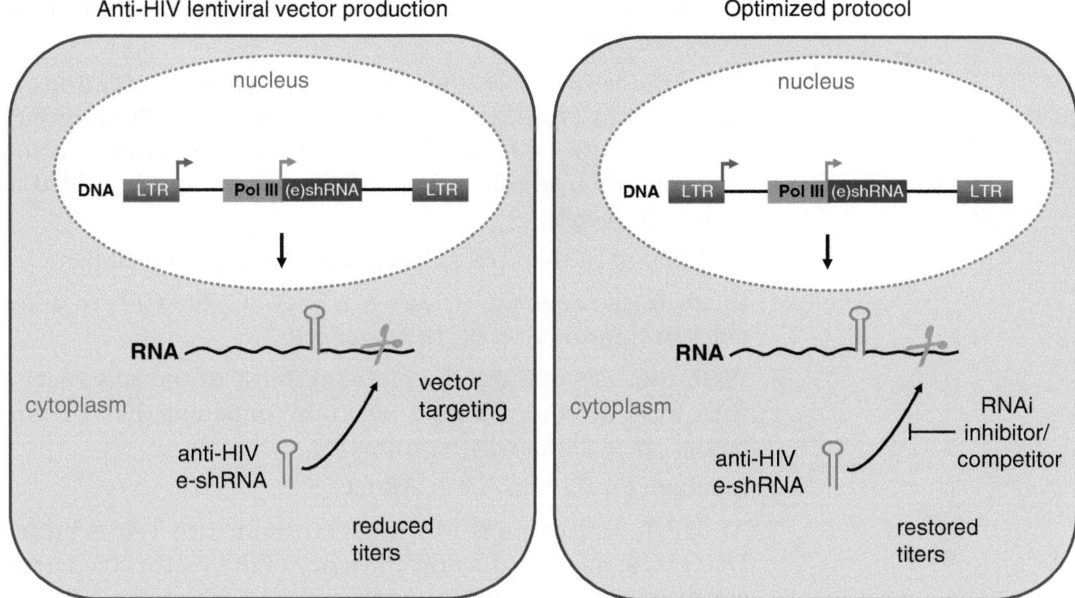

Fig. 5. Production of lentiviral vectors encoding anti-HIV e-shRNAs. *Left*: The titer of anti-HIV e-shRNA constructs can be reduced when the expressed e-shRNA targets HIV-derived sequences in the vector RNA genome during vector production in the producer cell. *Right*: This reduction of vector titer can be overcome by overexpressing an RNAi inhibitor or competitor during lentiviral vector production.

9. On day 4, harvest the produced lentiviral vector in the supernatant (see Notes 15 and 16). Spin down detached cellular debris by low speed centrifugation at 1,305 × *g* for 10 min and filter the supernatant over a 0.45 μm filter. The highest titers are obtained at day 4. Optionally, additional medium can be added on the cells to perform another harvest at day 5. However, the transduction titer of day 5 is generally about tenfold lower than that of day 4.
10. The cleared viral supernatant can be concentrated with a centrifugal filter (Amicon Ultra-15 centrifugal filter) at 4,000 × *g* (see Note 17), according to the instructions of the manufacturer. For a 10 ml vector preparation, we usually centrifuge for around 15–20 min to obtain a 250 μl end volume.
11. Aliquot the viral supernatants in cryovials and store the stocks at −80°C (see Note 18).
12. Determine the production of lentiviral vector particles (capsid titer) by CA-p24 ELISA as described previously (88).

3.6. Titration of Lentiviral Vectors

1. On day 1 in the morning, seed 50.000 SupT1 T cells in each well of a flat bottom 96-well plate in 100 μl of advanced RPMI supplemented with 1% FCS, L-glutamine, penicillin, and streptomycin. Place the plate in a humidified CO_2 chamber (37°C,

5% CO_2) to allow the cells to descend to the bottom of the wells (see Note 19).

2. Thaw the lentiviral vector stocks and make serial dilutions of the vector in Opti-MEM® I medium. We generally transduce the cells with 100, 10, 1, and 0.1 μl of lentiviral vector. Dilutions are made such that 100 μl of vector can be added to each well of cells.
3. Add 100 μl of the vectors to each well of the 96-well plate.
4. Incubate the cells for at least 6 h at 37°C, 5% CO_2 to allow efficient transduction of the target cells.
5. Wash the cells by carefully replacing most of the supernatant with fresh advanced RPMI medium containing FCS, L-glutamine, penicillin and streptomycin.
6. Incubate for 72 h at 37°C, 5% CO_2.
7. At day 4, replace most of the supernatant with FACS buffer. Determine the transduction efficiency of the vectors by detecting the percentage of GFP+ cells by FACS (see Note 20).
8. To calculate the transduction titer per ml of vector the following formula is used:

   Titer/ml = cell number on the day of transduction × ((percentage of GFP+ cells/100)/dilution factor calculated for 1 ml of vector). Thus, if 5% become GFP+ when a transduction was performed on 50,000 cells with 1 μl of vector the titer is calculated as follows: $50{,}000 \times (0.05/0.001) = 2.5 \times 10^6$ TU/ml. The titers should be calculated using a percentage of GFP+ cells within the 1–30% range (see Note 21).

4. Notes

1. Alternative polymerase III promoters could be used to express shRNAs or e-shRNAs. DNA plasmids containing the human U6 (90) or 7SK (91) polymerase III promoters, and the human U1 polymerase II promoter (92) are commercially available. pSilencer 2.0-U6 (Ambion Inc., Austin, TX, USA), psiRNA-h7SKhygro (Invivogen, San Diego, CA, USA), and the pGeneClip-BasicVector (Promega Corp., Madison, WI, USA) respectively contain the U6, 7SK, and U1 promoters (93).
2. Annealing of oligonucleotides can also be performed in a heat block at 94°C for 5 min, 80°C for 4 min, 75°C for 4 min, and 70°C for 10 min, and then turn off the heat block to slowly cool the samples to room temperature. In exceptional cases, annealing may be unsuccessful due to self-annealing of the oligonucleotides. To prevent self-annealing, lower the temperature

stepwise in a PCR machine (94°C for 5 min, 85°C for 4 min, 82°C for 4 min, 80°C for 4 min, 78°C for 4 min, etc.).

3. It is not necessary to CIP-treat the digested vector because the DNA ends are not compatible. But if the vector will be CIP-treated, one should phosphorylate the annealed oligonucleotides by kinase treatment.
4. Addition of 1 M Betaine in the sequence reaction will improve the quality of sequencing templates with hairpin structures and/or GC-rich sequences (94).
5. The pcDNA6.2-GW/EmGFP-miR vector is designed to express artificial pre-miRNAs in the murine miR-155 backbone. The vector is engineered in such a way that annealed oligonucleotides can be directly cloned into the linearized vector with miR-155 flanking sequences of the pre-miRNA. In our case, we used other miRNA backbones; therefore, we removed the miR-155 flanking sequences by digestion with BamHI and XhoI. Subsequently, we generated an amiRNA in the miRNA scaffold of interest and cloned the whole artificial pri-miRNA into the BamHI and XhoI site of the vector.
6. We removed the EmGFP encoding sequence from the pcDNA6.2-GW/EmGFP-miR vector because the JS1 lentiviral vector already encodes eGFP.
7. We use multiple oligonucleotides to construct e-shRNAs because there is a size limit for oligonucleotides of 80 nt. Longer oligonucleotides can be ordered too, but they are significantly more expensive and require a longer production time.
8. Destabilizing G–U base pairs can be introduced in the hairpin stem of the e-shRNA by modification of the passenger strand (by introducing A to G or C to T mutations), thus maintaining the guide strand sequence. We usually introduce three G–U base pairs per shRNA unit when we make e-shRNAs with two or more siRNA units to facilitate cloning and sequencing without affecting the RNAi activity. In the e3-63 molecule (Fig. 3c), we inserted G–U at base pair position 5, 14, and 17 in the base shRNA unit, positions 3, 8, and 16 in the middle unit, and positions 1, 5, and 13 in the top shRNA unit.
9. For lentiviral vector production, cells are seeded in medium without antibiotics. To diminish the chance of contamination, the plates can be covered in saran wrap.
10. To reduce pipetting variations, make a master mix of the packaging construct when performing the transfection to produce lentiviral vectors. Similarly, make a master mix for the Lipofectamine-Opti-MEM® I Medium mix.
11. A low lentiviral vector titer may be caused by poor quality of the plasmids, but it may also be affected by the inserted transgene.

To test this, prepare fresh plasmids and repeat the lentiviral vector production and titer measurement. For plasmid DNA isolation, we use the BIOKE NucleoBond Xtra Midi/Maxi kit. The DNA concentration and purity can be assessed by UV spectrophotometry and subsequent quantitative analysis on an agarose gel. The ratio of absorbance at 260 nm versus 280 nm of pure DNA is between 1.8 and 2.0. Lower ratios indicate protein contamination. Another measure for the purity of the DNA is the ratio of absorbance at 260 nm versus 230 nm. This ratio is generally between 2.0 and 2.2 for pure DNA. Lower ratios indicate a contamination with organic chemical compounds.

12. To inhibit or saturate the RNAi machinery we have used several competitor/inhibitor constructs: excess shRNAs, an shRNA against Drosha, or siRNA against Dicer. These reagents have been shown to boost the vector titers of vector-targeting e-shRNAs (95). Other constructs were also tested, including a luciferase reporter with the corresponding target sequence to provide an RNAi target decoy, a plasmid encoding adenovirus VA RNA to inhibit Dicer (96), or a plasmid encoding the RNAi suppressor protein VP35 of Ebola virus (97, 98). However, these constructs did not positively affect the vector titer.
13. To test whether inefficient nuclear transport of the vector RNA genome was causing the poor vector titer, we also overexpressed CRM1 (chromosome region maintenance 1) (99) during vector production. CRM1 is involved in nuclear transport of spliced and unspliced RNA and thus should enhance vector RNA genome transport in case this is impaired. However, CRM1 overexpression did not influence the vector titer, indicating that restricted RNA export is not the cause of the reduced vector titer.
14. Vector production in Opti-MEM® I Medium improves the titer about twofold as compared to standard DMEM with 10% serum. In addition, this medium improves the subsequent filtration and concentration steps.
15. Infectious material such as lentiviral vectors should be handled with care, e.g., always use gloves and filter tips to prevent contamination. The manipulations should be performed in a laboratory with the appropriate biosafety level.
16. When harvesting lentiviral vectors carefully remove the viral supernatant and avoid the inclusion of detached cells.
17. Ultracentrifugation can also be used to concentrate the lentiviral vector particles.
18. Freeze aliquots for single use to prevent multiple freeze–thaw cycles. Titers can drop as much as 2–4 fold upon freeze thawing (100, 101).

19. The vector titer should be determined on the eventual target cells because the transduction efficiency may vary for different cell types.
20. Another method to determine the vector titers is to assess DNA sequences in transduced cells using qPCR (102).
21. For SupT1 T cells, we generally obtain titers ranging from 1 to 10×10^6 transducing units/ml (TU/ml) depending on the lentiviral vector construct. For PM1 T cells, we generally obtain tenfold lower titers and for primary cells (e.g., PBMCs and CD4+ T cells) we obtain 100-fold lower titers.

Acknowledgements

We thank all the members of the RNAi group for stimulating discussions and useful suggestions. RNAi research in the Berkhout laboratory is sponsored by NWO-CW (Top grant) and ZonMw (Translational Gene Therapy program).

References

1. Waterhouse PM, Wang MB, Lough T (2001) Gene silencing as an adaptive defence against viruses. Nature 411:834–842
2. Voinnet O (2001) RNA silencing as a plant immune system against viruses. Trends Genet 17:449–459
3. Wilkins C, Dishongh R, Moore SC, Whitt MA, Chow M, Machaca K (2005) RNA interference is an antiviral defence mechanism in *Caenorhabditis elegans.* Nature 436:1044–1047
4. Wang XH, Aliyari R, Li WX et al (2006) RNA interference directs innate immunity against viruses in adult Drosophila. Science 312: 452–454
5. Segers GC, Zhang X, Deng F, Sun Q, Nuss DL (2007) Evidence that RNA silencing functions as an antiviral defense mechanism in fungi. Proc Natl Acad Sci U S A 104:12902–12906
6. Fire A, Xu S, Montgomery MK, Kostas SA, Driver SE, Mello CC (1998) Potent and specific genetic interference by double-stranded RNA in *Caenorhabditis elegans.* Nature 391:806–811
7. Ghildiyal M, Seitz H, Horwich MD et al (2008) Endogenous siRNAs derived from transposons and mRNAs in Drosophila somatic cells. Science 320:1077–1081
8. Sijen T, Plasterk RH (2003) Transposon silencing in the *Caenorhabditis elegans* germ line by natural RNAi. Nature 426:310–314
9. Malone CD, Hannon GJ (2009) Small RNAs as guardians of the genome. Cell 136: 656–668
10. Yang N, Kazazian HH Jr (2006) L1 retrotransposition is suppressed by endogenously encoded small interfering RNAs in human cultured cells. Nat Struct Mol Biol 13:763–771
11. Bernstein E, Caudy AA, Hammond SM, Hannon GJ (2001) Role for a bidentate ribonuclease in the initiation step of RNA interference. Nature 409:363–366
12. Nykanen A, Haley B, Zamore PD (2001) ATP requirements and small interfering RNA structure in the RNA interference pathway. Cell 107:309–321
13. Zamore PD, Tuschl T, Sharp PA, Bartel DP (2000) RNAi: double-stranded RNA directs the ATP-dependent cleavage of mRNA at 21 to 23 nucleotide intervals. Cell 101: 25–33
14. Elbashir SM, Lendeckel W, Tuschl T (2001) RNA interference is mediated by 21- and 22-nucleotide RNAs. Genes Dev 15: 188–200

15. Haase AD, Jaskiewicz L, Zhang H et al (2005) TRBP, a regulator of cellular PKR and HIV-1 virus expression, interacts with Dicer and functions in RNA silencing. EMBO Rep 6:961–967
16. Zhang H, Kolb FA, Brondani V, Billy E, Filipowicz W (2002) Human Dicer preferentially cleaves dsRNAs at their termini without a requirement for ATP. EMBO J 21: 5875–5885
17. Leuschner PJ, Ameres SL, Kueng S, Martinez J (2006) Cleavage of the siRNA passenger strand during RISC assembly in human cells. EMBO Rep 7:314–320
18. Matranga C, Tomari Y, Shin C, Bartel DP, Zamore PD (2005) Passenger-strand cleavage facilitates assembly of siRNA into Ago2-containing RNAi enzyme complexes. Cell 123:607–620
19. Rand TA, Petersen S, Du F, Wang X (2005) Argonaute2 cleaves the anti-guide strand of siRNA during RISC activation. Cell 123:621–629
20. Hammond SM, Caudy AA, Hannon GJ (2001) Post-transcriptional gene silencing by double-stranded RNA. Nat Rev Genet 2:110–119
21. Umbach JL, Cullen BR (2009) The role of RNAi and microRNAs in animal virus replication and antiviral immunity. Genes Dev 23:1151–1164
22. Cullen BR (2006) Is RNA interference involved in intrinsic antiviral immunity in mammals? Nat Immunol 7:563–567
23. Berkhout B, Jeang KT (2007) RISCy business: microRNAs, pathogenesis, and viruses. J Biol Chem 282:26641–26645
24. Haasnoot J, Westerhout EM, Berkhout B (2007) RNA interference against viruses: strike and counterstrike. Nat Biotechnol 25:1435–1443
25. Han J, Lee Y, Yeom KH, Kim YK, Jin H, Kim VN (2004) The Drosha-DGCR8 complex in primary microRNA processing. Genes Dev 18:3016–3027
26. Bohnsack MT, Czaplinski K, Gorlich D (2004) Exportin 5 is a RanGTP-dependent dsRNA-binding protein that mediates nuclear export of pre-miRNAs. RNA 10:185–191
27. Lund E, Guttinger S, Calado A, Dahlberg JE, Kutay U (2004) Nuclear export of microRNA precursors. Science 303:95–98
28. Yi R, Qin Y, Macara IG, Cullen BR (2003) Exportin-5 mediates the nuclear export of pre-microRNAs and short hairpin RNAs. Genes Dev 17:3011–3016
29. Chendrimada TP, Gregory RI, Kumaraswamy E et al (2005) TRBP recruits the Dicer complex to Ago2 for microRNA processing and gene silencing. Nature 436:740–744
30. Hutvagner G, McLachlan J, Pasquinelli AE, Balint E, Tuschl T, Zamore PD (2001) A cellular function for the RNA-interference enzyme Dicer in the maturation of the let-7 small temporal RNA. Science 293:834–838
31. Mourelatos Z, Dostie J, Paushkin S et al (2002) miRNPs: a novel class of ribonucleoproteins containing numerous microRNAs. Genes Dev 16:720–728
32. Brennecke J, Stark A, Russell RB, Cohen SM (2005) Principles of microRNA-target recognition. PLoS Biol 3:e85
33. Grimson A, Farh KK, Johnston WK, Garrett-Engele P, Lim LP, Bartel DP (2007) MicroRNA targeting specificity in mammals: determinants beyond seed pairing. Mol Cell 27:91–105
34. Yekta S, Shih IH, Bartel DP (2004) MicroRNA-directed cleavage of HOXB8 mRNA. Science 304:594–596
35. Krek A, Grun D, Poy MN et al (2005) Combinatorial microRNA target predictions. Nat Genet 37:495–500
36. Lewis BP, Shih IH, Jones-Rhoades MW, Bartel DP, Burge CB (2003) Prediction of mammalian microRNA targets. Cell 115:787–798
37. Liu J, Valencia-Sanchez MA, Hannon GJ, Parker R (2005) MicroRNA-dependent localization of targeted mRNAs to mammalian P-bodies. Nat Cell Biol 7:719–723
38. Sen GL, Blau HM (2005) Argonaute 2/RISC resides in sites of mammalian mRNA decay known as cytoplasmic bodies. Nat Cell Biol 7:633–636
39. Kim DH, Rossi JJ (2007) Strategies for silencing human disease using RNA interference. Nat Rev Genet 8:173–184
40. Brummelkamp TR, Bernards R, Agami R (2002) A system for stable expression of short interfering RNAs in mammalian cells. Science 296:550–553
41. Paddison PJ, Caudy AA, Bernstein E, Hannon GJ, Conklin DS (2002) Short hairpin RNAs (shRNAs) induce sequence-specific silencing in mammalian cells. Genes Dev 16:948–958
42. Siolas D, Lerner C, Burchard J et al (2005) Synthetic shRNAs as potent RNAi triggers. Nat Biotechnol 23:227–231
43. Kawasaki H, Taira K (2003) Short hairpin type of dsRNAs that are controlled by tRNA(Val) promoter significantly induce

RNAi-mediated gene silencing in the cytoplasm of human cells. Nucleic Acids Res 31:700–707

44. Lee Y, Kim M, Han J et al (2004) MicroRNA genes are transcribed by RNA polymerase II. EMBO J 23:4051–4060
45. van de Wetering M, Oving I, Muncan V et al (2003) Specific inhibition of gene expression using a stably integrated, inducible small-interfering-RNA vector. EMBO Rep 4:609–615
46. Gupta S, Schoer RA, Egan JE, Hannon GJ, Mittal V (2004) Inducible, reversible, and stable RNA interference in mammalian cells. Proc Natl Acad Sci U S A 101:1927–1932
47. Liu YP, Haasnoot J, Berkhout B (2007) Design of extended short hairpin RNAs for HIV-1 inhibition. Nucleic Acids Res 35:5683–5693
48. Liu YP, von Eije KJ, Schopman NC et al (2009) Combinatorial RNAi against HIV-1 using extended short hairpin RNAs. Mol Ther 17:1712–1723
49. Weinberg MS, Ely A, Barichievy S et al (2007) Specific inhibition of HBV replication in vitro and in vivo with expressed long hairpin RNA. Mol Ther 15:534–541
50. Saayman S, Barichievy S, Capovilla A, Morris KV, Arbuthnot P, Weinberg MS (2008) The efficacy of generating three independent anti-HIV-1 siRNAs from a single U6 RNA Pol III-expressed long hairpin RNA. PLoS One 3:e2602
51. Sano M, Li H, Nakanishi M, Rossi JJ (2008) Expression of long anti-HIV-1 hairpin RNAs for the generation of multiple siRNAs: advantages and limitations. Mol Ther 16:170–177
52. Wu Z, Asokan A, Samulski RJ (2006) Adeno-associated virus serotypes: vector toolkit for human gene therapy. Mol Ther 14:316–327
53. Kotin RM, Siniscalco M, Samulski RJ et al (1990) Site-specific integration by adeno-associated virus. Proc Natl Acad Sci U S A 87:2211–2215
54. Nakai H, Montini E, Fuess S, Storm TA, Grompe M, Kay MA (2003) AAV serotype 2 vectors preferentially integrate into active genes in mice. Nat Genet 34:297–302
55. Russell DW (2003) AAV loves an active genome. Nat Genet 34:241–242
56. Smith RH (2008) Adeno-associated virus integration: virus versus vector. Gene Ther 15:817–822
57. Donsante A, Miller DG, Li Y et al (2007) AAV vector integration sites in mouse hepatocellular carcinoma. Science 317:477
58. Cao H, Koehler DR, Hu J (2004) Adenoviral vectors for gene replacement therapy. Viral Immunol 17:327–333
59. Buchbinder SP, Mehrotra DV, Duerr A et al (2008) Efficacy assessment of a cell-mediated immunity HIV-1 vaccine (the Step Study): a double-blind, randomised, placebo-controlled, test-of-concept trial. Lancet 372:1881–1893
60. McElrath MJ, De Rosa SC, Moodie Z et al (2008) HIV-1 vaccine-induced immunity in the test-of-concept Step Study: a case-cohort analysis. Lancet 372:1894–1905
61. O'Brien KL, Liu J, King SL et al (2009) Adenovirus-specific immunity after immunization with an Ad5 HIV-1 vaccine candidate in humans. Nat Med 15:873–875
62. Lasaro MO, Ertl HC (2009) New insights on adenovirus as vaccine vectors. Mol Ther 17:1333–1339
63. Hartman ZC, Appledorn DM, Amalfitano A (2008) Adenovirus vector induced innate immune responses: impact upon efficacy and toxicity in gene therapy and vaccine applications. Virus Res 132:1–14
64. Baum C, Schambach A, Bohne J, Galla M (2006) Retrovirus vectors: toward the plentivirus? Mol Ther 13:1050–1063
65. Hacein-Bey-Abina S, Garrigue A, Wang GP et al (2008) Insertional oncogenesis in 4 patients after retrovirus-mediated gene therapy of SCID-X1. J Clin Invest 118:3132–3142
66. Mohamedali A, Moreau-Gaudry F, Richard E, Xia P, Nolta J, Malik P (2004) Self-inactivating lentiviral vectors resist proviral methylation but do not confer position-independent expression in hematopoietic stem cells. Mol Ther 10:249–259
67. Ellis J (2005) Silencing and variegation of gammaretrovirus and lentivirus vectors. Hum Gene Ther 16:1241–1246
68. Buchschacher GL Jr, Wong-Staal F (2000) Development of lentiviral vectors for gene therapy for human diseases. Blood 95:2499–2504
69. Laufs S, Guenechea G, Gonzalez-Murillo A et al (2006) Lentiviral vector integration sites in human NOD/SCID repopulating cells. J Gene Med 8:1197–1207
70. Montini E, Cesana D, Schmidt M et al (2006) Hematopoietic stem cell gene transfer in a tumor-prone mouse model uncovers low genotoxicity of lentiviral vector integration. Nat Biotechnol 24:687–696
71. Montini E, Cesana D, Schmidt M et al (2009) The genotoxic potential of retroviral vectors

is strongly modulated by vector design and integration site selection in a mouse model of HSC gene therapy. J Clin Invest 119: 964–975

72. Lois C, Hong EJ, Pease S, Brown EJ, Baltimore D (2002) Germline transmission and tissue-specific expression of transgenes delivered by lentiviral vectors. Science 295:868–872
73. Pfeifer A (2004) Lentiviral transgenesis. Transgenic Res 13:513–522
74. Grimm D (2009) Small silencing RNAs: state-of-the-art. Adv Drug Deliv Rev 61:672–703
75. Grimm D, Kay MA (2007) RNAi and gene therapy: a mutual attraction. Hematol Am Soc Hematol Educ Program 1:473–481
76. Mowa MB, Crowther C, Arbuthnot P (2010) Therapeutic potential of adenoviral vectors for delivery of expressed RNAi activators. Expert Opin Drug Deliv 7:1373–1385
77. Raoul C, Barker SD, Aebischer P (2006) Viral-based modelling and correction of neurodegenerative diseases by RNA interference. Gene Ther 13:487–495
78. Stewart SA, Dykxhoorn DM, Palliser D et al (2003) Lentivirus-delivered stable gene silencing by RNAi in primary cells. RNA 9:493–501
79. Chen J, Wall NR, Kocher K et al (2004) Stable expression of small interfering RNA sensitizes TEL-PDGFbetaR to inhibition with imatinib or rapamycin. J Clin Invest 113:1784–1791
80. Brummelkamp TR, Bernards R, Agami R (2002) Stable suppression of tumorigenicity by virus-mediated RNA interference. Cancer Cell 2:243–247
81. Liu YP, Berkhout B (2009) Lentiviral delivery of RNAi effectors against HIV-1. Curr Top Med Chem 9:1130–1143
82. Manjunath N, Wu H, Subramanya S, Shankar P (2009) Lentiviral delivery of short hairpin RNAs. Adv Drug Deliv Rev 61:732–745
83. Seppen J, Rijnberg M, Cooreman MP, Oude Elferink RP (2002) Lentiviral vectors for efficient transduction of isolated primary quiescent hepatocytes. J Hepatol 36:459–465
84. Liu YP, Haasnoot J, Ter Brake O, Berkhout B, Konstantinova P (2008) Inhibition of HIV-1 by multiple siRNAs expressed from a single microRNA polycistron. Nucleic Acids Res 36:2811–2824
85. Kotsopoulou E, Kim VN, Kingsman AJ, Kingsman SM, Mitrophanous KA (2000) A Rev-independent human immunodeficiency virus type 1 (HIV-1)-based vector that exploits a codon-optimized HIV-1 gag-pol gene. J Virol 74:4839–4852
86. Dull T, Zufferey R, Kelly M et al (1998) A third-generation lentivirus vector with a conditional packaging system. J Virol 72:8463–8471
87. Ter Brake O, Konstantinova P, Ceylan M, Berkhout B (2006) Silencing of HIV-1 with RNA interference: a multiple shRNA approach. Mol Ther 14:883–892
88. Jeeninga RE, Hoogenkamp M, Armand-Ugon M, de Baar M, Verhoef K, Berkhout B (2000) Functional differences between the long terminal repeat transcriptional promoters of HIV-1 subtypes A through G. J Virol 74:3740–3751
89. Schopman NC, Liu YP, Konstantinova P, Ter Brake O, Berkhout B (2010) Optimization of shRNA inhibitors by variation of the terminal loop sequence. Antiviral Res 86:204–211
90. Yu JY, DeRuiter SL, Turner DL (2002) RNA interference by expression of short-interfering RNAs and hairpin RNAs in mammalian cells. Proc Natl Acad Sci U S A 99:6047–6052
91. Koper-Emde D, Herrmann L, Sandrock B, Benecke BJ (2004) RNA interference by small hairpin RNAs synthesised under control of the human 7S K RNA promoter. Biol Chem 385:791–794
92. Denti MA, Rosa A, Sthandier O, De Angelis FG, Bozzoni I (2004) A new vector, based on the PolII promoter of the U1 snRNA gene, for the expression of siRNAs in mammalian cells. Mol Ther 10:191–199
93. Ter Brake O, 'tHooft K, Liu YP, Centlivre M, von Eije KJ, Berkhout B (2008) Lentiviral vector design for multiple shRNA expression and durable HIV-1 inhibition. Mol Ther 16:557–564
94. Haqqi T, Zhao X, Panciu A, Yadav SP (2002) Sequencing in the presence of Betaine: improvement in sequencing of the localized repeat sequence regions. J Biomol Tech 13:265–271
95. Liu YP, Vink MA, Westerink JT et al (2010) Titers of lentiviral vectors encoding shRNAs and miRNAs are reduced by different mechanisms that require distinct repair strategies. RNA 16:1328–1339
96. Andersson MG, Haasnoot PCJ, Xu N, Berenjian S, Berkhout B, Akusjarvi G (2005) Suppression of RNA interference by adenovirus virus-associated RNA. J Virol 79:9556–9565
97. de Vries W, Haasnoot J, van der Velden J et al (2008) Increased virus replication in mammalian cells by blocking intracellular innate defense responses. Gene Ther 15:545–552
98. Haasnoot J, de Vries W, Geutjes EJ, Prins M, de Haan P, Berkhout B (2007) The Ebola

virus VP35 protein is a suppressor of RNA silencing. PLoS Pathog 3:e86

99. Popa I, Harris ME, Donello JE, Hope TJ (2002) CRM1-dependent function of a cis-acting RNA export element. Mol Cell Biol 22:2057–2067

100. Higashikawa F, Chang L (2001) Kinetic analyses of stability of simple and complex retroviral vectors. Virol 280:124–131

101. Kwon YJ, Hung G, Anderson WF, Peng CA, Yu H (2003) Determination of infectious retrovirus concentration from colony-forming assay with quantitative analysis. J Virol 77: 5712–5720

102. Sastry L, Johnson T, Hobson MJ, Smucker B, Cornetta K (2002) Titering lentiviral vectors: comparison of DNA, RNA and marker expression methods. Gene Ther 9:1155–1162

Chapter 14

Bifunctional Short Hairpin RNA (bi-shRNA): Design and Pathway to Clinical Application

Donald D. Rao, Neil Senzer, Zhaohui Wang, Padmasini Kumar, Chris M. Jay, and John Nemunaitis

Abstract

The discovery of RNA interference (RNAi) engendered great excitement and raised expectations regarding its potential applications in biomedical research and clinical usage. Over the ensuing years, expanded understanding of RNAi and preliminary results from early clinical trials tempered enthusiasm with realistic appraisal resulting in cautious optimism and a better understanding of necessary research and clinical directions. As a result, data from more recent trials are beginning to show encouraging positive clinical outcomes. The capability of delivering a pharmacologically effective dose to the target site while avoiding adverse host reactions still remains a challenge although the delivery technology continues to improve. We have developed a novel vector-driven bifunctional short hairpin RNA (bi-shRNA) technology that harnesses both cleavage-dependent and cleavage-independent RISC loading pathways to enhance knockdown potency. Consequent advantages provided by the bi-shRNA include a lower effective systemic dose than comparator siRNA/shRNA to minimize the potential for off-target side effects, due to its ability to induce both a rapid (inhibition of protein translation) and delayed (mRNA cleavage and degradation) targeting effect depending on protein and mRNA kinetics, and a longer duration of effectiveness for clinical applications. Here, we provide an overview of key molecular methods for the design, construction, quality control, and application of bi-shRNA that we believe will be useful for others interested in utilizing this technology.

Key words: Bifunctional shRNA, Cancer, Gene therapy, Targeted, Clinical

1. Introduction

RNA interference (RNAi), the Nobel prize-winning discovery by Fire and Mello in 1998, has fostered an exponential number of studies and publications furthering our understanding of gene function (1, 2) and stimulating numerous phase I and II clinical trials (3–6). This naturally occurring gene-silencing mechanism by small RNAs, which includes endogenous microRNA (miRNA), is

Debra J. Taxman (ed.), *siRNA Design: Methods and Protocols*, Methods in Molecular Biology, vol. 942,
DOI 10.1007/978-1-62703-119-6_14, © Springer Science+Business Media, LLC 2013

highly dependent on gene sequence; thus the mechanism can, in theory, be used to inhibit the expression of any targeted gene(s) with strong specificity. RNAi is not limited by the pharmacologic constraints inherent to the development of small molecules, which creates an opportunity to access traditionally "undruggable" targets for disease treatment (7–9).

The central player of this mechanism is the RNA-Induced Silencing Complex (RISC). The process starts with double-stranded small RNA (composed of a passenger strand and a guide strand), which is incorporated into the pre-RISC, followed by the cleavage-dependent or cleavage-independent release of the passenger strand to form the guide strand containing RISC (10). The guide strand (antisense to mRNA) guides the RISC to recognize the target mRNA through sequence complementarity (full or extended partial) (11). A key component of the RISC is the family of Argonaute proteins (Ago), Ago 1, 2, 3, and 4 in mammalian systems, of which only Ago 2 has endonuclease activity so as to allow for cleavage of the target mRNA for further degradation (cleavage-dependent pathway) (10, 12); all the Ago containing RISCs can function through a cleavage-independent pathway resulting in translation repression and mRNA sequestration in *p*-bodies with subsequent degradation (13, 14). The cleavage-dependent process requires extensive homology between the guide strand and both the passenger strand and target mRNA, particularly in the central region; the cleavage-independent process, on the other hand, only requires partial homology between the guide strand and both the passenger strand and target mRNA (15–17).

RNAi can be triggered either by synthetic double-stranded small interfering RNA (siRNA) or by vector-driven short hairpin RNA (shRNA) (5, 18). Both siRNA and vector-driven shRNA have been demonstrated to be effective in in vitro and in vivo applications, each with their respective advantages. Most siRNAs are structurally designed to promote efficient incorporation into the Ago2 containing RISC; the RNase III containing Dicer-substrate design improves the efficiency of siRNA at least tenfold by initial association and processing at the pre-RISC (19). Vector-driven shRNA utilizes the host microRNA biogenesis pathway, which appears to be more efficient (20, 21). siRNA is more readily chemically modified while shRNA expression can be modulated and regulated by specific promoters.

We have recently developed a novel vector-driven shRNA technology, the bifunctional shRNA (bi-shRNA), to further improve the efficiency of RNAi by harnessing both cleavage-dependent and cleavage-independent pathways of RISC loading in one preprogramed molecule (Fig. 1) (18, 21). The vector-driven bi-shRNA consists of two stem–loop structures for each mRNA target sequence; one stem–loop shRNA has perfect complementarity at the stem and the second stem–loop shRNA contains mismatches on the passenger

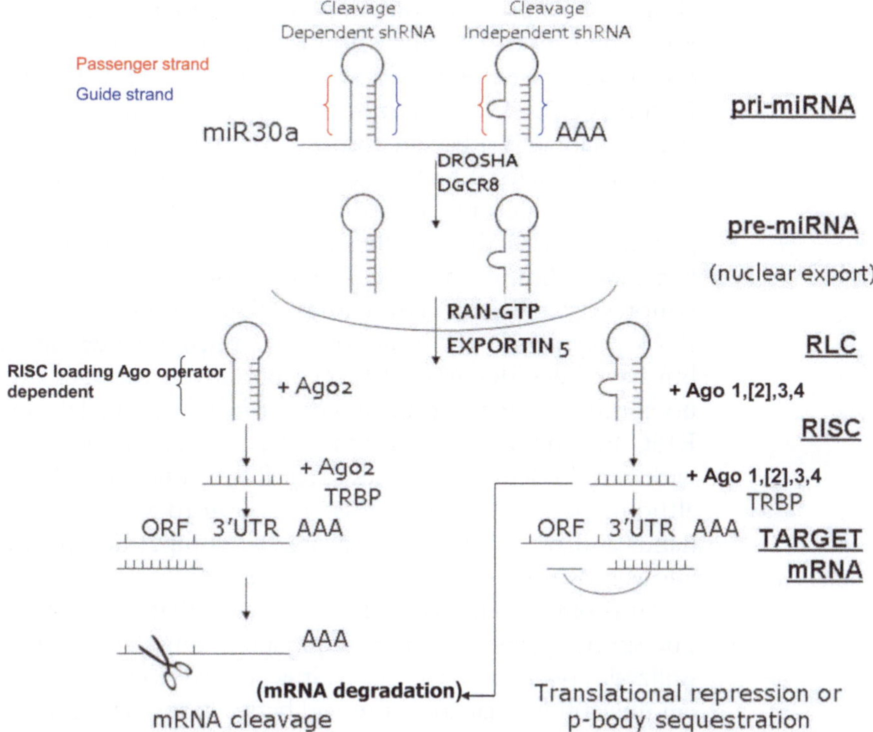

Fig. 1. Schematic showing the principle of bifunctional shRNA. Like a significant percentage of intronic miRNA, the primary transcripts of bi-shRNA are first synthesized in the nucleus by RNA polymerase II. The primary transcripts are quickly processed by the microprocessor, primarily consisting of Drosha and DGCR8, into individual stem–loop structures. The stem–loop structures are exported from the nucleus to the cytoplasm via Exportin 5 and/or Exportin 1 (CRM1), where upon association with Dicer and RISC loading complex (RLC), the loop of the stem–loop structures are removed. Upon loading onto multiple types of Ago containing RISCs, the passenger strands of bi-shRNA depart through either a cleavage-dependent or cleavage-independent process so that the guide strands are effectively loaded onto multiple types of Ago containing RISCs to more efficiently execute RNAi through either mRNA cleavage and degradation or through translational repression, p-body sequestration and degradation.

strand of the stem. Importantly, following incorporation into the RISC, the guide strands derived from each of the two structures are fully complementary to the mRNA target sequence but are associated with different Ago containing RISCs. The bi-shRNA design leads to more rapid onset of gene silencing, higher efficacy, and greater durability when compared with either siRNA or conventional shRNA (21). We are currently transitioning personalized cancer therapy with target-specific bi-shRNA into the clinic in Phase I studies using a modified bilamellar invaginated liposome delivery vehicle (22). Here, we provide key molecular methods involved in design, construction, and the implementation of bi-shRNA.

Once a target gene is selected, the initial step is to determine the objective of the study. Depending on that objective, there are many different vectors or plasmid backbones and delivery systems that can be used. It is a good idea to decide early in the process so the right expression cassette can be constructed. It is important to choose an expression vector with efficient transgene expression. We found that an expression vector with an extended CMV promoter containing immediate early (IE) 5′UTR and partial Intron A (pUMVC3) is more effective than those with a cloning site immediately adjacent to the CMV promoter (23, 24). It may be more beneficial to have a stretch of lead transcript before the stem–loop structures. In addition, if more than one vector usage is planned, an effective shuttle strategy should be worked out beforehand; modification by PCR amplification of the expressed cassette is not very efficient. The choice of promoter is also important; RNA polymerase III promoters are much stronger in expression but have been documented to competitively saturate the endogenous miRNA maturation process at both the nuclear export and RISC loading steps resulting in lethal toxicity in vitro and in vivo with certain delivery vehicles (25). RNA polymerase II promoters, although less strong in expression, appear to work efficiently in our hands and are much less toxic vis-à-vis competition for the endogenous miRNA pathway (26).

It is often useful to design a sequence that can act in more than one species particularly if multiple animal model systems are to be utilized. For most target genes, it is possible to find stretches of target nucleotides that are conserved between species; however, finding a sequence that is both conserved and optimum for knockdown may not be easy. One would have to painstakingly compare the homology matched sequence with the selected target site sequence.

Publicly accessible computer programs using differing algorithms (e.g., Dharmacon RNAi design center (www.dharmacon.com) and IDT (www.idtdna.com) are readily available and can be used to locate appropriate target sites within the targeted gene. A search with most computer programs will often yield a good set of targets for further analysis; additionally, many available publications offer worthwhile advice in the form of do and do-not suggestions (27–30). Special care must be taken to do a BLAST search for each target sequence to analyze potential cross homology with other mRNAs within the species of interest.

Once the target site sequence is selected, the bi-shRNA design process can begin; the design process is presented in Subheading 3.1. The bi-shRNA stem–loop structure routinely used in our facility employs the well analyzed miR-30a backbone (31, 32) although, in theory, one can use any functional miRNA backbone. A predicted stem–loop structure of a bifunctional construct is shown in Fig. 2a. The bi-shRNA consists of the two stem–loop structures on an miR-30a backbone located immediately adjacent to each other with a gap about 10 nucleotides long (Fig. 2a); however, a longer nucleotide gap can be used and multiple units of bi-shRNA can be designed to string together in a single transcript targeting either a single gene at multiple sites or multiple different genes simultaneously.

To construct the expression unit to be placed in the multiple cloning sites of an expression vector, we have developed an assembly

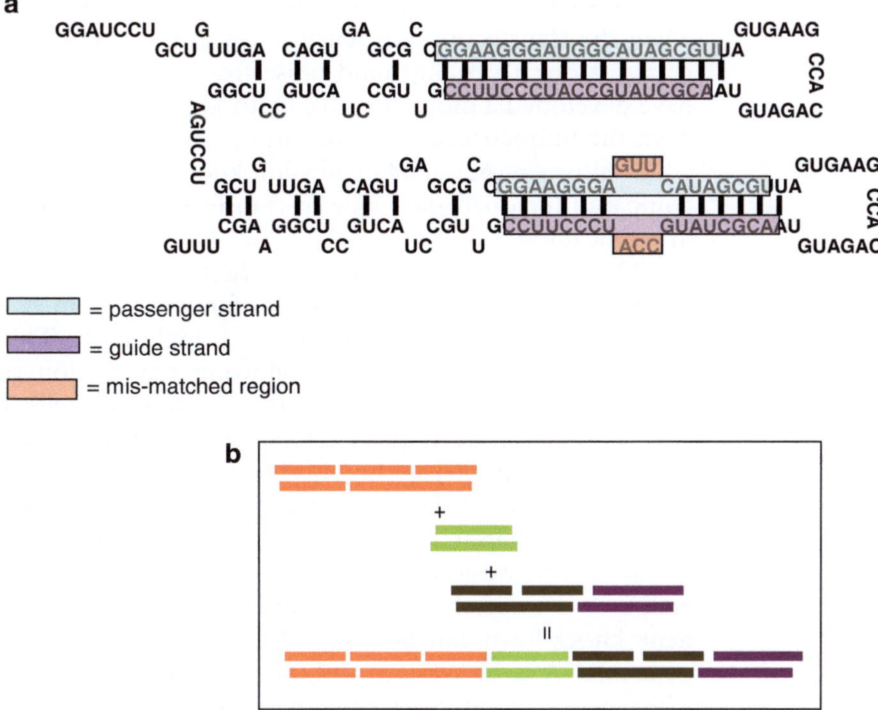

Fig. 2. Bi-shRNA structure and construction assembly process. (**a**) An example of predicted RNA structure of the bi-shRNA in the miR30a backbone. Two stem–loop structures are juxtaposed to each other each with the identical guide strand (*light purple*); the first one completely complementary to the passenger strand (*light blue box*), the second one with a mis-matched passenger strand. The mismatched region is in *red boxes*. (**b**) Schematic of the assembly process to generate the bi-shRNA expression unit. Overlapping fragments cover the entire expression unit. The fragments are designed with sticky ends to facilitate joining by ligation. The assembly process proceeds in stepwise fashion. The 5′ end fragments (*orange*) are ligated together. The 3′ fragments are ligated together (*black* and *purple*) before being joined together with the middle fragment (*green*).

strategy using synthetic oligonucleotides sequentially linked together (Fig. 2b). Alternatively, one can also outsource the synthesis of the gene construct with the specified sequence to one of many biotechnology service companies. For the oligonucleotide assembly process, overlapping DNA fragments were designed and synthesized; because of redundant sequences in the two stem–loop structures, we found it is necessary to initially ligate the 5′ fragments and 3′ fragments. The 5′ fragment and the 3′ fragment can then be purified on a gel and further ligated to the middle, linking fragments as illustrated in Fig. 2b and described in Subheading 3.2. This assembly process is efficient and, with careful design, many fragments can be repetitively used for different bifunctional constructs.

For each target, it is best to design and construct at least three bifunctional constructs to compare and from which to select the construct(s) with the most knockdown efficiency for further evaluation. Knockdown efficiency can be compared in vitro in tissue culture cells. It is often hard to compare the knockdown efficiency with endogenously expressed genes because in vitro transfection methods have widely different efficiencies. This is particularly so when the transfection efficiency is low as the knockdown is hard to assess due to background noise from untransfected cells. We have developed a more effective method in which we co-transfect both the bifunctional construct and transgene expression vector, which allows target gene expression knockdown to be effectively compared and quantified. The co-transfection method is described in Subheading 3.3.

The efficacy and efficiency of target gene knockdown by bi-shRNA can be tested with a variety of in vitro and in vivo systems depending on the target and planned application. This in vitro assessment can be conducted following transfection of the bi-shRNA expression plasmids in a variety of cultured cells. We found that transfection by both electroporation and by liposome (e.g., Lipofectamine 2000) are highly effective; however, the amount of plasmid DNA used should be carefully controlled using a control vector or universal random sequence. For Lipofectamine or a related agent, we found the reverse transfection method, in general, is less toxic than the forward transfection method. Target gene knockdown can be assessed by either qRT-PCR for target gene mRNA or by Western and/or ELISA for target gene protein. These assays are well described in many publications. We present two assay methods in detail here; one detects the expression of mature shRNA by stem–loop RT-PCR (Subheading 3.4), the other detects the target mRNA cleavage by 5′ RNA-Ligand Mediated RACE (5′ RLM-RACE; Subheading 3.5). We have successfully used both these methods to assess the efficacy of bi-shRNA both in vitro and in vivo. The stem–loop RT-PCR method is schematically illustrated in Fig. 3a and a typical result shown in Fig. 3b. Stem–loop RT-PCR is a sensitive method dependent on the specific probe primer used; in addition, one can specifically detect and quantify both the passenger strand and guide strand. For bi-shRNA, the method can differentially score both the fully complementary as well as the mismatched (partially complementary) passenger strand (21). The 5′ RLM-RACE method is schematically shown in Fig. 4a; the method requires ligation of an RNA oligomer onto the cleaved mRNA end, and consequently, the method is rendered less efficient. Insofar as a number of rounds of amplifications are often required, a nested primer design is essential to ensure specificity. A typical result is shown in Fig. 4b.

Evaluable functionality of bi-shRNA relies on effective plasmid delivery into target cells. Various in vitro transfection systems often do not translate to inherently more complex in vivo animal models (33). There are numerous delivery systems designed specifically for systemic applications in vivo (22, 33–35). We have utilized the fusogenic, cationic DOTAP:cholesterol bilamellar invaginated vesicle lipoplex (BIV) for our in vivo studies (22, 36) and have successfully translated it to the clinic (37). We are currently developing

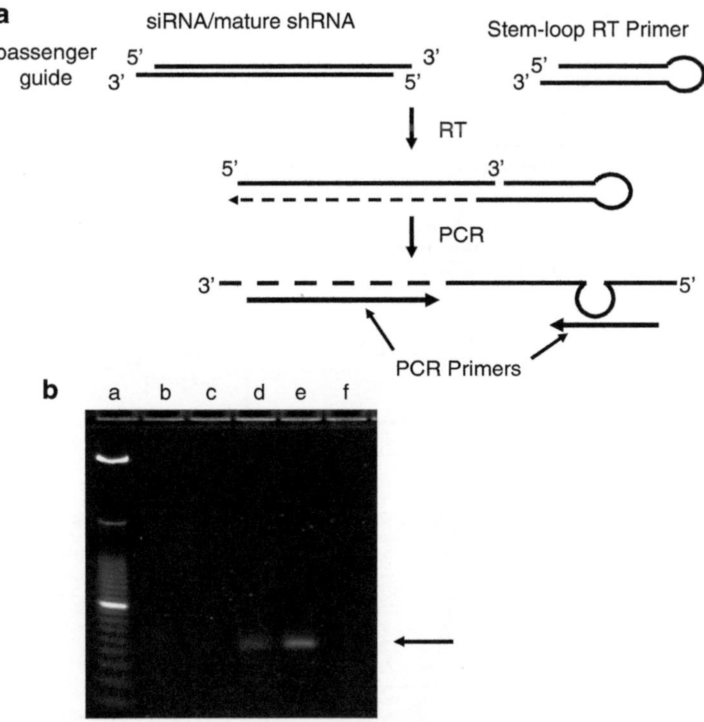

Fig. 3. Stem–loop RT-PCR to detect mature shRNAs. (**a**) Schematic depicting the stem–loop RT-PCR procedure. A stem–loop RT primer is designed with at least a 4-base overhang at the 3′ end to recognize the 3′ end of siRNA/mature shRNA. After RT, PCR amplification is accomplished with an siRNA/mature shRNA-specific PCR primer and a stem–loop RT primer-specific PCR primer. (**b**) A typical result of stem–loop RT-PCR. PCR products which were run onto a 3% agarose gel. Lane (a) is a 10 bp-size marker, (b) is a control with cellular RNA isolated from untransfected cells, (c) is a control with cellular RNA isolated from cells transfected with scrambled siRNA, (d) is with cellular RNA isolated from cells transfected with bi-shRNA, (e) is with cellular RNA isolated from cells transfected with siRNA, (f) is RNA isolated from bi-shRNA transfected cells without RT reaction.

modification strategies for more focused biodistribution, targeted delivery, and enhanced intracellular uptake. An effective lipoplex should use plasmids devoid of any contaminants from host *E. coli*. Although endo-free plasmid purification kit produced plasmids are generally used, GLP or GMP produced plasmids are more effective. Unfortunately, colanic acid and other non-endotoxin-associated polysaccharides co-purify with DNA by anion exchange chromatography and by cesium chloride density gradient centrifugation. Therefore, endotoxin removal does not remove these contaminants, and HPLC cannot detect these contaminants. To correct this, we recently developed the Superclean™ procedure to generate ultra-high quality plasmid DNA, cleansed of these contaminants, for in vivo and clinical applications (38). Liposome preparation involves highly specialized equipment; we routinely generate the DOTAP:cholesterol BIV in our GMP facility. It is advised that premade liposome be obtained from a collaborator or

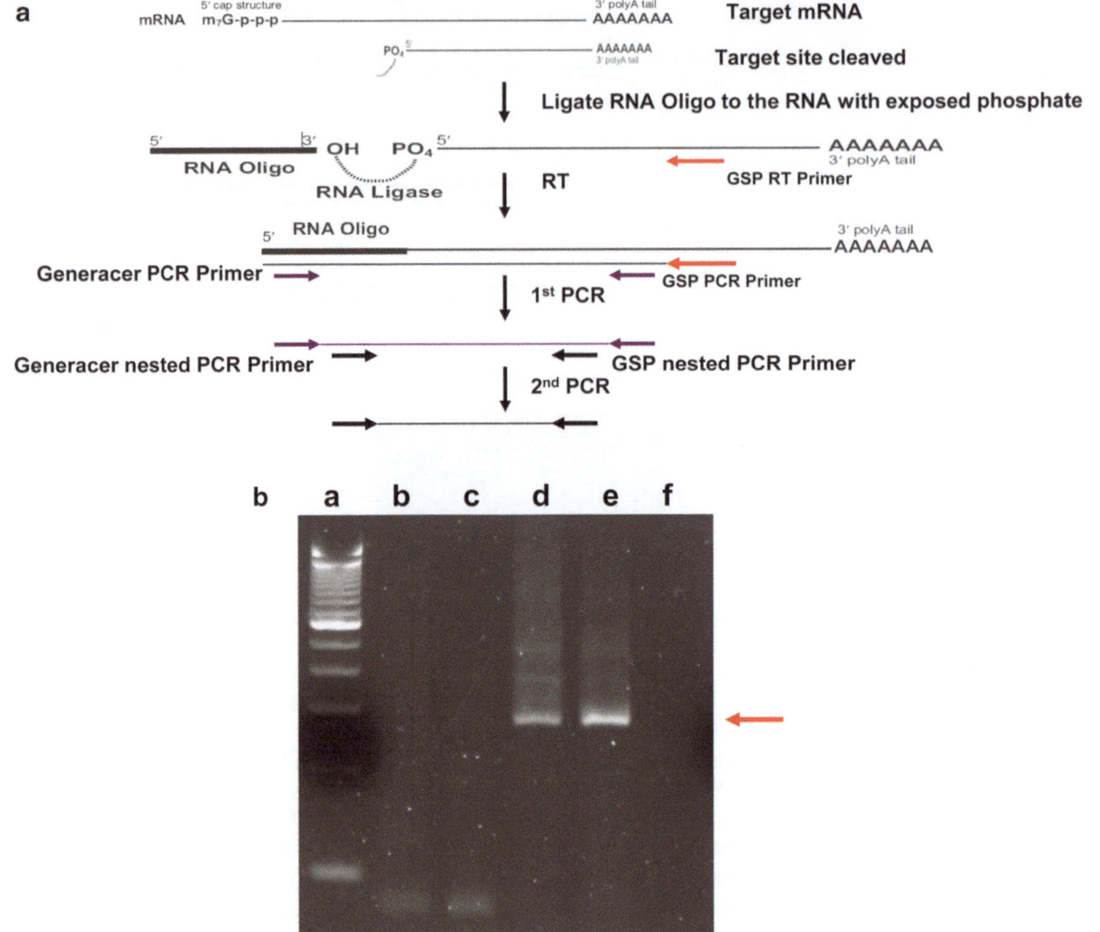

Fig. 4. 5′ RNA-ligand mediated RACE (5′RLM-RACE) to detect target site cleavage product. (**a**) Schematic depicting the 5′RLM-RACE method. Cleaved or partially degraded mRNA with exposed 5′ ends is first ligated with an RNA oligo via RNA ligase. The RNA oligo added RNA fragment is selectively copied into cDNA with a GSP primer and the cDNA is amplified by PCR with a GSP primer and an RNA oligo-specific primer, followed by further amplification with a nested GSP primer and a nested RNA oligo-specific primer. (**b**) A typical result of 5′RLM-RACE. PCR products which were run onto a 1.5% agarose gel. Lane (a) is a 100 bp size marker, (b) is a control with cellular RNA isolated from untransfected cells, (c) is a control with cellular RNA isolated from cells transfected with scramble siRNA, (d) is with cellular RNA isolated from cells transfected with bi-shRNA, (e) is with cellular RNA isolated from cells transfected with siRNA, (f) is RNA isolated from bi-shRNA transfected cells without RT reaction.

be purchased from a vendor. In Subheading 3.6, we describe the process of preparing lipoplex with high quality liposome and plasmid DNA. The lipoplex formulation can be achieved in most laboratory settings. Once the lipoplex is made, the formulation can be delivered systemically to experimental animals either through slow tail vein injection or with catheters. Target site vector expression can be analyzed using the PCR method for plasmid DNA and the stem–loop RT-PCR for mature bi-shRNA, respectively. bi-shRNA functionality can be assayed with the 5′ RLM-RACE for target mRNA cleavage and with western blot or IHC for target protein knockdown. It is advised that these analyses be performed at ~48 hours post treatment. For efficacy, repeated delivery into the experimental animal is often required; the dosing schedule needs to be experimentally determined and optimized.

2. Materials

2.1. Bifunctional shRNA Design

1. Computer with Internet access.

2.2. Construction of Bi-shRNA

1. DNA synthesis vendor or capacity.
2. Expression vectors.
3. Customized oligonucleotides.
4. T4 DNA ligase.
5. Agarose gel electrophoresis.
6. DNA gel extraction kit.
7. Restriction enzymes.
8. Competent *E. coli* cells.
9. Plasmid isolation kit.
10. Thermal cycler.

2.3. Assess Comparative Knockdown by Co-transfection

1. Gene Pulser XCell Electroporation apparatus (Bio-Rad).
2. Gene Pulser Cuvettes (Bio-Rad).
3. Tissue culture cells.
4. Target gene expression vector.
5. For protein extraction and estimation.
   - CelLytic™ M (Sigma).
   - Protease Inhibitor Cocktail (Sigma).
   - Cell scrapers.
   - Orbital shaker.
   - Microfuge (refrigerated).
   - Coomassie Plus—Bradford™ Assay Kit (Pierce).
   - Pre-diluted protein standards (Pierce).
   - 96-well round bottom plate.
   - Multichannel pipet.
   - ELISA plate reader (Molecular Devises).
6. For western immunoblot.
   - Mini-Protein II Cell: Mini vertical electrophoresis system (Bio-Rad).

- Mini-Protein II Ready Gels: 12% Tris–HCl, ten wells.
- Running buffer (TGS Buffer; Bio-Rad).
- Precision Plus All Blue standards (Bio-Rad).
- Laemmli sample loading buffer.
- β-Mercaptoethanol.
- Pre-prepared cell lysate with known protein concentration.
- Mini-Trans blot module (Bio-Rad).
- Trans-blot PVDF membrane Sandwich (Bio-Rad).
- Transfer buffer (TG buffer).
- Methanol.
- Stir plate.
- Power Pack.
- DPBS (Invitrogen).
- Nonfat dried milk, for blocking nonspecific binding.
- Tween-20.
- Primary antibody to specific protein.
- Primary antibody to loading control.
- HRP conjugated secondary antibody.
- SuperSignal West Dura Extended Duration Substrate (ThermoScientific, Pierce).
- Orbital shaker plate.
- Syngene G-Box.

2.4. Stem–Loop RT-PCR for Mature shRNA Detection

1. Thermal cycler.
2. Microcentrifuge.
3. SuperScript III RT (including reverse transcriptase, 5× first strand buffer, and 0.1 M DTT).
4. RNase H.
5. dNTP mix, 10 mM.
6. RNaseOut (40 U/μl).
7. siRNA for positive control.
8. Stem–loop RT primer.
9. Stem–loop PCR primer.
10. Mature shRNA-specific PCR primers.
11. Total cellular RNA including small RNA and miRNA extracted using the mirVana kit (Ambion, TX).
12. Agarose gel electrophoresis and TA cloning kit.

2.5. 5′ RLM-RACE for Cleavage Product Detection

1. Thermal cycler.
2. Microcentrifuge.
3. Microcon YM-100 centrifugal filter.
4. T4 RNA ligase and buffer.
5. SuperScript III RT (including reverse transcriptase, 5× first strand buffer, and 0.1 M DTT).
6. RNase H.
7. dNTP mix (10 mM).
8. RNaseOut (40 U/μl).
9. Platinum *Taq* Kit (including Platinum *Taq* Polymerase, 5× PCR buffer and 10 mM $MgCl_2$).
10. RNA oligo:

    (5′rCrGrArCrUrGrGrArGrCrArCrGrArGrGrArCrArCrUrGrArCrArUrGrGrArCrUrGrArArGrGrArGrUrArGrArArA3′).
11. 10 μM GeneRacer 5′ PCR primer: (5′-*CGACTGGAGCACGAGGACACTGA*-3′).
12. 10 μM GeneRacer 5′ nested PCR primer: (5′-*GGACACTGACATGGACTGAAGGAGTA*-3′).
13. 10 μM gene-specific (GSP) RT primer (3′ of the GSP PCR primer).
14. 10 μM each GSP PCR primer and nested GSP primer (at least 5 nt 5′ to the GSP primer).
15. Total RNA isolated with the mirVana miRNA isolation kit or RNeasy Mini kit.
16. Agarose gel electrophoresis and DNA gel extraction kit.

2.6. Preparation of DNA-DOTAP:Chol Lipoplexes

1. Certified Biological Safety Cabinet (BSC) (The Baker Company).
2. Spectrophotometer with standard and turbidity cell holder (Beckman Coulter).
3. ZetaSizer Nano (Malvern).
4. Quartz semi-microcell cuvette (Beckman Coulter).
5. Vortexer (VWR).
6. Pipettors (Gilson) and Pipet tips (Phenix).
7. Acrodisc filter 0.2 μm (Pall).
8. GMP (GLP) Grade plasmid DNA (≥2.0 μg/μl in water).
9. 5× DOTAP:Cholesterol Liposomes (Gradalis).
10. Sterile water for irrigation (Baxter).
11. Sterile D5W (5% dextrose in water) (Baxter).
12. Ultra High Pure Argon gas (Airgas).
13. Septihol (USP 70%, v/v, isopropyl alcohol + USP 30%, v/v, purified water) (Steris).

3. Methods

3.1. Bi-shRNA Design

1. Select target site sequences using available computer programs.
2. Do BLAST search to find target sites with lowest homology hits to other mRNA of the targeted species.
3. Plug the 19 nucleotides passenger strand (sense strand, blue boxed region in Fig. 2a) and guide strand (antisense strand, purple boxed region in Fig. 2a) sequences into both stems of the miR30a backbone sequence.
4. Replace nucleotides 9, 10 and 11 (counting from the 3′ end of the passenger strand) of the second stem–loop structure with nucleotides mismatched to the guide strand.
5. Run both stem–loop structure on mfold program (http://mfold.rna.albany.edu/?q=mfold), the ΔG free energy of the hairpin with mismatches should fall between −10 kcal/mol and equilibrium.
6. If the ΔG free energy is beyond −10 kcal/mol, additional mismatches should be introduced at the passenger strand of the second hairpin. The additional mismatch introduced is preferably at the 3′ half of the passenger strand.

3.2. Construction of Bi-shRNA Expression Vector

1. Design overlapping DNA oligonucleotides from both strands with sticky ends for ligation, four nucleotides overhang should be sufficient. The upper limit of oligonucleotides length for DNA synthesis is around 50 nucleotides, thus DNA fragments should be spaced evenly and be <50 nt each. Care should be taken to avoid any internal self-complementary potential for each fragment.
2. Remember to place a 5′ phosphate on each fragment for efficient ligation. Build the cloning site sequence into the 5′ and 3′ end of the expression unit to expedite cloning into the expression vector.
3. Order or synthesize oligonucleotides. Reconstitute synthesized oligonucleotides at 1 mg/ml.
4. Ligate the 5′ fragments together with one-step ligation of equal amount of oligonucleotides.
5. Ligate the 3′ fragments together with one-step ligation of equal amount of oligonucleotides.
6. Extract and purify the calculated full-length DNA fragments on an agarose gel.
7. Ligate 5′ fragment with linking fragment then 3′ fragment and purify the calculated full-length DNA fragment on an agarose gel.

8. Ligate the purified fragment into expression vector and transform competent *E. coli* cells.
9. Isolate colonies, and screen for clones with the appropriate size insert.
10. Plasmids isolated from positive clones are further confirmed by sequencing.

3.3. Compare Knockdown Efficiency by Co-transfection

1. Determine the most effective electroporation condition and the optimum amount of plasmid DNA for each cell types.
2. We usually use HEK-293 cells with electroporation conditions recommended by Bio-Rad.
3. Prepare the DNA mix for electroporation with appropriate controls. It is essential to have a control with the transgene expression vector and the empty vector for bi-shRNA expression because promoter competition is often observed.
4. We usually use the transgene expression vector and bi-shRNA expression vector at a 1:1 ratio. The experiment can also include a 1:2 ratio or 2:1 ratio to further validate the knockdown comparison at different doses.
5. After electroporation, cells are plated at 40–50% confluency. Cells are removed at various time points to analyze comparative knockdown. The most effective time point is 48 h post transfection.
6. Cells are removed from the culture plate using Trypsin EDTA 0.25%. Washed in DPBS two times to remove any residual culture media containing serum. After the final wash, the cells are pelleted.
7. Prepare the lysis buffer by adding 1% of protease inhibitor cocktail to the CelLyticM lysis buffer. For each pellet of 1.0×10^7 cells 500 μl of the lysis buffer is added. Pipette up and down to mix the cells with the lysis buffer. The lysis buffer plus the cell mixture should be cloudy but without being too dense. If the mixture is too dense another 100–200 μl of lysis buffer is added. Incubate at room temperature for 30 min on a slow shaker. Centrifuge the lysed cells for 20 min at 12,000–20,000 $\times g$ to pellet the cellular debris. Remove the protein containing supernatant to a chilled tube.
8. Protein estimation is done by the Coomassie Bradford Plus Assay.
9. Equal amount of total protein is loaded onto Ready Gel for PAGE for Western analysis.
10. Protein from the Gel is transferred to PVDF membrane.
11. Immuno probing is done with gene-specific antibody and anti body to loading control protein.

12. HRP conjugated secondary antibody is used followed by Super Signal Dura Extended Duration Substrate to detect the signal.
13. Images are captured using a G-Box and the band densities are quantified.
14. Side-by-side knockdown comparison of different bi-shRNA can be scored by semi-quantitative scan.

3.4. Demonstrate Bi-shRNA Expression by Stem–Loop RT-PCR (See Fig. 3)

1. Design and synthesize primers: the process essentially is the same as published (21, 39).
   (a) Design stem–loop RT primers for either the passenger strand or guide strand according to the predicted mature shRNA sequence following its processing into an siRNA. One can essentially use the same body of sequence as the published stem–loop RT primer sequence (21, 39) (see below) but change the final six 3′ nucleotides (the XXXXXX in sequence below) to the sequences complementary to the 3′ ends of either the guide strand or the passenger strand in query. Stem–loop RT primer sequences: 5′ GTCGTATCCAGTGCAGGGTCCGAGGTATTCGC ACTGGATACGACXXXXXX 3′.
   (b) The stem–loop PCR primer has the following sequence: 5′ GTGCAGGGTCCGAGGT 3′.
   (c) Design guide strand and passenger strand mature shRNA-specific PCR primers according to the predicted mature shRNA/siRNA sequence. The guide strand- and passenger strand-specific PCR primers are essentially the complete or shorter guide strand or passenger strand sequences pending on T_m calculation to match the T_m of the stem–loop PCR primer.
2. Reverse transcription (RT):
   (a) Add the following reagents to the tube containing 3.0 μg total RNA including small RNA (many RNA isolation methods or kits lose small RNAs; be sure to use an RNA isolation method or kit for which small RNAs are retained): 1 μl stem–loop RT primer (10 μM), 1 μl 10 mM dNTP mix (Invitrogen).
   (b) Incubate the mixture for 5 min at 65°C to disrupt the RNA secondary structure.
   (c) Keep on ice for 2 min and centrifuge briefly.
   (d) Add the following reagents to the mixture above and bring the volume to 20 μl by adding nuclease-free water: 4 μl 5× first strand buffer, 1 μl 0.1 M DTT, 1 μl RNaseOut (40 U/μl), 1 μl Superscript III RT (200 μ/μl).
   (e) Mix well and incubate at 55°C for 50 min.
   (f) Inactivate the RT reaction by incubating at 70°C for 15 min.

(g) Chill on ice for 2 min.

(h) Add 1 μl RNase H to the reaction mix and incubate at 37°C for 20 min. Centrifuge briefly and keep on ice or store in the –20°C freezer.

3. PCR setup

Make the master mix and follow the PCR parameters as below:

Reagent	**Volume (μl)**
Nuclease-free water	32.6
5× PCR buffer	10
25 mM $MgCl_2$	3
10 nM dNTP	1
Stem–loop PCR primer (10 μM)	1
Mature shRNA-specific PCR primer	1
GoTaq HotStart	0.4
cDNA generated in step 2	1
Total volume	50

Conditions

94°C	2 min	
94°C	1 min	
55°C	30 s	35 cycles
72°C	1 min	
72°C	9 min	
4°C	Hold	

4. PCR product analysis

After the PCR amplification is completed, run a 20 μl of sample on a 4% agarose gel to visualize the amplicon. The predicted PCR amplicon is TA cloned and then sequenced.

3.5. Demonstrate Target Gene Cleavage by 5′ RLM-RACE (See Fig. 4)

1. To ligate the RNA oligo to the RNA molecules with exposed 5′ phosphates, add 1–3 μg of total RNA from each sample to 250 ng of RNA oligo and mix well by pipetting several times.
2. Incubate the mixture at 65°C for 5 min in a thermal cycler and then hold at 4°C for 2 min.
3. Add the following reagents to the tube, mix well, and centrifuge briefly: 2 μl of 10× ligase buffer, 2 μl of RNaseOut (40 U/μl), 2 μl of nuclease-free water, and 1 μl of T4 RNA ligase (10 U/μl). Incubate the mixture at 37°C for 1 h in a thermal cycler.
4. RNA purification: After 37°C incubation, add 80 μl of nuclease-free water to the 20 μl ligation product, and load the 100 μl

mixture into a microcon YM-100 centrifugal filter. Centrifuge at 500×*g* for 15 min. Turn the column upside down in a new tube and centrifuge at 1,000×*g* for 3 min. The elution volume is about 10 μl.

5. RT: Add the following reagents to the tube containing 10 μl of ligated RNA from above: 1 μl of gene-specific (GSP) RT primer, 1 μl of dNTP mix, and 1 μl of nuclease-free water. Incubate the mixture for 5 min at 65°C, and then keep on ice for 1 min. Add the following reagents to the mixture of ligated RNA and RT primer: 4 μl of 5× first strand buffer, 1 μl of 0.1 M DTT, 1 μl of RNaseOut, and 1 μl of SuperScript III RT. Incubate at 55°C for 50 min, followed by incubating at 70°C for 15 min. Add 1 μl of RNaseH to the reaction mix and incubate at 37°C for 20 min.
6. Amplifying cDNA ends (first PCR, touch-down): Set up the reaction mix as listed below.

Reagent	Volume (μl)
Nuclease-free water	36.5
5× PCR buffer	5
25 mM $MgCl_2$	2
10 mM dNTP	1
Generacer 5′ PCR Primer (10 μM)	3
GSP PCR primer (10 μM)	1
Platinum *Taq*	0.5
cDNA	1.0
Total volume	50

Perform touch-down PCR following the parameters as shown below.

Conditions		
94°C	2 min	
94°C	30s	
72°C	1 min	5 cycles
94°C	30s	
70°C	1 min	5 cycles
94°C	1 min	
65°C	30s	25 cycles
72°C	1 min	
72°C	10 min	
4°C	Hold	

7. Nested PCR (second PCR): Nested PCR is employed to increase the specificity and sensitivity of the RACE product amplification. The reaction master mix is set up as listed below.

Reagent	Volume (μl)
Nuclease-free water	38.5
5× PCR buffer	5
25 mM $MgCl_2$	2
10 mM dNTP	1
Generacer 5′ nested PCR primer (10 μM)	1
Nested GSP PCR primer (10 μM)	1
Platinum Taq	0.5
DNA template from first PCR	1
Total volume	50

The PCR is performed following the parameters as shown below.

Conditions		
94°C	2 min	
94°C	30s	
65°C	30s	25 cycles
72°C	1 min	
72°C	10 min	
4°C	Hold	

8. Agarose gel electrophoresis and sequencing: 20 μl of the nested-PCR product is separated on a 2% agarose gel. The image of the agarose gel is captured by an imaging system. The identity of the PCR product is further confirmed by sequencing.

3.6. Preparation of DNA-DOTAP:Chol Lipoplexes

1. Perform the formulation in of the lipoplexes in a BSC. Do a test run before doing a bulk manufacture run.
2. Bring the liposomes, plasmid DNA, D5W, and water to room temperature.
3. Transfer D5W into 50 ml conical tubes using a 60 ml syringe and 16 G needle.
4. Measure the plasmid DNA concentration in a spectrophotometer by measuring the OD260 using a standard cuvette holder.
5. In a 1.5 ml tube, dilute the plasmid DNA to 1.0 μg/μl with D5W.
6. In a 1.5 ml tube, combine 3 volumes of D5W and 2 volumes of 5× DOTAP:Chol stock. Mix to make 2× liposomes.

7. Prepare the Lipoplex by transfering the diluted DNA (1.0 μg/μl) into an equal volume of 2× liposomes and mix by pipetting up and down.
8. Check for any signs of precipitation. If no precipitation is visible, proceed to the next step. If precipitation is visible, repeat steps 5–7 using less DNA and note the change.
9. Spectrophotometric analysis: Replace the cuvette holder with the turbidity cell holder. Dilute the DNA-Lipoplex 20-fold with sterile water. Blank the spectrophotometer at 400 nm with sterile water and then measure the OD400 of the diluted DNA-Lipoplex. The acceptable range is 0.65–0.95 OD units. If the OD400 reading is too high, repeat steps 5–7 using less DNA and note the changes in the operator comments section. If the OD400 reading is too low, repeat steps 5–7 using more DNA and note the changes.
10. After the test run is within the acceptable range, the same process is repeated to complete the manufacture run. Combine all DNA-Lipoplexes. Mix and measure the OD400 of the final pooled product. Aliquot samples for quality control studies. Layer argon gas (filtered through a 0.2 μm acrodisc filter) onto each aliquot and onto the bottle of remaining lipoplex to displace the ambient air and seal the lid.
11. Quality control tests include particle analysis (described in step 12), USP sterility, LAL endotoxin assay, residual chloroform analysis by GC/MS, and thin layer chromatography to identify lipids (QC tests service companies are listed below in step 13).
12. Particle analysis (size and zeta potential): Rinse the folded capillary cell with 5 ml 100% ethanol, and then with 5 ml USP water. Load 20-fold diluted DNA-Lipoplex sample into the capillary cell. Measure the average particle size (acceptable range <500 nm) and zeta potential (acceptable range >40 mV).
13. List of service providers:

    Sterility and endotoxin:

    AppTec.

    1265-B Kennestone Circle.

    Marietta, GA 30066.

    Residual chloroform:

    Exova Inc.

    9240 Santa Fe Springs Rd.

    Santa Fe Springs, CA 90670.

    Lipid analysis:

    Avanti Polar Lipids, Inc.

    700 Industrial Park Drive.

Acknowledgements

The authors gratefully acknowledge the generous support of the W.W. Caruth Jr. Foundation Fund of the Communities Foundation of Texas, the Jasper L. and Jack Denton Wilson Foundation, and the Marilyn Augur Family Foundation. They also wish to thank Susan W. Mill for her assistance in manuscript preparation and submission.

References

1. Fire AZ (2007) Gene silencing by double-stranded RNA (Nobel Lecture). Angew Chem Int Ed 46:6966–6984
2. Mello CC (2007) Return to the RNAi world: rethinking gene expression and evolution. Cell Death Differ 14:2013–2020
3. Burnett JC, Rossi JJ, Tiemann K (2011) Current progress of siRNA/shRNA therapeutics in clinical trials. Biotechnol J 6:1130–1146
4. Davidson BL, McCray PB Jr (2011) Current prospects for RNA interference-based therapies. Nat Rev 12:329–340
5. Wang Z, Rao DD, Senzer N, Nemunaitis J (2011) RNA interference and cancer therapy. Pharm Res (Epub ahead of print)
6. Phalon C, Rao DD, Nemunaitis J (2010) Potential use of RNA interference in cancer therapy. Expert Rev Mol Med 12:e26
7. Drews J (1996) Genomic sciences and the medicine of tomorrow. Nat Biotechnol 14: 1516–1518
8. Drews J (2000) Drug discovery: a historical perspective. Science 287:1960–1964
9. Verdine GL, Walensky LD (2007) The challenge of drugging undruggable targets in cancer: lessons learned from targeting BCL-2 family members. Clin Cancer Res 13:7264–7270
10. Matranga C, Tomari Y, Shin C, Bartel DP, Zamore PD (2005) Passenger-strand cleavage facilitates assembly of siRNA into Ago2-containing RNAi enzyme complexes. Cell 123:607–620
11. Preall JB, Sontheimer EJ (2005) RNAi: RISC gets loaded. Cell 123:543–545
12. Parker JS, Barford D (2006) Argonaute: a scaffold for the function of short regulatory RNAs. Trends Biochem Sci 31:622–630
13. Gregory RI, Chendrimada TP, Cooch N, Shiekhattar R (2005) Human RISC couples microRNA biogenesis and posttranscriptional gene silencing. Cell 123:631–640
14. Tang G (2005) siRNA and miRNA: an insight into RISCs. Trends Biochem Sci 30:106–114
15. Grimson A, Farh KK, Johnston WK, Garrett-Engele P, Lim LP, Bartel DP (2007) MicroRNA targeting specificity in mammals: determinants beyond seed pairing. Mol Cell 27:91–105
16. Ameres SL, Martinez J, Schroeder R (2007) Molecular basis for target RNA recognition and cleavage by human RISC. Cell 130:101–112
17. Parker JS, Roe SM, Barford D (2006) Molecular mechanism of target RNA transcript recognition by Argonaute-guide complexes. Cold Spring Harb Symp Quant Biol 71:45–50
18. Rao DD, Vorhies JS, Senzer N, Nemunaitis J (2009) siRNA vs. shRNA: similarities and differences. Adv Drug Deliv Rev 61:746–759
19. Kim DH, Behlke MA, Rose SD, Chang MS, Choi S, Rossi JJ (2005) Synthetic dsRNA Dicer substrates enhance RNAi potency and efficacy. Nat Biotechnol 23:222–226
20. McAnuff MA, Rettig GR, Rice KG (2007) Potency of siRNA versus shRNA mediated knockdown in vivo. J Pharm Sci 96:2922–2930
21. Rao DD, Maples PB, Senzer N et al (2010) Enhanced target gene knockdown by a bifunctional shRNA: a novel approach of RNA interference. Cancer Gene Ther 17:780–791
22. Templeton NS, Lasic DD, Frederik PM, Strey HH, Roberts DD, Improved PGN (1997) DNA: liposome complexes for increased systemic delivery and gene expression. Nat Biotechnol 15:647–652
23. Chapman BS, Thayer RM, Vincent KA, Haigwood NL (1991) Effect of intron A from human cytomegalovirus (Towne) immediate-early gene on heterologous expression in mammalian cells. Nucleic Acids Res 19:3979–3986
24. Simari RD, Yang ZY, Ling X et al (1998) Requirements for enhanced transgene expression by untranslated sequences from the human cytomegalovirus immediate-early gene. Mol Med (Cambridge, Mass) 4:700–706
25. Grimm D, Streetz KL, Jopling CL et al (2006) Fatality in mice due to oversaturation of cellular microRNA/short hairpin RNA pathways. Nature 441:537–541

26. Giering JC, Grimm D, Storm TA, Kay MA (2008) Expression of shRNA from a tissue-specific pol II promoter is an effective and safe RNAi therapeutic. Mol Ther 16:1630–1636
27. Walton SP, Wu M, Gredell JA, Chan C (2010) Designing highly active siRNAs for therapeutic applications. FEBS J 277:4806–4813
28. Hofacker IL, Tafer H (2010) Designing optimal siRNA based on target site accessibility. Methods Mol Biol 623:137–154
29. Moore CB, Guthrie EH, Huang MT, Taxman DJ (2010) Short hairpin RNA (shRNA): design, delivery, and assessment of gene knockdown. Methods Mol Biol 629:141–158
30. Dawson LA, Usmani BA (2008) Design, manufacture, and assay of the efficacy of siRNAs for gene silencing. Methods Mol Biol 439:403–419
31. Zeng Y, Cullen BR (2006) Recognition and cleavage of primary microRNA transcripts. Methods Mol Biol 342:49–56
32. Siolas D, Lerner C, Burchard J et al (2005) Synthetic shRNAs as potent RNAi triggers. Nat Biotechnol 23:227–231
33. Barteau B, Chevre R, Letrou-Bonneval E, Labas R, Lambert O, Pitard B (2008) Physicochemical parameters of non-viral vectors that govern transfection efficiency. Curr Gene Ther 8: 313–323
34. Chesnoy S, Huang L (2000) Structure and function of lipid–DNA complexes for gene delivery. Annu Rev Biophys Biomol Struct 29:27–47
35. Templeton NS (2010) Liposomes for gene transfer in cancer therapy. Methods Mol Biol 651:271–278
36. Phadke AP, Jay CM, Wang Z et al (2011) In vivo safety and antitumor efficacy of bifunctional small hairpin RNAs specific for the human Stathmin 1 oncoprotein. DNA Cell Biol 30:715–726
37. Nemunaitis G, Jay CM, Maples PB et al (2011) Hereditary inclusion body myopathy: single patient response to intravenous dosing of GNE gene lipoplex. Hum Gene Ther (Epub ahead of print)
38. Firozi P, Zhang W, Chen L, Quiocho FA, Worley KC, Templeton NS (2010) Identification and removal of colanic acid from plasmid DNA preparations: implications for gene therapy. Gene Ther 17:1484–1499
39. Chen C, Ridzon DA, Broomer AJ et al (2005) Real-time quantification of microRNAs by stem-loop RT-PCR. Nucleic Acids Res 33:e179

Chapter 15

Design and Chemical Modification of Synthetic Short shRNAs as Potent RNAi Triggers

Anne Dallas and Brian H. Johnston

Abstract

Synthetic shRNAs that are too short to be Dicer substrates (short shRNAs or sshRNAs) can be highly potent RNAi effectors when properly designed, with activities similar to or more potent than the more commonly used siRNAs targeting the same sequences. sshRNAs can be designed in two possible orientations: left- or right-hand loop, designated L-sshRNAs and R-sshRNAs, respectively. Because L- and R-sshRNAs are processed by the RNAi machinery in different ways, optimal designs for the two formats diverge in several key aspects. Here, we describe the principles of design and chemical modification of highly effective L- and R-sshRNAs.

Key words: shRNA, sshRNA, Dicer, Ago proteins

1. Introduction

RNA interference (RNAi)-inducing triggers such as small interfering RNAs (siRNAs) and small hairpin RNAs (shRNAs) have been widely used for gene function analysis, pathway mapping, and drug target validation (1–9). Because of their high specificity and potency, siRNAs and shRNAs also show promise as potential therapeutic agents, although effective delivery to target tissues and organs remains a challenge.

While the most commonly used design for siRNA consists of 19 base pair duplexes with 2-nucleotide 3′-overhangs on each strand, the design of shRNAs involves more parameters. The basic structure of a typical shRNA comprises a Watson-Crick base paired duplex connected by a "loop" or connecting sequence. The length of the duplex usually varies from 19 to 29 base pairs, and the loop can be almost any length or sequence. We can further classify shRNAs into two subgroups depending on whether or not they are substrates for the RNase III-family endonuclease Dicer. shRNAs

Debra J. Taxman (ed.), *siRNA Design: Methods and Protocols*, Methods in Molecular Biology, vol. 942,
DOI 10.1007/978-1-62703-119-6_15, © Springer Science+Business Media, LLC 2013

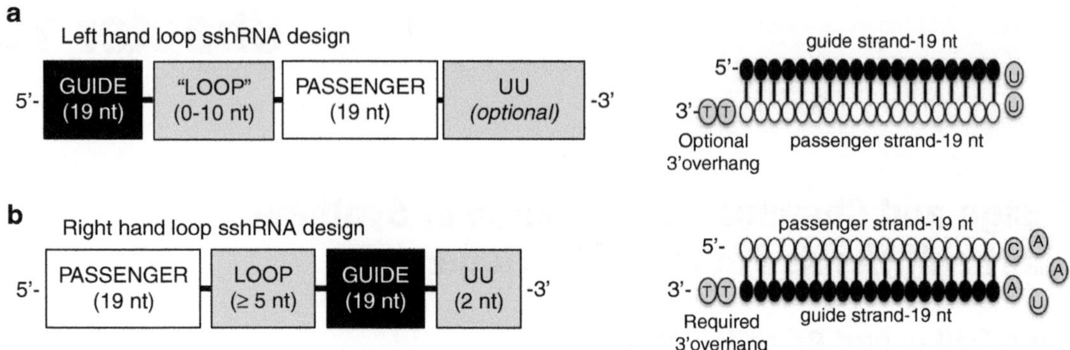

Fig. 1. Schematic diagram of sshRNA design. (**a**) Structural features and putative secondary structures of L-sshRNAs. (**b**) Structural features and putative secondary structures of R-sshRNAs.

having duplex lengths of ≥21 bp can be recognized and processed by Dicer before incorporation into the RNA-induced silencing complex (RISC) (10). In contrast, shRNAs with a duplex of 19 or fewer base pairs are not Dicer substrates (9–11). To distinguish shRNAs with 19 or fewer base pairs from longer, Dicer substrate shRNAs, we have designated the former as short shRNAs or sshRNAs (9). sshRNAs can be designed such that their efficacy may be as good as or, in some cases, better than siRNAs that target the same sequences (9, 10, 12–15). We have shown that sshRNAs can be extremely potent, with IC_{50}s in the low picomolar range, and that they are of interest for development as therapeutic agents (9).

sshRNAs have an intrinsic "handedness" to them because the guide sequence can be positioned either on the 5′ side of the loop (left-hand or L-type) or the 3′ side (right-hand or R-type) (Fig. 1) (16). Both L- and R-sshRNAs can be designed with high activity and specificity, but because their mechanisms differ, so do the structural aspects of their designs (17). L-sshRNAs can be loaded into RISC without any prior processing of their loops. Once loaded, RISC activation is completed by slicing of the passenger arm by Ago2 opposite nt 10–11 of the guide arm, measured from its 5′ phosphate (17). In contrast, potent R-sshRNAs need to have loops that can be cleaved prior to productive formation of an active RISC (17). Upon loop cleavage, a phosphate is produced at the 5′-end of the guide arm that can allow stable binding and accurate positioning of the guide arm in the RISC complex. The passenger arm may be sliced to facilitate its removal and leave the guide strand available for pairing with the target.

Although target site selection is critical to silencing activity, details of the structural design of sshRNAs also play a significant role. Effective target sites can be identified and characterized by a number of methods, including rational design, by the use of one of a number of available algorithms, or by library-based screening methods (18–23). We have studied the effects of design and various chemical modifications on the potency, stability, and immunostimulatory properties of sshRNAs. In this chapter, we describe the principles and guidelines for designing potent sshRNAs once an effective target site has been identified.

2. Materials

1. Commercial source of HPLC-purified synthetic sshRNA (e.g., IDT or ThermoFisher) or in-house equivalent.
2. Sterile, RNAse-free, pyrogen-free ddH_2O.
3. 1× sshRNA resuspension buffer: 20 mM KCl, 6 mM HEPES-KOH (pH 7.5), 0.2 mM $MgCl_2$.
4. An assay to measure the activity of sshRNAs in cultured cells expressing a target gene.
5. Microfuge tubes with low nucleic acid retention such as Eppendorf DNA LoBind tubes.
6. Disposable, sterile siliconized pipet tips.
7. Acrylamide solution (19:1 acryl:bis), TEMED, 10% ammonium persulfate.
8. Urea.
9. Formamide.
10. 10× TBE buffer: 890 mM Tris base, 890 mM boric acid, 20 mM EDTA, pH 8.0.
11. Apparatus for running polyacrylamide gels.
12. SYBR® Gold Nucleic Acid Gel Stain, 10,000× concentrate in DMSO (Life Technologies). Store at −20°C, protected from light. Dilute in 1× TBE for working solution.

3. Methods

3.1. Design of sshRNAs

Once a target sequence has been identified by the user's preferred method, highly potent non-Dicer substrate sshRNAs can be designed in either L or R orientation (Fig. 1). As noted above, the two orientations differ in the positions of the guide and passenger sequences with respect to the connecting loop sequence. The guide sequence of L-sshRNAs is located on the 5′-side of the loop, while the guide sequence of R-sshRNAs is placed to the 3′-side of the loop.

1. Design of L sshRNAs
   (a) Design L-sshRNAs to conform to the following general structure: 5′-19 nt guide sequence—"loop" connector sequence—19 nt passenger sequence—optional dinucleotide overhang-3′ (Fig. 1a).
   (b) L-sshRNAs should be designed such that they contain a 19 bp Watson-Crick base paired stem where the two strands (5′-guide arm and 3′-passenger arm) are joined by a connecting sequence that forms a hairpin loop. In addition, the guide sequence should be perfectly complementary to the target RNA to be silenced. While we have found that high silencing activity can occur with shorter stems (minimum of 16 bp), the reliability of such formats in many sequence contexts has not been explored in depth. By selecting a duplex length of 19 base pairs, the design of guide and passenger sequences is, in principle, similar to the standard design of siRNAs, and allows for molecules that are highly potent in their ability to silence their targets as well as maintaining a gene-specific effect.
   (c) While the connecting sequence that bridges the guide and passenger sequences can be varied from 0-to-10 nucleotides in length, we recommend a "loop" or connecting sequence length of 2 nucleotides with the sequence of UU (see Note 1).
   (d) Optionally include a 3′ overhang whose sequence may be either UU or TT. For some target sequences, we have found that L-sshRNAs have slightly higher activity if the molecules contain a 3′-overhang. It is recommended that both blunt and 3′-overhang-containing sshRNAs be synthesized and compared head to head in the gene knockdown assay of choice. One other point to consider is that blunt-ended hairpins induce expression of pro-inflammatory cytokines, which can lead to observation of nonspecific inhibition of gene expression. This interferon response can be abrogated by inclusion of 2′-*O*-methyl (2′-OMe) modifications at certain residues along the sshRNAs (see Subheading 3.2).
   (e) (Optional depending on application) We recommend that once a target site is selected, a sequence walk be performed around the target site. Several sshRNAs whose target sequences are positioned on either side of the target sequence should be synthesized and assayed in parallel to identify the sshRNAs with the highest activity. This step will be more important for development of sshRNAs to be used as therapeutic agents. Less optimization is required if sshRNAs are to be used as a research tool for inducing gene

knockdown. A scrambled-sequence sshRNA should also be included as a control for nonspecific gene knockdown.

2. Design of R sshRNAs
   (a) Design R-sshRNAs to conform to the following general structure: 5′-19 nt passenger sequence—loop sequence of at least 5 nt—19 nt guide sequence—dinucleotide overhang-3′ (Fig. 1b). The minimum length of R-sshRNAs of this design is 45 nucleotides.
   (b) The guide and passenger sequences should be fully complementary, and the guide sequence should have perfect complementarity to the target RNA to be silenced.
   (c) Many sequences are possible for the loop, but for optimal activity, the loop must have at least 5 nucleotides (15, 17) (see Note 2). Because we have found that R-sshRNAs are more potent if they can be cleaved by an endonuclease in the loop, we recommend including a 5′-Pyr-A-3′ sequence in the loop, which is the recognition motif for ribonuclease-A type endonucleases. In practice, we have found that the sequence 5′-CAAUA-3′ is a good choice for the loop as long as it is not complementary to either the guide or passenger sequence (9, 15). Whatever loop sequence is chosen, it is important to check that it does not have such complementarity, which could lead to misfolding of the hairpin.
   (d) It is essential to include a 3′-dinucleotide overhang for highly potent R-sshRNAs. The sequence can be either UU or dTdT.

3.2. Chemical Modification of sshRNAs

Chemical modification of shRNAs and siRNAs can be beneficial for several reasons: to enhance nuclease stability, to mitigate potential undesirable immune stimulatory effects, to reduce off-target effects, and to aid in conjugation to delivery agents (24–31). One of the factors that govern the overall pharmacokinetics of oligonucleotide-based drugs is their sensitivity to nucleases found in serum. Although dsRNAs are more stable than single-stranded RNAs (ssRNAs), sshRNAs without chemical modification are still relatively sensitive to nucleases. The inclusion of certain chemical modifications in sshRNAs can substantially increase their serum half-lives (16).

Numerous studies have demonstrated the capability of unmodified shRNAs and siRNAs to induce the undesired expression of proinflammatory cytokines such as type I interferon (IFN), IL-6, and TNF-α (28). Although sshRNAs have a duplex length of 19 bp or less, shorter than "ordinary" shRNAs, they may still be immune activators if they contain certain sequences (e.g., GU motifs) or structural features (e.g., blunt ends, which can stimulate RIG-I). Because L- and R-sshRNAs differ in their mechanisms of action and how they are processed in intracellular environments (16, 17), the rules governing chemical modification differ for each class of hairpin RNA.

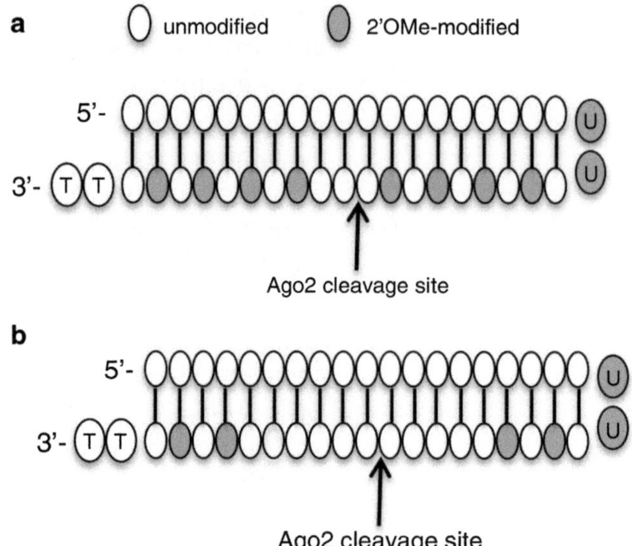

Fig. 2. Effective 2′-OMe-modification patterns for L-sshRNAs. *White ovals*: unmodified residues; *gray-ovals*: residues with 2′-OMe substitutions.

1. Modification of L sshRNAs
   (a) The connecting sequence between the guide and passenger arms may be modified with 2′-OMe groups, deoxy substitutions of the 2′-OH, or PS groups, or it may be completely substituted with nonnucleotide linkers such as C_3C_3 without loss of activity.
   (b) To improve serum stability while abrogating any induction of the innate immune system (see Note 3) without loss of gene silencing activity, the following chemical modification pattern is highly effective (shown schematically in Fig. 2a): place a 2′-OMe on each nucleotide of the loop and alternate nucleotides of the passenger strand, leaving an unmodified window of 3 nt at the slicer site. If this pattern of modification reduces potency of the sshRNAs beyond an acceptable level for the application of interest, we found that reducing the number of modified residues to the pattern shown in Fig. 2b was also highly effective in improving serum stability and eliminating immune response stimulation. As few as two 2′-OMe substitutions can reduce the innate immune response to background levels (16).
   (c) Avoid modification of the guide arm with 2′-OMe moieties (except at the last 2 residues at the 3′-end), as these reduce activity of the L-sshRNA.

(d) Phosphorothioate (PS) modifications at the 5′ and 3′ termini of the stem of L-sshRNAs do not reduce activity but have been found to be immune activating (16) (see Note 4). If PS modifications are desired, they should be combined with 2′-OMe modifications to avoid an immunostimulatory response.

(e) 3′-end conjugation with relatively bulky groups such as groups containing a disulfide linkage (e.g., 3′-S-S-C_6) does not adversely affect the activity of L-sshRNAs, whereas conjugation at the 5′-end does impair activity (e.g., 5′-S-S-C_6). These types of modifications can be useful for conjugation of delivery agents such as peptides, lipid nanoparticles, or antibodies that could enhance cellular uptake of sshRNAs (see Note 5). Again, L-sshRNAs are more permissive of modifications to the passenger strand than the guide strand and require a free 5′-phosphate on the guide strand. Because 5′-conjugation presumably blocks the phosphorylation of 5′-OH ends of synthetic RNAs that normally occurs upon their transfection into the cell, the loss of RNAi activity by this modification is not surprising.

2. Modification of R sshRNAs

(a) Do not modify the loop residues of R-sshRNAs

(b) R-sshRNAs are more tolerant than L-sshRNAs of 2′-OMe in guide strand (e.g., at position 2)

(c) Bulky groups can be conjugated to the 5′-end but not the 3′-end of R-sshRNAs without loss of activity.

3.3. Avoiding Multimeric Forms of shRNAs

Lyophilized synthetic sshRNAs have been found upon hydrating to comprise at least three major species that resolve in native polyacrylamide gels, regardless of whether they are resuspended in ddH_2O or a mildly buffered solution (9). In contrast, under denaturing conditions (12% polyacrylamide gel containing 8 M urea and 20% formamide), only a single band is usually observed (9). The three major bands behave in a manner consistent with their being monomer, dimer, and trimer forms of sshRNAs. We have seen this mix of structures with all the sshRNAs we have procured and purified, irrespective of stem length, loop size, or L vs. R loop orientation. In some cases, even higher-order multimeric complexes were observed. Although it will usually contain an internal loop at the middle (unless the loop is self-complementary), a dimer sshRNA has a duplex length of well over 30 bp and is thus a good candidate for protein kinase R (PKR) recognition (32). Before characterization of the functional activities of these sshRNAs in knockdown assays, the mixed population should be treated, so that it consists solely of monomeric hairpins, using a heating and quick-cooling procedure described as follows.

1. Handle all reagents, pipets, tubes, and other consumables with gloved hands.
2. We recommend the use of low nucleic acid binding microcentrifuge tubes and pipet tips especially when working with low concentrations of sshRNAs.
3. Dissolve sshRNA in either sterile, RNAse-free, pyrogen-free ddH_2O (commercially available from numerous suppliers) or a low ionic strength buffer such as 1× sshRNA resuspension buffer to a final concentration of 100 μM. In practice, we initially dissolve the RNAs to a relatively high concentration, which is more stable for long-term storage, and then make dilutions to working concentration. If resuspending in a different buffer than the one suggested, avoid pH > 8 and the inclusion of divalent cations in millimolar concentrations as these conditions will promote degradation of the RNA in subsequent steps.
4. Dilute to a working concentration if necessary. We typically dilute to 5 μM, but we have confirmed that the following steps can be performed at up to 150 μM concentration.
5. Heat the RNA at 95°C for 4 min.
6. Transfer the RNA immediately to an ice-water bath and let sit for 10–20 min until ready for further use.
7. sshRNAs can be stored at −20°C and can be thawed at room temperature and frozen repeatedly.
8. Confirm that the sshRNAs have been converted to monomeric hairpins by analyzing by both non-denaturing (Fig. 3) (see Note 6) and denaturing PAGE (see Note 7) for monomer formation. Stain with SYBR Gold to visualize RNA bands (see Note 8). In both non-denaturing and denaturing gels, only a single band should be observed. For comparison, load an aliquot of non-heat-treated RNA in an adjacent lane.

4. Notes

1. Dinucleotide UU connecting sequences, even when they are unmodified, dramatically improve resistance to serum nucleases compared to longer sequences (16). In addition, these L-sshRNAs can be loaded efficiently into Ago2-containing RISC complexes directly as intact hairpins (i.e., they are not cleaved by Dicer or some other cellular endonuclease to remove the loop) (17). Instead, the full-length hairpins are sliced in

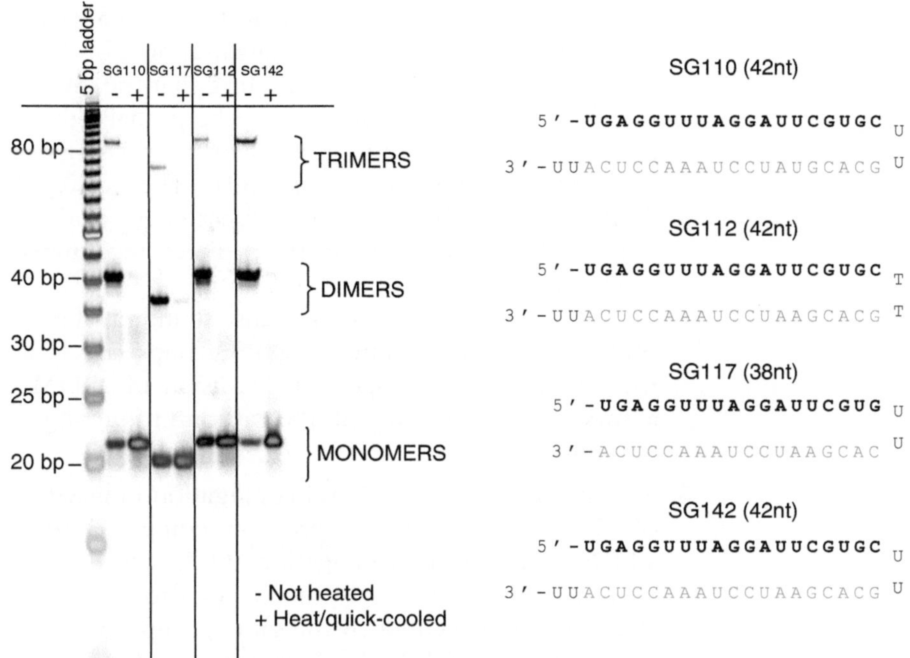

Fig. 3. Non-denaturing PAGE analysis of the conversion of mixed multimeric sshRNA species to monomeric hairpins. 10% non-denaturing PAGE of sshRNAs before (lanes labeled –) and after heating and quick-cooling treatment (lanes labeled +). The gel was stained with SYBR Gold. Sequences and putative secondary structures of sshRNAs analyzed are depicted to the *right* of the gel. Guide strand sequences are in *bold, black text*; passenger strand sequences are in *gray text*.

the passenger strand by Ago2 to generate the active molecule. sshRNAs with 2-nucleotide connectors will also be more cost-effective to synthesize since their overall length is shorter.

2. It is important to keep in mind that although R-sshRNAs may be highly potent, they are rapidly degraded in serum when they are unmodified because of the longer loop lengths required for activity (16).
3. 2′-OMe modification has been found to be particularly effective in preventing the recognition of siRNAs by TLR7/8 and RIG-I (28, 33–35). The modified groups interact with TLR7 without triggering signaling cascades, antagonizing TLR7-mediated activation by both RNA and small-molecule TLR7 agonists (34, 36).
4. The presence of many PS bonds in oligonucleotides can result in cytotoxicity, and the ability of PS oligonucleotides to nonspecifically bind to proteins may explain some if not all of its toxicity (37, 38). The addition of as few as three PS bonds in sshRNAs has caused significant induction of cytokines, particularly TNF. For blunt-ended hairpins, one possibility is that PS linkages may enhance the interaction with RIG-I. However,

some single-stranded oligonucleotides with PS backbone have been shown to inhibit the signaling of TLR3 and RIG-I induced by poly I:C (39), suggesting that the immune activation by PS-modified dsRNAs may be partially due to RIG-I having different recognition mechanisms for single- and double-stranded oligonucleotides (39). The up-regulation of TLR3 and RIG-I may be an indirect effect of the release of inflammatory cytokines or their direct recognition of the modification.

2′-OMe substitution in the sense strand (at alternate nucleotides) of sshRNAs containing PS groups largely eliminated the innate immune responses. Inclusion of 2′-OMe residues allows the substitution of PS without triggering an innate immune response.

5. The introduction of a 3′ end conjugation induced high levels of IFN-β and TNF-α expression compared to the same sshRNAs lacking the conjugation. TLR3 and RIG-I were also up-regulated. Interestingly, the up-regulation of cytokines and RIG-I was not seen when the same group was conjugated to the 5′ end of L-sshRNAs. It is not clear how the activity of RIG-I was affected differently by a disulfide group conjugated to the 3′ or 5′ ends of a blunt-ended RNA hairpin.

6. For sshRNAs with a typical length of 38–45 nucleotides, we recommend 10% acrylamide (19:1, acryl:bis), 1× TBE for non-denaturing gel electrophoresis, which will resolve monomers from dimers and higher-order multimeric species. Any commercially available native gel loading buffer can be used to prepare the samples, but we suggest avoiding the use of loading buffer containing bromophenol blue tracking dye as the dye may comigrate with RNA species of interest and can interfere with visualization of bands if imaging with a phosphorimager. Do not heat samples prior to loading on the gel. For sufficient sensitivity without overloading the gel, load approximately 100 ng per lane.

7. For sshRNAs with a typical length of 38–45 nucleotides, we recommend the following PAGE conditions for denaturing gel electrophoresis: 12% acrylamide (19:1, acryl:bis), 20% formamide, 8 M urea, 1× TBE. Samples should be diluted 1:1 with 2× Loading buffer (95% formamide, 50 mM EDTA, 0.015% xylene cyanol, and 0.015% bromophenol blue) and heated to 95°C for 2 min prior to loading on the gel. Because of the high degree of secondary structure in hairpin RNAs, it is essential to include formamide in the gel to provide complete denaturation. Run the gel at 45 W.

8. Do not substitute ethidium bromide stain for SYBR gold for visualization of small, denatured RNAs. Ethidium bromide

binds preferentially to double-stranded structures and does not accurately reflect the relative population of single-stranded RNA oligonucleotides. Because of the highly structured nature of hairpin RNAs, also do not rely on the intensity of 5′-end-labeling with a radioisotope such as ^{32}P to assess relative amounts of sshRNAs. The efficiency of 5′-end-labeling is very sensitive to the availability of the 5′-terminal residue, which, in the case of sshRNAs, may be either recessed if a 3′-overhang is included in the design, or blunt-ended. Consequently, even minor degradation products with comparatively available termini will be labeled with much higher efficiency than full-length products and will not accurately reflect the relative population of RNA species present in the sample.

References

1. Dorsett Y, Tuschl T (2004) siRNAs: applications in functional genomics and potential as therapeutics. Nat Rev Drug Discov 3:318–329
2. Chang K, Elledge SJ, Hannon GJ (2006) Lessons from nature: microRNA-based shRNA libraries. Nat Methods 3:707–714
3. Bernards R, Brummelkamp TR, Beijersbergen RL (2006) shRNA libraries and their use in cancer genetics. Nat Methods 3:701–706
4. Fewell GD, Schmitt K (2006) Vector-based RNAi approaches for stable, inducible and genome-wide screens. Drug Discov Today 11:975–982
5. Amarzguioui M, Lundberg P, Cantin E, Hagstrom J, Behlke MA, Rossi JJ (2006) Rational design and in vitro and in vivo delivery of Dicer substrate siRNA. Nat Protoc 1:508–517
6. Xia H, Mao Q, Eliason SL, Harper SQ, Martins IH, Orr HT, Paulson HL, Yang L, Kotin RM, Davidson BL (2004) RNAi suppresses polyglutamine-induced neurodegeneration in a model of spinocerebellar ataxia. Nat Med 10:816–820
7. Elbashir SM, Harborth J, Lendeckel W, Yalcin A, Weber K, Tuschl T (2001) Duplexes of 21-nucleotide RNAs mediate RNA interference in cultured mammalian cells. Nature 411:494–498
8. Wang Q, Contag CH, Ilves H, Johnston BH, Kaspar RL (2005) Small hairpin rnas efficiently inhibit hepatitis C IRES-mediated gene expression in human tissue culture cells and a mouse model. Mol Ther 12:562–568
9. Ge Q, Ilves H, Dallas A, Kumar P, Shorenstein J, Kazakov SA, Johnston BH (2010) Minimal-length short hairpin RNAs: the relationship of structure and RNAi activity. RNA 16: 106–117
10. Siolas D, Lerner C, Burchard J, Ge W, Linsley PS, Paddison PJ, Hannon GJ, Cleary MA (2005) Synthetic shRNAs as potent RNAi triggers. Nat Biotechnol 23:227–231
11. McManus MT, Petersen CP, Haines BB, Chen J, Sharp PA (2002) Gene silencing using micro-RNA designed hairpins. RNA 8:842–850
12. Harborth J, Elbashir SM, Vandenburgh K, Manninga H, Scaringe SA, Weber K, Tuschl T (2003) Sequence, chemical, and structural variation of small interfering RNAs and short hairpin RNAs and the effect on mammalian gene silencing. Antisense Nucleic Acid Drug Dev 13:83–105
13. McManus MT, Haines BB, Dillon CP, Whitehurst CE, van Parijs L, Chen J, Sharp PA (2002) Small interfering RNA-mediated gene silencing in T lymphocytes. J Immunol 169:5754–5760
14. Daly C, Coyle S, McBride S, O'Driscoll L, Daly N, Scanlon K, Clynes M (1996) mdrl ribozyme mediated reversal of the multi-drug resistant phenotype in human lung cell lines. Cytotechnology 19:199–205
15. Vlassov AV, Korba B, Farrar K, Mukerjee S, Seyhan AA, Ilves H, Kaspar RL, Leake D, Kazakov SA, Johnston BH (2007) shRNAs targeting hepatitis C: effects of sequence and structural features, and comparision with siRNA. Oligonucleotides 17:223–236
16. Ge Q, Dallas A, Ilves H, Shorenstein J, Behlke MA, Johnston BH (2010) Effects of chemical modification on the potency, serum stability, and immunostimulatory properties of short shRNAs. RNA 16:118–130

17. Dallas A, Ilves H, Ge Q, Kumar P, Shorenstein J, Kazakov SA, Cuellar TL, McManus MT, Behlke MA, Johnston BH (2012) Right- and left-loop short shRNAs have distinct and unusual mechanisms of gene silencing. Nucleic Acids Res. [Epub ahead of print]. doi:10.1093/nar/gks662.
18. Reynolds A, Leake D, Boese Q, Scaringe S, Marshall WS, Khvorova A (2004) Rational siRNA design for RNA interference. Nat Biotechnol 22:326–330
19. Seyhan AA, Vlassov AV, Ilves H, Egry L, Kaspar RL, Kazakov SA, Johnston BH (2005) Complete, gene-specific siRNA libraries: production and expression in mammalian cells. RNA 11:837–846
20. Shirane D, Sugao K, Namiki S, Tanabe M, Iino M, Hirose K (2004) Enzymatic production of RNAi libraries from cDNAs. Nat Genet 36:190–196
21. Yang D, Buchholz F, Huang Z, Goga A, Chen CY, Brodsky FM, Bishop JM (2002) Short RNA duplexes produced by hydrolysis with Escherichia coli RNase III mediate effective RNA interference in mammalian cells. Proc Natl Acad Sci USA 99:9942–9947
22. Amarzguioui M, Rossi JJ (2008) Principles of Dicer substrate (D-siRNA) design and function. Methods Mol Biol 442:3–10
23. Pei Y, Tuschl T (2006) On the art of identifying effective and specific siRNAs. Nat Methods 3:670–676
24. Behlke MA (2008) Chemical modification of siRNAs for in vivo use. Oligonucleotides 18:305–319
25. Watts JK, Deleavey GF, Damha MJ (2008) Chemically modified siRNA: tools and applications. Drug Discov Today 13:842–855
26. Judge A, MacLachlan I (2008) Overcoming the innate immune response to small interfering RNA. Hum Gene Ther 19:111–124
27. Morrissey DV, Lockridge JA, Shaw L, Blanchard K, Jensen K, Breen W, Hartsough K, Machemer L, Radka S, Jadhav V, Vaish N, Zinnen S, Vargeese C, Bowman K, Shaffer CS, Jeffs LB, Judge A, MacLachlan I, Polisky B (2005) Potent and persistent in vivo anti-HBV activity of chemically modified siRNAs. Nat Biotechnol 23:1002–1007
28. Robbins M, Judge A, MacLachlan I (2009) siRNA and innate immunity. Oligonucleotides 19:89–102
29. Jackson AL, Burchard J, Leake D, Reynolds A, Schelter J, Guo J, Johnson JM, Lim L, Karpilow J, Nichols K, Marshall W, Khvorova A, Linsley PS (2006) Position-specific chemical modification of siRNAs reduces "off-target" transcript silencing. RNA 12:1197–1205
30. Debart F, Abes S, Deglane G, Moulton HM, Clair P, Gait MJ, Vasseur JJ, Lebleu B (2007) Chemical modifications to improve the cellular uptake of oligonucleotides. Curr Top Med Chem 7:727–737
31. Bumcrot D, Manoharan M, Koteliansky V, Sah DW (2006) RNAi therapeutics: a potential new class of pharmaceutical drugs. Nat Chem Biol 2:711–719
32. Manche L, Green SR, Schmedt C, Mathews MB (1992) Interactions between double-stranded RNA regulators and the protein kinase DAI. Mol Cell Biol 12:5238–5248
33. Hornung V, Ellegast J, Kim S, Brzozka K, Jung A, Kato H, Poeck H, Akira S, Conzelmann KK, Schlee M, Endres S, Hartmann G (2006) 5′-Triphosphate RNA is the ligand for RIG-I. Science 314:994–997
34. Robbins M, Judge A, Liang L, McClintock K, Yaworski E, MacLachlan I (2007) 2′-O-methyl-modified RNAs act as TLR7 antagonists. Mol Ther 15:1663–1669
35. Judge AD, Bola G, Lee AC, MacLachlan I (2006) Design of noninflammatory synthetic siRNA mediating potent gene silencing in vivo. Mol Ther 13:494–505
36. Cekaite L, Furset G, Hovig E, Sioud M (2007) Gene expression analysis in blood cells in response to unmodified and 2′-modified siRNAs reveals TLR-dependent and independent effects. J Mol Biol 365:90–108
37. Levin AA (1999) A review of the issues in the pharmacokinetics and toxicology of phosphorothioate antisense oligonucleotides. Biochim Biophys Acta 1489:69–84
38. Kurreck J (2003) Antisense technologies. Improvement through novel chemical modifications. Eur J Biochem 270:1628–1644
39. Ranjith-Kumar CT, Murali A, Dong W, Srisathiyanarayanan D, Vaughan R, Ortiz-Alacantara J, Bhardwaj K, Li X, Li P, Kao CC (2009) Agonist and antagonist recognition by RIG-I, a cytoplasmic innate immunity receptor. J Biol Chem 284:1155–1165

Chapter 16

Production and Application of Long dsRNA in Mammalian Cells

Katerina Chalupnikova*, Jana Nejepinska*, and Petr Svoboda

Abstract

Double-stranded RNA (dsRNA) is involved in different biological processes. At least three different pathways can respond to dsRNA in mammals. One of these pathways is RNA interference (RNAi) where long dsRNA induces sequence-specific degradation of transcripts carrying sequences complementary to dsRNA. Long dsRNA is also a potent trigger of the interferon pathway, a sequence-independent response that leads to global suppression of translation and global RNA degradation. In addition, dsRNA can be edited by adenosine deamination, which may result in nuclear retention and degradation of dsRNA or in alteration of RNA coding potential. Here, we provide a technical review summarizing different strategies of long dsRNA usage. While the review is largely focused on long dsRNA-induced RNAi in mammalian cells, it also provides helpful information on both the in vitro production and in vivo expression of dsRNAs. We present an overview of currently available vectors for dsRNA expression and provide the latest update on oocyte-specific transgenic RNAi approaches.

Key words: dsRNA (double-stranded RNA), RNAi (RNA interference), Transgenic RNAi, IFN response, OAS (2′5′-oligoadenylate synthetase), siRNA (short interfering RNA), shRNA (short hairpin RNA), Mammalian somatic cells, Oocytes and embryos, IVT (in vitro transcription)

1. Introduction

Long dsRNA is a unique structure that plays various roles in different organisms, from antiviral defense to regulation of gene expression. Three pathways activated by long dsRNA are well characterized (Fig. 1): sequence-specific gene silencing by RNA interference (RNAi), sequence-independent silencing associated with the interferon (IFN) response, and RNA editing, which results either in alteration of the substrate specificity (hypoediting) of short interfering RNAs (siRNAs) or in nuclear retention of dsRNA making it

* Katerina Chalupnikova and Jana Nejepinska contributed equally to this article.

Debra J. Taxman (ed.), *siRNA Design: Methods and Protocols*, Methods in Molecular Biology, vol. 942,
DOI 10.1007/978-1-62703-119-6_16, © Springer Science+Business Media, LLC 2013

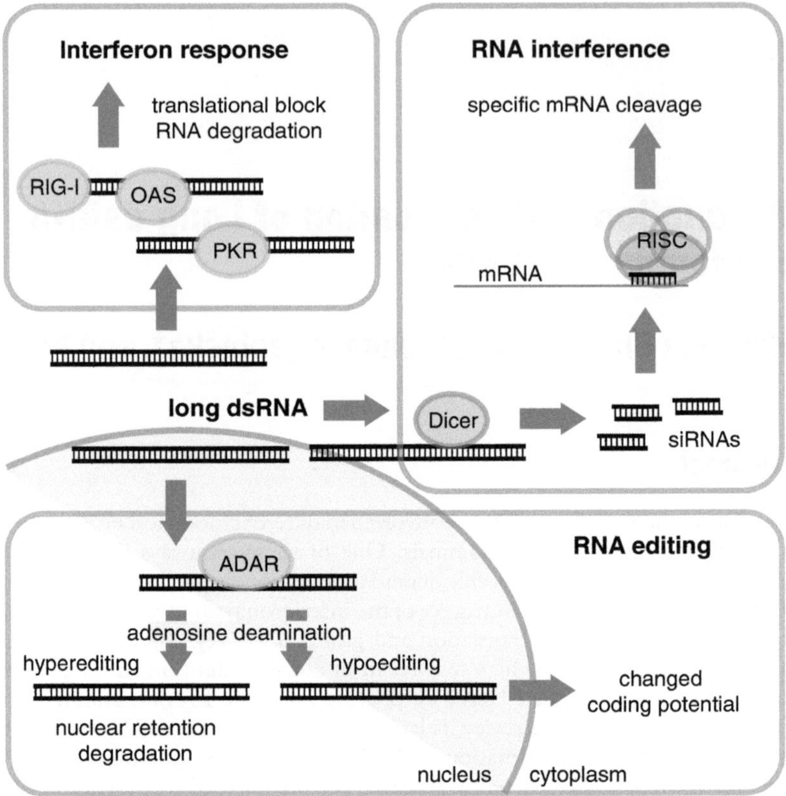

Fig. 1. Pathways activated by long dsRNA. In the RNAi pathway (*upper right frame*), long dsRNA is cleaved by Dicer into siRNAs. One of the strands is incorporated into the RNA-induced silencing complex (RISC) and acts as a guide for binding and cleavage of complementary mRNAs. In the interferon response (*upper left frame*), long dsRNA is recognized by dsRNA-binding proteins (PKR, OAS, RIG-I) and initiates global inhibition of proteosynthesis and RNA degradation. Long dsRNA can also undergo RNA editing (*bottom frame*) by ADARs. Hyperedited dsRNA is degraded in the nucleus, hypoedited RNA is exported into the cytoplasm and can cause alterations during translation.

unavailable for the cytoplasmic silencing machinery (hyperediting). Although all pathways share a common substrate-long dsRNA, they differ in their evolutionary conservation, cellular localization, and outcome. Therefore, the effect of specific long dsRNA depends on the combination of different factors.

The most conserved pathway is RNAi, which is present in almost all eukaryotes. Cytoplasmic long dsRNA is processed by the RNase-III enzyme Dicer into 21-nucleotide (nt)-long siRNAs with two nt 3′ overhangs (reviewed in ref. 1). One of the siRNA strands is incorporated into an RNA-induced silencing complex (RISC) and acts as a guide for binding to the complementary messenger RNA (mRNA). The cognate mRNA is subsequently cleaved in the middle of the basepairing region by a "slicer" (Ago2 protein) and rapidly degraded (reviewed in ref. 2).

Since RNAi discovery (3), dsRNA has been used as an experimental tool to selectively silence gene expression. Specific mRNA degradation mediated by an uptake of exogenous dsRNA is a common way of suppressing gene expression in nematodes (*Caenorhabditis elegans*) (3) and insects (*Drosophila*) (4–7), where RNAi serves as a form of antiviral response (reviewed in ref. 8). However, vertebrates evolved an independent complex response to viruses that includes global inhibition of proteosynthesis, expression of antiviral genes, RNA degradation, and apoptosis. This complex of pathways is known as the "IFN response" and it is activated by long dsRNA (which often serves as a marker of viral infection) (reviewed in ref. 9).

Cytoplasmic dsRNA-binding proteins, such as protein kinase R (PKR) and 2′5′-oligoadenylate synthetase (OAS), are the backbone of the IFN response. Activated PKR phosphorylates the alpha subunit of eukaryotic initiation factor 2 (eIF2α), resulting in a global inhibition of translation (10). The second part of the IFN response is represented by OAS/RNase L system. Upon binding of dsRNA to OAS, this enzyme is activated and converts adenosine triphosphate (ATP) to pyrophosphate and 2′5′-linked oligoadenylates (11). Oligoadenylates activate endoribonuclease RNase L, which nonspecifically degrades viral and cellular RNA (including mRNAs); thereby resulting in decreased proteosynthesis (12). Moreover, there are additional players involved in the activation or regulation of the IFN response (reviewed in ref. 9). The circumstances under which these molecules are activated depend on various factors, such as cell type, dsRNA origin, localization, length, structure, or termini (9, 13–18). This sophisticated system is essential to discriminate foreign dsRNAs from endogenous RNAs with intra or intermolecular complementarity in order to ignore harmless dsRNAs or initiate a deadly IFN response. The antiviral role of RNAi has been replaced by the IFN response in most, but not all, vertebrate cell types. In oocytes, where the IFN response is not functional (14), RNAi plays an important role by targeting repetitive elements and endogenous genes (reviewed in ref. 19).

In addition to RNAi and the IFN response, dsRNA can be modified by adenosine deaminases acting on RNA (ADARs), nuclear and cytoplasmic enzymes found in Metazoa (20, 21). The conversion of adenosines into inosines in dsRNA substrates can have a number of consequences. Hyperediting of dsRNA results in nuclear retention of the dsRNA and plays an important role in the defense against viruses producing dsRNA (22). In contrast, dsRNAs with a low level of editing are exported into the cytoplasm, but modifications in individual codons may affect splicing or cause alterations in translation. Specific modifications by ADARs increase diversity in protein sequences, which is essential for substrate specificity of neuronal channels (reviewed in ref. 23).

In organisms where more dsRNA-induced pathways are functional, individual pathways compete for the common substrate and interfere with each other. In the case of crosstalk between RNA editing and RNAi, ADAR can modify a dsRNA sequence, inducing a change in siRNA specificity and subsequent downregulation of different targets. Alternatively, edited siRNA is not perfectly complementary to its original target, which elicits a translational repression rather than a cleavage of the target mRNA. In the case of hyperediting, long dsRNA acquires excessive I–U pairs that make the duplex unstable; subsequently, edited dsRNA is degraded in the nucleus and unavailable for cytoplasmic RNAi machinery (24).

Opposing roles are played by RNAi and the IFN response. Although both pathways probably evolved as host defense mechanisms, RNAi induces silencing only of the sequences homologous to the original RNAi trigger, while the IFN response induces global RNA degradation and inhibition of translation. The interactions between these two pathways are still not well understood. In fact, while some groups described RNAi effects in mammals, others claim that RNAi is inefficient and masked by nonspecific inhibition of gene expression.

2. Long dsRNA Application in Mammals

In mammals, where dsRNA is part of three different pathways, diverse experimental setups are used to study various effects of dsRNA. Here, we focus primarily on the link between long dsRNA and the RNAi pathway. Sequence-specific mRNA degradation induced by exogenous dsRNA is a common strategy to suppress gene expression in invertebrates and plants. In mammals, however, RNAi induction by long dsRNA is rather uncommon (reviewed in ref. 25). The endogenous RNAi was clearly documented in mouse oocytes and embryonic stem cells (ESCs), where deep sequencing identified endogenous siRNAs (endo-siRNAs) produced from retrotransposons and processed pseudogenes (26–28). In these cell types, dsRNA is preferentially routed to the RNAi pathway while the IFN response is not readily induced (14). In addition, deep sequencing of hippocampal RNA revealed that small RNAs having features of endogenous siRNAs are expressed in the brain and possibly regulate synaptic plasticity of neuronal cells (29). In other cell types, endogenous siRNAs are either not detected or their levels are too low to be physiologically significant. However, experimental induction of RNAi with long dsRNA in different somatic cells indicates that the canonical RNAi pathway is latent or masked rather than absent in these cells.

3. Preparation of Long dsRNA for RNAi Experiments

The great advantage of using long dsRNA to induce RNAi is the simplicity, which can be appreciated only in permissive model systems. When siRNA and short hairpin RNA (shRNA) are used, their design and use have to follow several essential rules and experimental designs have to take off-target effects into consideration (30–32). In mammalian cells, off-target effects stem from the ability of siRNAs to suppress many other genes via their binding to mRNAs with partial complementarity (typically involving a perfect basepairing at the 5′ end of an siRNA). When siRNA is transfected, it causes detectable off-target effects because of the high abundance of siRNA molecules carrying the same 5′ end sequence, which can basepair with and repress a discrete set of cellular mRNAs (33). On the other hand, long dsRNA causes minimal off-targeting because it is processed into a pool of siRNAs with varying sequences, which binds to different mRNA targets. Thus, this effectively dilutes off-target effects. One should pay attention to select a nonredundant target sequence that would not show sequence identity in stretches longer than 20 nt with other transcripts. This can be easily resolved with a BLAST sequence search.

The position of dsRNA within the mRNA sequence is flexible. Often, dsRNAs derived from the coding sequence of the cognate transcript are used (15, 34–43). However, the 3′ untranslated region (3′UTR) of the transcript can be targeted by dsRNA as well (38). Using dsRNA against the 3′UTR, one can perform a rescue experiment by expressing the coding sequence of the cognate transcript without the 3′UTR or with another nontargeted 3′UTR. Notably, when designing long dsRNA against the 3′UTR, one should also test (e.g., by analyzing available ESTs) whether the dsRNA would target all possible alternative 3′UTRs.

While the relative position of dsRNA in the cognate transcript does not play a crucial role, dsRNA length may influence efficient silencing. dsRNA shorter than 500 base pair (bp) may not be effectively processed by Dicer as suggested by in vitro experiments (44). Nevertheless, it should also be noted that efficient RNAi was induced with 290 bp-long dsRNA (41). It is not clear if there is an upper limit for dsRNA length. However, for practical reasons, dsRNA length typically ranges from 500 to 800 bp and rarely exceeds 1,000 bp. If one cannot use a single unique sequence, two shorter sequences can be fused together. A continuous region of the cognate transcript is not critical for siRNA target recognition.

RNAi silencing can be triggered either by transfection of in vitro transcribed long dsRNA or by long dsRNA expression from vectors transfected in the cell. The next two sections summarize practical considerations for both strategies.

3.1. dsRNA Produced by In Vitro Transcription

A common in vitro dsRNA preparation involves in vitro transcription (IVT) of sense and antisense RNAs, which are annealed and purified. Thus, dsRNA preparation is usually accomplished in <1 week. However, the exact time depends on the availability of suitable sequences, primers, and other reagents. The starting step of dsRNA production, IVT, can be easily done in 1 day. Templates for IVT can either be directly produced from a target cDNA by PCR using primers carrying SP6, T3, or T7 promoter sequences at their 5′ ends or can be cloned into an appropriate vector. Before IVT, a template must be purified to remove primers and nucleotides. PCR-based preparation of templates for IVT is simple and fast, but when a large number of different templates needs to be prepared, necessary primers may be costly. Moreover, we have observed that RNA synthesis from different PCR-generated templates is more variable than IVT from a plasmid (45). Therefore, we recommend preparing a template for IVT by cloning the PCR product into an appropriate vector. We routinely amplify the desired sequence by reverse transcriptase PCR (RT-PCR) and insert it using the TOPO-TA Cloning Kit (Invitrogen) into the pCRII plasmid (Invitrogen), which carries T7 and SP6 promoters flanking a multiple cloning site. Alternatively, dsRNA can be prepared directly using a template carrying an inverted repeat (IR) of the target sequence (15, 46). However, it should be noted that cloning of IR may not be a trivial task (this issue is discussed in the next section).

A plasmid that is used for IVT needs to be linearized downstream of the transcribed sequence and purified (either by a standard phenol/chloroform extraction or using a DNA purification kit). The final DNA template concentration should be above 500 ng/μl, and 100–500 ng of the template DNA is generally required per IVT reaction. IVT protocols have been already described in details elsewhere, including standard laboratory protocol manuals, RNA polymerase manufacturers' manuals, or protocols available on the Internet. Some laboratories have their own protocols for IVT while others rely on IVT kits (e.g., MEGAscript kit from Ambion), which contain all necessary reagents and are optimized for a high yield. After IVT, which takes 2–4 h at 37°C, the template is removed by DNase treatment (e.g., by RQ1 DNase (Promega)). Then, one can purify sense and antisense strands, roughly estimate RNA yield by agarose electrophoresis and proceed to annealing. For annealing, approximately equimolar ratio of sense and antisense strands is mixed in up to 200 μl (the concentration of RNA should range between 200 and 500 ng/ml). We have successfully performed annealing of RNA dissolved in MilliQ pure water or in IVT reaction buffer. For efficient annealing, it is important to cool down the reaction slowly. Annealing can be performed in a boiling water bath, in a dry heatblock, or in a PCR machine. For the boiling water method, the 1.5 ml tube must be properly closed (we use safe-lock tubes) to prevent accidental opening when pressure builds inside the tube upon heating. The tube with the annealing mixture is placed into a large beaker (1 liter) with boiling water for 5 min. The beaker is then removed from the heat source and it is left to cool down slowly to room temperature. For even a slower cooling, one can just turn off the heating plate. Annealing in the heatblock is essentially the same: the annealing mixture is incubated for 5 min at 95°C before turning off the heatblock and letting the mixture cool down to room temperature. Alternatively, RNA can be annealed in a PCR machine programmed to cool down from 95 to 25°C for 30–60 min. When dsRNA is annealed from sense and antisense RNA strands, one can either remove the excess of single-stranded RNA and single-stranded overhangs by RNase T1 treatment prior to the purification step or proceed directly to purification. Annealed dsRNA is typically purified by phenol/chloroform extraction and dissolved in pure sterile water (do not use DEPC-treated water if dsRNA will be microinjected into oocytes). Produced dsRNA is fairly stable and can be stored at –20°C for a week or at –80°C for months. In vitro transcribed dsRNA can be microinjected into cells or transfected using Lipofectamine or Oligofectamine reagents.

3.1.1. dsRNA Produced from a Vector or a Transgene

There are several different strategies to express long dsRNA (Fig. 2). Sense and antisense RNA can be expressed separately, either from two different vectors (Fig. 2a), from one vector with two separately transcribed regions for the sense and antisense strands (Fig. 2b), or from one vector with one bidirectionally transcribed region (Fig. 2c). Alternatively, one can generate dsRNA as an intramolecular duplex—an RNA hairpin, which is formed upon transcription of an inserted IR (Fig. 2d). Even if plasmids expressing sense and antisense RNAs separately are easier to produce than plasmids with IR, hybridization of separately expressed sense and antisense RNAs in the cytoplasm may be inefficient. Moreover, the dose of dsRNA cannot be controlled. In contrast, IR forms dsRNA with a high probability and the amount of dsRNA is proportional to the strength of the promoter. Thus, the RNA hairpin expression from IR is a better choice than using the dual promoter plasmid. It should be noted that a dual promoter system (Fig. 2b) worked well in NIH 3T3 and HEK 293 cells (41). However, no systematic comparison of dual promoter and IR systems in mammals are currently available.

Expression of dsRNA can be controlled by various promoters. Polymerase II (pol II) promoters allow for tissue-specific expression or ubiquitous expression with diverse strength. When pol II transcribes IR, the product is typically spliced, capped, and polyadenylated. Such noncoding hairpin RNA is synthesized in the nucleus and presumably transported to the cytoplasm. On the other hand, polymerase III (pol III) induces nuclear production

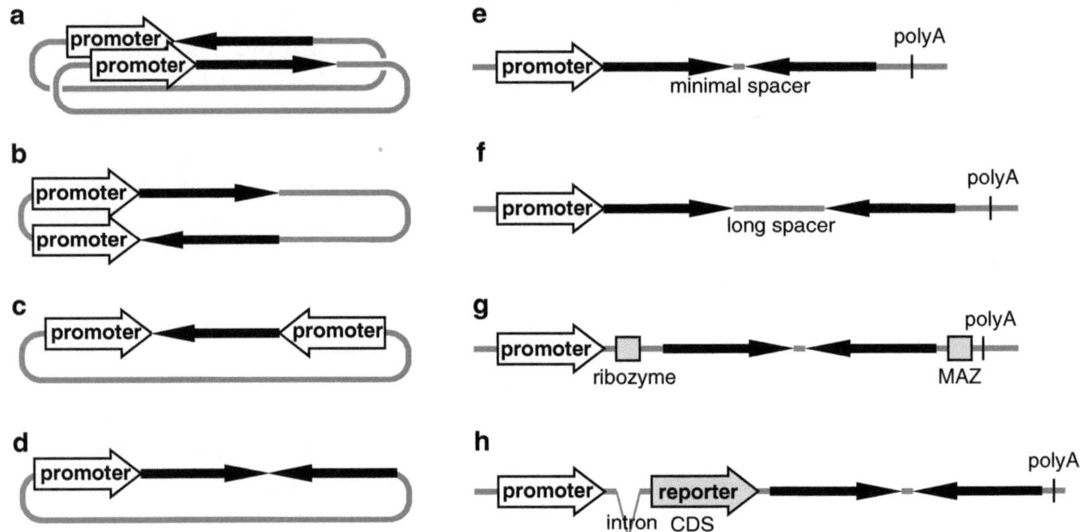

Fig. 2. General plasmid design for long dsRNA expression. (**a–d**) Long dsRNA can be produced by sense and antisense expression originating: (**a**) from two different plasmids, (**b**) from a single plasmid expressing independently sense and antisense RNA, (**c**) from a single plasmid where two promoters transcribe the same sequence in opposite directions, or (**d**) by transcription of an inverted repeat (IR). (**e–h**) Published hairpin-based expression systems include: (**e**) plasmids consisting of a promoter, an IR with a short spacer, and a polyA signal, (**f**) plasmids consisting of a promoter, an IR with a long spacer, and a polyA signal, (**g**) pDECAP derivates that produce decapped and non-polyadenylated dsRNA because of a presence of a ribozyme and MAZ attenuation site, (**h**) plasmids producing a spliced translatable mRNA carrying a long dsRNA stem in their 3′UTRs.

of non-spliced, non-capped, and non-polyadenylated RNA. Because pol III transcription prematurely terminates at TTTT, target long dsRNA sequence should be designed accordingly. Finally, it is possible to control dsRNA expression by ectopically expressed phage polymerase (15, 46). In this case, dsRNA synthesis can be targeted to the cytoplasm or the nucleus depending on the presence or absence of the nuclear localization signal in the phage polymerase.

Most of the published RNA hairpin expressing vectors fall into one of the four categories (Fig. 2e–h). The simplest two vector types are composed of the promoter, IR, and a polyA signal (Fig. 2e, f). The IR has either a minimal (tens of nucleotides) or a long (hundreds to kilobases) spacer and is followed by a polyA site (39, 46). The plasmid introduced by Shinagawa and Ishii produces decapped and non-polyadenylated dsRNA (Fig. 2g) that should be retained in the nucleus (40). The vector also does not contain an intron because splicing generally enhances the nuclear export of RNA (47). A vector expressing a translatable mRNA carrying a long dsRNA stem in its 3′UTR (Fig. 2h) was developed for transgenic RNAi experiments in mouse oocytes where it was used on numerous occasions (Table 3) (48).

3.2. Cloning of the Inverted Repeat

Several strategies for IR cloning are available (Fig. 3). One strategy uses PCR products with appropriate restriction sites that are ligated in vitro and then inserted into a plasmid in opposite directions forming the IR (Fig. 3a). It is important to avoid cryptic polyadenylation sites in PCR products in any direction, especially when a pol II promoter is used for expression. One of the PCR products can be longer at the end to form the center of the IR (Fig. 3a); this additional sequence will create the spacer, which enhances the efficiency of IR cloning. Based on our experience, the short (20–50 bp long) spacer is sufficient for successful cloning without any negative impact on RNAi efficiency. The spacer can be perhaps even longer, as successful RNAi was reported with a spacer as long as 700 bp (37). Another strategy uses sequential insertion of IR arms into the plasmid (Fig. 3b). One can also clone an IR with a long spacer first (1–2 kb), and then remove the spacer (Fig. 3c). Finally, one can clone a head to tail tandem array of two fragments, where one is flanked with loxP sites allowing for inverting it by Cre recombinase (37). This strategy (Fig. 3d) helps to minimize difficulties with IR insertion into the plasmid, which may become a frustrating task as transformation of ligated fragments frequently results in a large number of false positives. This is presumably because the presence of IR may interfere with plasmid's replication, thus creating a strong selection against the IR.

4. Effects of Long dsRNAs in Somatic Cells and Their Analysis

Since the 1970s, it has been known that long dsRNA is a potent trigger of the sequence-independent IFN response (49). Therefore, there was a strong skepticism about the existence of RNAi in mammals when the pathway was first identified in *C. elegans* and *Drosophila*. Mammalian RNAi was first demonstrated by microinjecting dsRNA into oocytes (46, 50, 51), which do not possess the IFN response. The problem of the IFN response was eventually minimized by introducing siRNAs and their derivatives (differently modified siRNAs) directly into somatic cells (34). Thus, siRNAs and shRNAs became the main RNAi tools in mammalian cells. It is generally accepted that in mammalian somatic cell types, long dsRNA readily triggers the IFN response causing sequence-independent silencing effects that mask a potential impact of RNAi-induced downregulation on cognate mRNAs (9, 37, 43, 52). The IFN pathway appears to become fully functional during differentiation as suggested by experiments in ESCs. While the specific RNAi effect was masked by sequence-independent silencing in differentiated ESCs, long dsRNA-induced RNAi was present in undifferentiated ESCs. This observation was interpreted as a manifestation of IFN effects although the activation of the IFN pathway was not studied

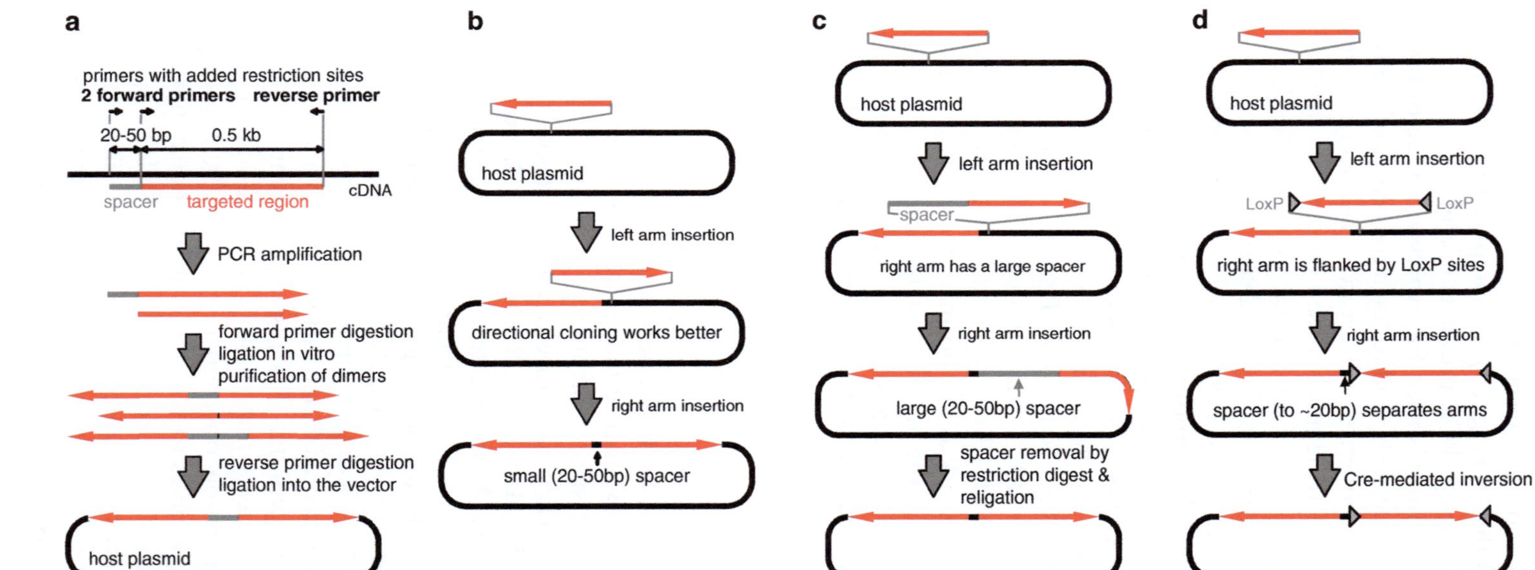

Fig. 3. Strategies for cloning an inverted repeat (IR) into a plasmid. (**a**) In vitro assembly of an IR, which is then inserted into a plasmid in a single cloning step. PCR products with appropriate restriction sites are digested at one end, ligated in vitro, digested at the other end, and the IR is then inserted into a plasmid. (**b**) A sequential insertion of arms (*red arrow*) of the IR. Arms can be prepared as PCR products carrying suitable restriction sites allowing for oriented insertion into the plasmid. (**c**) Spacer-in-and-out sequential insertion of IR arms into a plasmid. A large spacer is created either in an insert as shown here or by choosing an insertion site distant from the insertion of the first arm. In the final step, the spacer is removed and the plasmid is self-ligated. This strategy essentially eliminates all head-to-tail tandem insertions, which may cause problems when strategies **a** or **b** are used. (**d**) Production of an IR by Cre-mediated recombination. This approach also helps to minimize the previously mentioned difficulties sometimes experienced with strategies **a** and **b.**

in detail (15). A recent study of ubiquitous dsRNA expression in transgenic mice suggested a more complex routing of dsRNA into dsRNA-responding pathways in mouse somatic cells. It was shown that expression of long dsRNA embedded in a translatable mRNA (Fig. 2h) was a poor trigger of the IFN pathway; at the same time, it was also poorly processed into siRNAs (53). This suggests that somatic cells recognize different types of long dsRNA. In the following text, we will discuss contradictory results concerning long dsRNA-induced sequence-specific mRNA degradation in mammalian somatic cells.

4.1. dsRNA Transfection

Even if long dsRNA is of limited use for RNAi in mammalian cells, experiments employing long dsRNA are still a relevant tool to unravel how animal cells respond to foreign dsRNA. The ability of in vitro produced dsRNA to induce RNAi was addressed by several studies (Table 1). An initial study of long dsRNA as a potential RNAi trigger in mammalian somatic cells showed that, unlike siRNAs, long dsRNAs strongly and nonspecifically reduced reporter gene expression in NIH 3T3, COS-7, and HeLa S3 cells (34). The long dsRNAs, which ranged from 50 to 500 bp in length, were in vitro transcribed using T7 polymerase, annealed, and transfected into the cells. Long dsRNA targeting firefly (FF) or renilla (RL) luciferases were co-transfected together with luciferase reporter plasmids. Long dsRNA against GFP was used as an additional control for nonspecific repression of luciferase expression. RNAi efficiency was monitored by dual luciferase assay. The results showed that long dsRNA induced robust sequence-independent repression of luciferase reporters. The nature of the sequence-independent repression and the IFN pathway activation were not analyzed in further detail. A similar result was reported by Yang et al., who suggested that in mammalian cells long dsRNA may induce RNAi, but nonspecific activation of the IFN response may mask the specific effect (15). To repress EGFP reporter expression, they transfected either in vitro produced dsRNA (0.5 kb) or a plasmid harboring an IR (1.1 kb) dowstream of a T7 polymerase promoter, which was co-transfected with the plasmid expressing T7 polymerase (Tables 1 and 2). Delivery of dsRNA into undifferentiated ESCs resulted in a sequence-specific silencing of EGFP suggesting that undifferentiated ESCs may lack the IFN response to dsRNA. However, differentiated ESCs and mammalian cell lines HEK 293 and NIH 3T3 respond to dsRNA in a sequence-independent manner. In vitro assays showed that lysates of differentiated cells are capable of generating siRNAs from long dsRNA suggesting that RNAi effects may be masked by nonspecific effects (15). The ability of in vitro produced dsRNA to induce RNAi in

Table 1
Summary of experiments employing in vitro-transcribed long dsRNA in mammalian somatic cells

dsRNA length (bp)	dsRNA production	RNAi effect	Nonspecific effect	Cell line	Reference
50–500	IVT T7 promoter	Weak	+	HeLa S3 NIH 3T3 COS-7	Elbashir et al. (34)
547	IVT T7 promoter from vector	–	+	Diff. ESCs	Yang et al. (15)
		+	–	Drosophila S2 Undiff. ESCs	
123–771	IVT T7 promoter	+	– IFN NS	P19 F9 E14 ESCs	Billy et al. (35)
500	IVT T7 promoter	–	+	HeLa C2C12	Paddison et al. (37)
600–800	IVT T3 or T7 promoter	+	NS	Neuroblastama N2a (AGYNB010)	Gan et al. (38)
550–1,000	IVT T7 promoter	+	NS	Hypothalamic cells	Bhargava et al. (42)
472	IVT T7 promoter	+	+	KB-3-1 IMR-32 SK-N-MC	Kabilova et al. (43)

The dsRNA length, promoter used for dsRNA production, and the presence of sequence-specific (RNAi) effects or nonspecific effects as related to the cell lines used are discussed for each experiment. *Abbreviations*: *IVT* in vitro transcription, *bp* base pair, *NS* not studied, *IFN* interferon pathway

undifferentiated pluripotent cells was also reported by Billy et al., who showed sequence-specific downregulation of exogenous (EGFP) and endogenous (integrin a3 and b1) genes upon electroporation of dsRNA (30 μg of ~700 bp-long dsRNA per 1×10^7 cells) in undifferentiated ESCs and P19 embryonic carcinoma cells (35). Similarly, Paddison et al. showed that in vitro transcribed long dsRNA (~0.5 kb) induced sequence-specific silencing in pluripotent P19 cells but not in HeLa cells or C2C12 myoblasts, where general repression of reporter expression was observed (37). Sequence-independent effects upon dsRNA transfection were also observed in cancer cell lines KB-3-1 (epidermal oral carcinoma) and SK-N-MC (neuroblastoma) (43). In this study, the effects of Myc long dsRNA (~500 bp, sequence from exon 3 of c-myc gene) was compared with those of siRNAs and polyinosinic:polycytidylic acid (poly(I:C)). Long dsRNAs and poly(I:C) induced nonspecific suppression of *c-MYC* production and a transient increase in *PKR* and *OAS1* mRNA levels.

Table 2
Summary of different constructs used to express long dsRNA in mammalian somatic cells

dsRNA schema design	dsRNA length (bp)	dsRNA production	RNAi effect	Nonspecific effect	Cell line	Reference
T7 EGFP EGFP	547	Transfection linearized dsEGFP vector + T7 pol. expressing vector	–	+	Diff. ESCs	Yang et al. (15)
			+	–	Drosophila S2 Undiff. ESCs	
EF1α dEGFP/TP53 pA RSV LTR dEGFP/TP53 pA	1,083 1,195	Stable transfection episomal vectors	+	–	EcR293 HCT116	Tran et al. (36)
CMV FL/GFP Zeo LoxP FL/GFP LoxP	500	Stable transfection	+	–	P19	Paddison et al. (37)
CMV dyxin 700-bp spacer dyxin	1,090	Eletroporation Transfection	+	– IFN NS	C2C12	Yi et al. (39)
CMV ribozyme NO 5′cap G/L/S – S/L/G 12-bp spacer MAZ NO polyA CMV intron G – G 12-bp spacer pA	540 518 502	Transfection Injection of fer egg Transfection	+ weak	– IFN NS +	MEF Transgenic mice MEF	Shinagawa and Ishii (40)
GRx CMV TRE CMV GRx	289	Transfection	+	NS	NIH 3T3	Wang et al. (41)

(continued)

Table 2 (continued)

dsRNA schema design	dsRNA length (bp)	dsRNA production	RNAi effect	Nonspecific effect	Cell line	Reference
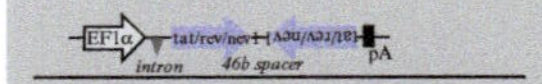	300	Transfection	+	–/NS	HEK 293T	Konstantinova et al. (46)
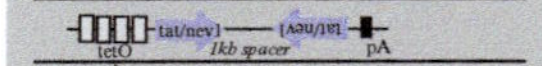			Weak	+	C33A	
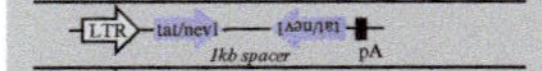			Weak	–/NS	Vero	
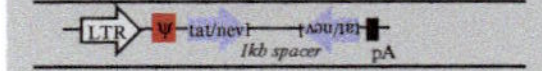			+	+		
	520	Transfection	–	–	Somatic cells HEK 293	Nejepinska et al. (53)
		Transgenic mice	+	–	Mouse oocytes	

The construct design, dsRNA length, method for dsRNA production, presence of sequence-specific (RNAi) effects, and nonspecific effects of long dsRNAs are shown for the cell lines used. *Drawing: light gray arrow* polymerase II or III promoter, *CMV* cytomegalovirus enhancer, *green arrow or box* enhance green fluorescent protein (EGFP) coding sequence, *violet arrow* targeting sequence, *light orange box* tetracycline responsive element (TRE), *dark gray box* ribozyme and binding site of the zinc finger protein MAZ, *yellow box* zeocin resistance (Zeo^r), *red triangle* LoxP site for Cre recombination, *dark red box* HIV-1 leader sequence (ψ). *Abbreviations: pA* polyadenylation signal, *bp* base pair, *NS* not studied, *IFN* interferon pathway, *ESC* embryonic stem cell, *MEF* mouse embryonic fibroblast, *NS* not studied

There are several observations suggesting that, in addition to ESCs and oocytes, long dsRNA may induce specific silencing in neuronal cells (38, 42). Using in vitro transcribed dsRNAs (~700 bp) in mouse neuronal cell, Gan et al. observed efficient silencing (50–60%) of exogenous GFP or endogenous proteins involved in apoptosis, signaling, and metabolism (38). In another report, injection of ~550 bp-long dsRNA into adult rat hypothalamic cells resulted in a successful sequence-specific knockdown of targeted genes in vivo (42). The observed effect of long dsRNA on two model neuroproteins, corticotrophin-releasing factor (CRF) and arginin vasopressin (VAP), was specific and restricted to injected hypothalamic cells and did not spread to other brain regions. In another set of experiments, 10–15 μg of long dsRNA (~0.5–1 kb) targeting CRF and VAP were mixed with 1 μl of Lipofectamine and injected into the brain. Sequence-specific effects were subsequently present for up to 1 week suggesting that long dsRNA can be applied for sequence-specific silencing in neural cells and for in vivo set studies. The notion that RNAi in neuronal cells may be more prominent than the IFN response is further supported by detection of endogenous siRNAs in the mouse hippocampus (29).

Experiments with delivering in vitro produced dsRNA into mammalian cells showed that its potential to induce RNAi is largely restricted to pluripotent cells. Differentiated cells (perhaps with some exceptions among neuronal cells) generally respond to dsRNA in a sequence-independent manner. This is attributed to the IFN pathway, in agreement with studies of poly(I:C) effects (54). Accordingly, in vitro produced long dsRNA is routinely used for inducing RNAi in invertebrate models, such as *C. elegans*, *Tribolium*, or *Drosophila*. However, it virtually disappeared from RNAi experiments in mammals except oocytes and early embryos. Mammalian RNAi is typically induced by siRNAs and shRNAs (reviewed in ref. 55), while long dsRNA remains a useful tool to study the IFN response and crosstalks between dsRNA-induced pathways.

4.2. dsRNA Expression

In contrast to in vitro-produced long dsRNA, several lines of evidence suggest that expressed long dsRNA more frequently induces sequence-specific silencing without provoking the IFN response in somatic cells (Table 2). For example, Tran et al. used episomal plasmids encoding long complementary RNAs to specifically suppress exogenous and endogenous gene expression in mammalian cells (36). In this system, co-expression of sense and antisense RNAs led to a specific inhibition of GFP or endogenous p53 transcriptional activator. Under these conditions, the IFN response was not triggered as evidenced by the lack of PKR phosphorylation. However, the mechanism of sequence-specific silencing may be different from RNAi, as Dicer knockdown did not influence the ability to reduce GFP expression. Furthermore, the gene silencing effect was transferable to other cells not previously exposed to the sense and antisense RNA via culture media, although mammalian cells do not exhibit transitive RNAi (reviewed in ref. 25).

Another system for long dsRNA expression was introduced by Paddison et al., who employed a plasmid expressing a self-complementary stem-loop against the first 500 nt of a GFP coding sequence (37). To produce long dsRNA hairpin, they used a FLIP cassette containing two sense-orientated fragments flanking the zeocin-resistance coding sequence. The second fragment was flanked by LoxP sites allowing its inversion into the antisense orientation by the Cre recombinase (Fig. 3d). Recombination resulted in an IR, which had the potential to form a long hairpin with a zeocin-resistance coding sequence as a loop. This system was used in the P19 embryonal carcinoma cell line and induced sequence-specific silencing of GFP accompanied with processing of GFP dsRNA into siRNAs (37). Whether this system could induce RNAi in differentiated somatic cell lines was not reported.

In contrast to the previously mentioned nonspecific effects of in vitro-produced dsRNA in C2C12 cells, Yi et al. demonstrated that long dsRNA could also trigger sequence-specific repression of Dyxin in differentiated C2C12 myoblast cells (39). To express dsRNA, they used a CMV-driven IR with a 700 bp-long spacer. In undifferentiated C2C12 cells, they observed sequence-specific downregulation of the ectopically expressed Dyxin. Moreover, the dsRNA also functioned in fully differentiated myotubes, where endogenous Dyxin (but not eIF2α, vinculin, or tubulin) was knocked down efficiently. Dyxin dsRNA-induced gene silencing of protein levels was accompanied by reduction of Dyxin mRNA suggesting that the silencing effect was caused by RNAi. However, sequence-independent effects including the activation of the IFN pathway was not studied in detail (39).

An interesting dsRNA expression system designed to prevent the IFN response by nuclear retention of dsRNA was developed by Shinagawa and Ishii (40). The pDECAP (deletion of Cap structure and polyA) vector contained a CMV promoter that controls expression of dsRNA hairpins lacking both the 5′-cap structure and the 3′-poly(A) tail (Fig. 2g). To remove the 5′-cap structure, a ribozyme cassette was added at a site downstream of the RNA start site. To eliminate the poly(A) tail, they inserted a MAZ site downstream of the IR. The zinc finger protein MAZ binding causes polymerase II to pause and fall off. As splicing generally enhances the nuclear export of RNA (47), the vector did not contain any intron. The pDECAP vector harbored an approximately 1 kb long IR containing the firefly luciferase, β-galactosidase, or the Ski gene sequence that was separated by a short sequence forming a small loop in expressed hairpin transcripts. This design should lead to the retention of long dsRNA in the nucleus, where it could be processed to siRNAs while preventing induction of the IFN response by dsRNA in the cytoplasm. Northern blot analysis of the nuclear and cytoplasmic fractions revealed that long dsRNA was not detected in either fraction suggesting that long dsRNA was rapidly processed to siRNAs. While it remains unclear whether dsRNA from pDECAP was exclusively processed in the nucleus, this system successfully induced sequence-specific silencing in HEK 293 cells. Transgenic mice carrying pDECAP vector expressing long dsRNA against the Ski gene showed similar developmental abnormalities as Ski knockout mice. Thus, specifically modified expressed long dsRNA can be routed into the RNAi pathway without provoking the IFN response. However, while the pDECAP vector appeared to be a suitable tool to efficiently generate tissue-specific knockdown mice, it did not get wider attention and was only sporadically used with variable success (56–60).

Another approach to overcome the IFN response was tested in Chock's laboratory (41). To express dsRNA, they used two opposite tetracycline-inducible CMV promoters driving expression of 290 nt sense and antisense sequences of glutaredoxin cDNA. After induction of sense and antisense RNA expression with doxycycline for 6 days in NIH 3T3 cells, glutaredoxin protein reduction was observed. Notably, the antisense RNA alone did not induce significant decrease of glutaredoxin.

Several different vectors containing expressed long hairpin RNAs of approximately 300 bp in length were used to inhibit human immunodeficiency virus type 1 (HIV-1) in infected mammalian cells (46). It was demonstrated that long dsRNAs are capable of inhibiting HIV-1 production in a sequence-specific manner without inducing expression of the class I IFN genes. The most effective constructs contained a viral RNA sequence Ψ cloned downstream of a promoter, or had a T7 promoter producing dsRNA directly in the cytoplasm. The viral Ψ sequence may provide the transcript with a nonself signature and thereby boost RNAi. The T7 promoter bypasses a possible problem with the nuclear export of expressed long dsRNAs, thus increasing HIV-1 repression. Interestingly, the study showed that the length of the loop between the IRs did not play a role in the efficiency of inhibiting HIV-1 production. As long dsRNA can target a large sequence of the viral genome, a major advantage of long dsRNA over shRNA use is the reduced chance of viral escape when mutations occur.

A comprehensive study of the effects of long dsRNA both in vivo and in vitro was done by Nejepinska et al. (53) who adapted a transgene design previously developed for mouse oocytes (48). An IR of ~500 bp was placed in the 3′UTR of spliced EGFP-encoding mRNA under the control of a chimeric CMV/β-actin promoter. This expression vector was used to generate transgenic mice and for the transfection of human cell lines. In transgenic mice, the IR was ubiquitously expressed and the formation of dsRNA was confirmed by detecting RNase T1-resistant RNA in cells. Interestingly, this long dsRNA did not activate the IFN response in somatic cells as evidenced by the lack of phosphorylation of PKR, analysis of several IFN-stimulated genes, and a microarray analysis in HEK 293 and HeLa cells transfected with the plasmid. There was a low level of RNA editing in mouse organs but a substantial editing in transfected HEK 293 cells, suggesting cell type-dependent activity of the ADAR enzyme. Moreover, edited dsRNA was detected in the cytoplasmic fraction, indicating that RNA editing did not prevent cytoplasmic localization of the dsRNA. Importantly, strong RNAi response was observed in oocytes derived from the transgenic mice but not in any other cell types. In addition, transfected HEK 293 or HeLa cells did not show strong RNAi. High-throughput sequencing showed that long dsRNA was poorly processed into siRNAs in the transgenic liver or brain, or in transfected HEK 293 cells. These data suggest that some long dsRNAs can be tolerated in mammalian somatic cells without induction of the IFN response and that siRNA biogenesis in somatic cells may be limited.

5. Effects of Long dsRNA in Mammalian Oocytes and Early Embryos

Mammalian oocytes and blastomeres of early embryos are the only mammalian cell types where long dsRNA is more frequently employed to induce RNAi. One of the reasons is that the IFN response is not functional in these cells (14). In addition, long dsRNA readily enters RNAi in these cells, presumably because of the presence of a robust endogenous RNAi (26, 27). In fact, mouse oocytes were the first cells where the existence of the mammalian RNAi pathway was demonstrated upon injection of long dsRNA (50, 51). Microinjection of in vitro-produced long dsRNA (typically ~10^6 molecules) offers a simple and robust method to induce RNAi. This method is still being used, although microinjection of siRNAs may be more prevalent nowadays because of the reduced price and improved efficiency of siRNAs. However, the induction of RNAi by long dsRNA microinjection is usually performed only in fully grown oocytes and 1-cell embryos, which are the stages most accessible to the injection. If one wants to target genes during oocyte growth and/or during oocyte-to-embryo transition, expression of dsRNA from a transgene is the method of choice.

5.1. Transgenic RNA

Transgenes for long dsRNA expression in oocytes mostly follow a previously developed design (Fig. 2h), where dsRNA is expressed as a part of EGFP-encoding mRNA (48). The RNA construct expression is specifically induced in oocytes at the post-meiotic stage with the ZP3 promoter or at the pre-meiotic stage with the GDF9 promoter. Detailed protocols for designing a transgene for RNAi in oocytes have been described elsewhere (19, 61). Here, we focus on additional considerations that have emerged recently. To date, 11 published transgenic RNAi experiments have been performed in oocytes (53, 62–71) (Table 3). None of the experiments encountered serious problems with cloning IRs and preparing transgenic constructs. Occasional delays in cloning IRs with a short spacer were solved by growing bacterial clones at room temperature in special bacterial strains (e.g., Stbl4 (Invitrogen) or Sure (Stratagene)). All transgenes yielded a strong mRNA knockdown in the best transgenic lines. In addition, we successfully used transgenic RNAi to silence two genes simultaneously (Flemr and Svoboda, unpublished). The average time from the transgene cloning to the first phenotype analysis typically ranged from 6 to 9 months. Phenotypes of knockdown mice were typically robust, except for the *Wee1B* knockdown, where only approximately 25% of the oocytes exhibited a phenotype and where *Wee1B* mRNA was not knocked down as efficiently as in other transgenic experiments (63).

Despite the successes, there are also several drawbacks that should be considered before deciding to perform a transgenic RNAi experiment. First, one has to select transgenic line(s) with a strong knockdown effect, which may be a problem when a small number of transgenic lines carrying randomly integrated transgenes is obtained. Second, randomly integrated transgenes may be subjected to epigenetic silencing. In our recent experience, some vectors are more prone to epigenetic silencing than others (72), and transgene silencing is more frequently observed if an inbred strain embryo is used to produce transgenic founder animals. Although, our vector for transgenic RNAi (65) is typically well expressed, it was found to be silenced in several C57Bl/6 animals (unpublished results). A hybrid cross of C57Bl/6 and BALB/c partially solved this problem. However, some RNAi transgenes still showed variable expression as judged by variable EGFP fluorescence in oocytes. Notably, RNAi transgenes targeting dormant maternal mRNAs (mRNAs, not translated until ovulation) have typically a uniform EGFP expression while RNAi transgenes targeting translated mRNAs, which may be of importance for oocyte growth, have a more variable EGFP expression. This suggests that some growing oocytes may escape from the silencing effect and fully mature with milder phenotype (Svoboda et al., unpublished observation). Selection against knockdown effects could also explain insufficient knockdown in the *Wee1B* experiment (63).

Table 3
Summary of experiments with transgenic RNAi mice expressing long dsRNA in oocytes

Gene	dsRNA length (bp)	Knockdown efficiency	Phenotype	Reference
Mos	535	90% mRNA reduction	Null phenotype parthenogenetic activation	Stein et al. (65)
Ctcf	700	70–99% mRNA reduction	Abnormal methylation decreased developmental competence	Fedoriw et al. (64)
Msy2	850	60% Protein reduction	Pleotropic effect	Yu et al. (66)
WeelB	566	50% mRNA reduction	Meiotic arrest	Han et al. (63)
Bnc (basonuclin)	~800	>90% mRNA reduction	Not identified	Ma et al. (62)
CPEB	612	~90% Reduction	Abnormal MII spindle parthenogenetic activation sterility	Racki et al. (67)
PLCb1	533	~75% mRNA reduction	Significant decrease in Ca^{2+} transient amplitude	Igarashi et al. (68)
SLBP	654	80% mRNA and >90% protein reduction	DNA replication defect at the 2-cell stage and developmental arrest	Arnold et al. (69)
CTCF	333	50–99% mRNA reduction	Meiotic defect in the egg and mitotic defect in the embryo	Wan et al. (70)
ATRX	450	80–90% Reduction	Abnormal MII spindle and increased aneuploidy	Baumann et al. (71)
Mos	520	90% mRNA reduction	Sterile or subfertile parthogenetic activation	Nejepinska et al. (53)

The table contains the name of the target genes, the length of the dsRNA stem, the knockdown efficiency of the transgene, and the phenotype observed in transgenic mice

6. Outlook

While long dsRNA may not be a perfect tool to induce RNAi in mammalian cells, it remains an important method to investigate mammalian responses to foreign nucleic acids. From this perspective, different strategies to produce and deliver diverse types of dsRNA molecules into mammalian cells are still highly relevant. Furthermore, long dsRNA has a great potential for nonmammalian model systems and many technical aspects discussed in this review are of general importance to anyone working with long dsRNA in any model systems.

Acknowledgements

Authors thank Radek Malik for help with preparation and Camille Du Roure for text revision of this manuscript. This work was supported by the GACR 204/09/0085, EMBO SDIG program #1488, and the Purkynje Fellowship to PS. JN is supported in part by Faculty of Science, Charles University in Prague.

References

1. Carthew RW, Sontheimer EJ (2009) Origins and mechanisms of miRNAs and siRNAs. Cell 136:642–655
2. Ghildiyal M, Zamore PD (2009) Small silencing RNAs: an expanding universe. Nat Rev Genet 10:94–108
3. Fire A, Xu S, Montgomery MK et al (1998) Potent and specific genetic interference by double-stranded RNA in *Caenorhabditis elegans*. Nature 391:806–811
4. Caplen NJ, Fleenor J, Fire A et al (2000) dsRNA-mediated gene silencing in cultured Drosophila cells: a tissue culture model for the analysis of RNA interference. Gene 252: 95–105
5. Clemens JC, Worby CA, Simonson-Leff N et al (2000) Use of double-stranded RNA interference in Drosophila cell lines to dissect signal transduction pathways. Proc Natl Acad Sci U S A 97:6499–6503
6. Hammond SM, Bernstein E, Beach D et al (2000) An RNA-directed nuclease mediates post-transcriptional gene silencing in Drosophila cells. Nature 404:293–296
7. Ui-Tei K, Zenno S, Miyata Y et al (2000) Sensitive assay of RNA interference in Drosophila and Chinese hamster cultured cells using firefly luciferase gene as target. FEBS Lett 479:79–82
8. Cullen BR (2006) Is RNA interference involved in intrinsic antiviral immunity in mammals? Nat Immunol 7:563–567
9. Gantier MP, Williams BR (2007) The response of mammalian cells to double-stranded RNA. Cytokine Growth Factor Rev 18:363–371
10. Garcia MA, Gil J, Ventoso I et al (2006) Impact of protein kinase PKR in cell biology: from antiviral to antiproliferative action. Microbiol Mol Biol Rev 70:1032–1060
11. Silverman RH (2007) A scientific journey through the 2–5A/RNase L system. Cytokine Growth Factor Rev 18:381–388
12. Silverman RH (2007) Viral encounters with 2′,5′-oligoadenylate synthetase and RNase L during the interferon antiviral response. J Virol 81:12720–12729
13. Judge AD, Sood V, Shaw JR et al (2005) Sequence-dependent stimulation of the mammalian innate immune response by synthetic siRNA. Nat Biotechnol 23:457–462
14. Stein P, Zeng F, Pan H et al (2005) Absence of non-specific effects of RNA interference triggered by long double-stranded RNA in mouse oocytes. Dev Biol 286:464–471

15. Yang S, Tutton S, Pierce E et al (2001) Specific double-stranded RNA interference in undifferentiated mouse embryonic stem cells. Mol Cell Biol 21:7807–7816
16. Hornung V, Guenthner-Biller M, Bourquin C et al (2005) Sequence-specific potent induction of IFN-alpha by short interfering RNA in plasmacytoid dendritic cells through TLR7. Nat Med 11:263–270
17. Sioud M (2005) Induction of inflammatory cytokines and interferon responses by double-stranded and single-stranded siRNAs is sequence-dependent and requires endosomal localization. J Mol Biol 348:1079–1090
18. Marques JT, Devosse T, Wang D et al (2006) A structural basis for discriminating between self and nonself double-stranded RNAs in mammalian cells. Nat Biotechnol 24:559–565
19. Svoboda P, Stein P (2009) RNAi experiments in mouse oocytes and early embryos. Cold Spring Harb Protoc. doi:10.1101/pdb.prot5134
20. Keegan LP, Leroy A, Sproul D et al (2004) Adenosine deaminases acting on RNA (ADARs): RNA-editing enzymes. Genome Biol 5:209
21. Bass BL (2002) RNA editing by adenosine deaminases that act on RNA. Annu Rev Biochem 71:817–846
22. Kumar A, Crawford K, Close L et al (1997) Rescue of cardiac alpha-actin-deficient mice by enteric smooth muscle gamma-actin. Proc Natl Acad Sci U S A 94:4406–4411
23. Nishikura K (2010) Functions and regulation of RNA editing by ADAR deaminases. Annu Rev Biochem 79:321–349
24. DeCerbo J, Carmichael GG (2005) Retention and repression: fates of hyperedited RNAs in the nucleus. Curr Opin Cell Biol 17:302–308
25. Nejepinska J, Flemr M, Svoboda P (2012) The canonical RNA interference pathway in animals. In: Mallick B, Ghosh Z (eds) Regulatory RNAs. Springer, Heidelberg, p 623
26. Tam OH, Aravin AA, Stein P et al (2008) Pseudogene-derived small interfering RNAs regulate gene expression in mouse oocytes. Nature 453:534–538
27. Watanabe T, Totoki Y, Toyoda A et al (2008) Endogenous siRNAs from naturally formed dsRNAs regulate transcripts in mouse oocytes. Nature 453:539–543
28. Babiarz JE, Ruby JG, Wang Y et al (2008) Mouse ES cells express endogenous shRNAs, siRNAs, and other microprocessor-independent, Dicer-dependent small RNAs. Genes Dev 22:2773–2785
29. Smalheiser NR, Lugli G, Thimmapuram J et al (2011) Endogenous siRNAs and noncoding RNA-derived small RNAs are expressed in adult mouse hippocampus and are up-regulated in olfactory discrimination training. RNA 17: 166–181
30. Fellmann C, Zuber J, McJunkin K et al (2011) Functional identification of optimized RNAi triggers using a massively parallel sensor assay. Mol Cell 41:733–746
31. Taxman DJ, Livingstone LR, Zhang J et al (2006) Criteria for effective design, construction, and gene knockdown by shRNA vectors. BMC Biotechnol 6:7
32. Olson A, Sheth N, Lee JS et al (2006) RNAi Codex: a portal/database for short-hairpin RNA (shRNA) gene-silencing constructs. Nucleic Acids Res 34:D153–D157
33. Birmingham A, Anderson EM, Reynolds A et al (2006) 3′ UTR seed matches, but not overall identity, are associated with RNAi off-targets. Nat Methods 3:199–204
34. Elbashir SM, Harborth J, Lendeckel W et al (2001) Duplexes of 21-nucleotide RNAs mediate RNA interference in cultured mammalian cells. Nature 411:494–498
35. Billy E, Brondani V, Zhang H et al (2001) Specific interference with gene expression induced by long, double-stranded RNA in mouse embryonal teratocarcinoma cell lines. Proc Natl Acad Sci U S A 98:14428–14433
36. Tran N, Raponi M, Dawes IW et al (2004) Control of specific gene expression in mammalian cells by co-expression of long complementary RNAs. FEBS Lett 573:127–134
37. Paddison PJ, Caudy AA, Hannon GJ (2002) Stable suppression of gene expression by RNAi in mammalian cells. Proc Natl Acad Sci U S A 99:1443–1448
38. Gan L, Anton KE, Masterson BA et al (2002) Specific interference with gene expression and gene function mediated by long dsRNA in neural cells. J Neurosci Methods 121:151–157
39. Yi CE, Bekker JM, Miller G et al (2003) Specific and potent RNA interference in terminally differentiated myotubes. J Biol Chem 278:934–939
40. Shinagawa T, Ishii S (2003) Generation of Ski-knockdown mice by expressing a long double-strand RNA from an RNA polymerase II promoter. Genes Dev 17:1340–1345
41. Wang J, Tekle E, Oubrahim H et al (2003) Stable and controllable RNA interference: investigating the physiological function of glutathionylated actin. Proc Natl Acad Sci U S A 100:5103–5106
42. Bhargava A, Dallman MF, Pearce D et al (2004) Long double-stranded RNA-mediated RNA interference as a tool to achieve site-specific

silencing of hypothalamic neuropeptides. Brain Res Brain Res Protoc 13:115–125

43. Kabilova TO, Vladimirova AV, Chernolovskaya EL et al (2006) Arrest of cancer cell proliferation by dsRNAs. Ann N Y Acad Sci 1091:425–436
44. Bernstein E, Caudy AA, Hammond SM et al (2001) Role for a bidentate ribonuclease in the initiation step of RNA interference. Nature 409:363–366
45. Stein P, Svoboda P, Anger M et al (2003) RNAi: mammalian oocytes do it without RNA-dependent RNA polymerase. RNA 9: 187–192
46. Konstantinova P, de Vries W, Haasnoot J et al (2006) Inhibition of human immunodeficiency virus type 1 by RNA interference using long-hairpin RNA. Gene Ther 13:1403–1413
47. Valencia P, Dias AP, Reed R (2008) Splicing promotes rapid and efficient mRNA export in mammalian cells. Proc Natl Acad Sci U S A 105:3386–3391
48. Svoboda P, Stein P, Schultz RM (2001) RNAi in mouse oocytes and preimplantation embryos: effectiveness of hairpin dsRNA. Biochem Biophys Res Commun 287: 1099–1104
49. Hunter T, Hunt T, Jackson RJ et al (1975) The characteristics of inhibition of protein synthesis by double-stranded ribonucleic acid in reticulocyte lysates. J Biol Chem 250: 409–417
50. Svoboda P, Stein P, Hayashi H et al (2000) Selective reduction of dormant maternal mRNAs in mouse oocytes by RNA interference. Development 127:4147–4156
51. Wianny F, Zernicka-Goetz M (2000) Specific interference with gene function by double-stranded RNA in early mouse development. Nat Cell Biol 2:70–75
52. Akimov IA, Kabilova TO, Vlassov VV et al (2009) Inhibition of human cancer-cell proliferation by long double-stranded RNAs. Oligonucleotides 19:31–40
53. Nejepinska J, Malik R, Filkowski J et al (2012) dsRNA expression in the mouse elicits RNAi in oocytes and low adenosine deamination in somatic cells. Nucleic Acids Res 40(1):399–413
54. Geiss G, Jin G, Guo J et al (2001) A comprehensive view of regulation of gene expression by double-stranded RNA-mediated cell signaling. J Biol Chem 276:30178–30182
55. Bantounas I, Phylactou LA, Uney JB (2004) RNA interference and the use of small interfering RNA to study gene function in mammalian systems. J Mol Endocrinol 33:545–557
56. Hou X, Omi M, Harada H et al (2011) Conditional knockdown of target gene expression by tetracycline regulated transcription of double strand RNA. Dev Growth Differ 53:69–75
57. Dai P, Nakagami T, Tanaka H et al (2007) Cx43 mediates TGF-beta signaling through competitive Smads binding to microtubules. Mol Biol Cell 18:2264–2273
58. Koster MI, Dai D, Marinari B et al (2007) p63 induces key target genes required for epidermal morphogenesis. Proc Natl Acad Sci U S A 104:3255–3260
59. Maekawa T, Shinagawa T, Sano Y et al (2007) Reduced levels of ATF-2 predispose mice to mammary tumors. Mol Cell Biol 27:1730–1744
60. Zipperlen PBaP (2005) Comparison of a range of approaches for RNAi in human cells. QIAGEN News, QIAGEN, Institute of Molecular Biology, University of Zurich, Zurich, Switzerland
61. Svoboda P (2009) Cloning a transgene for transgenic RNAi in mouse oocytes. Cold Spring Harb Protoc. doi:10.1101/pdb.prot5134
62. Ma J, Zeng F, Schultz RM et al (2006) Basonuclin: a novel mammalian maternal-effect gene. Development 133:2053–2062
63. Han SJ, Chen R, Paronetto MP et al (2005) Wee1B is an oocyte-specific kinase involved in the control of meiotic arrest in the mouse. Curr Biol 15:1670–1676
64. Fedoriw AM, Stein P, Svoboda P et al (2004) Transgenic RNAi reveals essential function for CTCF in H19 gene imprinting. Science 303: 238–240
65. Stein P, Svoboda P, Schultz RM (2003) Transgenic RNAi in mouse oocytes: a simple and fast approach to study gene function. Dev Biol 256:187–193
66. Yu J, Deng M, Medvedev S et al (2004) Transgenic RNAi-mediated reduction of MSY2 in mouse oocytes results in reduced fertility. Dev Biol 268:195–206
67. Racki WJ, Richter JD (2006) CPEB controls oocyte growth and follicle development in the mouse. Development 133:4527–4537
68. Igarashi H, Knott JG, Schultz RM et al (2007) Alterations of PLCbeta1 in mouse eggs change calcium oscillatory behavior following fertilization. Dev Biol 312:321–330
69. Arnold DR, Francon P, Zhang J et al (2008) Stem-loop binding protein expressed in growing oocytes is required for accumulation of mRNAs encoding histones H3 and H4 and for early embryonic development in the mouse. Dev Biol 313:347–358

70. Wan LB, Pan H, Hannenhalli S et al (2008) Maternal depletion of CTCF reveals multiple functions during oocyte and preimplantation embryo development. Development 135: 2729–2738

71. Baumann C, Viveiros MM, De La Fuente R (2010) Loss of maternal ATRX results in centromere instability and aneuploidy in the mammalian oocyte and pre-implantation embryo. PLoS Genet 6:e1001137

72. Sarnova L, Malik R, Sedlacek R et al (2010) Shortcomings of short hairpin RNA-based transgenic RNA interference in mouse oocytes. J Negat Results Biomed 9:8

Chapter 17

Design of RNAi Reagents for Invertebrate Model Organisms and Human Disease Vectors

Thomas Horn and Michael Boutros

Abstract

RNAi has become a very versatile tool to silence gene expression in a variety of organisms, in particular when classical genetic methods are missing. However, the application of this method in functional studies has raised new challenges in order to design RNAi reagents that minimize false positives and false negatives. Because the performance of reagents cannot be validated on a genome-wide scale, improved computational methods are required that consider experimentally derived quality measures. In this chapter, we describe computational methods for the design of RNAi reagents for invertebrate model organisms and human disease vectors, such as *Anopheles*. We describe procedures for designing short and long double-stranded RNAs for single genes, and evaluate their predicted specificity and efficiency. Using a bioinformatics pipeline we also describe how to design a genome-wide RNAi library for *Anopheles gambiae*.

Key words: RNAi, Invertebrates, Double-stranded RNA, Drosophila, Anopheles, High-throughput screening

1. Introduction

RNA interference (RNAi) screens have become an important tool for the identification and characterization of gene function on a large-scale and complement classic mutagenesis screens by providing a means to target almost every transcript in a sequenced and annotated genome (1, 2). RNAi is a posttranscriptional gene silencing mechanism conserved from plants to humans and relies on the delivery of exogenous short double-stranded (ds) RNAs that trigger the degradation of homologous mRNAs in cells (3, 4). As an experimental tool, RNAi is now widely used to silence the expression of genes in a broad spectrum of organisms (5).

RNAi was first observed in plants (6–8) and later mechanistically dissected in the nematode *Caenorhabditis elegans* (4). The introduction of dsRNAs into cells leads to their cleavage into

Debra J. Taxman (ed.), *siRNA Design: Methods and Protocols*, Methods in Molecular Biology, vol. 942, DOI 10.1007/978-1-62703-119-6_17, © Springer Science+Business Media, LLC 2013

short-interfering RNA (siRNA) duplexes of 21–23 nucleotides length that contain 2-nucleotide 3′ overhangs with 5′ phosphate and 3′ hydroxyl termini (9). This process (referred to as "dicing") is mediated by the RNase III-like endonuclease Dicer (10–13). The siRNAs are then incorporated into the RNA-induced silencing complex (RISC), and siRNA duplexes are unwound by the RISC's helicase activity. The crucial observation that each RISC complex contains only one of the two strands of an siRNA duplex (14), and that only the antisense strand of an siRNA can direct the cleavage of the sense mRNA target has provided important insights into the biochemical mechanism of RNAi-mediated gene silencing (15–20). Thus, it is hypothesized that the RISC preferentially accepts the strand of the siRNA that has the less stable 5′ end and that other biophysical, thermodynamic, and structural parameters as well as base preferences at specific positions in the sense strand influence the efficiency of incorporation. The siRNA incorporated into the RISC then binds to complementary mRNA that is consequently cut by the RISC RNase H-like nuclease activity at a defined position. Subsequently, the cleaved mRNA is recognized by the cell as aberrant, leading to its degradation (9). This leads to a depletion of the corresponding protein in the cell. The RNAi pathway is not only triggered by exogenous dsRNAs but also by sequences encoded in the organism's genome, including siRNAs, micro-RNAs (miRNA), and piwi-interacting RNAs. These play important roles in many fundamental biological and disease processes (21).

In *Drosophila*, RNAi can be triggered by 100–700 bp-long dsRNAs expressed as hairpins in a time- and tissue-specific manner using the UAS-Gal4 system in vivo (22) or added to the culture medium of *Drosophila* cells in vitro (23). The uptake of long dsRNAs in *Drosophila* cells in vitro is achieved by receptor-mediated endocytosis (24, 25). In addition, a genome-wide resource to silence transcripts during *Drosophila* oogenesis using short-hairpin RNAs (shRNAs) with the UAS-Gal4 system is being built (26). In contrast, RNAi-mediated silencing in mammalian cells is mediated through 21–23 nt-long siRNAs (27) mimicking Dicer-cleaved products to circumvent an interferon response triggered by long dsRNAs (28). Such short dsRNAs can be generated by different methods. Vectors transcribing short-hairpin RNAs (shRNAs) (29–31) or synthetic siRNAs (27) as well as endoribonuclease-prepared siRNAs (esiRNAs) (32) are commonly used.

The availability of genome-wide RNAi libraries for cell-based assays (5, 33) and whole organisms (34) has opened new avenues to query genomes for a broad spectrum of loss-of-function phenotypes. The number of sequenced invertebrate genomes is steadily rising, enabling reverse genetic approaches using RNAi in many novel model systems, including, e.g., the medically relevant vector *Anopheles gambiae* and species used to study evolutionary aspects of development such as *Tribolium castaneum*, *Acyrthosiphon pisum*, and *Schmidtea mediterranea*. RNAi libraries will facilitate the functional characterization of genes in these species, either through studying smaller subsets of candidates or on a genomic scale.

The design of RNAi reagents is key for obtaining reliable phenotypic data in large-scale RNAi experiments. Several recent studies demonstrated that the degradation of unintended transcripts (so-called off-target effects) (35, 36) and knockdown efficiency depend on the sequence of the RNAi reagent and have to be carefully monitored (15, 17–20, 37, 38). Based on experimental studies, rules for the design of RNAi reagents have been devised to improve knockdown efficiency and simultaneously minimize unspecific effects.

In this chapter, we provide protocols for the design of RNAi reagents for invertebrate model organisms. We describe a general workflow of how to identify suitable target regions that minimize the potential for off-target effects, increase the silencing capacity and allow an efficient synthesis of the reagents. We show how all these steps can be performed automatically, using computational tools that we have developed (39, 40). Specifically, we provide case studies for the design of long dsRNAs, siRNAs, and shRNAs. Finally, we outline how a genome-wide RNAi resource for a human disease vector genome can be generated.

2. Materials

2.1. E-RNAi

E-RNAi is a Web service available at http://www.e-rnai.org/. It currently enables the design and evaluation of RNAi reagents for 13 organisms, which can be extended upon the request.

2.2. NEXT-RNAi

NEXT-RNAi is a software package implemented in Perl and available for download at http://www.nextrnai.org/. It requires the installation of Bowtie (41) and Primer3 (42). To utilize all options of NEXT-RNAi, the BLAST (43), BLAT (44), RNAfold (45), and mdust (see Note 3) programs are also required. A platform-independent virtual machine (running on VirtualBox) with NEXT-RNAi and all dependencies preinstalled is also available. On a Linux server (two Intel Xeon Quad-Core CPU with 2.00 GHz, 16GB RAM) running Ubuntu 9.10 server edition, the design of a genome-wide RNAi library for the *Anopheles* genome took about 4 h.

3. Methods

3.1. A General Workflow for the Design of RNAi Reagents

3.1.1. Selection of Suitable RNAi Target Sequences

1. Obtaining sequences from databases: Potential target sequences (exon, transcript, or gene sequences) for RNAi can be downloaded from different genome databases, such as NCBI RefSeq (46) and ENSEMBL (47), or from model-organism databases, such as FlyBase for *Drosophila* (48), BeetleBase for *Tribolium*

castaneum (49), Wormbase for *Caenorhabditis elegans* (50), and many others. Table 1 lists invertebrate model organisms amenable to RNAi.

2. Sequences suitable for RNAi synthesis: We recommend using exon sequences as RNAi templates. Sequences containing exons interspersed by small introns also work efficiently. Care needs to be taken for the synthesis of long dsRNAs: if the reagent was designed based on a transcript sequence composed of multiple exons, which are interspersed by long introns in the gene sequences, it should also be amplified from cDNA (not from genomic DNA), because the PCR might fail due to the increased amplicon length. If the reagent was designed based on an exon sequence, it does not make a difference whether cDNA or genomic DNA is used as a template.
3. Targeting multiple transcripts of the same gene: If the target gene encodes for multiple predicted isoforms, it is recommended to select sequences of exons common to all annotated isoforms, except when knocking down a specific isoform is desired (see Note 1).
4. Minimal information about a genome required for RNAi: Novel sequencing technologies allow the rapid assembly of genomes and transcriptomes for emerging model organisms (51, 52). If the organism is amenable to RNAi, this is often the first method of choice for functional studies. How to use such assemblies for the design of RNAi reagents is discussed in detail in Note 2.

3.1.2. Assessing the Predicted RNAi Specificity

RNAi reagents have been shown to exert so-called off-target effects, which means they can target additional transcripts besides the desired one. Sequences that are prone to such effects should be excluded from the template sequence, including regions of low complexity, perfect or partial matches to unintended targets, and seed region targets:

1. Regions of low complexity: Tandem trinucleotide repeats of the type CA[ACGT) (CAN) are associated with promiscuous off-target effects (84) and longer stretches of such sequences (e.g., more than five contiguous CAN repeats) should be excluded from target regions. Additionally, simple nucleotide repeats and other poly-triplet sequences should also be avoided (see Note 3). Regions of low complexity are particularly critical during the design of long dsRNAs, which are usually longer than 100 bp making them more prone to contain such sequences.
2. Perfect homologies to unintended transcripts: siRNAs may cause unspecific gene silencing via short stretches of perfect homology with unintended mRNAs (84–86). To identify such homologies the siRNAs are aligned to the transcriptome of the

Table 1
Invertebrate model organisms amenable to RNAi

Organisms	Database	Web page	Database reference	RNAi reference
Acyrthosiphon pisum	AphidBase	http://www.aphidbase.com/aphidbase/	(53)	(54, 55)
Apis mellifera[a]	BeeBase	http://hymenopteragenome.org/beebase/	(56)	(57, 58)
Aplysia californica[b]	UCSC	http://genome.ucsc.edu/cgi-bin/hgGateway?org=Sea+here	–	(59)
Bombyx mori	SilkDB	http://silkworm.genomics.org.cn/	(60)	(61)
Brugia malayi	Brugia Genome Project	http://ghedinlab.csb.pitt.edu/GhedinLab/Brugia%20malayi	–	(62)
Caenorhabditis elegans[c]	WormBase	http://www.wormbase.org/	(50)	(4, 63)
Ciona intestinalis[d]	JGI Genome Portal	http://genome.jgi-psi.org/Coin2/	–	(64)
Ciona savignyi[e]	Ciona savignyi Database	http://www.broadinstitute.org/annotation/ciona/	–	(64)
Drosophila melanogaster[f]	FlyBase	http://flybase.org/	(48)	(65)
Hydra magnipapillata	Hydrazome	http://hydrazome.metazome.net	–	(66, 67)
Invertebrate vectors of human pathogens[g]	VectorBase	http://www.vectorbase.org/	(68)	(69)
Manduca sexta	Manduca Base	http://agripestbase.org/manduca/	–	(70)
Nasonia vitripennis	NansoniaBase	http://hymenopteragenome.org/nasonia/	(56)	(71)
Nematostella vectensis	StellaBase	http://nematostella.bu.edu/stellabase/	(72)	(73)

(continued)

Table 1 (continued)

Organisms	Database	Web page	Database reference	RNAi reference
Parhyale hawaiensis	JGI Genome Portal	http://genome.jgi-psf.org/parha/	–	(74)
Schistosoma mansoni	SchistoDB	http://schistodb.net/schistodb20/	(75)	(76)
Schmidtea mediterranea	SmedGD	http://smedgd.neuro.utah.edu/	(77)	(78, 79)
Six ant species[h]	Ant Genomes Portal	http://hymenopteragenome.org/ant_genomes/	(56)	(80)
Strongylocentrotus purpuratus	SpBase	http://www.spbase.org/	(81)	(82)
Tribolium casteneum	BeetleBase	http://beetlebase.org/	(49)	(83)

[a]Also including *Bombus terrestris* and *Bombus impatiens*
[b]Also available at BROAD http://www.broadinstitute.org/science/projects/mammals-models/vertebrates-invertebrates/aplysia/aplysia-genome-sequencing-project
[c]Also including *Caenorhabditis briggsae*
[d]Also available via ENSEMBL
[e]Also available via ENSEMBL and http://mendel.standford.edu/sidowlab/ciona.html
[f]Also including 11 other Drosophila species (ananassae, erecta, grimshawi, mojavensis, persimilis, pseudoobscura, sechellia, simulans, virillis, willistoni, yakuba)
[g]*Anopheles gambiae, Aedes aegypti, Ixodes scapularis, Culex quinquefasciatus, Pediculus humanus, Rhodnius prolixus, Glossina morsitans*
[h]*Atta cephalotes, Camponotus floridanus, Harpegnathos saltator, Linepithema humile, Pogonomyrmex barbatus, Solenopsis invicta*

organism using alignment programs such as BLAST (87) or Bowtie (41) (see Note 4). For the design of a long dsRNA all possible siRNAs that can be generated need to be aligned. Usually, the long dsRNA is cut into all possible 19 bp-long siRNAs (with an offset of 1 base, so that, e.g., a 300 bp-long sequence produces 282 siRNAs) to mimic the function of Dicer. An siRNA is predicted to be specific if it only aligns to transcripts of the intended target gene.

3. Partial homologies to unintended transcripts: siRNAs can also recognize targets through partial sequence homologies (88, 89). BLAST searches allowing partial alignments mapping the RNAi reagent to the transcriptome can identify such homologies (see Note 5). Significant homologies of an siRNA or a long dsRNA to unintended transcripts should be avoided.
4. miRNA-like seed matches in 3'-UTRs of unintended transcripts: siRNAs can cause unspecific transcript silencing by a route similar to microRNA-mediated silencing through sequence similarity in positions 2–9 (hexamer, heptamer or octamer) of the siRNA guide strand to the 3′-untranslated region (UTR) of unintended transcripts (90, 91). The number of siRNA seed matches (seed complement frequency) can be determined by aligning the siRNA seed-sequence to a database of 3′-UTR sequences, using, e.g., Bowtie. The more targets a seed has, the more likely an off-target silencing effect is (see Note 6). Thus, siRNA sequences with low seed complement frequencies should be prioritized.

3.1.3. Assessing the Predicted RNAi Efficiency

Factors influencing the efficiency of RNAi reagents depend very much on the RNAi species. Long dsRNAs, siRNAs, and shRNAs are processed in different ways by the cellular RNAi machinery (92). In *Drosophila*, long dsRNAs are first cleaved into 21–23 bp-long siRNAs by Dicer2, before they can form a functional RISC. Exogenously applied siRNAs mimic these Dicer2 products and are directly integrated into the RISC. shRNA constructs are designed to enter the miRNA biogenesis pathway, including a two-step processing by Drosha/Pasha complexes and Dicer 1 into mature shRNAs, before functional RISCs can form.

Every enzymatic step during the processing of exogenous RNAi reagents might influence the overall efficiency. Many studies have been performed in human cell lines to find the best nucleotide composition for siRNAs. Because siRNAs are not processed before they are integrated into the RISC, most siRNA properties identified might reflect the preferences of the RISC. Several criteria can be taken into account to determine the predicted efficiency of an siRNA:

1. Asymmetric thermodynamic properties: According to the asymmetry rule, the strand (of the double-stranded siRNA)

with the less stably paired 5′ end is incorporated into the RISC and becomes the guide strand (17, 19, 93). This should be the strand complementary to the target mRNA, which is called the "antisense strand." To increase integration of the antisense strand into the RISC, its 5′ end should contain A and U bases (in positions 1–5).

2. G/C content: It has been experimentally determined that the G/C content of a functional siRNA should be between 30 and 52% (18, 94). An mRNA target site with a high G/C content can form stable secondary structures that are not accessible by the siRNA (95).
3. Internal repeats: siRNAs containing internal repeats or palindromes may form internal fold-back structures, which may exist in equilibrium with the duplex form (96) and decrease the effective siRNA concentration and silencing potential. Thus, a functional siRNA should not contain internal repeats that cannot be resolved at temperatures lower or equal to 20°C (18).
4. Base preferences: Different studies identified specific base preferences at several positions required for the optimal efficiency of an siRNA (18, 20, 37). Referring to the sense strand, the following bases are preferred: G/C at position 1, A at position 3, 6, and 19, U at position 10 and 13. The following bases are disadvantaged: A/U at position 1, G at position 13, G/C at position 19.
5. Structural properties of the target: mRNA target sites can take secondary structures with different stabilities making one site more accessible to an siRNA than another (97). Easy accessible target sites should be prioritized.

Two main methods for scoring the overall efficiency of an siRNA were developed from these criteria, here referred to as the "Rational" design by Reynolds et al. (18) and the "Weighted" design by Shah et al. (38) (see Note 7). One of the main differences between both methods is that Reynolds et al. suggests the absence of "internal repeats" a significant feature, whereas Shah et al. does not. Both methods do not consider the structural properties of the target mRNA, which can be calculated using the RNAxs online tool (97).

Recent studies suggest further sequence features influencing RNAi efficiency that could be considered during the reagent design. Argonaut (Ago) proteins form the core of the RISC and possess the endonuclease activity required for target-mRNA cleavage (98). siRNAs bind to a specific Ago family member, Ago2, to mediate RNAi (99). But they can also interact with other Ago proteins (Ago1, Ago3, and Ago4) that usually bind miRNAs. This binding can induce miRNA-like silencing of transcripts by partial homology with their 3′-UTRs, leading to unwanted off-target effects (86, 91, 100–102). It was recently reported that this could be overcome by increasing the siRNA duplex stability, which selectively blocks Ago1, -3 and -4 action and even increases silencing efficiency via Ago2 (103). Duplex stability could be increased by selecting target sites with higher G/C content or by using chemically modified nucleotides for the passenger strand, to increase thermodynamic stability. However, increasing the G/C content could interfere with target site accessibility. Two other recent studies found siRNA guide strands starting with C preferentially binding to *Drosophila* Ago2, which also increases silencing efficiency and circumvents off-target activities via the other Ago family members (104, 105).

In the case of long dsRNAs, it is not understood which siRNAs are produced by Dicer2. Although there are studies monitoring siRNAs generated from long dsRNAs to study Dicer specificity and efficiency (106), no predictive algorithm is available. However, long dsRNAs have the potential of generating multiple siRNAs, increasing the likelihood that one or more of these can efficiently silence the target transcript. Their efficiency could be predicted using the criteria for siRNAs described above and summarized for the long dsRNA (e.g., by counting the number of efficient siRNAs per long dsRNA or by averaging the efficiency of all siRNAs).

Commonly, the same algorithms used to predict the efficiency of siRNAs are used to predict the efficiency of shRNAs (26, 30, 31) and computational tools have been made available (107). Using an experimental approach in mammalian cells, it has been shown recently that siRNA algorithms are poor predictors for shRNA efficiency (108). This is most likely because of the multistep process of shRNA biogenesis, which introduces additional structural constraints. For shRNAs, the following efficiency criteria were identified (referring to a 22 bp-long sense sequence):

1. An overall high A/U content of 9–18 bases with at least 7 A/U bases between positions 9 and 22.
2. The ratio of percent A/U bases from positions 9–22 and positions 1–8 should be higher than 1 to ensure the thermodynamic asymmetry.
3. For antisense strand integration into the RISC, there should be an A or U in position 22 of the sense strand and no U in position 3.
4. There should be either a U or A base at positions 10 or an A base at position 9. At positions 2 and 3 C bases are preferred. This increases Drosha cleavage efficiency.
5. There should be no G in position 3 to increase Drosha cleavage accuracy.
6. An A/U is preferred at position 13 for efficient target cleavage.
7. The sequence should not contain simple nucleotide repeats, such as AAAAAA, UUUUU, CCCC, or GGGG.

In addition, the efficiency of shRNA constructs largely depends on the design of the vectors containing the shRNA sequences. These should enable an efficient processing via the miRNA biogenesis pathway (30).

3.1.4. Primer Design for Long dsRNAs

The amplification of DNA templates for long dsRNAs by PCR from genomic or cDNA sources is a required step during the synthesis of long dsRNAs. Once suitable target sequences with optimal predicted specificity and efficiency (see Subheadings 3.1.1–3.1.3) have been identified, optimized primers need to be designed.

1. We recommend using the program Primer3 (42) with default parameters, which is also available as an online tool (see Note 8). The desired "amplicon size range" (length of dsRNA, e.g., 100–700 bp) should be adjusted. All other parameters can be used with the default settings.
2. Primer3 also reports a primer penalty, which scores the overall quality of the primer pair. According to our experience, this penalty should be below 10 to enable an efficient PCR synthesis.
3. We have also found that the standardized primer design with Primer3 facilitates similar PCR synthesis efficiency and that the selection of smaller windows for the "amplicon size range" (resulting in designs of comparable lengths) facilitates similar in vitro transcription reactions (see Note 9).
4. For in vitro transcription reactions, suitable bacteriophage polymerase promoter sequences, such as T7 (TAATACGACTC ACTATAGGG) or SP6 (ATTTAGGTGACACTATAG), must be added to the 5′ end of both primers (109).

3.1.5. Ranking of RNAi Reagents

The quality of an RNAi target site is determined by the predicted specificity and efficiency as described above. We recommend different strategies for ranking long dsRNAs and siRNAs (for the same target gene) based on the quality measures described.

1. Filtering and ranking of long dsRNAs: Filtering potential target sequences for regions of low sequence complexity should be a first step during the ranking of long dsRNAs. Further, all siRNAs (e.g., 19-bp sequences) contained in a potential RNAi target site should be specific and not show perfect homology to other, unintended target transcripts. These specific regions can then be used as templates for primer designs with Primer3. Each primer pair will define a potential dsRNA targeting the desired gene. As a last step, the number of predicted efficient siRNAs contained in these dsRNAs should be calculated and the dsRNA with the highest number of efficient silencers should be prioritized. For some genes it might not be possible to find longer sequence stretches containing only specific

siRNAs. In this case, dsRNAs should be first sorted for specificity (prioritizing dsRNAs containing low numbers of unspecific siRNAs) before sorting for efficiency occurs.

2. Filtering and ranking of siRNAs: siRNAs should not contain low-complexity regions and should not show perfect homology to any other than the intended target transcripts. The remaining sequences that pass these filters should be further sorted in the following way: first, for predicted efficiencies (prioritizing efficient silencers) and second for seed complement frequencies (prioritizing siRNAs with small numbers of seed matches to 3′-UTRs of unintended transcripts).

3.2. Case Studies

We developed computational tools that enable a straightforward design of optimized long dsRNA, siRNA, and shRNA sequences factoring in the quality measures discussed in Subheading 3.1 (Fig. 1). E-RNAi (39) is an online tool which facilitates the design of RNAi reagents currently for 13 different organisms. NEXT-RNAi (40) is a stand-alone software that was specifically developed for the design of large-scale RNAi libraries for any sequenced and annotated organism. In the following sections we will provide detailed methods on designing different types of RNAi reagents using E-RNAi. In the last section, we will show how to use NEXT-RNAi for designing novel genome-wide RNAi resources.

3.2.1. Design of Long dsRNAs

The design of a long dsRNA needs to take into account both the properties of the target sequence, e.g., its sequence complexity, as well as the properties of all potential siRNAs contained within the long dsRNA, such as their predicted target specificity and efficiency. Because long dsRNAs are often generated by in vitro transcription (IVT, see ref. 109), the design of suitable primer pairs to amplify IVT templates through PCR from genomic DNA or cDNAs must be implemented.

Here, we show how to use E-RNAi (http://www.e-rnai.org/) for the design of long dsRNAs targeting the *Drosophila* gene *flw* (CG2096), which encodes a serine/threonine phosphatase (110).

1. E-RNAi facilitates the selection of target sequences by several means: using a gene identifier, a gene sequence, or by visual selection via the generic genome browser (111). The option "ID or sequence input" (on the start page) enables identifier and sequence inputs and provides examples for all available organisms. Sequences can be pasted in raw or FASTA formats (see Note 10). In this example, we searched via the identifier *flw*, which is the official gene name. We selected "long dsRNA" as the reagent type and "*D. melanogaster*" as the organism.
2. E-RNAi identifies the gene model for *flw* as well as all annotated isoforms (FBtr0071446, FBtr0071447) and exons (flw:1–7). From here, target sequences can be selected. We want to

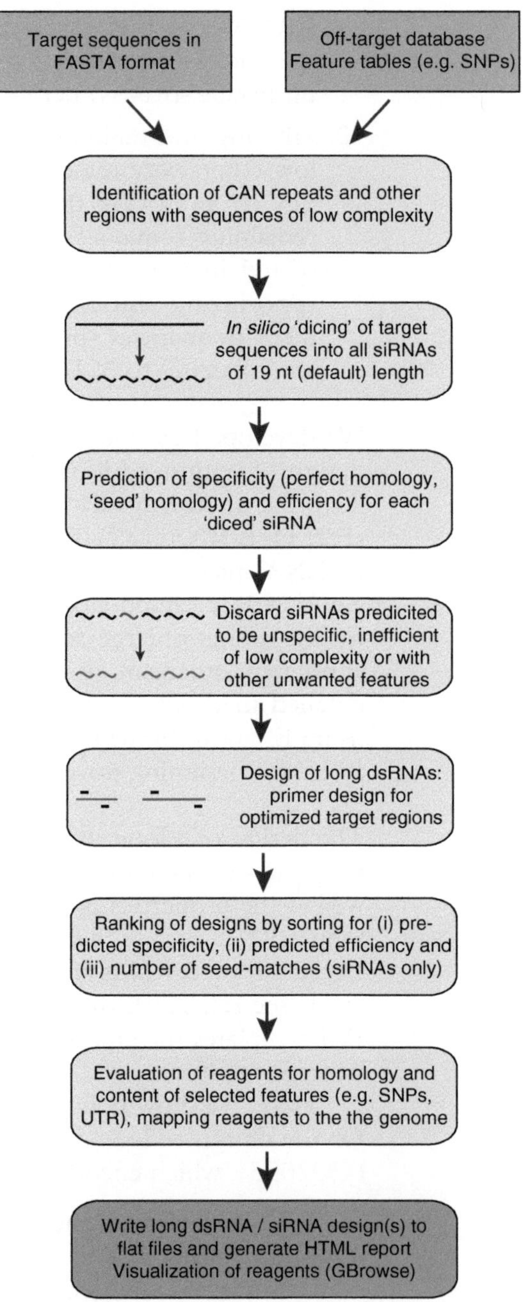

Fig. 1 Computational workflow of RNAi design. Target sequence files and sequence databases are provided in FASTA format. The target sequences are first filtered for six (default) or more contiguous CAN repeats and for other regions of low complexity (e.g., simple nucleotide repeats) using mdust (see Note 3). Sequences are then "diced" to generate all possible siRNA sequences with a default length of 19 bp and an offset of 1 bp. Subsequently, each siRNA is aligned to a user-defined "off-target" database (e.g., the whole transcriptome) to determine its specificity using Bowtie. The specificity is set to 1 if the siRNA targets a single gene or to 0 otherwise. In the next step, the predicted efficiency of each 19 bp siRNA is computed using either the "Rational" method according to Reynolds et al. (18) or the "weighted" method according to Shah et al. (38), assigning each siRNA an efficiency

target both *flw* isoforms via the exons flw:1 and flw:3–6 (flw:7 equals flw:6 and is not selected) as well as the FBtr0071446-specific isoform flw:2 (Fig. 2a).

3. Now E-RNAi allows the adjustment of several design options (Fig. 2c). The default settings E-RNAi uses result from our experience in designing long dsRNAs for *Drosophila* and should be a good starting point for other invertebrates, too.
Settings (I)–(VII) (Fig. 2c) are important to assess the quality of the input sequences. E-RNAi will use these settings as filters:

   I. The siRNA length for specificity prediction is set to 19 bp by default. E-RNAi will align all 19-bp siRNAs derived from the selected exon sequences (with an offset of 1 bp) to the *Drosophila* transcriptome to assess their specificity using Bowtie.

   II. The analysis for regions of low sequence complexity and sequences with more than 5 contiguous CAN repeats are enabled by default. Sequences of low complexity are identified using the mdust program (see Note 3).

   III. The siRNA seed-match analysis is disabled by default, but allows the user to upload a FASTA file containing 3′-UTR sequences, to adjust the seed length (6, 7, or 8 bases starting at position 2 in the antisense strand) and to define the seed-match cutoff (default 500), which is the maximum allowed number of siRNA seed matches to unintended 3′-UTRs. The seed complement frequency is then determined for every siRNA calculated from option (I) by aligning the siRNA seeds to uploaded 3′-UTRs using Bowtie. We usually do not use this filter during the design of long dsRNAs because it will yield hundreds or thousands of predicted 3′-UTR seed matches. It is not possible to predict the effect that they will have on the protein expression for their target.

   IV. The efficiency is predicted for all siRNAs calculated from option (I) and can be calculated using the "Weighted" or "Rational" methods (see above). The minimum required

Fig. 1. (continued) score between 0 and 100. Optionally, the seed complement frequency for each siRNA can be computed for any FASTA file provided (e.g., a file containing 3′-UTR sequences). siRNAs that do not pass the low-complexity filters, show perfect homology to multiple target genes or do not meet the user-defined cutoffs for efficiency or seed complement frequency are excluded from the queried target sequences. Remaining sequences are used as templates for primer design with Primer3 for long dsRNAs or are directly subjected to the final ranking for the design of siRNAs. Designs are ranked by (1) their predicted specificity, (2) their predicted efficiency and, in the case of siRNA designs, (3) their calculated seed complement frequency (preferring a low number of seed matches). Sequences can also be evaluated for additional features, such as homology to unintended transcripts, or SNP- and UTR-contents. Final designs can be visualized using GBrowse. All results are presented in a comprehensive HTML report and are also exported as text files.

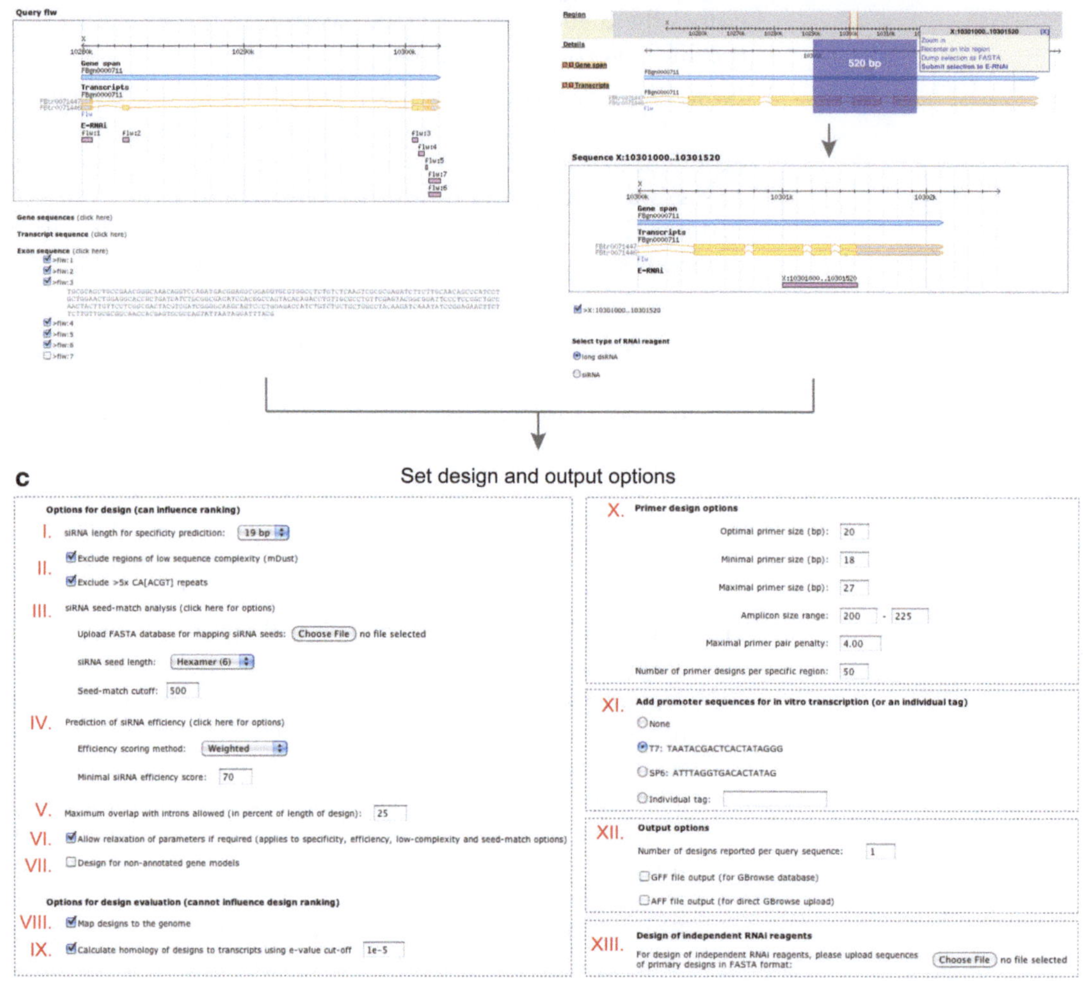

Fig. 2. Sequence selection and design options. (**a**) After querying E-RNAi with gene identifiers, gene sequences, transcript sequences, or exon sequences can be selected for design. (**b**) Individual sequences can be selected as design templates visually from GBrowse (*upper panel*). After sequence submission to E-RNAi, the type of reagent (long dsRNA or siRNA) can be selected (*lower panel*). (**c**) Options available for the design of long dsRNAs include settings for the prediction of specificity, efficiency, and low-complexity regions (*left panel*) as well as primer design settings (*upper right panel*). In addition, different output options, such as the number of designs per query sequence and different report formats, can be adjusted. The sequence of a previous design can also be uploaded in FASTA format to exclude it from a new (independent) design (*lower right panel*). Details for certain options are provided in the text.

efficiency score per siRNA can be adjusted between 0 and 100 to filter for efficient silencers. According to the "Weighted" method a score ≥63 defines an efficient siRNA, for the "Rational" method the score should be ≥66.7 (see Note 7). We selected the "Weighted" method and a score of 70 (default is 20). However, it is very unlikely to find a region longer than 100 bp that contains only efficient siRNAs. To maximize the efficiency, E-RNAi will sort designs according to the number of efficient siRNAs they contain, which is also reported in the output (Fig. 3a).

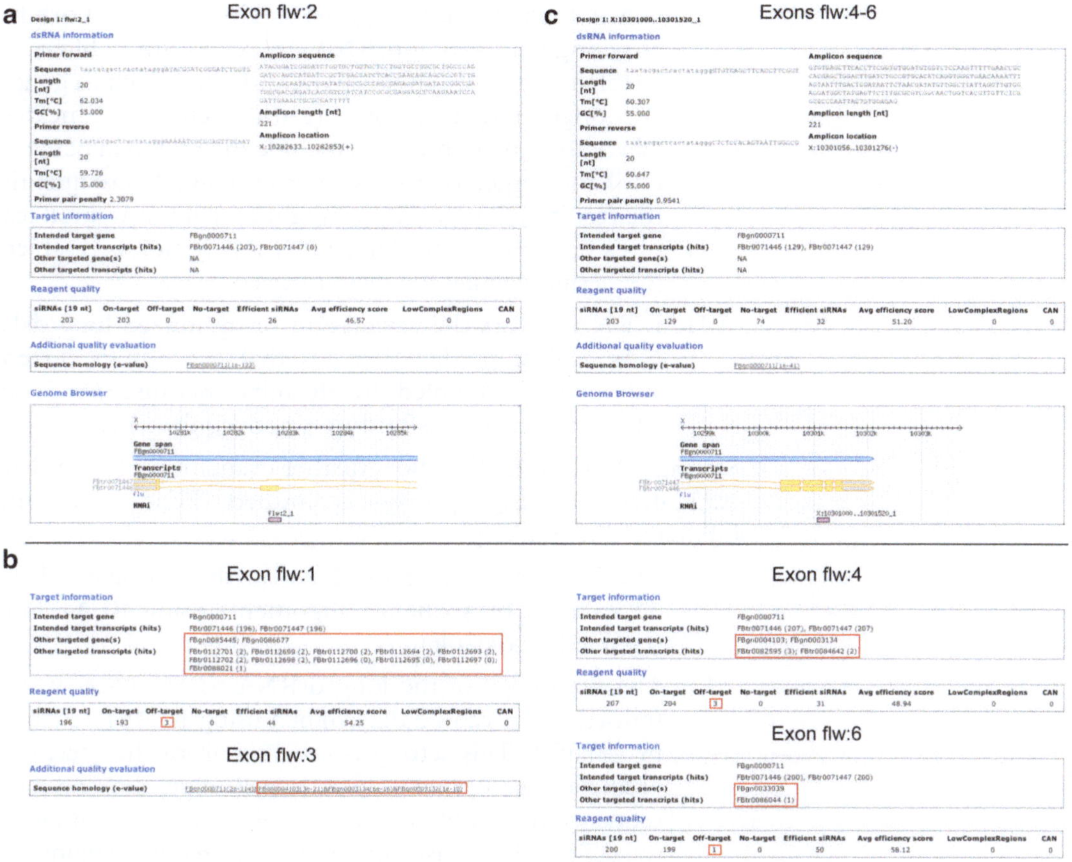

Fig. 3. Report for the design of a long dsRNA against the gene *flw*. (**a**) Optimal design for exon flw:2: "dsRNA information" summarizes the properties of the primers required to amplify the reagent from genomic or cDNA sources and also shows the full sequence of the construct and its length and location in the genome. The primers shown here were also tagged with the T7 promoter sequence (*lower letters*) at the 5′ end. "Target information" lists all "intended" and "other" target genes found as well as all transcripts belonging to them (including the number of siRNA "hits" to each transcript). "Reagent quality" summarizes the analyzed quality parameters and shows the number of specific ("on-target") siRNAs contained within the dsRNA, the number of "off-target" and "no-target" siRNAs as well as the number of "efficient siRNAs," the average siRNA efficiency and low-complexity information. In addition, all targets with significant "sequence homology" to the long dsRNA are listed. GBrowse visualizes the location of the designed reagent in the gene model. (**b**) Suboptimal designs for exons flw:1, -3, -4, -6: for these designs only parts of the reports are shown. Designs for exons flw:1, -4 and -6 all have predicted 19-bp homologies to unintended transcripts (*red boxes*), which can be read from the "Target information" and "Reagent quality" boxes. The design for flw:3 shows significant overall homology to unintended transcripts (*red box*), which can be read from the "Additional quality evaluation" box. (**c**) Optimal design for the region spanning *flw* exons 3–6: the design spans the intron between exon 4 and 5 (74 19-bp siRNAs have "No target") and has no predicted off-target effects.

V. The maximum allowed overlap of the long dsRNA with introns (by default 25% of its length in bp) is important only in the case that genomic sequences containing exons and introns are queried (see step 5 below). It is not important, for our example, where exons are selected as target sites.

VI. Settings (I)–(IV) define "favorable" target regions within the queried *flw* exons. Depending on the stringency applied, these regions might get too short for targeting with a long dsRNA. If "relaxation of filters" is enabled, E-RNAi can use an iterative approach to generate longer target regions. It will identify the nearest "favorable" neighbors and include the region in between until the target region is long enough. Consequently, the resulting long dsRNAs might contain unwanted features (e.g., off-target effects). If this option is disabled, the design will fail for these queries.

VII. E-RNAi allows the calculation of optimized long dsRNAs against predicted genes that are not in current annotations (disabled by default). Because *flw* is an annotated gene, this setting is not required here.

Settings (VIII) and (IX) allow further evaluation of long dsRNAs after the design process is finished. Thus, it will not influence their ranking:

VIII. The location of long dsRNAs on the genome will be determined by sequence alignments using BLAT (44) (enabled by default).

IX. The homology of the long dsRNA to sequences in the transcriptome can be determined using BLAST (enabled by default). This setting is different from the specificity prediction in option (I), as it will align the complete sequence to identify partial homologies with unintended transcripts. The sensitivity can be adjusted by defining the E-value cutoff for the BLAST search (see Note 5), which is set to 0.00001 by default.

The primer design options define the quality of primer pairs and the desired length of the long dsRNAs. E-RNAi will design primers for all optimal target regions identified from settings (I)–(VII):

X. According to our experience, long dsRNAs as small as 60 bp and as large as 800 bp can be synthesized efficiently and also trigger sufficient knockdowns. If multiple long dsRNAs are designed we recommend using the same primer size settings (e.g., the defaults) and a small amplicon size-range (e.g., 200–225 bp). This facilitates comparable PCR and IVT yields, as well as transfection efficiencies, when long dsRNAs are transfected during the RNAi experiment (see Note 9). The primer pair penalty is a measure of the overall quality of the primer pair. Pairs with higher penalty are predicted to perform less efficiently during the PCR. According to our experience, penalties up to 10 still allow sufficient PCR yields. The number of primer designs per specific region (default 50) defines the number of primer pairs suggested by Primer3 for each optimal target region. Increasing this value will yield more long dsRNAs covering the target region and also increase the likelihood of identifying better designs ultimately because E-RNAi will find the best prediction by sorting them according to their overall quality (see Subheading 3.1.5).

Settings (XI) and (XII) define additional output options:

XI. T7 or SP6 promoters (required for IVT reactions) or any individual tag can be added to the 5′-end of both primers (by default no tag is added).

XII. The number of designs reported per query can be adjusted (default 1). If multiple designs are requested, these do not necessarily target independent regions of the query sequence. Independent designs can be calculated using option (XIII). Further, generic feature files (GFF) and annotation feature files (AFF) for the designed long dsRNA might be requested. These are tab-delimited files that can be used to visualize the location of the long dsRNAs in a genome browser, which is, e.g., provided by FlyBase (and other model organism databases).

XIII. Long dsRNAs targeting the *flw* exons designed during this run can be forced to be independent of previously designed reagents targeting *flw*. This requires the upload of FASTA sequences for the available designs. E-RNAi will exclude the regions targeted by these reagents. For this case study no sequences are uploaded.

4. E-RNAi designs long dsRNAs targeting the *flw* exons according to the settings selected. A comprehensive report for designs will be generated, which can also be downloaded and locally viewed in any Web browser. However, the only design without predicted homologies to unintended targets is the design targeting exon flw:2, which is specific for transcript FBtr0071446. The main output for this design is shown and described in Fig. 3a. In addition to the HTML report, a tab-delimited file (compatible with any spreadsheet program) containing the same information is also available for download. Sequences are also downloadable in FASTA format. No long dsRNA targeting flw:5 could be calculated, because this exon was too short for the selected settings. Long dsRNAs targeting exons flw:1, flw:4, and flw:6 have unwanted 19-bp homologies to unintended transcripts (Fig. 3b), including other members of the PP1 family of phosphatases (*Pp1-87B* and *Pp1α-96A*) and ankyrin repeat-containing proteins (*Ank2*). The design for exon flw:3 has no such perfect 19-bp homologies to unintended transcripts, but shares a significant sequence homology with more than 80% sequence identity over stretches of at least 77 bp with all three members of the PP1 family of phosphatases (*Pp1-87B*, *Pp1α-96A*, *Pp1-13C*) (Fig. 3b). This can lead to

unspecific knockdowns of these phosphatases via partial homology. How an optimal long dsRNA targeting both isoforms of *flw* can be designed by refining the E-RNAi settings is explained in step 5 below.

5. The exons flw:3–6 are either too short for a long dsRNA or contain short regions of homology to unintended transcripts. One way to circumvent these issues is to decrease the desired length of the long dsRNA by setting the minimum to 80 bp (for the amplicon size-range in the primer settings), and follow steps 1–4 again as described above. However, we sometimes observe decreased synthesis efficiencies for very small amplicons. Because the *flw* exons flw:3–6 are interspersed only by three very small introns, it is possible to include these for the design. Using the GBrowse option of E-RNAi we selected a sequence spanning exons flw:4–6 including the introns in between (Fig. 2b). To enable targeting introns, option (V) was modified to allow up to 50% intron content; all other settings were used as described in step 3. This enables E-RNAi to design a specific long dsRNA of suitable size (203 bp) spanning exons flw:4 and flw:5 (Fig. 3c). In E-RNAi, the introns are subjected to the same quality evaluations as the exons to prevent off-target effects.

3.2.2. Design of siRNAs

Long dsRNAs are the preferred reagents used for RNAi in invertebrate model organisms because they can be synthesized quite easily (and cheaply) in the lab (109) and show high knockdown efficiencies. However, as the synthesis of siRNAs becomes more cost-effective they offer a good alternative for long dsRNA and under certain experimental conditions they even perform more efficiently (112). Using siRNAs enables targeting of very short sequences, which might be valuable for transcript-specific knockdowns or when very short exons need to be covered. Because siRNA experiments usually use just a single siRNA or a pool of three to four siRNAs, the strength of the knockdown is compromised by inefficient sequences and the phenotype can be easily affected by targeting of unintended transcripts. Here, we use E-RNAi to design siRNAs targeting a specific isoform of the *Drosophila* gene *wls/evi* (CG6210), which encodes a transmembrane protein required for Wnt secretion (113).

1. Using FlyBase (48), we downloaded the sequence for *wls* exon 2 (wls:2, 96 bp long), an exon specific to the transcript FBtr0076218, which is one of the two annotated transcripts of *wls*. The sequence was pasted in FASTA format in the sequence box using the "ID or sequence input option," "siRNA" was selected as reagent type and "*D. melanogaster*" as organism.
2. E-RNAi aligns the input sequence to the *Drosophila* genome using BLAT and displays the mapping in GBrowse (Fig. 4a).
3. Now E-RNAi allows adjustment of the design options (see Fig. 2c for long dsRNA designs). All settings described for

a **Query evi:2**

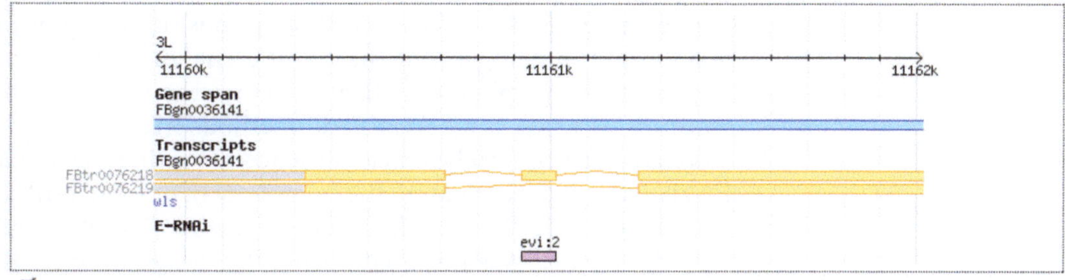

>evi:2
TTCCTTTAGACCAGAAAGTTGAAGATTGGGCCGGAATCGTATATTTTTACACAAAAGCTTTTTTTTTTCAATTGCATAAGGCAAATGAAAGTAAAG

b

Design 3: wls:2_3

siRNA information

siRNA sequence	GATTGGGCCGGAATCGTAT
Position in queried target	24
Length [nt]	19
siRNA location(s)	3L:11160973..11160991(-)

Target information

Intended target gene	FBgn0036141
Intended target transcripts (hits)	FBtr0076218 (1), FBtr0076219 (0)
Other targeted gene(s)	NA
Other targeted transcripts (hits)	NA

Reagent quality

siRNAs [19 nt]	On-target	Off-target	No-target	SCF	Efficiency score	LowComplexRegions	CAN
1	1	0	0	150	82.01	0	0

Additional quality evaluation

Sequence homology (e-value)	FBgn0036141(9e-04)

Genome Browser

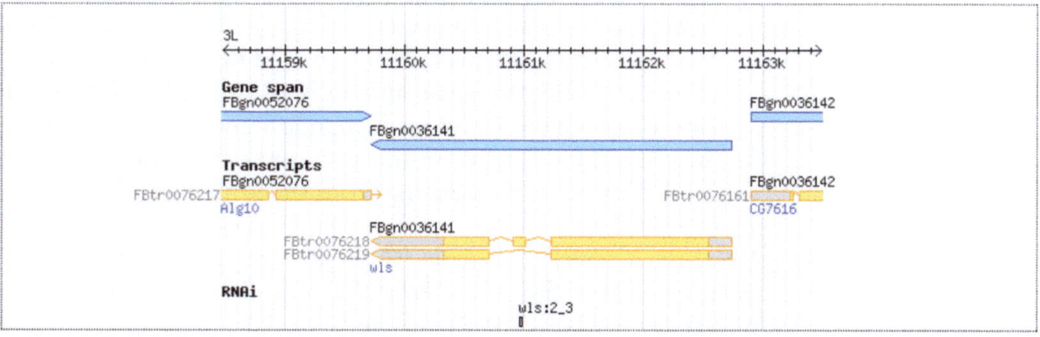

Fig. 4. Report for the design of siRNAs against the gene *wls*. (**a**) Selection of *wls* exon 2 as the target site, which is specific for the transcript FBtr0076218. (**b**) Optimal siRNA design for the *wls* exon: "siRNA information" shows the 19 bp-long siRNA sequence, its position in the input sequence as well as its position in the genome. "Target information" lists all "intended" and "other" target genes found. This siRNA is specific for *wls* and targets the transcript FBtr0076218, but not FBtr0076219. "Reagent quality" summarizes the analyzed quality parameters and shows that this siRNA is specific ("on-target"), has 150 seed matches ("SCF," seed complement frequency) in annotated 3′-UTRs, an efficiency score of ~82 and contains no low-complexity sequences ("LowComplexRegions" or "CAN" repeats). Further, there is no "sequence homology" to unintended genes. The genome browser image visualizes its location.

the design of long dsRNAs in Subheading 3.2.1 also apply for the design of siRNAs (except for the options regarding primer design and tagging), but have different default settings. Importantly, the minimum efficiency in option (IV) is set to a score of 70 using the "Weighted" method. siRNAs with this score are predicted to be efficient silencers. Further, the default E-value cutoff for evaluating the homology of designed siRNAs in option (IV) is 0.1, to adjust for the smaller size of siRNAs. One option that was not used during the design of long dsRNAs is the siRNA seed match analysis, which is more important during the design of siRNAs and was applied in step III below.

III. We downloaded a pre-build file containing 18,807 annotated 3′-UTRs from FlyBase. This file was uploaded to E-RNAi as FASTA database to identify siRNA seed matches. The seed match cutoff was set to 500 for aligning octamers (positions 2-9 in the siRNA antisense strand). It has been shown that siRNAs with low seed complement frequencies induce fewer off-target phenotypes than siRNAs with more abundant 3′-UTR complements (90) (see Note 6).

IV. We requested four siRNAs to be designed. They will be independent and could be used in a pool for an experiment.

4. E-RNAi designs four siRNAs targeting exon 2 of *wls*. The output is structured in a similar way as for long dsRNAs and exemplified for one siRNA in Fig. 4b. All four siRNAs are specific for their intended target and have an efficiency score of at least 75, and three out of four siRNAs have seed complement frequencies of less than 300.
5. E-RNAi reports the "raw" siRNA sense sequence. For RNA synthesis, "Ts" would need to be changed to "Us" and 2-nucleotide overhangs (usually "TT") would need to be added to the 3′-end of both the sense and antisense strands. Biotechnology companies have developed chemical modifications that may increase the efficiency of the antisense strand further or diminish off-target effects arising from the sense strand.

3.2.3. Design of shRNAs

Recently, a novel transgenic RNAi resource based on shRNAs has been generated for *Drosophila* (26) using a design method optimized for siRNAs (107) to predict efficient sequences. In this study, the shRNAs tested did not only prove to be very effective at silencing gene expression, but they also could be used to knockdown transcripts in the female germline, where resources based on long dsRNAs do not function. shRNA libraries are more widely used in mammalian systems than siRNA libraries because they can be delivered into any cell type using lentiviral transfection, which can greatly facilitate pooled screening approaches (30, 31). However, recent studies found that shRNAs have sequence requirements different from siRNAs making siRNA efficiency rules poor predictors for shRNA efficiency (108, 114, 115).

E-RNAi in its current version does not implement an shRNA-specific efficiency predictor. The sequence preferences derived from a current study (108) are described in Subheading 3.1.3 and could be used for identifying efficient sequences. Sequences that pass the efficiency filter could then be evaluated for their specificity using the "Evaluation" option of E-RNAi. This requires just pasting the sequences in FASTA format, selecting "siRNA" as the reagent type and selecting the desired organism. Similarly to the design of RNAi reagents, E-RNAi offers different options for sequence evaluations. However, default settings will work fine. Only the length of the siRNA/shRNA for specificity evaluation needs to be adjusted properly (19 bp by default).

Depending on the application, shRNAs are usually cloned into specific vectors. Consequently, any restriction sites required for cloning must be excluded from the shRNA sequence.

3.2.4. Design of Multiple Independent RNAi Reagents for the Same Target Gene

In order to validate phenotypes from RNAi experiments, validation experiments using independent RNAi reagents are recommended (35). E-RNAi provides different ways to design independent RNAi reagents:

1. The sequence of a previously designed reagent can be submitted as a FASTA file in the process of designing a new reagent (see Subheading 3.2.1, option (XIII)). E-RNAi will exclude this sequence during the design.
2. E-RNAi can be queried with multiple sequences for one run. As an example, multiple exons of the same transcript can be queried and the resulting designs will be independent. Similarly, after querying E-RNAi with a sequence identifier, all annotated exon sequences for this gene are shown. Multiple sequences can be selected resulting in independent designs (see Subheading 3.2.1). For the design of independent siRNAs, selecting more than one design for the output under design option (VII) will be sufficient to obtain independent reagents (see Subheading 3.2.2).
3. The possibility of submitting sequences from GBrowse to E-RNAi allows a precise selection of potential target sequences (see Subheading 3.2.1 and Fig. 2b). Multiple, independent sequences can be submitted this way resulting in independent designs.

3.2.5. Design of Genome-Wide RNAi Libraries

We developed the software package NEXT-RNAi that allows the design and evaluation of genome-wide RNAi libraries using the workflow shown in Fig. 1 (40). NEXT-RNAi is a stand-alone Perl program and depends on several additional programs (see Subheading 2). Here, we describe how NEXT-RNAi can be used to design a novel library of long dsRNAs for the mosquito *A. gambiae*, which is widely studied to analyze the mechanism of innate immunity as a vector for *Plasmodium falciparum*. RNAi by long dsRNAs has been demonstrated in *A. gambiae* in vitro and in vivo and leads to efficient depletion of mRNAs (69).

1. Obtaining sequence and sequence-feature files: VectorBase is the model organism database for invertebrate vectors of human pathogens, including *A. gambiae* (68). NEXT-RNAi requires a defined set of input files with fixed formats. Those can either be downloaded directly from VectorBase or generated via its database management system BioMart (116), a tool that is becoming available for an increasing number of model organism databases to facilitate sequence and feature downloads. The following files are required:

   (a) A file containing target sequences: we recommend using exon sequences as target sites. This facilitates amplification of templates for long dsRNAs from genomic DNA and cDNA. If whole transcripts are used as target sites, cDNA libraries are required for the initial amplification step. Exon sequences in FASTA format can be downloaded via the BioMart implementation of VectorBase. We identified ~56,000 exon sequences with an average length of 418 bp, offering suitable target sites. To obtain even more potential target sequences, we split all exons longer than 560 bp into two halves, resulting in ~76,600 sequences.

   (b) An off-target transcript database: In order to assess the specificity of target sequences, NEXT-RNAi requires a database containing all annotated transcripts. This file (in FASTA format) can be directly downloaded from VectorBase. It contains ~15,000 transcript sequences. NEXT-RNAi uses Bowtie for aligning siRNAs to the transcriptome. Bowtie requires building of an index from the transcriptome FASTA file. This can be done using the *bowtie_build* script included in the Bowtie software. The Bowtie Web page also offers many pre-built indices that can be downloaded and directly used.

   In order to evaluate the designed reagents for partial homologies to unintended transcripts a BLAST index needs to be built from the transcriptome FASTA file using the *formatdb* script included in the BLAST program.

   (c) A file linking transcripts and genes: NEXT-RNAi needs to know, which annotated transcripts belong to the same gene in order to discriminate between on- and off-target effects. This requires a simple tab-delimited file with two columns: the column "Target" contains the transcript identifier and the column "TargetGroup" contains the corresponding gene identifier (for *Anopheles,* e.g., the transcript "AGAP004677-RB" belongs to the gene AGAP004677). This file can be generated using BioMart, querying for "Ensembl Transcript ID" and "Ensembl

Gene ID." Only the header names need to be changes to "Target" and "TargetGroup," respectively.

These are the minimal input requirements for NEXT-RNAi. In order to align reagents to the genome, a FASTA file containing the *Anopheles* genome would be required (and would require building it into a Bowtie index), which is also available from VectorBase. Additional sequence features could also be analyzed, e.g., whether the sequences contain annotated SNPs or conserved miRNA seeds. Examples of how to prepare such files are available from the NEXT-RNAi wiki (see Note 11).

2. Adjusting the design options: The user can modify basically every step during the design of RNAi reagents implemented in NEXT-RNAi. Although most of these can be used with default values, we recommend adjusting the options as follows:
   (a) Set the desired minimum and maximum length of the long dsRNA (we recommend using a rather small window, e.g., 200–225 bp).
   (b) Enable the analysis for low-complexity regions (simple nucleotide repeats and ≥6x CAN repeats).
   (c) Select the siRNA length for the off-target analysis (we recommend 19 bp).
   (d) Set the definition of the efficiency algorithm ("Weighted" or "Rational") used for assessing the siRNA efficiency and the efficiency cutoff. For long dsRNAs, we recommend to set the cutoff to 0. This will not set any restrictions to the efficiency of individual siRNAs, but NEXT-RNAi will finally sort all potential designs according to their average efficiency.
   (e) Enable the redesign option, which allows NEXT-RNAi to relax specificity and efficiency parameters (see Subheading 3.2.1).
   (f) Enable the analysis for homology to unintended transcripts.
   (g) Adjust the number of long dsRNAs to be designed per input sequence (in this case per exon). We recommend just designing one sequence per exon.
   (h) Provide the location of the file linking transcripts and genes.

   These options can be set using an options file (with an *OPTION=VALUE* structure) and can be provided to NEXT-RNAi during its start. Alternatively, NEXT-RNAi offers an interactive starting mode, which helps to define these settings. Further options and examples are explained in the NEXT-RNAi wiki (see Note 11).

3. Starting the design: NEXT-RNAi is a Perl script that can be started, e.g., from a command-line. It requires some additional information for the startup, including (a) the location of the

sequence input-file and (b) the knowledge of whether these sequences are templates for a *de novo* design or for the evaluation of RNAi reagents, (c) the type of reagents to be designed, (d) the location of the Bowtie index of the transcriptome, (e) the location of the options file, and (f) an identifier for the design process (e.g., "agambiae"). Help about how to start NEXT-RNAi and all available options can be obtained by starting NEXT-RNAi with "–h."

4. The output: NEXT-RNAi produces a comprehensive HTML output providing links to detailed reports for every long dsRNA designed (see e.g., Fig. 3 and Note 12). Further, a tab-delimited file (compatible with virtually any spreadsheet software) containing the same information for all designs is provided.

5. Working with the output: Overall, NEXT-RNAi designed 48,128 long dsRNAs covering 12,286 of the 13,319 annotated genes (92%). About 89% of the targeted genes are covered by a long dsRNA with no predicted off-target effects, and 86% are covered by at least two independent reagents. We recommend identifying the best two independent designs per gene by sorting the long dsRNAs for their target gene, followed by sorting steps for specificity and efficiency. In addition, reagents that target other, unintended genes (either via 19-bp homologies or significant overall homology) or that contain low-complexity regions should be filtered out. These steps could be easily accomplished using spreadsheet software. Next, to reach a nearly complete coverage, genes that are not covered so far or are just covered by a long dsRNA with predicted off-target effects as well as genes that are covered only by one long dsRNA could be repeated in a second run. For this run we recommend decreasing the minimum amplicon length allowed (e.g., to 80 bp). This will allow NEXT-RNAi to target smaller regions. Further, for genes for which a second, independent design is desired, the first design should be provided to NEXT-RNAi. These sequences will be excluded for targeting by a second reagent. After this run, a library with nearly complete genome coverage by two independent designs per gene will be achieved.

6. Synthesizing the library: The long dsRNAs designed for the *Anopheles* genome could be synthesized by amplification of DNA templates from genomic DNA or cDNA using PCR, followed by in vitro transcription of the DNA using T7 polymerase. Primers designed by NEXT-RNAi all have similar properties (e.g., melting temperatures and GC content) facilitating efficient synthesis in high-throughput formats, such as 96-well plates. To this end, we recommend using a two-step PCR approach for DNA amplification, which requires tagging of primers with re-amplifiable tags (109).

4. Notes

1. If genes have multiple annotated isoforms, regions common to all of them are very likely to be real. We identify such regions by comparing the absolute chromosomal locations of all exons (and UTRs) of all transcripts of a certain gene. Parts of the exons that exist in all isoforms are considered potential target sites. Common regions that are too short for the design of a long dsRNA (e.g., <120 bp) can also be extended on both ends (e.g., by 70 bp). Very long common regions (e.g., >700 bp) can be split into two halves to provide more independent target sites. Targeting specific isoforms of a gene requires the identification of exons or UTRs that are unique for this transcript.
2. The optimal design of RNAi reagents requires annotations for the genome and the transcriptome, including the information about which transcripts belong to the same gene. Gene functions can then be annotated by BLAST searches to other, well-annotated organisms, preferentially related species. This allows the identification of target genes with desired functions. Then, based on the transcript sequences, RNAi reagents can be designed. However, in the case of long dsRNAs, which requires synthesis by PCRs, DNA templates would need to be amplified from cDNA (not from genomic DNA), since regions in the transcripts might be interspersed by large introns. If assemblies of the genome (contigs or even chromosomes) are available this could be tested by aligning the long dsRNAs using BLAT.
3. We analyze sequences for low-complexity regions using the so-called mdust filter. This filter is available as BioPerl Module (at http://www.bioperl.org/wiki/Mdust) and as stand-alone C script (at http://compbio.dfci.harvard.edu/tgi/software/ or via our NEXT-RNAi packed archives at http://www.nextrnai.org/). We use the software with default settings.
4. For aligning many siRNA sequences to the transcriptome or genome we recommend to use Bowtie (http://bowtie-bio.sourceforge.net/), which was developed for aligning short sequences from next-generation sequencing data. We use it with the options "–f" (reading from FASTA files, see Note 10), "–v 0" (allowing no mismatches). BLAST could also be used; however, Bowtie is significantly faster for aligning short sequences.
5. To determine the overall homology of an RNAi reagent to all annotated transcripts we use BLAST to align the sequence to the transcriptome. This will uncover partial homologies to unintended transcripts. Except excluding low-complexity sequences ("–F F") and setting an E-value cutoff (long dsRNAs: 0.00001, siRNAs: 0.1) we use default BLAST parameters.

The low E-value cutoffs we use avoid reporting of insignificant homologies. However, other tools such as WU-BLAST (now AB-BLAST) might be even more sensitive for alignments of very short sequences considering mismatches.

6. We consider only seed complement frequencies during the design of siRNAs, because analyzing the seed matches for all siRNAs contained in a long dsRNA will be very time-consuming and will result in an immense amount of potential off-target effects. But even for single siRNAs the predictive value of the seed complement frequency is rather weak. Anderson et al. (90) found a correlation of low seed complement frequencies (<350) with a small number of off-target effects and medium seed complement frequencies (~2,500) with a high number of off-target effects. However, for high seed complement frequencies (>3,800) the number of off-target effects decreased again. Further, even if seed complement frequencies are low, it is not clear which of the targeted transcripts are really silenced.

7. The "Rational" prediction includes eight criteria to score the efficiency of an siRNA (sense strand): (a) G/C content (+1 if between 30% and 52%), (b) A/U bases at positions 15–19 (+1 each), (c) absence of internal repeats (+1 if melting temperature of potential internal hairpin <20°C), (d) A base at position 19 (+1), (e) A base at position 3 (+1), (f) U base at position 10 (+1), (g) G/C base at position 19 (−1), and (h) G base at position 13 (−1). This scoring provides a score range from −2 to 10. In the studies done by Reynolds et al. (18) siRNAs with scores ≥6 were found to be efficient silencers. We normalized this score to obtain a range between 0 and 100 using the following formula:

   FinalScore = ((Score of siRNA − minimal possible score)/(maximal possible score − minimal possible score)) × 100

   For an siRNA to be considered an efficient silencer, its score must be higher than 66.67.

   The "weighted" prediction includes twelve criteria to predict the efficiency of an siRNA (sense strand): (a) A/U base at position 1 (−1.4), (b) G/C base at position 1 (+1.11), (c) A base at position 6 (+0.7), (d) U base at position 10 (+0.25), (e) G base at position 13 (−1.66), (f) U base at position 13 (+0.31), (g–j) A/U bases at position 16–19 (+0.74, +1.2, +1.44, +0.87), (k) G/C base at position 19 (−1.02), and (l) G/C content (+0.42 if between 30% and 55%). This scoring provides a score range from −4.08 to 7.04. We applied the same normalization to obtain a score range between 0 and 100 as described for the "Rational" score. Shah et al. (38) found in their study that the average score of the most potent siRNAs was 63.

8. The Primer3 online tool is available at http://frodo.wi.mit.edu/primer3/.

9. We currently design long dsRNAs in the size range of 200 and 225 bp if possible (and decrease the minimum allowed size otherwise). We found that this results in more similar PCR and IVT yields. If transfection reagents (e.g., lipid-based transfection) are used for RNAi delivery into the cells, this also facilitates similar transfection efficiencies.
10. The FASTA format is a general sequence format composed of a header starting with ">" followed by the sequence identifier. The next line contains the sequence. More detailed information is available at http://en.wikipedia.org/wiki/FASTA_format. "Raw" sequence means that the sequence is provided without any identifier.
11. Detailed descriptions of how to use the E-RNAi online service (http://www.e-rnai.org) are available via the wiki at http://b110-wiki.dkfz.de/signaling/wiki/display/ernai/. The NEXT-RNAi software, installation packages and instructions for Linux and Mac operation systems, and further documentations are accessible via http://www.nextrnai.org/.
12. A detailed description of the preparation of files required for designing a genome-wide library of long dsRNAs for *Anopheles* and instructions for running NEXT-RNAi with these files is available at http://b110-wiki.dkfz.de/signaling/wiki/display/nextrnai/Anopheles+gambiae.

References

1. Jorgensen EM, Mango SE (2002) The art and design of genetic screens: caenorhabditis elegans. Nat Rev Genet 3:356–369
2. St Johnston D (2002) The art and design of genetic screens: Drosophila melanogaster. Nat Rev Genet 3:176–188
3. Chapman EJ, Carrington JC (2007) Specialization and evolution of endogenous small RNA pathways. Nat Rev Genet 8: 884–896
4. Fire A, Xu S, Montgomery MK, Kostas SA, Driver SE, Mello CC (1998) Potent and specific genetic interference by double-stranded RNA in Caenorhabditis elegans. Nature 391:806–811
5. Boutros M, Ahringer J (2008) The art and design of genetic screens: RNA interference. Nat Rev Genet 9:554–566
6. van der Krol AR, Mur LA, Beld M, Mol JN, Stuitje AR (1990) Flavonoid genes in petunia: addition of a limited number of gene copies may lead to a suppression of gene expression. The Plant cell 2:291–299
7. van der Krol AR, Mur LA, de Lange P, Mol JN, Stuitje AR (1990) Inhibition of flower pigmentation by antisense CHS genes: promoter and minimal sequence requirements for the antisense effect. Plant Mol Biol 14:457–466
8. Napoli C, Lemieux C, Jorgensen R (1990) Introduction of a chimeric chalcone synthase gene into petunia results in reversible co-suppression of homologous genes in trans. Plant Cell 2:279–289
9. Zamore PD, Tuschl T, Sharp PA, Bartel DP (2000) RNAi: double-stranded RNA directs the ATP-dependent cleavage of mRNA at 21 to 23 nucleotide intervals. Cell 101:25–33
10. Bernstein E, Caudy AA, Hammond SM, Hannon GJ (2001) Role for a bidentate ribonuclease in the initiation step of RNA interference. Nature 409:363–366
11. Bernstein SL, Guo Y, Kelman SE, Flower RW, Johnson MA (2003) Functional and cellular responses in a novel rodent model of anterior ischemic optic neuropathy. Investig Ophthalmol Vis Sci 44:4153–4162
12. Hammond SM, Bernstein E, Beach D, Hannon GJ (2000) An RNA-directed nuclease mediates post-transcriptional gene silencing in Drosophila cells. Nature 404:293–296

13. Ketting RF, Fischer SE, Bernstein E, Sijen T, Hannon GJ, Plasterk RH (2001) Dicer functions in RNA interference and in synthesis of small RNA involved in developmental timing in C. elegans. Genes Dev 15:2654–2659
14. Martinez J, Patkaniowska A, Urlaub H, Luhrmann R, Tuschl T (2002) Single-stranded antisense siRNAs guide target RNA cleavage in RNAi. Cell 110:563–574
15. Chiu YL, Rana TM (2002) RNAi in human cells: basic structural and functional features of small interfering RNA. Mol Cell 10:549–561
16. Chiu YL, Rana TM (2003) siRNA function in RNAi: a chemical modification analysis. RNA (New York, NY) 9:1034–1048.
17. Khvorova A, Reynolds A, Jayasena SD (2003) Functional siRNAs and miRNAs exhibit strand bias. Cell 115:209–216
18. Reynolds A, Leake D, Boese Q, Scaringe S, Marshall WS, Khvorova A (2004) Rational siRNA design for RNA interference. Nat Biotechnol 22:326–330
19. Schwarz DS, Hutvagner G, Du T, Xu Z, Aronin N, Zamore PD (2003) Asymmetry in the assembly of the RNAi enzyme complex. Cell 115:199–208
20. Ui-Tei K, Naito Y, Takahashi F, Haraguchi T, Ohki-Hamazaki H, Juni A, Ueda R, Saigo K (2004) Guidelines for the selection of highly effective siRNA sequences for mammalian and chick RNA interference. Nucleic Acids Res 32:936–948
21. Liu Q, Paroo Z (2010) Biochemical principles of small RNA pathways. Annu Rev Biochem 79:295–319
22. Fischer JA, Giniger E, Maniatis T, Ptashne M (1988) GAL4 activates transcription in Drosophila. Nature 332:853–856
23. Clemens JC, Worby CA, Simonson-Leff N, Muda M, Maehama T, Hemmings BA, Dixon JE (2000) Use of double-stranded RNA interference in Drosophila cell lines to dissect signal transduction pathways. Proc Natl Acad Sci U S A 97:6499–6503
24. Ulvila J, Parikka M, Kleino A, Sormunen R, Ezekowitz RA, Kocks C, Ramet M (2006) Double-stranded RNA is internalized by scavenger receptor-mediated endocytosis in Drosophila S2 cells. J Biol Chem 281: 14370–14375
25. Saleh MC, van Rij RP, Hekele A, Gillis A, Foley E, O'Farrell PH, Andino R (2006) The endocytic pathway mediates cell entry of dsRNA to induce RNAi silencing. Nat Cell Biol 8:793–802
26. Ni JQ, Zhou R, Czech B, Liu LP, Holderbaum L, Yang-Zhou D, Shim HS, Tao R, Handler D, Karpowicz P, Binari R, Booker M, Brennecke J, Perkins LA, Hannon GJ, Perrimon N (2011) A genome-scale shRNA resource for transgenic RNAi in Drosophila. Nat Methods 8:405–407
27. Elbashir SM, Harborth J, Lendeckel W, Yalcin A, Weber K, Tuschl T (2001) Duplexes of 21-nucleotide RNAs mediate RNA interference in cultured mammalian cells. Nature 411:494–498
28. Sledz CA, Holko M, de Veer MJ, Silverman RH, Williams BR (2003) Activation of the interferon system by short-interfering RNAs. Nat Cell Biol 5:834–839
29. Bernards R, Brummelkamp TR, Beijersbergen RL (2006) shRNA libraries and their use in cancer genetics. Nat Methods 3:701–706
30. Chang K, Elledge SJ, Hannon GJ (2006) Lessons from Nature: microRNA-based shRNA libraries. Nat Methods 3:707–714
31. Root DE, Hacohen N, Hahn WC, Lander ES, Sabatini DM (2006) Genome-scale loss-of-function screening with a lentiviral RNAi library. Nat Methods 3:715–719
32. Buchholz F, Kittler R, Slabicki M, Theis M (2006) Enzymatically prepared RNAi libraries. Nat Methods 3:696–700
33. Fuchs F, Boutros M (2006) Cellular phenotyping by RNAi. Brief Funct Genomics Proteomics 5:52–56
34. Dietzl G, Chen D, Schnorrer F, Su KC, Barinova Y, Fellner M, Gasser B, Kinsey K, Oppel S, Scheiblauer S, Couto A, Marra V, Keleman K, Dickson BJ (2007) A genome-wide transgenic RNAi library for conditional gene inactivation in Drosophila. Nature 448:151–156
35. Echeverri CJ, Beachy PA, Baum B, Boutros M, Buchholz F, Chanda SK, Downward J, Ellenberg J, Fraser AG, Hacohen N, Hahn WC, Jackson AL, Kiger A, Linsley PS, Lum L, Ma Y, Mathey-Prevot B, Root DE, Sabatini DM, Taipale J, Perrimon N, Bernards R (2006) Minimizing the risk of reporting false positives in large-scale RNAi screens. Nat Methods 3:777–779
36. Perrimon N, Mathey-Prevot B (2007) Off Targets and Genome Scale RNAi Screens in Drosophila. Fly 1:5
37. Amarzguioui M, Prydz H (2004) An algorithm for selection of functional siRNA sequences. Biochem Biophys Res Commun 316:1050–1058
38. Shah JK, Garner HR, White MA, Shames DS, Minna JD (2007) sIR: siRNA information resource, a web-based tool for siRNA sequence design and analysis and an open access siRNA database. BMC Bioinformatics 8:178

39. Horn T, Boutros M (2010) E-RNAi: a web application for the multi-species design of RNAi reagents–2010 update. Nucleic Acids Res 38:W332–339
40. Horn T, Sandmann T, Boutros M (2010) Design and evaluation of genome-wide libraries for RNA interference screens. Genome Biol 11:R61
41. Langmead B, Trapnell C, Pop M, Salzberg SL (2009) Ultrafast and memory-efficient alignment of short DNA sequences to the human genome. Genome Biol 10:R25
42. Rozen S, Skaletsky H (2000) Primer3 on the WWW for general users and for biologist programmers. Methods Mol Biol 132:365–386
43. Altschul SF, Gish W, Miller W, Myers EW, Lipman DJ (1990) Basic local alignment search tool. J Mol Biol 215:403–410
44. Kent WJ (2002) BLAT–the BLAST-like alignment tool. Genome Res 12:656–664
45. Hofacker IL (2003) Vienna RNA secondary structure server. Nucleic Acids Res 31: 3429–3431
46. Pruitt KD, Tatusova T, Klimke W, Maglott DR (2009) NCBI reference sequences: current status, policy and new initiatives. Nucleic Acids Res 37:D32–36
47. Flicek P, Aken BL, Ballester B, Beal K, Bragin E, Brent S, Chen Y, Clapham P, Coates G, Fairley S, Fitzgerald S, Fernandez-Banet J, Gordon L, Graf S, Haider S, Hammond M, Howe K, Jenkinson A, Johnson N, Kahari A, Keefe D, Keenan S, Kinsella R, Kokocinski F, Koscielny G, Kulesha E, Lawson D, Longden I, Massingham T, McLaren W, Megy K, Overduin B, Pritchard B, Rios D, Ruffier M, Schuster M, Slater G, Smedley D, Spudich G, Tang YA, Trevanion S, Vilella A, Vogel J, White S, Wilder SP, Zadissa A, Birney E, Cunningham F, Dunham I, Durbin R, Fernandez-Suarez XM, Herrero J, Hubbard TJ, Parker A, Proctor G, Smith J, Searle SM (2010) Ensembl's 10th year. Nucleic Acids Res 38:D557–562
48. Tweedie S, Ashburner M, Falls K, Leyland P, McQuilton P, Marygold S, Millburn G, Osumi-Sutherland D, Schroeder A, Seal R, Zhang H (2009) FlyBase: enhancing Drosophila gene ontology annotations. Nucleic Acids Res 37:D555–559
49. Kim HS, Murphy T, Xia J, Caragea D, Park Y, Beeman RW, Lorenzen MD, Butcher S, Manak JR, Brown SJ (2010) BeetleBase in 2010: revisions to provide comprehensive genomic information for Tribolium casttaneum. Nucleic Acids Res 38:D437–442
50. Harris TW, Antoshechkin I, Bieri T, Blasiar D, Chan J, Chen WJ, De La Cruz N, Davis P, Duesbury M, Fang R, Fernandes J, Han M, Kishore R, Lee R, Muller HM, Nakamura C, Ozersky P, Petcherski A, Rangarajan A, Rogers A, Schindelman G, Schwarz EM, Tuli MA, Van Auken K, Wang D, Wang X, Williams G, Yook K, Durbin R, Stein LD, Spieth J, Sternberg PW (2010) WormBase: a comprehensive resource for nematode research. Nucleic Acids Res 38:D463–467
51. Wang Z, Gerstein M, Snyder M (2009) RNA-Seq: a revolutionary tool for transcriptomics. Nat Rev Genet 10:57–63
52. Metzker ML (2010) Sequencing technologies—the next generation. Nat Rev Genet 11: 31–46
53. Legeai F, Shigenobu S, Gauthier JP, Colbourne J, Rispe C, Collin O, Richards S, Wilson AC, Murphy T, Tagu D (2010) AphidBase: a centralized bioinformatic resource for annotation of the pea aphid genome. Insect Mol Biol 19(Suppl 2):5–12
54. Mutti NS, Park Y, Reese JC, Reeck GR (2006) RNAi knockdown of a salivary transcript leading to lethality in the pea aphid, Acyrthosiphon pisum. J Insect Sci 6:1–7
55. Jaubert-Possamai S, Le Trionnaire G, Bonhomme J, Christophides GK, Rispe C, Tagu D (2007) Gene knockdown by RNAi in the pea aphid Acyrthosiphon pisum. BMC Biotechnol 7:63
56. Munoz-Torres MC, Reese JT, Childers CP, Bennett AK, Sundaram JP, Childs KL, Anzola JM, Milshina N, Elsik CG (2011) Hymenoptera genome database: integrated community resources for insect species of the order Hymenoptera. Nucleic Acids Res 39:D658–662
57. Jarosch A, Moritz RF (2011) Systemic RNA-interference in the honeybee Apis mellifera: tissue dependent uptake of fluorescent siRNA after intra-abdominal application observed by laser-scanning microscopy. J Insect Physiol 57:851–857
58. Dearden PK, Duncan EJ, Wilson MJ (2009) RNA interference (RNAi) in honeybee (Apis mellifera) embryos, Cold Spring Harb Protoc 6:1–3
59. Lee JA, Kim HK, Kim KH, Han JH, Lee YS, Lim CS, Chang DJ, Kubo T, Kaang BK (2001) Overexpression of and RNA interference with the CCAAT enhancer-binding protein on long-term facilitation of Aplysia sensory to motor synapses. Learn Mem 8:220–226
60. Duan J, Li R, Cheng D, Fan W, Zha X, Cheng T, Wu Y, Wang J, Mita K, Xiang Z, Xia Q (2010) SilkDB v2.0: a platform for silkworm (Bombyx mori) genome biology. Nucleic Acids Res 38:D453–456

61. Masumoto M, Yaginuma T, Niimi T (2009) Functional analysis of Ultrabithorax in the silkworm, Bombyx mori, using RNAi. Dev Genes Evol 219:437–444
62. Song C, Gallup JM, Day TA, Bartholomay LC, Kimber MJ (2010) Development of an in vivo RNAi protocol to investigate gene function in the filarial nematode, Brugia malayi. PLoS Pathog 6:e1001239
63. Maine EM (2008) Studying gene function in Caenorhabditis elegans using RNA-mediated interference. Brief Funct Genomics Proteomics 7:184–194
64. Nishiyama A, Fujiwara S (2008) RNA interference by expressing short hairpin RNA in the Ciona intestinalis embryo. Dev Growth Differ 50:521–529
65. Neumuller RA, Perrimon N (2011) Where gene discovery turns into systems biology: genome-scale RNAi screens in Drosophila. Wiley Interdiscip Rev Syst Biol Med 3:471–478
66. Lohmann JU, Endl I, Bosch TC (1999) Silencing of developmental genes in Hydra. Dev Biol 214:211–214
67. Galliot B, Miljkovic-Licina M, Ghila L, Chera S (2007) RNAi gene silencing affects cell and developmental plasticity in hydra. C R Biol 330:491–497
68. Lawson D, Arensburger P, Atkinson P, Besansky NJ, Bruggner RV, Butler R, Campbell KS, Christophides GK, Christley S, Dialynas E, Hammond M, Hill CA, Konopinski N, Lobo NF, MacCallum RM, Madey G, Megy K, Meyer J, Redmond S, Severson DW, Stinson EO, Topalis P, Birney E, Gelbart WM, Kafatos FC, Louis C, Collins FH (2009) VectorBase: a data resource for invertebrate vector genomics. Nucleic Acids Res 37:D583–587
69. Levashina EA, Moita LF, Blandin S, Vriend G, Lagueux M, Kafatos FC (2001) Conserved role of a complement-like protein in phagocytosis revealed by dsRNA knockout in cultured cells of the mosquito, Anopheles gambiae. Cell 104:709–718
70. Eleftherianos I, Millichap PJ, ffrench-Constant RH, Reynolds SE (2006) RNAi suppression of recognition protein mediated immune responses in the tobacco hornworm Manduca sexta causes increased susceptibility to the insect pathogen Photorhabdus. Dev Comp Immunol 30:1099–1107
71. Werren JH, Loehlin DW, Giebel JD (2009) Larval RNAi in Nasonia (parasitoid wasp), Cold Spring Harb Protoc 10:1358–1361
72. Sullivan JC, Reitzel AM, Finnerty JR (2008) Upgrades to StellaBase facilitate medical and genetic studies on the starlet sea anemone, Nematostella vectensis. Nucleic Acids Res 36: D607–611
73. Pankow S, Bamberger C (2007) The p53 tumor suppressor-like protein nvp63 mediates selective germ cell death in the sea anemone Nematostella vectensis. PLoS One 2:e782
74. Rehm EJ, Hannibal RL, Chaw RC, Vargas-Vila MA, Patel NH (2009) The crustacean Parhyale hawaiensis: a new model for arthropod development. Cold Spring Harb Protoc 1:1–10
75. Zerlotini A, Heiges M, Wang H, Moraes RL, Dominitini AJ, Ruiz JC, Kissinger JC, Oliveira G (2009) SchistoDB: a Schistosoma mansoni genome resource. Nucleic Acids Res 37:D579–582
76. Bhardwaj R, Krautz-Peterson G, Skelly PJ (2011) Using RNA interference in Schistosoma mansoni. Methods Mol Biol 764:223–239
77. Robb SM, Ross E, Sanchez Alvarado A (2008) SmedGD: the Schmidtea mediterranea genome database. Nucleic Acids Res 36:D599–606
78. Newmark PA, Reddien PW, Cebria F, Sanchez Alvarado A (2003) Ingestion of bacterially expressed double-stranded RNA inhibits gene expression in planarians. Proc Natl Acad Sci U S A 100(Suppl 1):11861–11865
79. Sanchez Alvarado A, Newmark PA (1999) Double-stranded RNA specifically disrupts gene expression during planarian regeneration. Proc Natl Acad Sci U S A 96:5049–5054
80. Lu HL, Vinson SB, Pietrantonio PV (2009) Oocyte membrane localization of vitellogenin receptor coincides with queen flying age, and receptor silencing by RNAi disrupts egg formation in fire ant virgin queens. FEBS J 276:3110–3123
81. Cameron RA, Samanta M, Yuan A, He D, Davidson E (2009) SpBase: the sea urchin genome database and web site. Nucleic Acids Res 37:D750–754
82. Song JL, Wessel GM (2007) Genes involved in the RNA interference pathway are differentially expressed during sea urchin development. Dev Dyn 236:3180–3190
83. Posnien N, Schinko J, Grossmann D, Shippy TD, Konopova B, Bucher G (2009) RNAi in the red flour beetle (Tribolium), Cold Spring Harb Protoc 8:1–8
84. Ma Y, Creanga A, Lum L, Beachy PA (2006) Prevalence of off-target effects in Drosophila RNA interference screens. Nature 443: 359–363
85. Kulkarni MM, Booker M, Silver SJ, Friedman A, Hong P, Perrimon N, Mathey-Prevot B (2006) Evidence of off-target effects associated with

long dsRNAs in Drosophila melanogaster cell-based assays. Nat Methods 3:833–838

86. Lin X, Ruan X, Anderson MG, McDowell JA, Kroeger PE, Fesik SW, Shen Y (2005) siRNA-mediated off-target gene silencing triggered by a 7 nt complementation. Nucleic Acids Res 33:4527–4535
87. Camacho C, Coulouris G, Avagyan V, Ma N, Papadopoulos J, Bealer K, Madden TL (2009) BLAST+: architecture and applications. BMC Bioinformatics 10:421
88. Du Q, Thonberg H, Wang J, Wahlestedt C, Liang Z (2005) A systematic analysis of the silencing effects of an active siRNA at all single-nucleotide mismatched target sites. Nucleic Acids Res 33:1671–1677
89. Elbashir SM, Martinez J, Patkaniowska A, Lendeckel W, Tuschl T (2001) Functional anatomy of siRNAs for mediating efficient RNAi in Drosophila melanogaster embryo lysate. EMBO J 20:6877–6888
90. Anderson EM, Birmingham A, Baskerville S, Reynolds A, Maksimova E, Leake D, Fedorov Y, Karpilow J, Khvorova A (2008) Experimental validation of the importance of seed complement frequency to siRNA specificity. RNA (New York, NY) 14:853–861.
91. Birmingham A, Anderson EM, Reynolds A, Ilsley-Tyree D, Leake D, Fedorov Y, Baskerville S, Maksimova E, Robinson K, Karpilow J, Marshall WS, Khvorova A (2006) 3' UTR seed matches, but not overall identity, are associated with RNAi off-targets. Nat Methods 3:199–204
92. Czech B, Hannon GJ (2011) Small RNA sorting: matchmaking for Argonautes. Nat Rev Genet 12:19–31
93. Tomari Y, Matranga C, Haley B, Martinez N, Zamore PD (2004) A protein sensor for siRNA asymmetry. Science 306:1377–1380
94. Elbashir SM, Harborth J, Weber K, Tuschl T (2002) Analysis of gene function in somatic mammalian cells using small interfering RNAs. Methods 26:199–213
95. Chan CY, Carmack CS, Long DD, Maliyekkel A, Shao Y, Roninson IB, Ding Y (2009) A structural interpretation of the effect of GC-content on efficiency of RNA interference. BMC Bioinformatics 10(Suppl 1):S33
96. Kirchner R, Vogtherr M, Limmer S, Sprinzl M (1998) Secondary structure dimorphism and interconversion between hairpin and duplex form of oligoribonucleotides. Antisense Nucleic Acid Drug Dev 8:507–516
97. Tafer H, Ameres SL, Obernosterer G, Gebeshuber CA, Schroeder R, Martinez J, Hofacker IL (2008) The impact of target site accessibility on the design of effective siRNAs. Nat Biotechnol 26:578–583
98. Ender C, Meister G (2010) Argonaute proteins at a glance. J Cell Sci 123:1819–1823
99. Liu J, Carmell MA, Rivas FV, Marsden CG, Thomson JM, Song JJ, Hammond SM, Joshua-Tor L, Hannon GJ (2004) Argonaute2 is the catalytic engine of mammalian RNAi. Science 305:1437–1441
100. Aleman LM, Doench J, Sharp PA (2007) Comparison of siRNA-induced off-target RNA and protein effects. RNA (New York, NY) 13:385–395.
101. Jackson AL, Bartz SR, Schelter J, Kobayashi SV, Burchard J, Mao M, Li B, Cavet G, Linsley PS (2003) Expression profiling reveals off-target gene regulation by RNAi. Nat Biotechnol 21:635–637
102. Jackson AL, Burchard J, Schelter J, Chau BN, Cleary M, Lim L, Linsley PS (2006) Widespread siRNA "off-target" transcript silencing mediated by seed region sequence complementarity. RNA (New York, NY) 12: 1179–1187.
103. Petri S, Dueck A, Lehmann G, Putz N, Rudel S, Kremmer E, Meister G (2011) Increased siRNA duplex stability correlates with reduced off-target and elevated on-target effects. RNA (New York, NY) 17:737–749.
104. Ghildiyal M, Xu J, Seitz H, Weng Z, Zamore PD (2010) Sorting of Drosophila small silencing RNAs partitions microRNA* strands into the RNA interference pathway. RNA (New York, NY) 16:43–56.
105. Ameres SL, Hung JH, Xu J, Weng Z, Zamore PD (2011) Target RNA-directed tailing and trimming purifies the sorting of endo-siRNAs between the two Drosophila Argonaute proteins. RNA (New York, NY) 17:54–63.
106. Vermeulen A, Behlen L, Reynolds A, Wolfson A, Marshall WS, Karpilow J, Khvorova A (2005) The contributions of dsRNA structure to Dicer specificity and efficiency. RNA (New York, NY) 11:674–682.
107. Vert JP, Foveau N, Lajaunie C, Vandenbrouck Y (2006) An accurate and interpretable model for siRNA efficacy prediction. BMC Bioinformatics 7:520
108. Fellmann C, Zuber J, McJunkin K, Chang K, Malone CD, Dickins RA, Xu Q, Hengartner MO, Elledge SJ, Hannon GJ, Lowe SW (2011) Functional identification of optimized RNAi triggers using a massively parallel sensor assay. Mol Cell 41:733–746
109. Steinbrink S, Boutros M (2008) RNAi screening in cultured Drosophila cells. Methods Mol Biol 420:139–153

110. Raghavan S, Williams I, Aslam H, Thomas D, Szoor B, Morgan G, Gross S, Turner J, Fernandes J, VijayRaghavan K, Alphey L (2000) Protein phosphatase 1beta is required for the maintenance of muscle attachments. Curr Biol 10:269–272

111. Stein LD, Mungall C, Shu S, Caudy M, Mangone M, Day A, Nickerson E, Stajich JE, Harris TW, Arva A, Lewis S (2002) The generic genome browser: a building block for a model organism system database. Genome Res 12:1599–1610

112. Yamaguchi J, Mizoguchi T, Fujiwara H (2011) siRNAs induce efficient RNAi response in Bombyx mori Embryos. PLoS One 6, 1–7 (e25469)

113. Bartscherer K, Pelte N, Ingelfinger D, Boutros M (2006) Secretion of Wnt ligands requires Evi, a conserved transmembrane protein. Cell 125:523–533

114. Bassik MC, Lebbink RJ, Churchman LS, Ingolia NT, Patena W, LeProust EM, Schuldiner M, Weissman JS, McManus MT (2009) Rapid creation and quantitative monitoring of high coverage shRNA libraries. Nat Methods 6:443–445

115. Li L, Lin X, Khvorova A, Fesik SW, Shen Y (2007) Defining the optimal parameters for hairpin-based knockdown constructs. RNA (New York, NY) 13:1765–1774.

116. Guberman JM, Ai J, Arnaiz O, Baran J, Blake A, Baldock R, Chelala C, Croft D, Cros A, Cutts RJ, Di Genova A, Forbes S, Fujisawa T, Gadaleta E, Goodstein DM, Gundem G, Haggarty B, Haider S, Hall M, Harris T, Haw R, Hu S, Hubbard S, Hsu J, Iyer V, Jones P, Katayama T, Kinsella R, Kong L, Lawson D, Liang Y, Lopez-Bigas N, Luo J, Lush M, Mason J, Moreews F, Ndegwa N, Oakley D, Perez-Llamas C, Primig M, Rivkin E, Rosanoff S, Shepherd R, Simon R, Skarnes B, Smedley D, Sperling L, Spooner W, Stevenson P, Stone K, Teague J, Wang J, Whitty B, Wong DT, Wong-Erasmus M, Yao L, Youens-Clark K, Yung C, Zhang J, Kasprzyk A (2011) BioMart Central Portal: an open database network for the biological community. Database (Oxford), bar041:1–9

Chapter 18

Construction of shRNA Expression Plasmids for Silkworm Cell Lines Using Single-Stranded DNA and *Bst* DNA Polymerase

Hiromitsu Tanaka

Abstract

Transfection of short hairpin RNA (shRNA) expression plasmids is conventionally performed for gene-specific knockdown in cultured mammalian and insect cells. Here, I describe a simple method to synthesize an inverted repeat DNA in a U6 small nuclear RNA promoter-based parent vector using a single-stranded inverted repeat DNA and *Bst* DNA polymerase. The shRNA expression plasmids constructed by this method were confirmed to promote efficient RNA interference knockdown in silkworm cell lines. This method may be useful for constructing a relatively large number of shRNA expression plasmids.

Key words: shRNA expression plasmids, *Bst* DNA polymerase, Single-stranded inverted repeat DNA, Silkworm cell line

1. Introduction

Posttranscriptional silencing by RNA interference (RNAi) is widely used as a technique for suppressing the expression of specific genes in many organisms (1–4). A conventional procedure for inducing RNAi knockdown in cultured mammalian and insect cells is the direct transfection of 21–23 nucleotides (nt) of small interfering RNA (siRNA) (5, 6) or over-expression of short hairpin RNA (shRNA) composed of 19–29 nt of stem regions and 4–23 nt of loop sequences by the transfection of an RNA polymerase III-dependent promoter-driven shRNA expression plasmid (7, 8). Knockdown by transfection of an shRNA expression plasmid has some advantages over knockdown by siRNA transfection (9). First, the RNAi effect may be more stable because of the sustained production of shRNA. Second, the transfected cells can be selected by antibiotics when the shRNA expression plasmid possesses

Debra J. Taxman (ed.), *siRNA Design: Methods and Protocols*, Methods in Molecular Biology, vol. 942, DOI 10.1007/978-1-62703-119-6_18, © Springer Science+Business Media, LLC 2013

antibiotic-resistance genes. Furthermore, inducible shRNA expression is available. In general, shRNA expression plasmids can be generated by two methods. One method is the insertion of a double-stranded inverted repeat (IR) DNA that is obtained by annealing of two complementary oligonucleotides into a parent vector (9). The second method is a polymerase chain reaction (PCR)-based strategy in which the promoter sequence serves as the template (10). We developed another simple method to create IR DNA in the parent vector using a single-stranded DNA possessing a short hairpin structure and *Bst* DNA polymerase, which has strand displacement activity (11). This method comprises the following steps: (a) linearization of the plasmid with the 5′ end of one terminus dephosphorylated by treating stepwise with one restriction enzyme, alkaline phosphatase, and then a second restriction enzyme; (b) ligation of a hairpin oligonucleotide to one end of the linear plasmid; (c) execution of the strand displacement reaction by *Bst* DNA polymerase; and (d) self-ligation of the linear plasmid (Fig. 1). This method reduces the cost of unique oligonucleotides compared with the conventional method. Therefore, it is useful for constructing relatively large numbers of shRNA expression plasmids. We further demonstrated that the shRNA expression plasmid constructed by this method effectively induces target-specific RNAi a silkworm cell line (11).

2. Materials

2.1. Oligonucleotide Annealing

1. Synthesized oligonucleotides: 53 mer; 21 mer of IR structure separated by 11 mer of spacer DNA. These oligonucleotides may be obtained from most custom oligonucleotide-synthesizing facilities and companies. For a discussion of the oligonucleotide design, see Notes 1–5. Store at −20°C.
2. 10× M buffer: 100 mM Tris–HCl (pH 7.9), 100 mM $MgCl_2$, 500 mM NaCl, and 10 mM DTT. Store at −20°C.
3. Ultrapure water: Milli Q grade; sterilized by autoclaving.

2.2. Construction of shRNA Expression Plasmids

1. Parent plasmid for constructing an shRNA expression plasmid: We used pIEx-4-BmU6M, which contains the enhancer and promoter region between *Sma*I and *Nco*I. Multicloning sites between the *Nco*I and *Dra*III sites of pIEx-4 were substituted by 467 bp of the promoter region of the *Bombyx mori* U6-2 small nuclear RNA gene (12) and the sequence "5′-CCATGG CTGCAG**AGGCCT**TTTTCACTAAGTG-3′" (underlining indicates the *Nco*I site; bold letters indicate the *Stu*I site), respectively.
2. Restriction endonucleases: *Nco*I and *Stu*I at 10 U/μL. Store at −20°C.

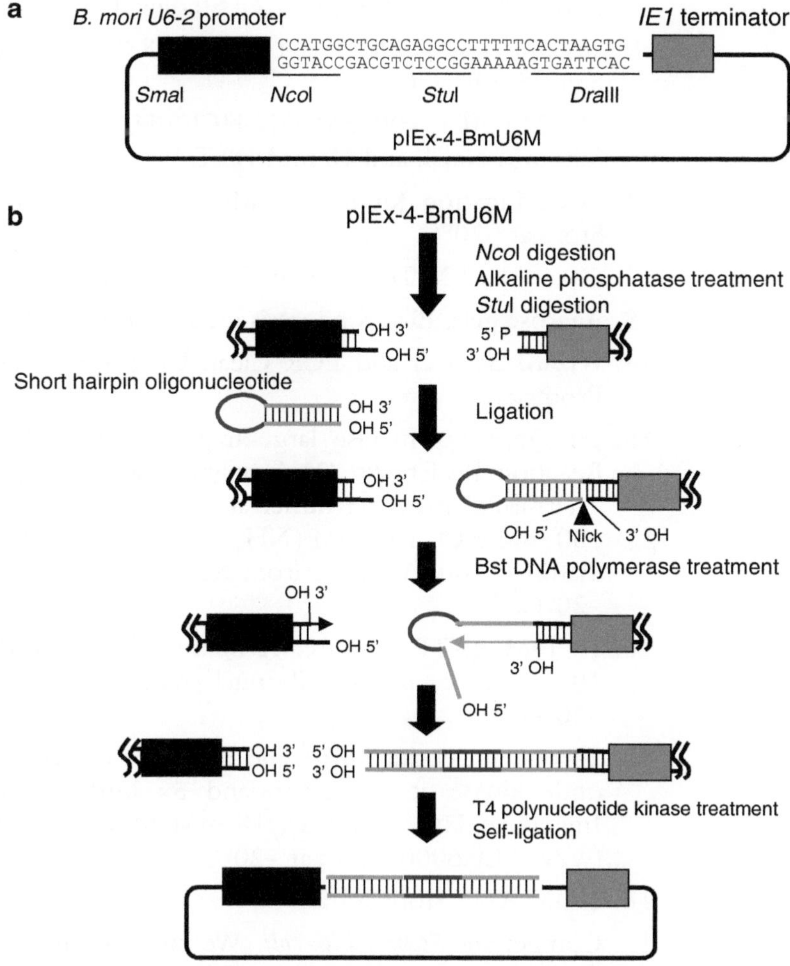

Fig. 1. The structure of pIEx-4-BmU6M and the procedure for construction of an shRNA expression plasmid. (**a**) Diagram of pIEx-4-BmU6M. The nucleotide sequences possessing the *Stu*I recognition site (*Stu*I) and a T cluster were inserted into the *Nco*I and *Dra*III sites of pIEx-4 (Novagen); the enhancer and promoter region between the *Sma*I and *Nco*I sites of pIEx-4 was replaced by 467 bp of a promoter region of *Bombyx mori* U6-2 small nuclear RNA gene (*black box*). Gray box indicates the terminator region from the *Autographa californica* nucleopolyhedrovirus-derived immediate early 1 gene. (**b**) Strategy to create the IR DNA in pIEx-4-BmU6M. A short hairpin oligonucleotide is ligated with the *Stu*I-digested blunt end of linear pIEx-4-BmU6M. *Bst* DNA polymerase extends the 3′ end of the *Nco*I-digested terminus and 3′ end at the nick followed by the displacement of the 5′ portion of the hairpin oligonucleotide. Kinase reaction and self-ligation yield a circular shRNA expression plasmid.

3. 10× H buffer: 500 mM Tris–HCl (pH 7.9), 100 mM $MgCl_2$, 1 M NaCl, and 10 mM DTT. Store at −20°C.
4. Alkaline phosphatase: 10 U/μL. Store at −20°C.
5. CIA: 24:1 (v/v) mixture of chloroform and isoamyl alcohol.
6. Phenol/chloroform: 1:1 (v/v) mixture of Tris–HCl (pH 8.0) buffered phenol and CIA. Store at 4°C.
7. Ethanol: 100% and 70% (v/v) solution.

8. 3 M Sodium acetate (pH 5.2): Sterilized by autoclaving.
9. TE buffer: 10 mM Tris–HCl (pH 8.0) and 1 mM EDTA. Sterilized by autoclaving.
10. 10× M buffer: 100 mM Tris–HCl (pH 7.9), 100 mM $MgCl_2$, 500 mM NaCl, and 10 mM DTT.
11. DNA Ligation Kit Mighty Mix: Available from Takara Bio. Store at −20°C.
12. 50× TAE: 2 M Tris–acetate, 50 mM EDTA.
13. Agarose gels: Electrophoresis grade agarose in 1× TAE.
14. Wizard SV Gel and PCR Clean-Up System: Available from Promega.
15. *Bst* DNA polymerase large fragment and 10× ThermoPol Reaction Buffer: *Bst* DNA polymerase at 8 U/μL and 10× ThermoPol Reaction Buffer at 200 mM Tris–HCl (pH 8.8), 100 mM KCl, 100 mM $(NH_4)_2SO_4$, 20 mM $MgSO_4$, and 1% Triton X-100. Available from New England Biolabs. Store at −20°C.
16. 10 mM dNTP mixture: A mixture in water that contains 10 mM of each deoxyribonucleoside triphosphate. Store at −20°C.
17. T4 polynucleotide kinase and 5× kinase buffer: T4 polynucleotide kinase at 10 U/μL and 5× buffer at 50 mmol/L Imidazole–HCl (pH 6.4), 18 mM $MgCl_2$, 5 mM DTT, 6% (w/v) PEG6000. Store at −20°C.
18. 2 mM ATP: Store at −20°C.
19. Competent *Escherichia coli*: We successfully used both the Sure2 Supercompetent Cells (Stratagene) and DH5α (Takara Co. Ltd) strains. Store at −80°C.
20. 2× YT agar plate: To make 1 L, add 16 g of polypeptone, 10 g of yeast extract, 5 g of NaCl, and 15 g of agar to 900 mL of water. Fill to 1 L with water and autoclave. After cooling, add ampicillin to a final concentration of 100 μg/mL. Pour into plates and store the plates at 4°C.

2.3. Confirmation of Insert Size by Colony PCR

1. Forward and reverse primers: Dilute each synthetic oligonucleotide to 10 μM with water. Store at −20°C.
2. 10× PCR buffer: 100 mM Tris–HCl (pH 8.3), 500 mM KCl, and 15 mM $MgCl_2$. Store at −20°C.
3. Taq polymerase (5 U/μL): Store at −20°C.
4. 2× YT medium: 1.6% polypeptone, 1.0% yeast extract, and 85 mM NaCl. Sterilize by autoclaving.

3. Methods

3.1. Oligonucleotide Annealing

1. Oligonucleotides were suspended in water to a concentration of 100 pmol/μL.
2. Mix 32 μL of oligonucleotide solution (100 pmol/μL), 32 μL of 10× M buffer, and 40 μL of water in a 0.2 mL tube.
3. Heat at 95°C for 5 min and gradually cool to 30°C (1–2°C/min). Annealed oligonucleotides should form a hairpin structure.
4. Store at −20°C if the annealed oligonucleotides are not to be used immediately.

3.2. Construction of shRNA Expression Plasmids

The construction method using pIEx-4-BmU6M (11) is as follows:

1. Digest 10 μg of pIEx-4-BmU6M with 25 units of *Nco*I at 37°C for 1–12 h in a 400 μL reaction volume containing 40 μL of 10×H buffer. Heat DNA at 65°C for 5 min to inactivate the enzymes.
2. Add 2 μL of alkaline phosphatase and incubate the solution at 37°C for 30 min.
3. Extract the reaction solution with phenol/chloroform and then CIA. Add 1 mL of absolute ethanol and 40 μL of 3 M sodium acetate to the upper phase solution. Centrifuge for 12,000×*g* at 4°C for 10 min. Discard the supernatant, wash the pellet in 70% ethanol, and recentrifuge for 5 min. Dissolve the pellet with 357 μL of water and then add 40 μL of 10× M buffer and 2 μL of *Stu*I. Incubate at 37°C for 1–12 h. Extract the reaction solution with phenol/chloroform and then CIA. Precipitate the reactant DNA with ethanol. The pellet is dissolved with TE buffer at a concentration of 0.25 μg/μL. The product can be stored at −20°C.
4. Mix 1.5 μL of linear plasmid, 1 μL of annealed oligonucleotide, 5 μL of DNA Ligation Kit Mighty Mix, and 2.5 μL of water. Incubate at 16°C for 30 min.
5. Load 10 μL of ligated DNA solution onto a 1% agarose gel in 1× TAE gel running buffer. After electrophoresis is performed, remove the desired bands from the gel. See Note 6.
6. Recover DNA from the gel slice using a Wizard SV Gel and PCR Clean-Up System according to the instruction manual. Finally, elute DNA with 50 μL of water.
7. Mix 43 μL of recovered DNA, 5 μL of 10× ThermoPol Reaction Buffer, 1 μL of 10 mM dNTP mixture, and 1 μL of *Bst* DNA polymerase. Incubate at 50°C for 2 min, and then move to 62.5°C for 30 min.
8. Extract the reaction solution with phenol/chloroform and then CIA. Precipitate the reactant DNA with ethanol. Dissolve

the pellet with 43 μL of water. Add 11.3 μL of 5× kinase buffer, 1 μL of 2 mM ATP, and 1 μL of T4 polymerase kinase. Incubate at 37°C for 30 min.

9. Extract the reaction solution with phenol/chloroform and then CIA. Precipitate the reactant DNA with ethanol. Dissolve the pellet with 20 μL of water.
10. Mix 5 μL of reactant solution and 5 μL of DNA Ligation Kit Mighty Mix. Incubate at 16°C for 30 min.
11. Transform 10 μL of ligation reaction into 100 μL of a competent strain of *E. coli*. Plate the appropriate amount of cells onto 2× YT agar plates. Incubate at 37°C overnight.

3.3. Confirmation of Insert Size by Colony PCR

1. Prepare the PCR reaction mixture. For 500 μL, mix 50 μL of 10× PCR buffer, 10 μL of 10 mM dNTP mixture, 10 μL of each of forward and reverse primers (10 μM), and 5 μL of Taq polymerase. Then, bring to 500 μL with water. Dispense 10 μL of the PCR reaction mixture to each 0.2 mL PCR tube.
2. Pick each colony with a sterile toothpick, and swirl it into the PCR reaction mixture in a tube.
3. Place each PCR tube in a thermal cycler. Heat at 95°C for 2 min, and then subject to 35 cycles as follows:

Denaturation	95°C for 0.5 min
Primer annealing	50–55°C for 0.5 min
Primer extension	72°C for 0.5–3 min

4. Run on a 1.5–2% agarose gel to analyze the insert size (See Notes 7 and 8).
5. Culture positive colonies in 2 mL of 2× YT medium at 37°C overnight.
6. Prepare a plasmid from the cultured *E. coli*, and confirm that the nucleotide sequence of the inserted DNA is correct.

4. Notes

1. GC contents in the stem region should be less than 55%; the extension by *Bst* polymerase may not be completed if the GC contents are higher.
2. Stretches of four or more T nt should not be included in the stem region because RNA polymerase III may terminate transcription by recognizing it as a terminator in the transfected cells.
3. For the spacer sequence of oligonucleotides, we used "5-GTGT GCTGTCC-3′," which was derived from human microRNA

mir26b and has been reported to be an effective spacer sequence in mammalian and *Drosophila* cell lines (13, 14). We confirmed that the shRNA possessing this spacer sequence also effectively knocked down the target gene in silkworm cells. Furthermore, we confirmed that a randomly designed spacer "5′-AGTCCAACAGG-3′" functioned efficiently in the silkworm cell lines.

4. shRNA with 19 or 17 nt stem regions were inefficient in silkworm cell lines (11). Therefore, the length of the stem region should be at least 21 nt.
5. In general, RNAi efficiency in the cells is known to depend on the sequence of the stem region, and only approximately 30% of random siRNAs have been reported to show highly effective RNAi in cultured mammalian cells (15). However, in our experiment, six of eight shRNA expression plasmids—each having randomly designed nucleotide sequences at the stem region—suppressed the expression of the reporter gene by more than 95% in silkworm cells (11). Another two constructs also showed 75–80% reductions. These results suggest that sequence preference in silkworm cell lines is much lower than that in mammalian cell lines.
6. Agarose gel electrophoresis should be performed after ligation of a hairpin oligonucleotide to the linear plasmid to remove free hairpin oligonucleotides (Fig. 2).
7. IR DNA-inserted plasmids can be easily distinguished from empty plasmids by colony PCR (Fig. 3). However, incomplete IR DNA is sometimes inserted into the vector. Therefore, confirmation of the nucleotide sequence of each plasmid is necessary.
8. An examination of nine constructions using oligonucleotides with 21 nt of stem regions and 11 nt of the spacer sequence "5′-GTGTGCTGTCC-3′" revealed that 20–70% of the transformed clones contained correctly sized inserts by colony PCR. The efficiency of creating an expected DNA insert in the plasmid would be dependent on the nt sequences of the stem region. Analysis of nt sequences revealed that 68% of the recombinant clones possessing correctly sized inserts had correct nt sequences, and that one additional nt was created in 80% of the clones at the junction between the *Nco*I site and the oligonucleotide (11).

Acknowledgments

This work was supported by a grant from Promotion of Basic Research Activities for Innovative Biosciences (PRO-BRAIN).

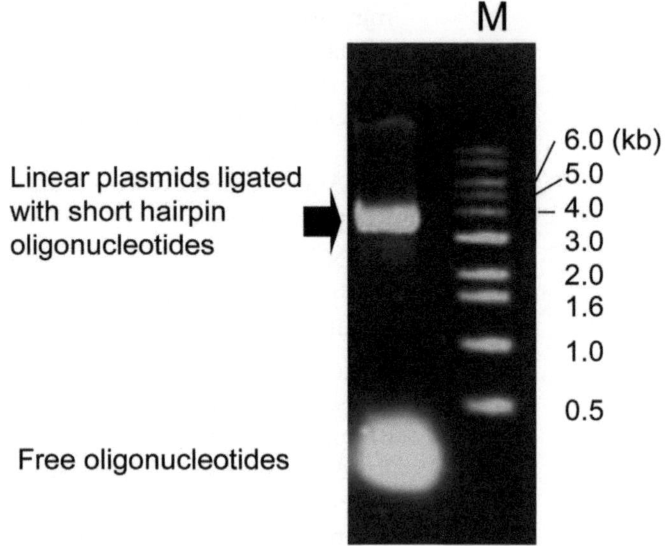

Fig. 2. Agarose gel electrophoresis to separate the short hairpin oligonucletotide-ligated plasmid and free oligonucleotides. The short hairpin oligonucleotide-ligated plasmid (*arrow*) is recovered by Wizard SV Gel and the PCR Clean-Up System. M; 1 kb DNA ladder.

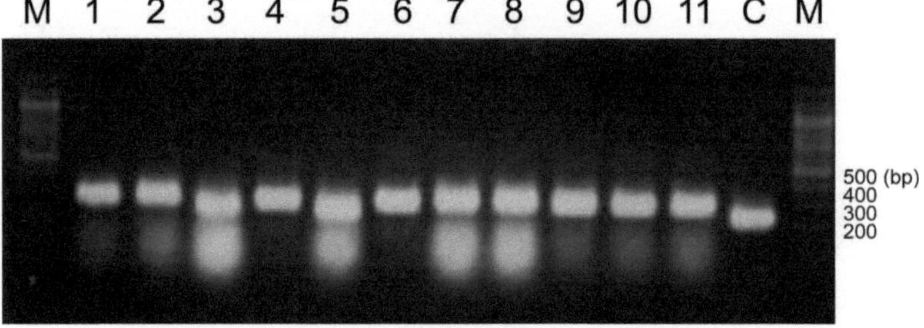

Fig. 3. Confirmation of insert size in pIEx-4-BmU6M by colony PCR. Colony PCR was performed in a total volume of 10 μL that contained 200 nM of each forward "5′-TGTAAAGTCGAGTGTTGTTGTA-3′" and reverse "5′-CAAAACCCCACACCAACAAC-3′" primer. In this experiment, an oligonucleotide, "5′-TCATTCCTGAAGACAGCTGAGGTGTGCTGTCCCTCAGCTGTCTTCAGGAATGA-3′," was used for construction of an shRNA expression plasmid. Two different sizes of bands were detected. The band that was amplified from empty plasmids (Lines 3 and 5) was 221 bp long, and the band that was amplified from shRNA plasmids was 275 bp long (Lines 1, 2, 4, 6, 7, 8, 9, 10, 11). Nucleotide sequences of these plasmids were confirmed. C; pIEx-4-BmU6M. M; 100 bp DNA ladder.

References

1. Agrawal N, Dasaradhi PV, Mohmmed A, Malhotra P, Bhatnagar RK, Mukherjee SK (2003) RNA interference: biology, mechanism, and applications. Microbiol Mol Biol Rev 67: 657–685
2. Dawe RK (2003) RNA interference, transposons, and the centromere. Plant Cell 15:297–301
3. Fire A, Xu S, Montgomery MK, Kostas SA, Driver SE, Mello CC (1998) Potent and specific genetic interference by double-stranded RNA in *Caenorhabditis elegans*. Nature 391: 806–811
4. Hannon GJ (2002) RNA interference. Nature 418:244–251

5. McManus MT, Sharp PA (2002) Gene silencing in mammals by small interfering RNAs. Nat Rev Genet 3:737–747
6. Hammond SM, Bernstein E, Beach D, Hannon GJ (2000) An RNA-directed nuclease mediates post-transcriptional gene silencing in *Drosophila* cells. Nature 404:293–296
7. Brummelkamp TR, Bernards R, Agami R (2002) A system for stable expression of short interfering RNAs in mammalian cells. Science 296:550–553
8. Dykxhoorn DM, Novina CD, Sharp PA (2003) Killing the messenger: short RNAs that silence gene expression. Nat Rev Mol Cell Biol 4: 457–467
9. Sano M, Kato Y, Akashi H, Miyagishi M, Taira K (2005) Novel methods for expressing RNA interference in human cells. Methods Enzymol 392:97–112
10. Castanotto D, Scherer L (2005) Targeting cellular genes with PCR cassettes expressing short interfering RNAs. Methods Enzymol 392: 173–185
11. Tanaka H, Fujita K, Sagisaka A, Tomimoto K, Imanishi S, Yamakawa M (2009) shRNA expression plasmids generated by a novel method efficiently induce gene-specific knockdown in a silkworm cell line. Mol Biotechnol 41:173–179
12. Hernandez G Jr, Valafar F, Stumph WE (2007) Insect small nuclear RNA gene promoters evolve rapidly yet retain conserved features involved in determining promoter activity and RNA polymerase specificity. Nucleic Acids Res 35:21–34
13. Miyagishi M, Sumimoto H, Miyoshi H, Kawakami Y, Taira K (2004) Optimization of an siRNA-expression system with an improved hairpin and its significant suppressive effects in mammalian cells. J Gene Med 6:715–723
14. Wakiyama M, Matsumoto T, Yokoyama S (2005) *Drosophila* U6 promoter-driven short hairpin RNAs effectively induce RNA interference in Schneider 2 cells. Biochem Biophys Res Commun 331:1163–1170
15. Ui-Tei K, Naito Y, Takahashi F, Haraguchi T, Ohki-Hamazaki H, Juni A, Ueda R, Saigo K (2004) Guidelines for the selection of highly effective siRNA sequences for mammalian and chick RNA interference. Nucleic Acids Res 32:936–948

Chapter 19

Designing Effective amiRNA and Multimeric amiRNA Against Plant Viruses

Muhammad Fahim and Philip J. Larkin

Abstract

RNA-mediated virus resistance is increasingly becoming a method of choice for antiviral defense in plants when effective natural resistance is unavailable. In this chapter we discuss the design principles of artificial micro RNA (amiRNA), in which a natural miRNA precursor gene is modified to target a different species of RNA, in particular viral RNA. In addition, we explore the advantages and effectiveness of multiple amiRNAs within one polycistronic amiRNA precursor against a virus, as illustrated with *Wheat streak mosaic virus*, WSMV. The judicious selection of amiRNAs, which are sequences of short length as compared to other related methodologies of RNA interference, greatly assists in avoiding unintended off-targets in the host plant. The viral sequences targeted can be genomic or replicative and should be derived from conserved regions of the published WSMV genome. In short, using published folding and miRNA selection rules and algorithms, candidate miRNA sequences are selected from conserved regions between a number of WSMV genomes, and are BLASTed against wheat TIGR ESTs. Five miRNAs are selected that are least likely to interfere with the expression of transcripts from the wheat host. Then, the natural miRNA in each of the five arms of the polycistronic rice miR395 is replaced in silico with the chosen artificial miRNAs. This artificial precursor is transformed into wheat behind a ubiquitin promoter, and its integration into transformed wheat plants is confirmed by PCR and Southern blot analysis. We have demonstrated the effectiveness of this methodology using an amiRNA precursor that we have termed Fanguard. The processing of amiRNAs in transgenic leaves is verified through splinted ligation assay, and the functionality of the transgene in preventing viral replication is verified by virus bioassay. Resistance is confirmed using mechanical virus inoculation over two subsequent generations. This example demonstrates the potential of polycistronic amiRNA to achieve stable immunity to economically important viruses.

Key words: Multiplex virus resistance, Artificial microRNA design, Multimeric amiRNA, Polycistronic amiRNA, virus immunity

1. Introduction

Plant viruses pose a major threat to crop production and are responsible for substantial economic losses to important crops across the globe. Despite strict bio-security practices, plant viruses

Debra J. Taxman (ed.), *siRNA Design: Methods and Protocols*, Methods in Molecular Biology, vol. 942, DOI 10.1007/978-1-62703-119-6_19, © Springer Science+Business Media, LLC 2013

sometimes escape into new agro-ecosystems and become a major limiting factor in crop production in the absence of durable natural resistance. The widespread occurrence of their natural vectors in such areas may further exacerbate the potential impact on yield. A number of factors including the impracticality of virus and vector management have added to the priority of breeding virus-resistant varieties and developing alternative methods of virus control including recombinant DNA methods. Natural resistance genes, when available, rarely confer immunity and are often overcome by different strains of the viral species. In the mid 1980s the first uses of pathogen-derived resistance (PDR) provided some niche economic benefit to agriculture; however public perceptions of safety and concerns expressed by some virologists of the risk of new viral recombinants emerging helped constrain commercialization of such genetically modified crops. The challenge being addressed by this chapter is how to design synthetic artificial microRNA (amiRNA), especially against viruses of economic importance that confer broad-spectrum resistance against multiple strains or even multiple virus species with negligible risk of unintended consequences for the plants and negligible risk of new types of viral pathogen emerging.

The first-generation transgenic plants with resistance against viruses utilized the sense sequence of the virus genome, especially the coat protein (CP) gene (1). Despite extensive studies, the molecular mechanisms that govern CP-mediated resistance (CPMR) are not fully understood and appear to be different for different viruses (2). It was assumed that somehow the expression of the transgenic viral protein disrupted the life cycle of the virus (1, 3, 4). Later, various labs around the world showed that in addition to viral CP genes, the expression of functional and dysfunctional sequences derived from elsewhere in the viral genome, i.e., replicase, movement protein, or protease genes, can also confer various degrees of protection characterized by symptom delay, reduced symptom severity, partial resistance, and immunity (5, 6). The term *PDR* was coined for this phenomenon and the technique became widely used for the development of transgenic virus-resistant plants (7–10).

In late 1990s, the RNA-based PDR strategies converged on RNA interference (RNAi) while leaving behind the CP-mediated resistance as a complex mystery, yet to be completely resolved (2, 4, 11). This second generation of transgenic plants expressed the viral sequence as an inverted repeat instead of sense or antisense orientation alone. This approach allows the formation of double-stranded RNA (dsRNA) molecules in the form of hairpin RNA if there is intervening sequence between the sense and antisense portions. Transgenic plants of this type display a more efficient and effective protection against viruses in plants (12) utilizing induced RNAi. There are now numerous examples of the development of virus-resistant transgenic plants through expression of virus-derived dsRNA or hairpin RNAs (hpRNAs) (13–16). These hpRNAs are processed into small interfering RNAs (siRNAs) by Dicer-like enzymes. siRNAs are then incorporated into the RNA-Induced Silencing Complex (RISC), and are then primed to recognize and degrade complementary viral sequence resulting in resistance or immunity (17). However, the long hpRNA from conventional RNAi vectors increases the probability of "off-target" effects, i.e., silencing of unintended host plant genes (18–21). Despite this risk a number of transgenic plants with antiviral RNAi constructs have been produced and perform successfully (22). The likelihood of off-target silencing problems for the host plant seems remote and could be further reduced by judicious selection of smaller sequences for the hpRNA. Nevertheless, this risk is one of the reasons experimentation began with microRNAs (miRNAs), which might be manipulated and used to achieve transgenic virus resistance. Gene silencing mediated by amiRNA may be considered the third-generation transgenic strategy for virus resistance, which promises to address the issues associated with off-target effects and the hypothetical generation of novel recombinant virus biotypes (23, 24).

miRNAs encoded by plant genes are similar to siRNAs in their biogenesis. However, they are derived from noncoding RNA (ncRNA) transcripts that fold into imperfect hairpin loop structures (pre-miRNAs). This characteristic structure is then processed by Dicer-like enzymes to produce the mature miRNA, a short (about 21 nt) dsRNA which is then loaded into the RISC. The miRNA–RISC then mediates complementarity-dependent repression or degradation of target mRNA (25, 26). It is possible to alter the sequence of the 21-nt mature miRNAs within the natural miRNA precursor, without affecting miRNA biogenesis and maturation, as long as the secondary structure of the pre-miRNA is maintained. This opens up the prospect of creating amiRNAs with new targets by design (23, 27–29).

The first successful amiRNA-based virus resistance was achieved in *Arabidopsis thaliana* (30) using two plant virus systems. Here, amiRNA amiR-p69-159 was engineered into the backbone of miR159, targeting the p69 gene of *Turnip yellow mosaic virus* (TYMV). Furthermore, amiR-HC-Pro-159 was derived from *Turnip mosaic virus*, TuMV, and targeted the HC-Pro gene of TuMV, which encodes the viral suppressor of RNA silencing. Transgenic lines expressing these two amiRNAs specifically conferred resistance to TYMV or TuMV (30). Since then, a similar strategy has been successfully applied to other virus–plant systems using a number of different miRNA precursors as backbones for delivery of amiRNAs (27, 30–36).

Previously, we developed virus-resistant transgenic wheat plants based on second-generation constructs, capable of forming dsRNA transcripts and resulting in total immunity against *Wheat streak mosaic virus* (WSMV) (17). This approach was demonstrably more effective against WSMV than either of the previous two strategies that involved sense expression of the nuclear inclusion "b" or coat protein genes (37, 38). Many species of siRNA are generated, which is useful to protect against the invading virus because it is unable to mutate in the many positions required to avoid this highly diverse attack. However, the use of long hpRNA from conventional RNAi vectors as in the 696 bp segment of WSMV genome used in our previous work (17) theoretically entails an increased risk of "off-target" effects, i.e., silencing of unintended genes (18). Furthermore, some express concern that agricultural scale deployment of antivirus hpRNA-expressing transgenic plants might lead to evolution of new virus biotypes via heterologous recombination or complementation between the relatively long viral sequences expressed from the transgene and RNA from a nontarget virus infecting the same plant.

In this chapter we provide practical information for designing and using RNAi vectors that are based on the third-generation approach of amiRNA, and in particular polycistronic amiRNA with multiple targets in the viral genome. First, we summarize the relative merits and demerits of amiRNA-dependent virus resistance compared to the long hairpin (or double stranded) RNA approach:

- *Frequency of useful events.* It is our experience with WSMV and wheat transformation that the long hpRNA approach had the advantage over amiRNA in that a much higher frequency of insertion events results in stable heritable immunity. It was possible to select stable immune events by both approaches, but the frequency was higher with hpRNA. Similarly, it has been revealed in other systems that not all amiRNAs targeting the viral CP sequence of Potato virus Y (PVY) were equally effective in preventing virus infection. Certain regions exhibited high virus resistance compared to other regions (39, 40). Furthermore, northern blots revealed a positive correlation between the resistance level and the accumulation of amiRNA molecules (39, 40). Other factors that contribute to successful transgenic events might include target accessibility, where targeting sequences of loose molecular structure may enhance virus resistance. Therefore, along with selection of typical amiRNA, the selection of an appropriate target sequence based on the secondary structure is crucial for transgenic plants (39). Experience in many other systems will be needed before generalizations can be contemplated. Furthermore, it would be interesting to explore the effect of ploidy level on amiRNA effectiveness in other wheat species such as durum wheat.
- *Off-target effects.* There is a greater chance that the abundant and varied siRNAs derived from long dsRNAs may have off-target effects, e.g., on essential endogenous genes of the host plant. Monomeric amiRNA produces one stable small RNA

(21 nt) making it relatively easy to choose that sequence to avoid off-target effects by screening potential amiRNAs against genomic or EST databases (20, 21, 23, 41–44). More recently, it has been shown that siRNAs derived from satellite-associated Cucumber mosaic virus (CMV)-induced yellowing symptoms in tobacco by targeting chlorophyll biosynthetic gene (CHLI) (45, 46). Our knowledge is still limited to a certain level when it comes to the interaction between small RNAs and their targets. Many variables play an important role and there is more to discover about how and to what extent virus-derived siRNAs play a role in mis-regulating endogenous host gene expression for their successful replication and infectivity.

- *Risk of trans-encapsidation and new pathogens.* Concern has been expressed that deploying virus sequences in transgenic crops engenders the risk that new and more destructive viruses could emerge through trans-encapsidation (24). Put simply, it is envisioned that RNA from the targeted virus could recombine with a second virus to achieve new functionalities. Irrespective of whether this risk is judged feasible, long hpRNA approaches increase the probability because they are more likely to include functional domains. While it is possible to design hpRNA sequences which do not represent functional domains, it is certainly the case that the short sequences used in the amiRNA strategy virtually eliminate this perceived risk.
- *Environmental effects on resistance efficacy.* Plant–virus interactions are usually influenced by environmental factors such as temperature. There is now conclusive evidence that certain siRNAs and their silencing efficiency are drastically affected by several factors in insect, plant, and mammalian cells (47–51). It has been established that low temperatures such as 15°C lead to the inhibition of RNA silencing resulting in susceptibility and loss of silencing-mediated transgenic phenotypes resulting from the low level of virus- or transgene-derived siRNAs (48). However, the accumulation of miRNA is believed to be independent of changing temperatures and the transgenic lines expressing virus-derived amiRNA retain their resistance (30, 48).
- *The likelihood of transgene stability.* The amiRNA approach uses the existing endogenous pre-miRNA genes as templates whereas hpRNA transgenes are entirely artificial. It can be argued that this may have impacts on transgene silencing and stability of expression, with amiRNA being more likely to be stable. More extensive experience in using diverse systems such as endogenous and tissue-specific promoters will be needed before this generalization can be substantiated.
- *Opportunity to target multiple viruses.* There are demonstrated opportunities to generate resistance to multiple viruses using chimeric genes which fuse virus-derived sequences behind a

single promoter to produce multiplex RNAi derived from an hpRNA transgene. Chimeric hpRNA transgenes targeting several viruses have been demonstrated in a number of plant transgenic systems (9, 52, 53). However, polycistronic amiRNA precursor genes (discussed below) might also be harnessed to attack multiple viruses.

- *Virus escape*. Viruses have a great capacity to generate biological variation by mutation. Documented examples of this include the selection of HIV escape mutants that avoid RNAi in animals (54–57), and the emergence of *Plum pox virus* (PPV) escape mutants in plants (58, 59). The disadvantage of amiRNA for protection against biological agents, especially viruses, is that the pathogen may evolve to avoid the surveillance of the amiRNA because it is only a short sequence and minor base changes might avoid detection and degradation.

The latter point potentially is the only major disadvantage of amiRNA compared to hpRNA. For this reason we advocate the use of multimeric (polycistronic) amiRNA, where multiple sequences in the pathogen are targeted simultaneously, greatly reducing the opportunity for natural diversity or mutation in the pathogen to escape surveillance and degradation (30, 59). In 2009, a similar rationale was followed in the work of Israsena and colleagues (27) in which they designed precursor genes encoding three amiRNA against rabies virus and tested them in cell culture. Likewise, multiple siRNAs were introduced in the backbone of a multiplex miRNA, directed against HIV in cultured cells (56, 60). The amiRNA design principles provided within this chapter are based on a study in which we chose five independent targets in the WSMV genome, incorporated all five into a polycistronic amiRNA precursor, and demonstrated that all five were being processed and presumably are active against the virus (33). For the virus to avoid this resistance, and a new virulent biotype to emerge, it would need to evolve mutations simultaneously in all five target sequences and without loss of functionality in the various genes involved. The rice polycistronic miRNA used in our experiments, pri-miR395 from *Oryza sativa* (61), has seven natural miRNA arms, making it feasible to engineer resistance to two or three viruses targeting multiple sequences in each. All the approaches discussed here circumvent the need to combine multiple transgenic insertions which would add significantly to breeding difficulties and regulatory compliance issues.

The first attempt to provide miRNA-mediated resistance against chimeric PPV failed as the virus readily evolved to escape the negative pressure of silencing mechanisms (58). However, given that the target accessibility affects the efficacy of miRNA-dependent RNA silencing (51, 62), the escaped resistance may have been due to suboptimal target accessibility.

A number of research studies have broadened our view of miRNA biogenesis and therefore are assisting in the design of more efficient amiRNAs. However, there are currently limitations in the available algorithms for predicting of RNA secondary structure which also limit the ability to exploit amiRNA strategy. We utilized the available guidelines summarized online at Web MicroRNA Design versions 2 and 3 (WMD3 at http://wmd3.weigelworld.org/cgi-bin/webapp.cgi) and took advantage of open source algorithms and software available online. The polycistronic amiRNA against WSMV which resulted was tested in transgenic wheat where it successfully conferred immunity in two subsequent generations (33).

The information presented in this chapter provides basic guidelines for developing amiRNA-mediated resistant transgenic plants against plant viruses and can be extended to develop the preferred multiplex resistance conferred by polycistronic amiRNA genes. The guidelines for amiRNA design will improve with time and as more information is revealed concerning miRNA–target interactions.

2. Materials

1. Competent cells suitable for the transformation protocols routine in your lab (e.g., DH alpha *E. coli*).
2. Suitable Monocot/Dicot transformation vector (e.g., pWubi (33)).
3. amiRNA carrier plasmid (*see* Subheadings 3.1–3.3).
4. Reagents for cloning into the destination vector. (Alternatively, cloning can be done commercially.)
5. Tissue culture media (67).
6. Gene Gun (for biolistic transformation).
7. Reagents for PCR, Northern blotting (optional), and Southern blotting.
8. An abrasive such as diatomaceous earth or carborundum (63), used together with leaf sap from infected plants for mechanical inoculation of the virus on leaves.
9. Small RNA isolation kits such as miRVana (or Trizol).
10. Small RNA detection kit, for the detection of specific amiRNA, such as miRTect IT based on splinted ligation (33). Alternatively, we believe that normal northern blot analysis especially 1-ethyl-3-(3-dimethylaminopropyl) carbodiimide (EDC) method (68) should be sufficient and less expensive with similar or improved sensitivity.

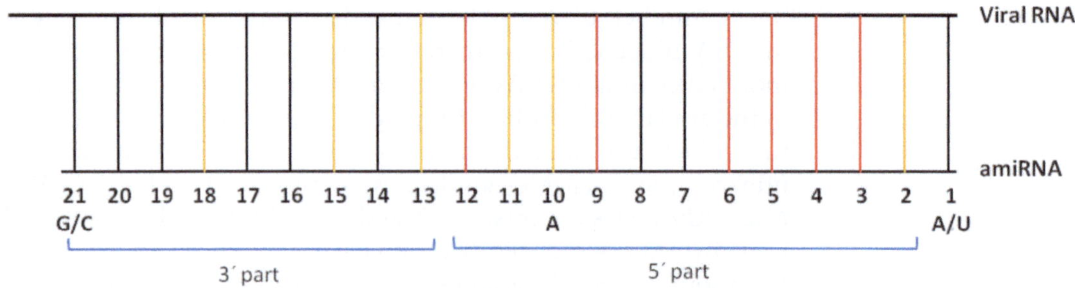

Fig. 1. Architecture of a typical artificial microRNA: Widely adopted rules in amiRNA design describe position 9–11 as having a vital role in miRNA-mediated gene silencing and repression. However, additional positions have been identified that play a vital role in determining the efficacy of amiRNA against plant viruses. Mismatches at positions 3–6, 9, and 12 (*red*) are more critical (i.e., failure of amiRNA-mediated resistance). Positions 2, 10, 11, 13, 15, and 18 (*orange*) are of intermediate importance, while the remaining positions (*black*) were classified as least critical for resistance. Thus, mismatches at the 5′ portion of the miRNA and the center are more critical as compared to the 3′ end of miRNA.

3. Methods

The following instructions are basic guidelines modified from the WMD3 Web site (29) http://wmd3.weigelworld.org/ and a number of other publications cited where appropriate. The basic structure of a microRNA is given in Fig. 1.

3.1. Choosing Conserved Regions for Selecting amiRNA Targeting Sites

1. Download the virus sequences from NCBI or *VirusFasta* developed by Rothamsted Research, UK, from http://www.dpvweb.net/analysis/fasta/index.php.
2. Download all available sequences of the virus species of interest. In cases such as PVY and CMV where there are very many variations in the sequence, choose the sequences closest to your agro-climate, with consideration of the virus epidemiology. Include the most virulent strains in the analysis to ensure that the conferred protection will also be against those strains. The analysis of potential targets can use either full viral genomes or just preselected regions or genes. VirusFasta gives the option to download sequences of individual genes.
3. Align the downloaded sequences both in sense and antisense orientation (see Note 1) using appropriate software. We used AlignX (Vector NTi 10, newer versions are available). Other alignment programs are available including *MEGA* http://www.megasoftware.net/. From the alignment of virus sequences choose conserved regions that span a minimum of 20 nucleotides (see Note 2). Note the coordinates on the consensus sequence and save each conserved region as a separate FASTA file.
4. These conserved regions are then processed based on published natural miRNA characteristics. For further details see WMD3

at http://wmd3.weigelworld.org/, which is a comprehensive site for Web miRNA design on various platforms, primarily against endogenous plant genes. We have provided the algorithm of miRNA Mate (Fig. 2), developed under Microsoft .NET framework. This algorithm may be freely modified according to the need and freely distributed. This package utilizes the Vienna RNA package (http://www.tbi.univie.ac.at/RNA/) and automatically picks potential amiRNA based on the A/U at position 1, A at position 10, and C/G at position 21 (see Note 3 and Fig. 1).

5. As a next selection criterion, pass your amiRNA sequences through RNAfold.exe, available as an open source package from *Vienna RNA Package* (http://www.tbi.univie.ac.at/RNA/). The algorithm we provided does this automatically by utilizing RNAfold.exe with the ΔG calculated and the amiRNA then ranked based on the minimum free energy. Enough flexibility should be available at various stages to make the conditions more or less stringent. The interaction between a short sequence miRNA (or amiRNA in this case) and the longer target sequence mRNA (the virus sequence in this case) can be calculated using *RNAhybrid* at http://bibiserv.techfak.uni-bielefeld.de/rnahybrid/ or *RNAup* at http://www.tbi.univie.ac.at/~ulim/RNAup/.
6. Screen amiRNA candidates against the particular host species genome or EST databases to search for possible off-targets. If the genomic resources are scarce for the actual host, BLAST the potential amiRNA sequences against the closest plant relative with a more extensive genomic resource. A plant small RNA target analysis server is available at http://plantgrn.noble.org/psRNATarget/. amiRNAs with potential targets in the host should be discarded. Here, the amiRNA selection criterion should be kept in mind, but the amiRNAs that have best matches in the target host should be discarded.
7. If after applying these selection filters, the final list of amiRNAs is greater than the desired number, the list may be further culled, for example by choosing amiRNAs directed against the most conserved regions in the viral genome or by biasing the selection towards the most virulent isolates. If a polycistronic amiRNA is being produced, consider choosing amiRNAs on both strands of the genome and targeting multiple genes within the virus.

The targets chosen in our example with WSMV are shown in Table 1, where the conservation across different isolates of the virus is evident. The positions of these target sequences in the WSMV genome and the structure of the transformation plasmids are shown in Fig. 3.

```
print "enter file number ";
$num=<STDIN>;
chomp $num;

$dg = "deltaG" . $num . '.txt';

open(OUTFILE, "> $dg") or die("could not open file out \n");

$seqName = "";
$silseq = "1";
$numSeqs = 0;
$targseq = "GCGCCGAGGTGAAGTTCGA";
$revt = "CGCGGCTCCACTTCAAGCT";

while ($silseq){

print "      TARGET SEQUENCE = $targseq \n";
print "REV C TARGET SEQUENCE = $revt \n";
print "      Enter  sequence : ";

$silseq=<STDIN>;
chomp $silseq;

if ($silseq eq "exit"){exit}
$silseq =~ tr/acgtuU/ACGTTT/;
$revtarg = reverse($revt);

print "      SilenceSEQUENCE = $silseq \n";


            open(TMPFILE, "> minFE.tmp") or die("could not open file
minFE.tmp\n");

            $silseqcomp = reverse($silseq);
            #$silseqcomp =~ tr/ACGT/TGCA/;

            $hp = $targseq . "AAAA" . $revtarg;
print "  perfect match hp  = $hp\n";
            print TMPFILE "$hp\n";
            close TMPFILE;

open(TMPFILE2, "> minFE2.tmp") or die("could not open file minFE2.tmp\n");

            $hp2 = $targseq . "AAAA" . $silseqcomp;
print "imperfect match hp2 = $hp2\n";

            print TMPFILE2 "$hp2\n";
            close TMPFILE2;
$results = `C:/Perl/RNAfold.exe < minFE.tmp`;
$results =~ s/\n/ /;

if ($results=~ /(\(\-)/) {
my $match = $&;

$mfe = $';
chomp $mfe;
chop $mfe;

$results2 = `C:/Perl/RNAfold.exe < minFE2.tmp`;
```

Fig. 2. The miR Mate algorithm (for Microsoft .Net framework)

```
$results2 =~ s/\n/ /;
#print $results2;

#print "\n $results2 \n";
$res1 = substr($results2, 0,42);
$res2 = substr($results2, 43,43);
print "         tested hp2 = $res1 \n";
print "         tested hp2 = $res2 \n";


if ($results2=~ /(\(\-)/) {

print " Perfect Free energy : $mfe";


$mfe2 = $';
chomp $mfe2;
chop $mfe2;
print "Free energy of new miR: $mfe2\n";


$ratio = ($mfe2/$mfe);

print OUTFILE " $hp \n $hp2 \n $res1 \n $res2 \n $mfe \n $mfe2 \n $ratio
\n\n\n";

}
}
}

print "why are you stopping here? \n";
close OUTFILE;
```

Fig. 2. (continued)

3.2. Selecting the Pre-miRNA Backbone

For our experiments we used osa-miR395 from rice. Osa-miR395 is induced by sulfur starvation to regulate a low-affinity sulfate transporter and two ATP sulfurylases (61, 64–66). The rice mi395 is a single ~1 kb transcript that generates a convoluted RNA structure that produces seven fully processed miRNA (Belide et al., unpublished). This gave us the option of replacing up to seven natural miRNA sites with antiviral amiRNAs. In our case we truncated the miR395 backbone so that it folded to only five duplex arms and replaced the five natural miRNAs with our five chosen amiRNAs.

3.3. Cloning In Silico

The final list of amiRNAs should replace the natural/endogenous miRNAs and miRNA* that are derived from the miR395 precursor. We replaced the native sequences with these amiRNA and amiRNA* sequences to conserve the secondary structure of the transcript. The predicted secondary structures of the polycistronic miR395 and the modified artificial miR395 were almost identical (Fig. 4), presumably enhancing the prospect of the predicted biogenesis of mature amiRNAs. This artificial polycistronic precursor was named

Table 1
Conservation of amiRNA targets in the WSMV genome

Accession	Origin	amiRNA-1 AGCTCTCGCATAGAGATAAGC	amiRNA-2 TCGAGCAAGATCTTTCACACG
a.			
AF057533	Nebraska	AGCTCTCGCATAGAGATAAGC	TCGAGCAAGATCTTTCACACG
AF285169	Kansas	AGCTCTCGCATAGAGATAAGC	TCGAGCAAGATCTTTCACACG
AF285170	Mexico	AGCTCTCGCATAGAGATAAGC	TCGAGCAAGATCTTTCACACG
AF454454	Czech	AGCTCTCGCATAGAGATAAGC	TCGAGCAAGATCTTTCACACG
AF454455	Turkey	AGCTCTCGCATAGAGATAAGC	TCGAGCAAGATCTTTCACACG
b.			
AF511614	Kansas	AGCTCTCGCATAGAGATAAGC	TCGAGCAAGATCTTTCACACG
AF511615	Kansas	AGCTCTCGCATAGAGATAAGC	TCGAGCAAGATCTTTCACACG
AF511618	Idaho	AGCTCTCGCATAGAGATAAGC	TCGAGCAAGATCTTTCACACG
AF511619	Idaho	AGCTCTCGCATAGAGATAAGC	TCGAGCAAGATCTTTCACACG
AF511630	Montana	AGCTCTCGCATAGAGATAAGC	TCGAGCAAGATCTTTCACACG
EU914917	Iran	AGCTCTCGCATAGAGATAAGC	TCGAGCAAGATCTTTCACACG
EU914918	Iran	AGCTCTCGCATAGAGATAAGC	TCGAGCAAGATCTTTCACACG
F511643	WA, USA	AGCTCTCGCATAGAGATAAGC	TCGAGCAAGATCTTTCACACG
FJ348358	WA, USA	AGCTCTCGCATAGAGATAAGC	TCGAGCAAGATCTTTCACACG
FJ348359	Argentina	AGCTCTCGCATAGAGATAAGC	TCGAGCAAGATCTTTCACACG
c.			
Unpublished	Canberra, Australia	AGCTCTCGCATAGAGATAAGC	TCGAGCAAGATCTTTCACACG

Accession	amiRNA-3 GAAGATTCCATTATGTGCCGA	amiRNA-4 CCAGGAAGCATTTTCTGGTCA	amiRNA-5 CCGCGAACGTCTTGCAAGTTA
a.			
AF057533	GAAGATTCCATTATGTGCCGA	CCAGGAAGCATTTTCTGGTCA	CCGCGAACGTCTTGCAAGTTA
AF285169	GAAGATTCCATTATGTGCCGA	CCAGGAAGCATTTTCTGGTCA	CCGCGAACGTCTTGCAAGTTA
AF285170	GAAGATACCACTATGTACCGA	CCAGGAAGCATTTTCTGGTCA	CCGCGAACGTCTTGCAAGTTA
AF454454	GAAGATTTCATTATGTGCCGA	CCAGGAAGCATTTTCTGGTCA	CCGCGAACGTCTTGCAAGTTA
AF454455	GAAGATTCCATTATGTGCCGA	CCAGGAAGCATTTTCTGGTCA	CCGCGAACGTCTTGCAAGTTA
b.			
AF511614	GAAGGTTCCATTATGTGCCGA	CCAGGAAGCATTTTCTGGTCA	CCGCGAACGTCTTGCAAGTTA
AF511615	GAAGATTCCATTATGTGCCGA	CCAGGAAGCATTTTCTGGTCA	CTGCGAACGTCTTGCAAGTTA
AF511618	GAAGATTCCATTATGTGCCGA	CCAGGAAGCATTTTCTGGTCA	CCGCGAACGTCTTGCAAGTTA
AF511619	GAAGATTCCATTATGTGCCGA	CCAGGAAGCATTTTCTGGTCA	CCGCGAACGTCTTGCAAGTTA
AF511630	GAAGATTCCATTATGTGCCGA	CCAGGAAGCATTTTCTGGTCA	CCGCGAACGTCTTGCAAGTTA
EU914917	GAAGATTTCATTATGTACCAA	CCAGGAAGCATTTTCTGGTCA	CCGCGAACGTCTTGCAAGTTA
EU914918	GAAGATTCCATTATGTGCCGA	CCAGGAAGCATTTTCTGGTCA	CCGCGAACGTCTTGCAAGTTA
F511643	GAAGATTTCATTATGTGCCGA	CCAGGAAGCATTTTCTGGTCA	CCGCGAACGTCTTGCAAGTTA
FJ348358	GAAGATTTCATTATGTGCCGA	CCAGGAAGCATTTTCTGGTCA	CCGCAAACGTCTTGCAAGTTA
FJ348359	GAAGATTCCATTATGTGCCGA	CCAGGAAGCATTTTCTGGTCA	CCGCGAACGTCTTGCAAGTTA
c.			
Unpublished	GAAGATTCCATTATGTGCCGA	CCAGGAAGCATTTTCTGGTCA	CCGCGAACGTCTTGCAAGTTA

Shown is the alignment and conservation of chosen target sequences in the WSMV genome. AlignX was used in Vector NTi10. (a) Shows the alignment of the five chosen target regions in the five published WSMV genomes available at the time of the design of the amiRNA. (b) Shows the alignment in ten new WSMV isolates that became available subsequently. (c) Shows alignment with the WSMV-ACT isolate used in our experiments (unpublished). The mismatched nucleotides with other isolates are highlighted. amiRNA-1 and -2 target the replicating strand; amiRNA-3, -4, and -5 target the genomic strand of the WSMV. The chosen targets remain absolutely conserved in all five target regions of the Australian isolate.

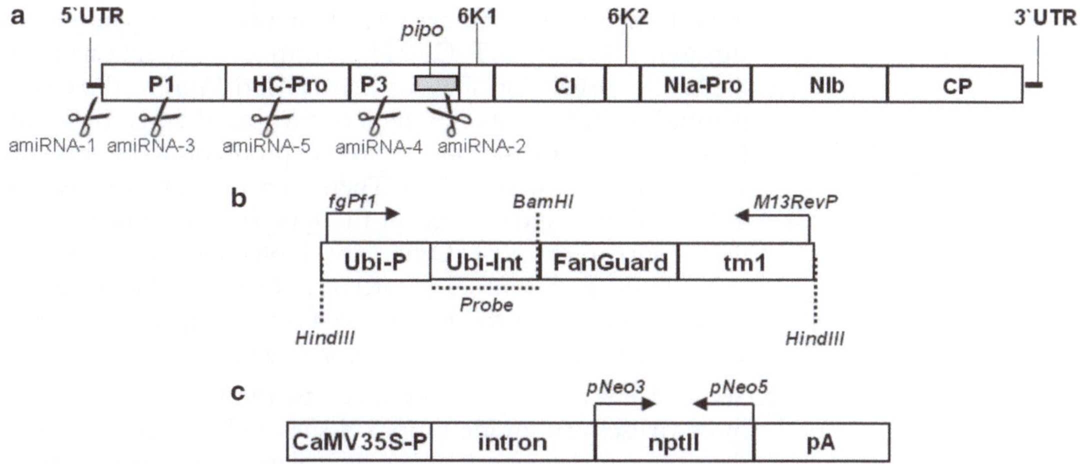

Fig. 3. Structure of the *Wheat streak mosaic virus* (WSMV) genome (approx 9,400 nt), with the target sites for amiRNAs and the FanGuard395 (*FGmiR395*) transgene. (**a**) Genome map of WSMV showing the five conserved regions (indicated by scissors) targeted by amiRNAs, amiRNA-1 to -5. (**b**) Design of the *FGmiR395* construct (1,400 nt) used to transform wheat using biolistics; shown are the probe region for Southern blot and primer sequences FgPf1 and M13RevP used in PCR. *Ubi-P* the maize ubiquitin promoter, *tm1* a plant gene terminator. (**c**) Diagram of pCMneoSTLS2 containing the *nptII* gene for geneticin resistance, used in the co-transformation of immature wheat embryos. CaMV35S-P is the cauliflower mosaic virus 35S promoter. The positions of the PCR primers pNeo3 and pNeo5 are shown.

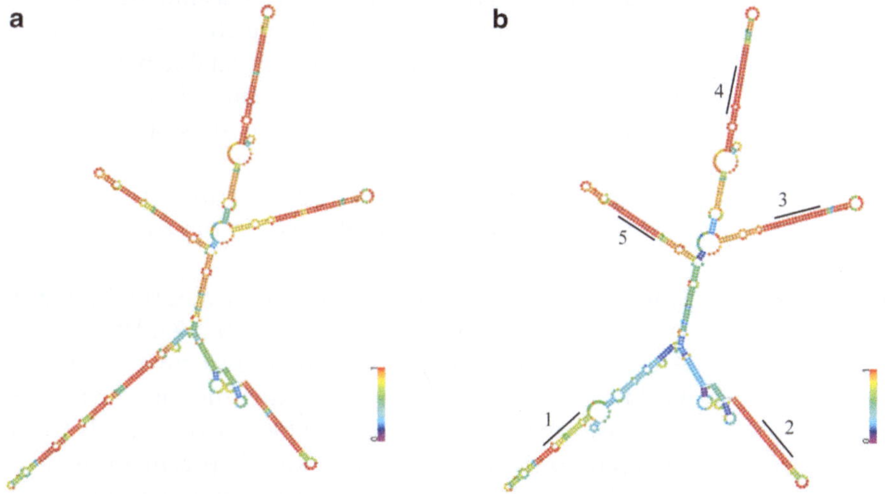

Fig. 4. amiRNA secondary structure. The structures of the truncated *Osa-miR395* and *FGmiR395* are shown. (**a**) Predicted secondary structure of *miR395* truncated to include only the first five native miRNAs. (**b**) Predicted secondary structure of *FGmiR395* replacing the first five natural miRNA sequences with amiRNAs designed against WSMV, numbered 1–5 and with bars showing regions corresponding to the amiRNA guide strand. These are the predicted fold structures of transcripts using *RNAfold*. The heat map indicates structure probabilities (*Blue*—weak, *Red*—strong).

FanGuard395 (FGmiR395). The following considerations should be used in the design of the vector:

1. Because of the complexity as a manual cloning exercise and the difficulties likely imposed by the secondary structures, it is strongly recommended that this type of gene construction be

done by synthesis. The designed FGmiR395 was synthesized through GENEART_GmbH (http://www.geneart.com) flanked by restriction sites for BamHI and KpnI in the carrier plasmid to facilitate its cloning into the final destination vector (pWubi). You can use routine cloning protocols to increase the vector and store it as backup. Then carry out sub-cloning into a destination vector based on the type of plant under study, i.e., monocot or dicot. Otherwise, biotechnology companies also will do such syntheses quickly and with quality assurance by sequencing the product and even offer to ligate the synthesized construct into the final vector as supplied.

2. Some attention needs to be given to the promoter; because most viruses can invade various tissues and are active at many developmental stages, it generally makes sense to use a constitutive type promoter (e.g., 35S CaMV promoter). For our experiments, the FG gene was excised from the carrier plasmid using appropriate restriction enzymes and ligated between the Ubiquitin promoter and the tml terminator of the vector pWubi-tml (67) generating cereal transformation plasmid FG-pWubi (see Note 4).

3. In the case of wheat transformation by biolistics, we use a selectable marker on a separate plasmid which is mixed with the transformation vector during the microparticle bombardment process. Often a suitable selectable marker gene will be a part of the T-DNA of the *Agrobacterium* binary plasmid (see Note 5). Appropriate restriction sites should be introduced into the construct to facilitate manipulation for gene fusion with reporter genes and Southern analysis to determine both insert size and copy number (see Note 6).

3.4. Transformation

Transformation should be carried out using the best available protocol for the host plant species (see Note 7). We use a well-established protocol for wheat biolistic transformation (68) with slight modification (17). We suggest experimenting with the so-called co-transformation protocols in which two plasmids (or fragments) are used: one carrying the amiRNA precursor gene without a selectable marker, and a second plasmid (or fragment) carrying the selectable marker (we used *npt*II). Because the pri-amiRNA and selectable marker transgenes sometimes insert in independent locations, this opens the opportunity to select segregant plants in subsequent generations carrying only the pre-amiRNA transgene conferring resistance against the virus and null for the selectable marker (17) (see Note 5).

3.5. Analysis of Transgenic Insertion

1. T_0 plants should be screened using PCR to select plants carrying the transgene of choice with pre-amiRNA against the virus. Obviously only T_0 plants which prove to have a full-length transgene insert will be of interest to phenotype in subsequent generations.

2. We also recommend performing northern blots using the EDC method involving chemical cross-linking (69). This method requires very little amount of RNA (up to 2 μg) and has 100-fold high sensitivity over the normal southern blot.
3. Plants should also be analyzed for the presence and stable insertion of the transgene via Southern blot (both for copy number and copy size) (see Note 6). Often the condition of T_0 plants does not warrant collecting enough tissue for a Southern blot; in this case we recommend collecting tissue from the family of T_1 plants which can be pooled to simulate T_0 because all the insertions of the T_0 plant can be assumed to be represented in the T_1 pool. Alternatively Southern blotting can be performed separately on each of the T_1 plants, which are also being screened for virus resistance, and this way initial assessment can be made of which inserts are associated with resistance.

3.6. Analysis of amiRNA Functionality via Virus Bioassay

T_1 families should be inoculated with virus using methods suitable for the virus. In the case of WSMV, mechanical inoculation is possible with rubbing or using a spray gun. The efficacy of the virus preparation should be established with control sensitive plants. The multiplication of the virus and its spread to new leaves over time can be monitored in a number of ways including Enzyme-linked Immunosorbent Assay (ELISA), reverse transcription-PCR (RT-PCR), and observation of symptoms. With ELISA, the presence of virus protein is directly correlated to the presence of virus particle. RT-PCR or quantitative RT-PCR is extremely sensitive in detecting the presence of viral RNA in tissues developing after the plant is inoculated. The utility of symptoms will vary with the virus of interest. Where immunity is suspected, it can be informative to attempt test inoculations from the putative immune plants as a control for susceptible plants; this is used to determine if the new tissues of the inoculated transgenic plant harbor any infectious virus. The absence of infectious virus from the sap of inoculated transgenic plant, absence of symptoms, background ELISA values, and RT-PCR for a nontarget region of the virus genome are usually sufficient to classify resistance as immunity (17, 33).

3.7. Analysis of Small RNA

Leaf samples should be collected from T_0, T_1, or T_2 plants without or with virus inoculation and analyzed for small RNA expression either using miRTect IT (USB, Cleveland, OH, USA) utilizing splinted ligation (70, 71) or the EDC method of Northern hybridization (69) (see Note 8). It has been observed that if transgenic plants are accumulating low levels of amiRNAs, then the pri-amiRNA transgene may not be transcribed or processed efficiently, leading to accumulation of virus and typical symptom manifestation.

3.8. Characterizing Resistance

Virus accumulation and symptoms should be monitored at 7, 14, 21, and 28 days post-infection. At each time point we recommend using a virus back-inoculation assay, especially if immunity is suspected (see Note 9). This is a very useful indirect method to determine infectious virus particles in the inoculated transgenic plants, although it might only be practical when the virus can be mechanically transmitted from sap, such as with WSMV. Various infective sap dilutions can further decipher the level of resistance or susceptibility. The resistance level is categorized as follows:

1. *Immunity:* It is permissible to characterize challenged plants as immune if at all time points they develop no virus symptoms, do not accumulate virus as determined by immunological assays such as ELISA, do not evidence the presence of viral RNA using RT-PCR, and leaf samples are shown not to harbor infective virus by back-inoculation to susceptible controls.
2. *Resistance*: If the virus-challenged plants show no symptoms, accumulate much reduced virus through ELISA, but do have infectious sap, then the plants may be characterized as resistant. Various degrees of resistance can be defined based on parameters such as intermediate levels of virus, or intermediate symptoms including effects on plant height compared to healthy and infected susceptible controls.
3. *Susceptibility*: Plants are categorized as susceptible if they show symptoms at all time points, accumulate virus through ELISA, and have infectious leaf sap. However, the susceptible transgenic plants should be analyzed in the next generation to see if immune plants can be recovered.
4. *Resistance breakdown/recovery*: It is a common phenomenon that viruses encounter initial resistance, have their rate of replication and accumulation suppressed, but overcome the resistance mechanism after an initial delay. A number of factors may account for such resistance breakdown: it could be attributed to mutation in the inoculated virus such that it escapes the amiRNA surveillance; and it could be due to the inoculum pressure overwhelming amiRNA-mediated resistance. Therefore, care should be taken to differentiate between the two phenomena. This can be achieved by sequencing the virus particles from challenged plants and then comparing the target sequence with the wild type (used in primary inoculation). Alternatively, a known load of infectious clones could be used so that one can easily follow the virus evolution, if any, in subsequent generations. However, such plants do yield resistant progeny in subsequent generations, and that could be attributed to several factors that are beyond the scope of this chapter.
5. *Tolerance*: This is a category that is probably best defined as a susceptible plant which performs well despite the infection.

It is most commonly applied to genotypes which allow virus replication and accumulation but which in the field continue to yield well despite the infection. This is not a category expected of amiRNA transgenics.

4. Notes

1. It may be helpful with a polycistronic pre-amiRNA to target both invading virus (genomic sequence) and the subsequent replicative strand (–RNA). In the life cycle of the virus one strand may be more available for the amiRNA surveillance than the other. By choosing amiRNAs against a mix of targets on the genomic and replicative strands you enhance the prospect of successfully disrupting the life cycle of the virus.
2. Conserved sequences in plant virus genomes are the preferred targets; however, it is ideal if such conserved sequences are located in functionally or structurally conserved regions. This would minimize the chances for virus to mutate and overcome the resistance. It has been shown that mutation in such regions often affects the virus replication and movement within the plant. Furthermore, mutations in positions 3–6, 9, and 12 resulted in disease symptoms (i.e., failure of amiRNA-mediated resistance). Positions 2, 10, 11, 13, 15, and 18 were of intermediate importance as mutations at these positions resulted in viral pathogenicity in 36 % of inoculated transgenic plants. The remaining positions were classified as noncritical for resistance as mutation in these positions resulted in disease in less than 7 % of inoculated plants (59) (see Fig. 1).
3. Although early reports suggested that nucleotide positions at 1, 10, and 21 are critical for effective amiRNAs, recent studies have shown that various other positions are important for amiRNA effectiveness. Therefore, we suggest choosing a perfect match between amiRNA and its target. You can choose a pri-miRNA backbone of your choice from miRBase (http://www.mirbase.org/) or Arabidopsis small RNA project ASRP (http://asrp.cgrb.oregonstate.edu/db/). Care should be taken to consider the use of the pri-miRNA backbone—this could depend upon the choice of host under study. We also recommend that a parallel analysis be carried out so as to see if there is any effect on the expression of the endogenous miRNA.
4. It may sometimes be preferable to use tissue-specific promoters. For example when dealing with viruses transmitted by nematodes or soil fungi, root-specific promoters may give better

results. Likewise leaf-specific promoters may be indicated for viruses transmitted by the feeding activity of vectors such as whiteflies or aphids. Perhaps phloem active promoters will be more effective for phloem-limited viruses. Tissue-specific expression may also serve to limit the presence of the amiRNA in the food product.

5. Marker-free transgenics will be preferred for deregulation and commercial release. When using biolistics for transformation, marker-free immune segregants could be recovered from some of the transformation events when the selectable marker was on a second plasmid from the pri-amiRNA carrying plasmid (co-transformation). The ratio of the plasmid vector mix in biolistics can be optimized to achieve this goal. With co-transformation we have been able to recover transgenic plants that lack the selectable marker nptII. We believe it would be possible to recover marker-free transgenic plants because of the possibility that the selectable marker and transgene may occupy different loci in the wheat genome.
6. Analysis of the transgene both for copy size and copy number, especially in biolistically transformed plants, is important. This should help identify single copy carrying events or segregants and can help characterize the resistance and various other phenotypes, when combined with other molecular data.
7. Before stable transformation, the individual or multiplex amiRNA construct can be expressed in vivo using a transient assay to verify if it is precisely processed, and co-infiltration with the relevant virus infectious clone should determine its efficacy. Transient expression of individual amiRNA constructs can be used as a positive control to examine the expression of the corresponding amiRNA in a complex multiplex pre-miRNA in transgenic plants.
8. It is possible that replacing the natural miRNA sequences in pre-miRNA in a complex miRNA cluster such as miR395 can affect the amiRNA biogenesis. This may lead to phasing of amiRNA in a manner similar to tasiRNAs, after initial cleavage by dcl1. The choice of the miRNA probe (LNA or simple oligos) can affect the detection of amiRNA in Northern blots. We also recommend that using EDC method could give higher sensitivity of detection.
9. Inoculum pressure is very important as high dose and repeated inoculation may overcome resistance mediated by amiRNAs. Care should be taken to simulate the natural field situation, when using mechanical inoculation or vector transmission for these studies. Mechanical inoculation may represent virus loads and intensity of challenge that greatly exceeds anything that is likely to be encountered in the field.

References

1. Abel PP et al (1986) Delay of disease development in transgenic plants that express the tobacco mosaic virus coat protein gene. Science 232(4751):738–743
2. Bendahmane M et al (2007) Coat protein-mediated resistance to TMV infection of Nicotiana tabacum involves multiple modes of interference by coat protein. Virology 366(1): 107–116
3. Register JC 3rd, Beachy RN (1988) Resistance to TMV in transgenic plants results from interference with an early event in infection. Virology 166(2):524–32
4. Nejidat A, Beachy RN (1990) Transgenic tobacco plants expressing a coat protein gene of tobacco mosaic virus are resistant to some other tobamoviruses. Mol Plant Microbe Interact 3(4):247–251
5. Lindbo JA, Dougherty WG (2005) Plant pathology and RNAi: a brief history. Annu Rev Phytopathol 43:191–204
6. Lomonossoff GP (1995) Pathogen-derived resistance to plant viruses. Annu Rev Phytopathol 33:323–343
7. Baulcombe DC (1996) Mechanisms of pathogen-derived resistance to viruses in transgenic plants. Plant Cell 8(10):1833–1844
8. Beachy RN (1997) Mechanisms and applications of pathogen-derived resistance in transgenic plants. Curr Opin Biotechnol 8(2):215–220
9. Bucher E et al (2006) Multiple virus resistance at a high frequency using a single transgene construct. J Gen Virol 87(Pt 12):3697–3701
10. Goldbach R, Bucher E, Prins M (2003) Resistance mechanisms to plant viruses: an overview. Virus Res 92(2):207–212
11. Nejidat A, Beachy RN (1989) Decreased levels of TMV coat protein in transgenic tobacco plants at elevated temperatures reduce resistance to TMV infection. Virology 173(2):531–538
12. Waterhouse PM, Graham HW, Wang MB (1998) Virus resistance and gene silencing in plants can be induced by simultaneous expression of sense and antisense RNA. Proc Natl Acad Sci USA 95(23):13959–13964
13. Kalantidis K et al (2002) The occurrence of CMV-specific short RNAs in transgenic tobacco expressing virus-derived double-stranded RNA is indicative of resistance to the virus. Mol Plant Microbe Interact 15(8):826–833
14. Di Nicola-Negri E et al (2005) Hairpin RNA-mediated silencing of Plum pox virus P1 and HC-Pro genes for efficient and predictable resistance to the virus. Transgenic Res 14(6): 989–994
15. Tougou M et al (2006) Development of resistant transgenic soybeans with inverted repeat-coat protein genes of soybean dwarf virus. Plant Cell Rep 25(11):1213–1218
16. Fuentes A et al (2006) Intron-hairpin RNA derived from replication associated protein C1 gene confers immunity to tomato yellow leaf curl virus infection in transgenic tomato plants. Transgenic Res 15(3):291–304
17. Fahim M et al (2010) Hairpin RNA derived from viral NIa gene confers immunity to wheat streak mosaic virus infection in transgenic wheat plants. Plant Biotechnol J 8(7):821–834
18. Jackson AL et al (2003) Expression profiling reveals off-target gene regulation by RNAi. Nat Biotechnol 21(6):635–637
19. Jackson AL, Linsley PS (2004) Noise amidst the silence: off-target effects of siRNAs? Trends Genet 20(11):521–524
20. Jackson AL et al (2006) Widespread siRNA "off-target" transcript silencing mediated by seed region sequence complementarity. RNA 12(7):1179–1187
21. Jackson AL et al (2006) Position-specific chemical modification of siRNAs reduces "off-target" transcript silencing. RNA 12(7):1197–1205
22. Fuchs M, Gonsalves D (2007) Safety of virus-resistant transgenic plants two decades after their introduction: lessons from realistic field risk assessment studies. Annu Rev Phytopathol 45:173–202
23. Schwab R et al (2006) Highly specific gene silencing by artificial microRNAs in Arabidopsis. Plant Cell 18(5):1121–1133
24. Schnippenkoetter WH et al (2001) Forced recombination between distinct strains of Maize streak virus. J Gen Virol 82:3081–3090
25. Brodersen P et al (2008) Widespread translational inhibition by plant miRNAs and siRNAs. Science 320(5880):1185–1190
26. Meister G et al (2004) Sequence-specific inhibition of microRNA- and siRNA-induced RNA silencing. RNA 10(3):544–550
27. Israsena N et al (2009) Inhibition of rabies virus replication by multiple artificial microRNAs. Antiviral Res 84(1):76–83
28. Khraiwesh B et al (2008) Specific gene silencing by artificial microRNAs in Physcomitrella patens: an alternative to targeted gene knockouts. Plant Physiol 148(2):684–693
29. Ossowski S, Schwab R, Weigel D (2008) Gene silencing in plants using artificial microRNAs and other small RNAs. Plant J 53(4):674–690
30. Niu QW et al (2006) Expression of artificial microRNAs in transgenic Arabidopsis thaliana

confers virus resistance. Nat Biotechnol 24(11): 1420–1428

31. Duan CG et al (2008) Artificial MicroRNAs highly accessible to targets confer efficient virus resistance in plants. J Virol 82(22): 11084–11095
32. Ai T et al (2011) Highly efficient virus resistance mediated by artificial microRNAs that target the suppressor of PVX and PVY in plants. Plant Biol 13(2):304–316
33. Fahim M et al (2012) Resistance to wheat streak mosaic virus generated by expression of an artificial polycistronic microRNA in wheat. Plant Biotechnol J 10(2):150–163
34. Kung YJ et al (2012) Multiple artificial microRNAs targeting conserved motifs of the replicase gene confer robust transgenic resistance to negative-sense single-stranded RNA plant virus. Mol Plant Pathol 13(3):303–317
35. Qu J, Ye J, Fang R (2007) Artificial microRNA-mediated virus resistance in plants. J Virol 81(12):6690–6699
36. Zhang X et al (2011) Expression of artificial microRNAs in tomato confers efficient and stable virus resistance in a cell-autonomous manner. Transgenic Res 20(3):569–581
37. Sivamani E et al (2000) Resistance to wheat streak mosaic virus in transgenic wheat expressing the viral replicase (NIb) gene. Mol Breed 6(5):469–477
38. Sivamani E et al (2002) Resistance to wheat streak mosaic virus in transgenic wheat engineered with the viral coat protein gene. Transgenic Res 11(1):31–41
39. Jiang F et al (2011) The choice of target site is crucial in artificial miRNA-mediated virus resistance in transgenic Nicotiana tabacum. Physiol Mol Plant Pathol 76(1):2–8
40. Jiang F et al (2011) Special origin of stem sequence influence the resistance of hairpin expressing plants against PVY. Biol Plantarum 55(3):528–535
41. Anderson E et al (2008) Identifying siRNA-induced off-targets by microarray analysis. Methods Mol Biol 442:45–63
42. Bartz SR et al (2002) Off-target activity of siRNA oligos in mammalian cells. Mol Biol Cell 13:409a–410a
43. Jackson AL, Linsley PS (2010) Recognizing and avoiding siRNA off-target effects for target identification and therapeutic application. Nat Rev Drug Discov 9(1):57–67
44. Petri S et al (2011) Increased siRNA duplex stability correlates with reduced off-target and elevated on-target effects. RNA 17(4):737–749
45. Smith NA, Eamens AL, Wang MB (2011) Viral small interfering RNAs target host genes to mediate disease symptoms in plants. PLoS Pathog 7(5):e1002022
46. Shimura H et al (2011) A viral satellite RNA induces yellow symptoms on tobacco by targeting a gene involved in chlorophyll biosynthesis using the RNA silencing machinery. PLoS Pathog 7(5):e1002021
47. Fortier E, Belote JM (2000) Temperature-dependent gene silencing by an expressed inverted repeat in Drosophila. Genesis 26(4): 240–244
48. Szittya G et al (2003) Low temperature inhibits RNA silencing-mediated defence by the control of siRNA generation. EMBO J 22(3):633–640
49. Kameda T et al (2004) A hypothermic-temperature-sensitive gene silencing by the mammalian RNAi. Biochem Biophys Res Commun 315(3):599–602
50. Chen J et al (2004) Viral virulence protein suppresses RNA silencing-mediated defense but upregulates the role of MicroRNA in host gene expression. Plant Cell 16(5):1302–1313
51. Schubert S et al (2005) Local RNA target structure influences siRNA efficacy: systematic analysis of intentionally designed binding regions. J Mol Biol 348(4):883–893
52. Praveen S, Mishra AK, Antony G (2006) Viral suppression in transgenic plants expressing chimeric transgene from tomato leaf curl virus and cucumber mosaic virus. Plant Cell Tissue Organ Cult 84(1):47–53
53. Antony G, Mishra AK, Praveen S (2005) A single chimeric transgene derived from two distinct viruses for multiple virus resistance. J Plant Biochem Biotechnol 14(2):101–105
54. Lafforgue G et al (2011) Tempo and mode of plant RNA virus escape from RNA interference-mediated resistance. J Virol 85(19): 9686–9695
55. Rodrigo G et al (2011) Optimal viral strategies for bypassing RNA silencing. J R Soc Interface 8(55):257–268
56. Liu YP et al (2008) Inhibition of HIV-1 by multiple siRNAs expressed from a single microRNA polycistron. Nucleic Acids Res 36(9):2811–2824
57. Westerhout EM et al (2005) HIV-1 can escape from RNA interference by evolving an alternative structure in its RNA genome. Nucleic Acids Res 33(2):796–804
58. Simon-Mateo C, Garcia JA (2006) MicroRNA-guided processing impairs Plum pox virus replication, but the virus readily evolves to escape this silencing mechanism. J Virol 80(5):2429–2436
59. Lin SS et al (2009) Molecular evolution of a viral non-coding sequence under the selective

pressure of amiRNA-mediated silencing. Plos Pathogens 5(2):e1000312

60. ter Brake O et al (2006) Silencing of HIV-1 with RNA interference: a multiple shRNA approach. Mol Ther 14(6):883–892
61. Guddeti S et al (2005) Molecular evolution of the rice miR395 gene family. Cell Res 15: 631–638
62. Overhoff M et al (2005) Local RNA target structure influences siRNA efficacy: a systematic global analysis. J Mol Biol 348(4): 871–881
63. Fahim M et al (2011) Resistance to wheat streak mosaic virus – a survey of resources and development of molecular markers. Plant Pathol 61(3):425-440
64. Kawashima CG et al (2009) Sulphur starvation induces the expression of microRNA-395 and one of its target genes but in different cell types. Plant J 57(2):313–321
65. Chiou TJ (2007) The role of microRNAs in sensing nutrient stress. Plant Cell Environ 30(3):323–332
66. Jones-Rhoades MW, Bartel DP (2004) Computational identification of plant MicroRNAs and their targets, including a stress-induced miRNA. Mol Cell 14(6):787–799
67. Wang M-B, Abbott DC, Waterhouse PM (2000) A single copy of a virus-derived transgene encoding hairpin RNA gives immunity to barley yellow dwarf virus. Mol Plant Pathol 1(6):347–356
68. Pellegrineschi A et al (2002) Identification of highly transformable wheat genotypes for mass production of fertile transgenic plants. Genome 45(2):421–430
69. Pall GS, Hamilton AJ (2008) Improved northern blot method for enhanced detection of small RNA. Nat Protoc 3(6):1077–1084
70. Maroney PA et al (2008) Direct detection of small RNAs using splinted ligation. Nat Protoc 3(2):279–287
71. Maroney PA et al (2007) A rapid, quantitative assay for direct detection of microRNAs and other small RNAs using splinted ligation. RNA 13(6):930–936

Chapter 20

Downregulation of Plant Genes with miRNA-Induced Gene Silencing

Felipe Fenselau de Felippes

Abstract

In plants, some microRNAs (miRNAs) can trigger the production of secondary small interfering RNAs (siRNAs) from their targets. miRNA-induced gene silencing (MIGS) exploits this unique feature to efficiently downregulate gene expression. The simple flanking of a sequence of interest with the target site for the miR173 (an miRNA able to trigger transitivity) is sufficient to start the production of secondary siRNAs and, consequently, silencing of the target gene. This technique can be easily adapted to promote gene silencing of more than one gene, even with those that share no sequence similarities. This chapter describes the necessary steps for designing and implementing the use of MIGS in plants.

Key words: Functional genomics, RNA silencing, RNA interference, miR173, tasiRNA, Gene downregulation

1. Introduction

One of the most important mechanisms of gene regulation in plants is the one mediated by 19–24 nt long RNA molecules, also known as small RNAs (sRNAs). This process, which is usually referred as sRNA silencing or RNA interference, results in the downregulation of gene expression, either at the transcription or post-transcription level (transcription gene silencing (TGS) and post-transcription gene silencing (PTGS), respectively). In plants, sRNAs can be divided into microRNAs (miRNAs) and small interfering RNAs (siRNAs). In both cases, the sRNAs are produced from double-stranded RNA (dsRNA) through the action of DICER-LIKE (DCL) enzymes. The resulting sRNAs are loaded into ARGONAUT (AGO) proteins to form the main component of the RNA-induced silencing complex (RISC). RISC will then lead to downregulation of gene expression by promoting DNA

Debra J. Taxman (ed.), *siRNA Design: Methods and Protocols*, Methods in Molecular Biology, vol. 942,
DOI 10.1007/978-1-62703-119-6_20, © Springer Science+Business Media, LLC 2013

methylation, cleavage of the target transcript, and/or inhibition of translation (1–3).

Trans-acting small interfering RNA (tasiRNA) is a plant-specific class of endogenous siRNA, which is produced from *TAS* gene-derived transcripts after they are targeted by an miRNA. The miRNA-dependent cleavage of the *TAS* transcript leads to the recruitment of SUPPRESSOR OF GENE SILENCING 3 (SGS3) and RNA-DEPENDENT RNA POLYMERASE 6 (RDR6). The latter uses the *TAS* transcript as a template for the synthesis of a dsRNA, which is processed by DICER-LIKE 4 (DCL4) into 21 nt long tasiRNAs. tasiRNAs are loaded in ARGONAUTE 1 (AGO1) and promote the downregulation of genes in trans (4, 5).

In the model plant *Arabidopsis thaliana*, four *TAS* families can be recognized (*TAS1-4*). The generation of tasiRNAs from *TAS1* and *TAS2* is triggered by miR173-directed cleavage. In both cases, tasiRNAs are produced from the sequence located downstream to the miR173 target site, with the cleavage position setting the beginning of this region (4). It has been shown that the ability of miR173 to trigger tasiRNA production is due to its size of 22 nt, unlike the majority of plant miRNAs, which are 21 nt in length and do not induce production of secondary siRNAs (6, 7). Moreover, the cleavage mediated by miR173 is sufficient to start transitivity, and targeting of a given gene by miR173 results in the production of secondary sRNAs originated from the target sequence (8, 9).

miRNA-Induced Gene Silencing (MIGS) is a technique to drive downregulation of gene expression. It is based on the unique feature of miR173 to trigger production of secondary sRNAs from its targets (10). The concept behind MIGS is very simple; it consists of targeting the miR173 to a sequence of the gene for which silencing is desired. This can be easily done with the addition of the miRNA target site directly upstream of the sequence of interest (Fig. 1a). When the transcript of the MIGS construct (sequence of interest flanked by miR173 target site) is cleaved by the miR173, a process similar to the one described for tasiRNA generation starts, resulting in the production of secondary sRNAs. These secondary sRNAs originating from the sequence of interest will be loaded into AGO1 and promote silencing of the endogenous target gene.

MIGS can also be used to simultaneously silence multiple genes. In a strategy similar to that discussed above, MIGS modules (miR173 target site plus sequence of interest) are produced for each gene of interest and fused together to generate a single MIGS construct, which can be then cloned into the binary vector of preference (Fig. 1b) (10). This construct will be capable of efficiently silencing different genes simultaneously. This approach has many advantages including the following: (1) sequence similarity is not required between the different target genes because a different, specific MIGS module will be generated for each intended target; (2) sRNAs targeting each one of the target genes are ensured

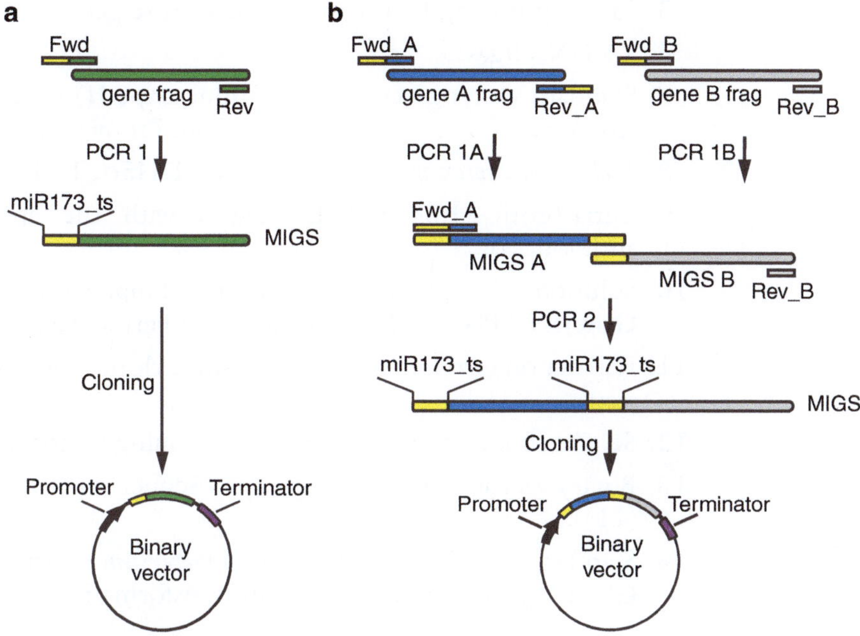

Fig. 1. Schematic representation of MIGS generation. (**a**) MIGS targeting a single gene. miR173 target site (miR173_ts/*yellow bar*) is incorporated upstream of a sequence of the target gene (gene frag) by PCR with specific forward (Fwd, modified for miRNA target site incorporation) and reverse (Rev) primers. The resulting MIGS construct is cloned into a binary vector containing the regulatory sequences (promoter and terminator) for expression in plants. (**b**) Simultaneous silencing of two genes with MIGS. In independent PCR reactions using the respective specific primers (Fwd_A and Rev_A for gene A; Fwd_B and Rev_B for gene B, see text), an miR173 target site is placed flanking both sides of gene A fragment (gene A frag) and upstream of a sequence of gene B (gene B frag). The two MIGS modules (MIGS A and MIGS B) are fused by PCR into a single MIGS construct and cloned into a binary vector containing regulatory sequences.

to be produced due to the presence of an miR173 target site in front of every single targeted gene sequence; and (3) regardless of the number of genes to be silenced, only one MIGS construct is generated, saving time and facilitating the generation of transgenic plants with multiple genes downregulated.

2. Materials

1. Sequence of the target gene(s) that will be silenced. For *A. thaliana*, gene sequences can be obtained at *The Arabidopsis Information Resource* (TAIR, http://www.arabidopis.org).
2. Primers as described in step 2 of Subheadings 3.1 and 3.2.
3. Proofreading DNA polymerase (e.g., Phusion® High-Fidelity DNA polymerase from Fermentas), dNTPs, and cDNA to be used as template in the PCR reaction.
4. Equipment and reagents to perform agarose gel electrophoresis.

5. Kit for purifying PCR bands from agarose gels.
6. T4 DNA ligase.
7. Standard cloning vector (e.g., pBluescript II) or commercial cloning kit (e.g., pGEM®-T Easy from Promega).
8. *Escherichia coli* competent cells (e.g., DH5α, TOP10).
9. Luria-broth (LB) and LB plates with the appropriated antibiotic(s).
10. Solutions for plasmid extraction (Miniprep) or kit (e.g., GeneJET™ Plasmid Miniprep kit from Fermentas).
11. Restriction enzymes to test for positive clones and sub-cloning steps.
12. Sequencing primers (specific to the cloning vector used).
13. Binary vector containing regulatory sequences (e.g., pGreenII (11)).
14. Standard competent strain of *Agrobacterium tumefaciens* (e.g., GV3101) and solutions for plant transformation.

3. Methods

In this section, the steps necessary for the design and utilization of MIGS will be presented. The first part focuses on the generation of MIGS mediating the silencing of a single gene, while in the second part, the strategy to generate MIGS targeting multiple genes is described.

3.1. Single Gene Silencing by MIGS

The first step in the generation of a MIGS construct is the choice of the sequence that will be flanked by an miR173 target site. This sequence should be part of the target gene cDNA and should be between 200 and 500 nt in length (see Note 1). Often multiple genes have shared or highly similar sequences; therefore, it is recommended to select a gene fragment with low sequence similarity to other regions of the plant transcriptome to minimize the risk of off targeting. The following steps are suggested:

1. Use BLAST (http://blast.ncbi.nlm.nih.gov/) to look for conserved regions in the target gene cDNA. At the tool Webpage, choose the genome database of the species for which silencing is desired (see Note 2). Using standard settings, proceed with the alignment of the target gene cDNA. Select a region presenting either no matches or minimal conservancy.
2. Design specific primers for the amplification of the selected region. The utilization of primer design tools, such as Primer3 (http://frodo.wi.mit.edu/primer3/), is recommended for best results. For the forward primer, include the miR173

target site sequence as follows: *GTGATTTTTCTCTACAA GCGAA*$N_{(17–24)}$, where *italicized* letters represent the miR173 target site and $N_{(17–24)}$ denotes the gene-specific sequence, which usually range from 17 to 24 nt in length. After PCR, the miR173 target site will be incorporated directly upstream of the gene fragment (Fig. 1a; see Note 3). The reverse primer should be a standard-specific oligonucleotide.

3. Carry out a PCR reaction using the designed primers, a proofreading DNA polymerase, and cDNA as template. The typical PCR cycle conditions include the following: (a) initial denaturation; (b) 30 cycles of denaturation, primer annealing (temperature depending on the oligonucleotide melting temperature (Tm), and extension; and (c) a final extension step. Check the polymerase instructions for recommended temperatures and incubation times.
4. Check the PCR product for size, quantity, and intactness by running an agarose gel, and purify the amplified band with standard gel extraction procedures.
5. Clone the purified PCR fragment into a standard cloning vector by using commercial kits such as pGEM®-T Easy (see Notes 4 and 5).
6. Check for mutations and for the correct incorporation of the miR173 target site upstream of the sequence of interest by sequencing the plasmid insert with specific primers. Most standard cloning vectors have sequencing primer-binding sites directly outside the multi-cloning region specifically for this purpose (T7 and SP6 primers for the pGEM®-T Easy vector).
7. Before proceeding with plant transformation, the MIGS construct needs to be provided with the proper regulatory sequences, i.e., a promoter and a terminator, which should be added by subsequent sub-cloning steps. For applications where near-complete reduction in gene expression is desired, strong constitutive promoters, such as the cauliflower mosaic virus 35S promoter (CaMV35S) (12), should be used. However, if modulation of the gene expression is the main purpose, the utilization of a weak or moderate promoter might be more appropriate. Tissue-specific promoters should be used with caution. tasiRNAs have been shown to act in a non-cell autonomous fashion (13–15); therefore it is possible that secondary sRNAs derived from MIGS constructs may not be confined to the tissue where they have been produced. In this case, silencing tools based on less mobile sRNAs, such as miRNAs, are recommended (16).
8. For plant transformation protocols relying on *Agrobacterium* infection, it is necessary to clone the MIGS construct into a final binary vector.
9. Finally, the generation of transgenic plants carrying the MIGS construct is done using standard transformation protocols.

3.2. Multi-Gene Silencing with MIGS

The design of MIGS mediating the downregulation of multiple genes follows the same principles applied for that of single gene silencing; for each gene to be silenced a fragment is selected and the miR173 target site is inserted upstream of this sequence. These MIGS modules (mir173 target site plus the sequence of interest) are then fused by PCR into a single MIGS construct and cloned into the vector of choice. Here the steps are presented for generating a MIGS construct targeting two distinct genes (gene A and gene B) simultaneously:

1. For each gene to be silenced, select a fragment as described in step 1 of Subheading 3.1 (single gene MIGS).
2. Design specific primers for the amplification of target genes A and B fragments as instructed in step 2 of Subheading 3.1. To both forward primers (Fwd_A and Fwd_B; designed for the amplification of gene A and gene B fragments, respectively) include the miR173 target site directly upstream of their sequences, as described in step 2 of Subheading 3.1. The reverse primer for amplifying the gene A fragment (Rev_A) should contain the reverse complementary sequence of miR173 target site as follows: *TTCGCTTGTAGAGAAAATCA* *C*$N_{(17-24)}$, where $N_{(17-24)}$ stands for the gene-specific sequence and the *italicized* letters represent the reverse complementary sequence of miR173 target site. This modification in the Rev_A primer will allow the pairing between MIGS module A and B during the final PCR step, aiding in the fusion of both fragments into a single MIGS construct (Fig. 1b). Finally, the reverse primer for amplification of gene B fragment (Rev_B) should contain no additional sequence.
3. Using the respective primers, individually amplify each of the fragments originated from either target gene A or B (PCR 1A and 1B, respectively), as instructed in step 3 of Subheading 3.1.
4. Analyze the PCR result by running an agarose gel and purify the PCR products using standard methods. The products from both PCR reactions can be pooled at this step.
5. To generate a single MIGS construct targeting both gene A and gene B, carry out a PCR reaction as follows: in the same reaction, use products from PCR 1A and 1B as template (0.1–1 μl of the purified product) and primers Fwd_A and Rev_B. By employing this strategy, a PCR product will only be generated when the miR173 target site located downstream of gene A fragment pairs with the same site located upstream to gene B fragment, resulting in the fusion of both MIGS modules (Fig. 1b). As suggested in step 3 of Subheading 3.1, use a proofreading DNA polymerase, following the manufacturer's instructions (see Note 6).

6. Check the PCR reaction in an agarose gel and purify the amplified band.
7. Follow with the cloning, sequencing analysis, sub-cloning, and plant transformation as described in steps 5–9 of Subheading 3.1.

4. Notes

1. In principle, the entire gene could be used in the generation of a MIGS construct. This is particularly relevant in cases where the target gene is already cloned in Gateway® entry vectors and cDNA libraries (see Note 3). However, working with small gene fragments is usually easier and allows the selection of a region in which off targeting can be avoided. Gene fragments smaller than 200 bp can also be used; however loss of efficiency can happen due to the lower amount of sRNAs produced. It has been reported that sequences as short as 98 nt were effective in triggering gene silencing through hairpin-based RNAi methods (hpRNAi) (17). Based on the similarities shared between hpRNAi and MIGS, it is likely that the same could apply for the latter technique.
2. MIGS is based on the miR173, which is a non-conserved miRNA present only in *A. thaliana* and a few close relatives. Therefore, gene silencing by MIGS in species that are not *A. thaliana* should be done together with the co-expression of miR173. A collection of Gateway®-compatible binary plasmids is available (10), where the MIGS and the miR173 expression cassette are both located in the same T-DNA, allowing the straightforward generation of MIGS silenced plants in species other than *A. thaliana*.
3. An alternative strategy for ligating the miR173 target site upstream of the sequence of interest is to clone this target site into the final binary vector, just after the promoter and directly before the fragment insertion site. Those working with Gateway®-compatible vectors will find this strategy very convenient because any given gene already cloned in a Gateway® entry vector can be readily recombined into the final binary vector containing the miR173 target site, thus generating a MIGS construct without the need for PCR and all subsequent steps. In addition, this strategy can also be used in high-throughput silencing screens of Gateway®-based gene libraries. The plasmid collection mentioned in Note 2 also includes vectors in which the miR173 target site is placed directly before the recombination site specifically for this purpose.

4. Because most proofreading DNA polymerases generate blunt-ended products, it might be necessary to add an adenine overhang to the PCR product before using some cloning kits, which rely on a TA strategy (e.g., pGEMT-easy from Promega). This is easily done by incubating the purified PCR product with ordinary standard DNA polymerase or Klenow-fragment in the presence of dATP. For more details, refer to enzymes and cloning kits instructions.
5. The PCR product can also be cloned directly into a binary vector or any other plasmid of choice (for instance, Gateway® entry vectors). An efficient method for cloning the blunt-ended PCR product generated by the proofreading polymerases is by using plasmids linearized with enzymes generating blunt-end extremities. It is required that the enzyme employed does not cut the PCR product. A good candidate is SmaI, which is present in the multi-cloning sites of most plasmids. Do not dephosphorylate the linearized vector, because the PCR product does not contain a 5' phosphorylated end. Perform the ligation in SmaI buffer, by adding ATP (final concentration of 0.5 mM) and SmaI (3 units) into the reaction. Incubate the reaction overnight at room temperature. Adding the enzyme in the ligation reaction will avoid plasmid re-ligation, highly increasing cloning efficiency.
6. If the fusion PCR does not yield any product, try using a different DNA polymerase and/or increase the annealing incubation time. If this does not succeed, it might be necessary to increase the hybridization area between both MIGS modules. In this case, a new primer needs to be generated and should include (a) the last 10–20 nt of the gene A fragment; (b) the miR173 target site; and (c) 4–6 nt of gene B fragment, as follows: ($A_{(10-20)}$ *GTGATTTTTCTCTACAAGCGAA*$B_{(4-6)}$). Carry out the first PCR step as indicated in step 3 of Subheading 3.2. Using the purified PCR product originating from PCR B as template, proceed with a second PCR reaction using the newly designed primer and Rev_B. The purified product of this reaction should be used now together with the PCR 1A product in the fusion PCR reaction (step 5, Subheading 3.2).

Acknowledgments

The author would like to thank Dr. Michael Johnston and Dr. Samanta B. de Campos for the critical reading and suggestions on the preparation of this chapter.

References

1. Carthew RW, Sontheimer EJ (2009) Origins and mechanisms of miRNAs and siRNAs. Cell 136:642–655
2. Ghildiyal M, Zamore PD (2009) Small silencing RNAs: an expanding universe. Nat Rev Genet 10:94–108
3. Chapman EJ, Carrington JC (2007) Specialization and evolution of endogenous small RNA pathways. Nat Rev Genet 8:884–896
4. Allen E, Howell MD (2010) miRNAs in the biogenesis of trans-acting siRNAs in higher plants. Semin Cell Dev Biol 21:798–804
5. Vazquez F, Legrand S, Windels D (2010) The biosynthetic pathways and biological scopes of plant small RNAs. Trends Plant Sci 15:337–345
6. Cuperus JT et al (2010) Unique functionality of 22-nt miRNAs in triggering RDR6-dependent siRNA biogenesis from target transcripts in Arabidopsis. Nat Struct Mol Biol 17:997–1003
7. Chen H-M et al (2010) 22-Nucleotide RNAs trigger secondary siRNA biogenesis in plants. Proc Natl Acad Sci USA 107:15269–15274
8. Felippes FF, Weigel D (2009) Triggering the formation of tasiRNAs in Arabidopsis thaliana: the role of microRNA miR173. EMBO Rep 10:264–270
9. Montgomery TA et al (2008) AGO1-miR173 complex initiates phased siRNA formation in plants. Proc Natl Acad Sci USA 105: 20055–20062
10. de Felippes FF, Wang JW, Weigel D (2011) MIGS: miRNA-induced gene silencing. Plant J 70:541–547
11. Hellens RP et al (2000) pGreen: a versatile and flexible binary Ti vector for Agrobacterium-mediated plant transformation. Plant Mol Biol 42:819–832
12. Odell JT, Nagy F, Chua NH (1985) Identification of DNA sequences required for activity of the cauliflower mosaic virus 35S promoter. Nature 313:810–812
13. de Felippes FF, Ott F, Weigel D (2011) Comparative analysis of non-autonomous effects of tasiRNAs and miRNAs in Arabidopsis thaliana. Nucleic Acids Res 39:2880–2889
14. Schwab R et al (2009) Endogenous TasiRNAs mediate non-cell autonomous effects on gene regulation in Arabidopsis thaliana. PLoS One 4:e5980
15. Chitwood DH et al (2009) Pattern formation via small RNA mobility. Genes Dev 23:549–554
16. Schwab R et al (2010) Directed gene silencing with artificial microRNAs. Methods Mol Biol 592:71–88
17. Wesley SV et al (2001) Construct design for efficient, effective and high-throughput gene silencing in plants. Plant J 27:581–590

Index

A

Ago. *See* Argonaut (Ago)
Ago2 ... 3, 8, 88, 90, 93–94, 122, 136–139, 141, 145–146, 153, 220, 260, 280, 286–287, 292, 322–323
asiRNAs. *See* Asymmetric siRNAs (asiRNAs)
amiRNAs. *See* Artificial microRNAs (amiRNAs)
Anopheles gambiae ... 316, 317, 320, 335–338, 341
Antisense oligonucleotides (ASOs) ... 87, 88, 90
Antiviral defense ... 235, 291
Arabadopsis thaliana ... 359, 373, 380–382, 385
Argonaut (Ago) ... 3, 194, 220, 234, 235, 260, 261, 322, 323, 379
Artificial microRNAs (amiRNAs) ... 208–212, 220, 225, 235–237, 239–241, 243, 245–246, 251, 357–374
ASOs. *See* Antisense oligonucleotides (ASOs)
Asymmetric siRNAs (asiRNAs) ... 89, 100, 101, 121, 135–151
Asymmetry ... 3–4, 6–9, 57–58, 88–89, 91, 94, 100–102, 121, 129–130, 135, 153–154, 162, 174, 180–181, 199, 321–323
Average silencing probability ... 18, 23, 32–33, 35, 47–49

B

Base preferences. *See* Nucletide preferences
Bayes' theorem ... 18, 23–32, 37–47
Bifunctional shRNA (bi-shRNA) ... 210, 259–277
Bifunctional siRNAs ... 180, 183, 185
Bioinformatics ... 1, 3–4, 6, 112, 117–121, 197
Biolistic transformation. *See* plant transformation
bi-shRNA. *See* Bifunctional shRNA
BLAST searching ... 13, 74, 83, 130, 262, 270, 295, 321, 330, 339
Bst DNA polymerase ... 347–354

C

Caenorhabditis elegans ... 197, 293, 299, 315, 317–319
Chemical engineering ... 87, 88, 90–99, 101
Chemical modification ... 10, 87–102, 113, 121, 122, 129, 136, 137, 179–180, 211, 279–289, 323, 334
Clinical trials ... 87, 88, 111, 219, 259
CMV promoter ... 166, 216, 246, 262, 306–308
Combinatorial RNAi (Co-RNAi). *See* Multiple gene RNAi
Co-RNAi. *See* Combinatorial RNAi (Co-RNAi)
Crop virus resistance ... 357–358

D

Decision tree learning ... 18, 23–24, 37
Delivery. *See* RNAi delivery
Dicer ... 1, 2, 5, 6, 66, 88, 89, 94, 101, 117, 130, 153, 171, 172, 174, 175, 180, 181, 194, 207–211, 234, 235, 237, 241, 248, 252, 260, 261, 279–281, 286, 292, 295, 305–306, 316, 321, 323, 359, 379, 380
Dicer-substrate siRNAs (D-siRNAs) ... 89, 100, 171–172, 175, 181–183
Drosha ... 5, 207–208, 234, 235, 241, 252, 261, 321, 323
Drosophila melanogaster ... 136, 197, 319, 325, 332
Dual-targeting siRNAs ... 169–176
Duplex stability ... 3–4, 6–10, 58, 66, 71, 94, 170–172, 175, 323

E

Efficacy ... 18, 21–23, 57, 65, 70, 74, 82–83, 113, 117, 129, 145, 154, 157, 170–173, 175, 180–182, 205, 206, 210–211, 216, 220, 224, 242, 261, 264, 267, 280, 361, 362, 364, 371, 374
Embryos ... 206, 224, 294, 302, 304–306, 308–310, 369
Endoribonuclease prepared siRNAs (esiRNAs) ... 193–202, 316
E-RNAi ... 317, 325, 327, 328, 330–332, 334–335, 341
e-shRNAs. *See* Extended shRNAs (e-shRNAs)
esiRNAs. *See* Endoribonuclease prepared siRNAs (esiRNAs)
Exportin 5 (Exp5) ... 207–208, 220, 234–235, 261
Extended shRNAs (e-shRNAs) ... 222, 223, 236, 237, 240, 242, 243, 247–252

Debra J. Taxman (ed.), *siRNA Design: Methods and Protocols*, Methods in Molecular Biology, vol. 942,
DOI 10.1007/978-1-62703-119-6, © Springer Science+Business Media, LLC 2013

F

Fanguard (FGmiR395)....369–370
Fork-like siRNAs (fsiRNAs)....89, 94, 100, 101, 117, 153–167
Fork-siRNAs. *See* Fork-like siRNAs (fsiRNAs)
fsiRNAs. *See* Fork-like siRNAs (fsiRNAs)

G

Gene expression profiling....98, 116–117, 131
Gene silencing....1–3, 9, 10, 18, 19, 21, 23–27, 32, 33, 35–37, 39–42, 45, 47–50, 63, 87–88, 93, 95, 101, 111, 121, 135–151, 153, 154, 161, 162, 195, 205–225, 233, 235, 259–261, 284, 291, 306, 309, 315, 316, 318, 359, 361, 364, 379–386
Gene targeting....9, 17, 18, 22, 23, 25, 26, 31, 32, 35–36, 39, 48, 49, 53, 63, 64, 67, 69, 79, 80, 95, 111, 112, 115, 117–119, 121, 129, 141, 153, 154, 157, 166, 170, 171, 173, 174, 180, 195, 218, 221, 261, 262, 264, 267, 273–275, 281, 308, 310, 318, 321, 324, 327, 329, 333, 335, 338, 339, 353, 380–382, 384, 385
Genome-wide siRNA libraries. *See* RNAi libraries
Guide strand....2, 3, 7–13, 58, 59, 61, 65, 69, 89, 90, 92–95, 97–99, 101, 102, 112, 119, 121, 122, 130, 136, 169, 182, 194, 209, 234, 235, 243, 245, 247, 251, 260, 261, 263, 264, 270, 272, 280, 284, 285, 287, 321–323, 369

H

Hamming distances....171–172, 174–175
HCV. *See* Hepatitis C virus (HCV)
Hepatitis C virus (HCV)....71, 77–82, 130, 218, 221, 224
hpRNAs. *See* long hairpin RNAs (lhRNAs)
Hidden Markov Model (HMM)....18, 33–36, 49–53
Highly structured target RNA....69–85
High-throughput screening....71, 385
HIV-1. *See* Human immunodeficiency virus type 1 (HIV-1)
HMM. *See* Hidden Markov Model (HMM)
Human disease vectors....315–341
Human immunodeficiency virus type 1 (HIV-1)....216, 219, 221, 237, 240–243, 245, 248, 304, 307

I

Immunogenicity....88, 89, 91, 96–99, 102, 238
Immune responses....10, 12, 92, 96, 97, 101, 136, 179,180, 183, 211, 219, 284, 288
Immunostimulatory siRNA....179–189
Inducible promoters....206, 216, 307
Interferon response....96, 97, 101, 237, 282, 292, 316
Invertebrates....194, 294, 305, 315–341
Inverted repeat (IR)....4, 70, 233–234, 296–301, 306–308, 348, 349, 353, 358
IR. *See* Inverted repeat (IR)

K

Knockdown....145, 195, 206, 236, 241, 271, 282, 317, 332, 334

L

Lentiviruses....216–219, 222, 233–253
lhRNAs. *See* Long hairpin RNAs (lhRNAs)
Liposome-mediated transfection....264–266, 269, 275–276
LNAs. *See* Locked nucleic acids (LNAs)
Locked nucleic acids (LNAs)....87, 89, 91, 92, 94–97, 99, 101, 102, 121, 129, 374
Long double-stranded RNA (dsRNA)....1, 153, 194
Long hairpin RNAs (lhRNAs)....215, 216, 222–224, 236, 237, 240, 242, 243, 247, 307, 359–362, 385
Loop region....347

M

Mammalian oocytes....308–310
Mammals....1–3, 17, 19, 22, 24–25, 57, 88, 111, 112, 116, 122, 128, 135–151, 180, 181, 194, 195, 205–207, 209, 216–219, 235, 237, 260, 291–311, 316, 323, 334, 347, 352–353, 361
Microarray....9, 58, 59, 112, 117, 121, 127–128, 131, 140–142, 150–151, 308
MicroRNA (miRNA)-like translational suppression. *See* translational suppression
MIGS. *See* miRNA induced gene-silencing
miR395....362, 367, 369, 370, 374
miRNA induced gene-silencing (MIGS)....379–386
miRNA pathway competition. *See* RNAi machinery saturation
Mismatch....5, 7, 13, 62–65, 89, 118, 154, 162, 170, 181, 188, 199, 207, 209, 210, 220, 234, 235, 245, 260–261, 263, 264, 270, 339, 340, 364
Multiple gene RNAi....169–176, 201, 215–216, 221–224, 236–237, 240–252, 359–363, 365, 367–369, 373, 384–385
Multi-gene silencing. *See* Multiple gene RNAi
Multimeric amiRNA. *See* polycistonic amiRNA
Multiplex amiRNA (Multi-amiRNA). *See* polycistronic amiRNA
Multiple promoter shRNA....221, 222, 224, 236, 242–245
Multi-shRNA. *See* multiple promoter shRNA

N

NEXT-RNAi....317, 325, 335–339, 341
Nuclease sensitivity....83, 90–92, 95, 100–102, 154, 157, 159–160, 162–165
Nucleotide preferences....3–6, 9, 17–53, 58–63, 70, 101, 118–119, 198–199, 316, 322, 323, 335, 340
Nucleotide substitution....154

O

Off-target effects 10, 57–67, 74, 83, 88, 98–99, 117, 121, 142, 150, 151, 194, 195, 197, 199, 283, 295, 317, 318, 321, 322, 329, 330, 332, 334, 336, 338
Oligonucleotide hybridization kinetics................. 71, 76–78
2'-O-methyl (2'-OMe) modification 91, 93, 96, 97, 99, 102, 113, 122, 129, 154, 162, 282, 284, 285, 287
Oocytes.................... 293, 294, 297–299, 304, 305, 307–310

P

Passenger strand 2, 3, 7, 9, 58, 61, 63, 65, 70, 88, 89, 91, 92, 94, 95, 98–102, 112, 117, 119, 121, 122, 136, 169, 181, 182, 209, 222, 234, 243, 245, 247, 251, 260–261, 264, 270, 272, 284–287, 323
Passenger strand competition 7–8
P-bodies 112, 235, 260
Plants...................................... 1, 2, 194, 207, 233, 294, 315, 357–374, 379–386
Plant transformation 363, 369–370, 374, 382, 383, 385
Polycistronic amiRNA.............................215, 216,222-224, 236, 242–243, 245–246, 360, 362, 363, 365, 367–369, 373, 374
Polymerase III (Pol III) promoters 65–66, 213–217, 223, 235–237, 242, 250, 262, 304
Polymerase II (Pol II) promoters 212, 213, 215–217, 223, 235, 236, 250, 262, 297, 299
Precursor miRNAs (pre-miRNAs)......................2, 207-208, 234–235, 251, 359, 361, 367, 374
pre-miRNAs. *See* Precursor miRNAs (pre-miRNAs)
pri-miRNAs 207–208, 215, 224–234, 235, 243, 245–246, 251, 373

R

RACE.....................................137, 139–141, 146–149, 264, 266–267, 269, 273–275
Reporter assays 59, 61, 115, 140–142, 148–149
Retinoic acid-inducible gene I (RIG-I)...................... 98, 99, 101, 112–113, 180, 182, 183, 185, 283, 287, 288, 292
Retroviruses 217–219, 237, 238
RIG-I. *See* Retinoic acid-inducible gene I (RIG-I)
RISC. *See* RNA-induced silencing complex (RISC)
RNA editing...291, 292, 294, 308
RNAfold................................. 170, 172, 174, 317, 365, 369
RNAi delivery 2, 88, 90, 91, 95–97, 99, 101, 129, 131, 166, 179, 182, 185, 206, 216, 218–219, 220, 221, 224, 225, 237, 261, 264–265, 267, 275–276, 279, 283, 285, 301, 315, 341, 359
RNAi machinery saturation 135–137, 140–142, 149–150, 215, 220, 252, 262
RNAi libraries 21, 98, 218, 280–281, 316–317, 318, 325, 334, 335–338, 341
RNA-induced silencing complex (RISC).............. 2, 3, 5–9, 18, 57, 69, 70, 83, 88–96, 99, 101, 113, 121, 136, 153, 169, 170, 194, 196, 207–210, 220, 234, 235, 241, 248, 260–262, 280, 286, 292, 316, 321–323, 359, 379–380
RNAse III 5, 88, 207, 260, 279, 292, 316

S

Secondary structure 6, 9–12, 18, 21, 70, 74, 83, 93, 101, 114, 119, 130, 146, 207, 210, 221, 272, 280, 287, 288, 322, 359, 360, 363, 367, 369
Seed region.................................. 10, 13, 58, 59, 61, 65, 90, 92–94, 98–99, 112, 113, 122, 123, 130, 162, 175, 200, 318
Sequence selection. *See* Target sequence selection
Short hairpin RNAs (shRNAs)............................. 1–3, 5–6, 8–9, 10, 13, 65–66, 89, 100, 101, 117, 205–225, 233–252, 259–276, 279–289, 295, 299, 305, 307, 316–317, 321, 323–325, 334–335, 347–354
Short small hairpin RNAs (sshRNAs) 89, 210, 211, 279–289
shRNAmirs 209, 210, 212
shRNAs. *See* Short hairpin RNAs (shRNAs)
siDirect 2.0 .... 58, 63–66
Silkworms..347–354
siRNA design software 58
siRNA pools 99, 195, 197, 199
siRNA and shRNA libraries. *See* RNAi libraries
siRNA backbone modification 90, 136–151, 153–167
sshRNAs. *See* Short small hairpin RNAs (sshRNAs)
Stable integration 206, 218
Statistical modelling 5
Stem-loop.............................. 210, 212, 243, 245, 260–266, 268, 270, 272–273, 306
Strand selection bias......... 7–8, 70, 89, 91–94, 113, 121, 196

T

Target sequence selection 1–13, 17–53, 56–67, 73–74, 101, 117–119, 157, 170–176, 180, 197–199, 317–323
Target site accessibility 70, 71, 77, 78, 83–84, 93, 323
tasiRNAs. *See* Trans-acting small interfering RNAs (tasiRNAs)
Therapeutics.................................. 57, 88, 91, 94, 111, 125, 136, 137, 206, 207, 211, 216, 225, 280, 282
Thermodynamics................................ 12, 58, 59, 61–62, 71, 88, 91, 93–95, 101, 102, 112, 114, 118–119, 153, 154, 170, 194, 196, 210, 211, 316, 321–323
TLRs. *See* Toll-like receptors (TLRs)
Toll-like receptors (TLRs)..........................12, 96, 112–113, 179, 186–188
Toxicity.. .... 90, 117, 121, 212, 215, 220–221, 224, 225, 262, 287

Trans-acting small interfering RNAs (tasiRNAs)....374, 380
Translational suppression.... 2, 10, 13, 96, 112, 118, 194, 208–210, 234–235, 259–261, 291–294, 380
Transduction....219, 248–250, 253
Transgenic plants.... 358–360, 363, 371–374, 381, 383
Transitivity....306, 380

U

UNAs. *See* Unlocked nucleobase analogs (UNAs)
Unlocked nucleobase analogs (UNAs)....111–131
U6 promoter....9, 212–215, 217, 219, 250

V

Validated targets....279
Viral vectors.... 2, 206, 218, 234, 236, 237, 242
Viral Titer.....237, 249–253

W

Web microRNA design versions 3 (WMD3)....363
Wheat streak mosaic virus (WSMV)....359–360, 362, 363, 365, 369, 371, 372
WMD3. *See* Web microRNA design versions 3 (WMD3)
WSMV. *See Wheat streak mosaic virus* (WSMV)

MIX
Papier aus verantwortungsvollen Quellen
Paper from responsible sources
FSC® C105338

If you have any concerns about our products,
you can contact us on
ProductSafety@springernature.com

In case Publisher is established outside the EU,
the EU authorized representative is:
Springer Nature Customer Service Center GmbH
Europaplatz 3, 69115 Heidelberg, Germany

Printed by Libri Plureos GmbH
in Hamburg, Germany